SCIENCEWORLD

SW•9

> Explore
> Question
> Imagine

Peter Saffin
BSc, BEd, GD Env. Sci

Peter Stannard
BSc, DipEd

Ken Williamson
BSc (Hons), DipEd

VICTORIAN CURRICULUM

This edition published in 2021 by

Matilda Education Australia, an imprint
of Meanwhile Education Pty Ltd
Level 1/274 Brunswick St
Fitzroy, Victoria Australia 3065
T: 1300 277 235
E: customersupport@matildaed.com.au
www.matildaeducation.com.au

First edition published in 2017 by Macmillan Science and Education Australia Pty Ltd

National Library of Australia
cataloguing in publication data

Creator:	Saffin, Peter, author
Title:	Scienceworld Victorian Curriculum 9 / Peter Saffin, Peter Stannard, Ken Williamson.
ISBN:	978 1 4202 3736 8 (paperback)
Target Audience:	For secondary school age
Subjects:	Science--Textbooks. Science--Study and teaching (Secondary)--Victoria
Other creators/contributors:	Stannard, Peter, author. Williamson, Ken, author.

Publisher: Emma Cooper
Project editors: Barbara Delissen, Diana Saad and Lachlan Smith
Proofreader: Marta Veroni
Illustrators: Andrew Craig, Chris Dent, Brent Hagen, Guy Holt and Nives Porcellato
Cover and text designer: Dimitrios Frangoulis
Production control: Katherine Fullagar
Photo research and permissions clearance: Annthea Lewis and Vanessa Roberts
Typeset in Merriweather Light 9.5/13.5 by Dim Frangoulis
Cover image: Shutterstock/WAYHOME studio
Indexer: Karen Gillen

August 2020
Printed in Australia by SOS Print + Media
1 2 3 4 5 6 7 25 24 23 22 21

Warning: It is recommended that Aboriginal and Torres Strait Islander peoples exercise caution when viewing this publication as it may contain images of deceased persons.

Contents

Using ScienceWorld

ScienceWorld takes a constructivist approach to learning, helping students explore what they already know, then building on that knowledge as they progress. This guide will show the key features of ScienceWorld, and help you understand how best to use it in the classroom.

Each chapter starts with a **Chapter opening page** containing an engaging photo for discussion. It shows the main content and skills to be covered in that chapter.

Students' prior knowledge is explored with a **Get started** activity, each with an **Explore**, **Question** or **Imagine** task that introduces the chapter topic.

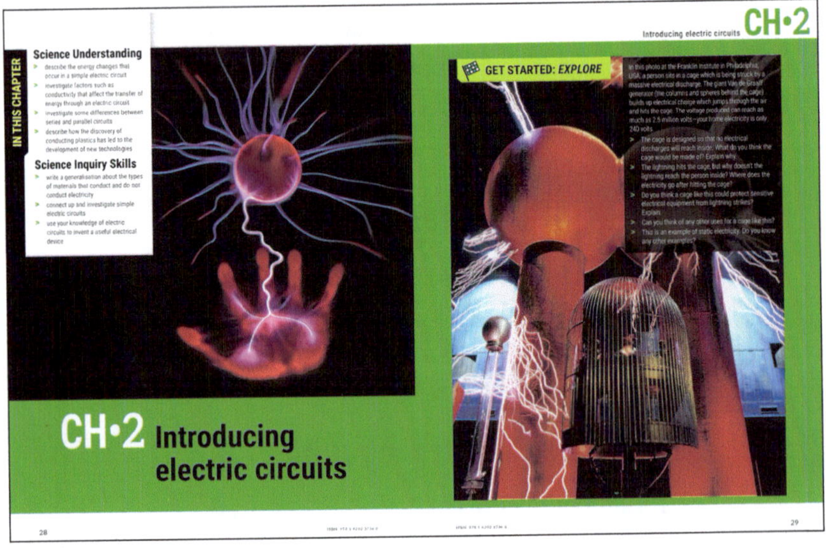

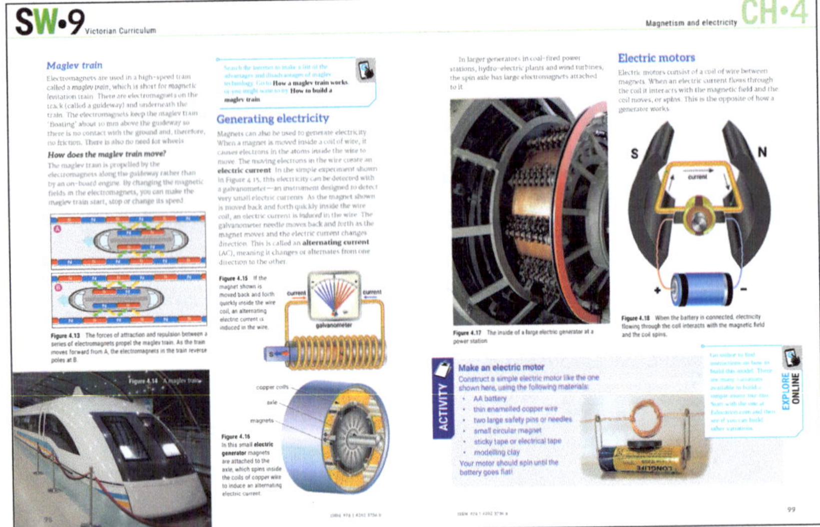

Each chapter is broken into sections focusing on certain aspects of theory and skills covered. *ScienceWorld* uses engaging photos, illustrations, cartoons and contexts to make science accessible to all students.

Activities are dispersed throughout each section to reinforce concepts and provide hands-on learning opportunities for each lesson.

Explore online boxes provide opportunity for wider research.

ISBN 978 1 4202 3736 8

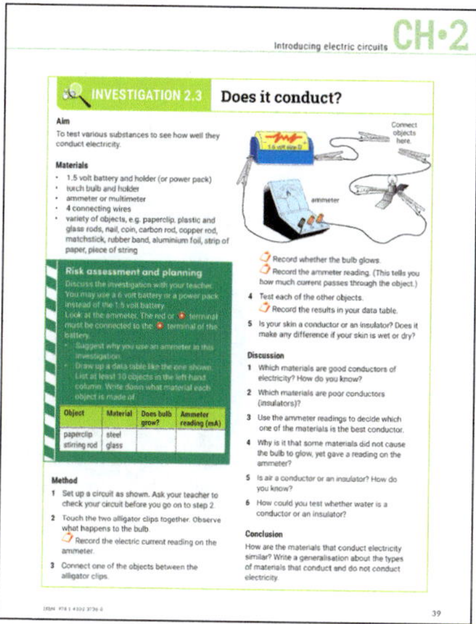

Investigations provide opportunities for students to apply Science Inquiry skills, while exploring key concepts.

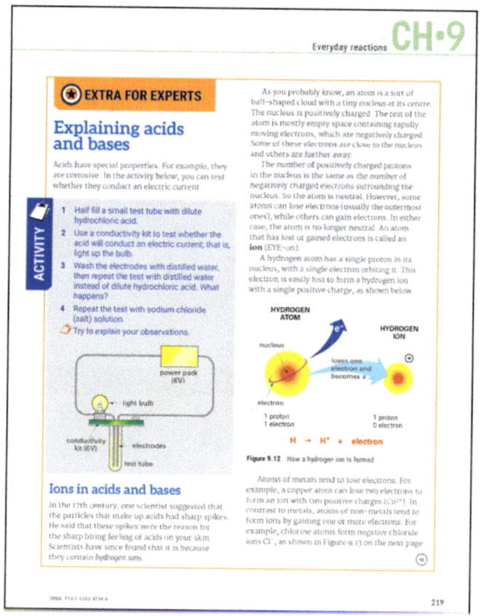

Experiments allow students to design their own experiments and inventions. Students explore and apply science skills and method while solving problems. This allows students to discover and engage with science concepts for themselves.

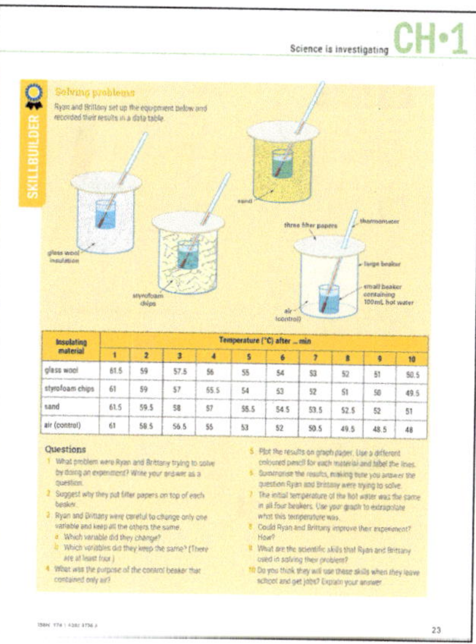

Skillbuilders teach key skills explicitly, supporting a clear progression of skill development throughout the book.

Science as a human endeavour features bring science to life; putting science in context historically, for today, and for the future. These features include activities that allow students to understand the nature and context of science and to imagine the future.

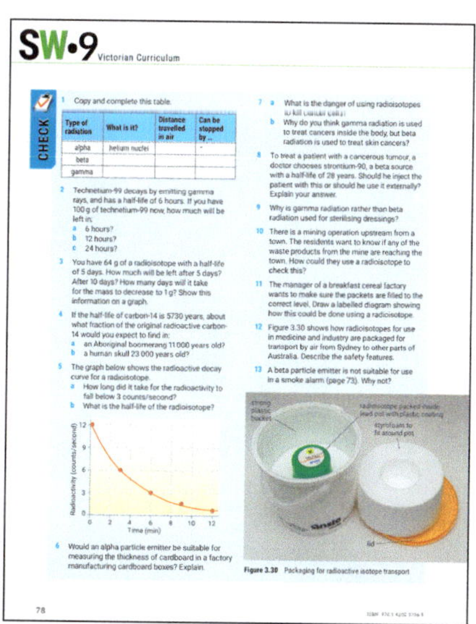

Check questions review students' understanding of concepts for each section.

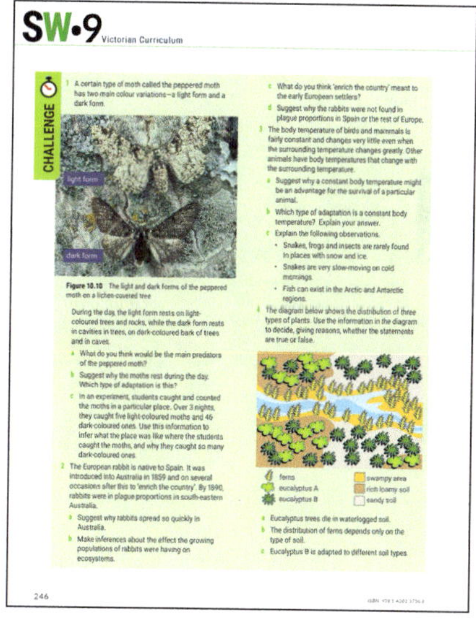

Challenge questions provide an opportunity for students to apply their knowledge and skills in context, and challenge students to think at a higher level.

ISBN 978 1 4202 3736 8

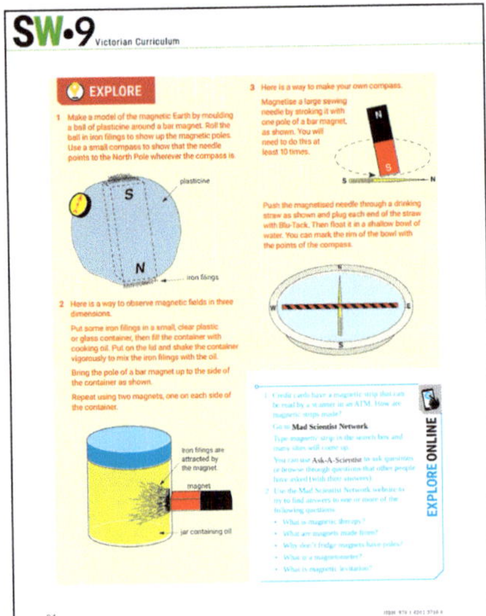

Explore sections allow students to create, invent and inquire into concepts learnt throughout the chapters.

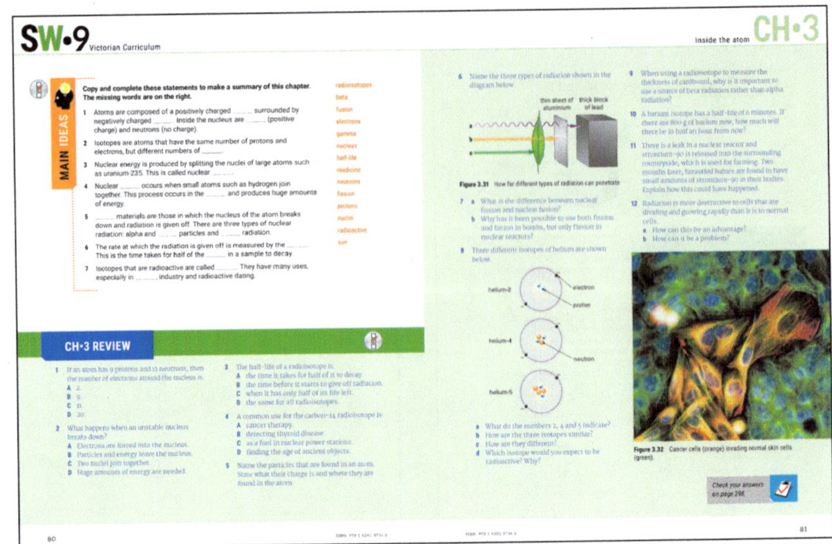

Each chapter ends with a **Main ideas** cloze exercise to test students' understanding of the key chapter concepts through comprehension.

The **Chapter review** then provides an opportunity for students to revise key science knowledge and skills developed through each chapter.

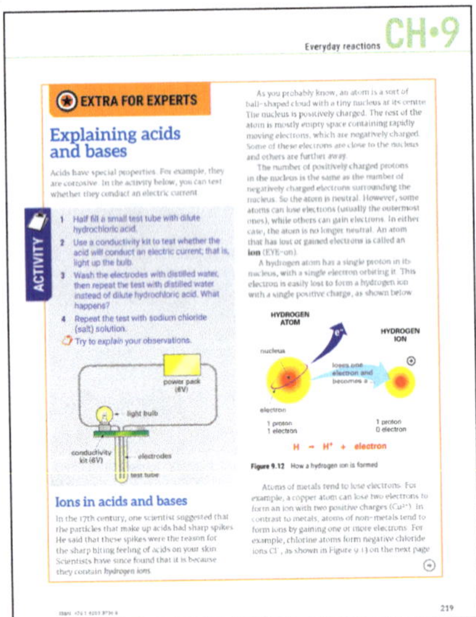

Extra for experts features provide opportunities for students to engage at a higher level, to challenge their scientific thinking and understanding.

ISBN 978 1 4202 3736 8

Foreword

As you probably know, I'm mad about science. Every day I learn something new about the world around me—dark matter and dark energy, living creatures of all shapes and sizes, the amazing irrationalities and untapped abilities of our human brain etc. My *Great Moments in Science* radio series/podcast is one way in which I explore these things and try to make them easier for people to understand. In this book you will explore all these things and learn how to think scientifically, asking questions about the world and imagining new solutions to these questions.

Doing science can lead to so many different and fascinating careers where you can design intelligent robots, use giant telescopes in space, produce food in a world where global warming is real, or go as an astronaut to planet Mars! It's fun to apply your knowledge of science to the real world and let your imagination run riot. After all, it's not the answer that gets you the Nobel Prize, it's the question. You could be the next Elizabeth Blackburn who won a Nobel Prize for her work on the telomere—a previously unexplored section of the human chromosome that gives us new and deep insights into aging.

And remember the words of Richard Feynman, 'Science is a way to not get fooled' ...

Dr Karl

ISBN 978 1 4202 3736 8

Links to the Victorian Curriculum

This scope and sequence provides an overview of how *ScienceWorld* 9 covers the Victorian Curriculum. The focus is on the Science Understanding strand, although only some of the Science as a Human Endeavour content and elaborations are covered in this version of the scope and sequence. Included online in the teacher support are curriculum scope and sequence guides that detail how *ScienceWorld* covers the Victorian Curriculum content descriptions across all four books, and these also include a full mapping of the Science as Human Endeavour sub-strand, and the Science Inquiry Skills.

Abbreviations:
SHE: Science as a Human Endeavour
BS: Biological Sciences
CS: Chemical Sciences
ESS: Earth and Space Sciences
PS: Physical Sciences

ScienceWorld 9

Chapter & Unit titles	Science Understanding	Elaborations
1 Science is investigating		
1.1 Steps in investigating	Science Inquiry Skills	
1.2 Collecting data	Science Inquiry Skills	
1.3 Processing data	Science Inquiry Skills	
2 Introducing electric circuits		
2.1 Electric charges	PS: Electric circuits can be designed for diverse purposes using different components; the operation of circuits can be explained by the concepts of voltage and current (VCSSU130)	
2.2 Electric currents	PS: Electric circuits can be designed for diverse purposes using different components; the operation of circuits can be explained by the concepts of voltage and current (VCSSU130)	• investigating parallel and series circuits and measuring voltage drops across and currents through various components
2.3 Electric circuits	PS: Electric circuits can be designed for diverse purposes using different components; the operation of circuits can be explained by the concepts of voltage and current (VCSSU130)	• investigating parallel and series circuits and measuring voltage drops across and currents through various components
3 Inside the atom		
3.1 Atomic structure	CS: All matter is made of atoms which are composed of protons, neutrons and electrons natural radioactivity arises from the decay of nuclei in atoms (VCSSU122)	• describing and modelling the structure of atoms in terms of the nucleus, protons, neutrons and electrons • comparing the mass and charge of protons, neutrons and electrons

ISBN 978 1 4202 3736 8

3.2 Nuclear reactions	CS: All matter is made of atoms which are composed of protons, neutrons and electrons; natural radioactivity arises from the decay of nuclei in atoms (VCSSU122) CS: Chemical reactions involve rearranging atoms to form new substances; during a chemical reaction mass is not created or destroyed (VCSSU124)	• describing in simple terms how alpha and beta particles and gamma radiation are released from unstable atoms • recognising that the conservation of mass in a chemical reaction can be demonstrated by simple chemical equations
3.3 Radioactivity	CS: All matter is made of atoms which are composed of protons, neutrons and electrons; natural radioactivity arises from the decay of nuclei in atoms (VCSSU122)	• describing in simple terms how alpha and beta particles and gamma radiation are released from unstable atoms
4 Magnetism and electricity		
4.1 Investigating magnets	PS: The interaction of magnets can be explained by a field model; magnets are used in the generation of electricity and the operation of motors (VCSSU131)	
4.2 Magnetic fields	PS: The interaction of magnets can be explained by a field model; magnets are used in the generation of electricity and the operation of motors (VCSSU131)	• investigating the action at a distance or the field model around magnets of different shapes
4.3 Electricity and magnets	PS: The interaction of magnets can be explained by a field model; magnets are used in the generation of electricity and the operation of motors (VCSSU131)	• investigating the movement of a magnet and a wire to produce electricity • investigating the effect of a magnet on a current from a battery to produce movement
5 Microbes		
5.1 Microscopic life	BS: Ecosystems consist of communities of interdependent organisms and abiotic components of the environment; matter and energy flow through these systems (VCSSU121)	• exploring interactions between organisms, for example, predator/prey, parasites, competitors, pollinators and disease vectors
5.2 Helpful microbes	BS: Ecosystems consist of communities of interdependent organisms and abiotic components of the environment; matter and energy flow through these systems (VCSSU121)	• exploring interactions between organisms, for example, predator/prey, parasites, competitors, pollinators and disease vectors
5.3 Microbes and disease	BS: Multicellular organisms rely on coordinated and interdependent internal systems to respond to changes to their environment (VCSSU117)	• investigating the response of the body to changes as a result of the presence of micro-organisms
6 Everyday substances		
6.1 Metals	CS: Different types of chemical reactions are used to produce a range of products and can occur at different rates; chemical reactions may be represented by balanced chemical equations (VCSSU125)	• investigating how chemical reactions result in the production of a range of useful substances, for example, fuels, metals and pharmaceuticals • using word or symbol equations to represent chemical reactions

ISBN 978 1 4202 3736 8

6.2 Carbon compounds	CS: Different types of chemical reactions are used to produce a range of products and can occur at different rates; chemical reactions may be represented by balanced chemical equations (VCSSU125)	• investigating how chemical reactions result in the production of a range of useful substances, for example, fuels, metals and pharmaceuticals • using word or symbol equations to represent chemical reactions
6.3 Plastics and fibres	CS: Different types of chemical reactions are used to produce a range of products and can occur at different rates; chemical reactions may be represented by balanced chemical equations (VCSSU125)	• investigating how chemical reactions result in the production of a range of useful substances, for example, fuels, metals and pharmaceuticals • using word or symbol equations to represent chemical reactions
7 Body balance		
7.1 The nervous and endocrine systems	BS: Multicellular organisms rely on coordinated and interdependent internal systems to respond to changes to their environment (VCSSU117)	• describing how the requirements for life (oxygen, nutrients, water and removal of waste) are provided through the coordinated function of body systems, for example, the respiratory, circulatory, digestive, nervous and excretory systems
	BS: An animal's response to a stimulus is coordinated by its central nervous system (brain and spinal cord); neurons transmit electrical impulses and are connected by synapses (VCSSU118)	• identifying functions for different areas of the brain • modelling the 'knee jerk' reaction and explaining why it is a reflex action • identifying responses involving the nervous and endocrine systems • researching the causes and effects of spinal cord damage
7.2 Plant responses	BS: Multicellular organisms rely on coordinated and interdependent internal systems to respond to changes to their environment (VCSSU117)	• describing how the requirements for life (oxygen, nutrients, water and removal of waste) are provided through the coordinated function of body systems, for example, the respiratory, circulatory, digestive, nervous and excretory systems
7.3 Keeping the balance	BS: Multicellular organisms rely on coordinated and interdependent internal systems to respond to changes to their environment (VCSSU117)	• explaining (using models, flow diagrams or simulations) how body systems work together to maintain a functioning body
8 Dynamic Earth		
8.1 Inside the Earth	ESS: The theory of plate tectonics explains global patterns of geological activity and continental movement (VCSSU127)	
8.2 Earthquakes	ESS: The theory of plate tectonics explains global patterns of geological activity and continental movement (VCSSU127)	• relating the occurrence of earthquakes and volcanic activity to constructive and destructive plate boundaries
8.3 Earth plates	ESS: The theory of plate tectonics explains global patterns of geological activity and continental movement (VCSSU127)	• recognising the major plates on a world map • considering the role of heat energy and convection currents in the movement of tectonic plates • relating the occurrence of earthquakes and volcanic activity to constructive and destructive plate boundaries

ISBN 978 1 4202 3736 8

9 Everyday reactions		
9.1 Acids and bases	CS: Chemical reactions, including combustion and the reactions of acids, are important in both non-living and living systems and involve energy transfer (VCSSU126)	• investigating reactions of acids with metals, bases, and carbonates
9.2 The pH scale	CS: Chemical reactions, including combustion and the reactions of acids, are important in both non-living and living systems and involve energy transfer (VCSSU126)	• investigating reactions of acids with metals, bases, and carbonates
	CS: Chemical reactions involve rearranging atoms to form new substances; during a chemical reaction mass is not created or destroyed (VCSSU124)	• modelling chemical reactions in terms of rearrangement of atoms
9.3 Reactions of acids and bases	CS: Chemical reactions, including combustion and the reactions of acids, are important in both non-living and living systems and involve energy transfer (VCSSU126)	• investigating reactions of acids with metals, bases, and carbonates
	CS: Chemical reactions involve rearranging atoms to form new substances; during a chemical reaction mass is not created or destroyed (VCSSU124)	• modelling chemical reactions in terms of rearrangement of atoms
9.4 Energy in reactions	CS: Chemical reactions, including combustion and the reactions of acids, are important in both non-living and living systems and involve energy transfer (VCSSU126)	• investigating a range of different reactions to classify them as exothermic or endothermic • comparing respiration and photosynthesis and their role in biological processes • investigating how chemical reactions result in the production of a range of useful substances, for example, fuels, metals and pharmaceuticals
	CS: Chemical reactions involve rearranging atoms to form new substances; during a chemical reaction mass is not created or destroyed (VCSSU124)	• modelling chemical reactions in terms of rearrangement of atoms • considering the role of energy in chemical reactions
	ESS: Global systems, including the carbon cycle, rely on interactions involving the atmosphere, biosphere, hydrosphere and lithosphere (VCSSU128)	• investigating how human activity affects global systems
	SHE: The values and needs of contemporary society can influence the focus of scientific research (VCSSU116)	• investigating how social actions have led to changed government policies and social behavioural change in relation to the use of chlorofluorocarbons (CFCs) in aerosol spray cans

ISBN 978 1 4202 3736 8

10 Energy in ecosystems		
10.1 Living in ecosystems	BS: Ecosystems consist of communities of interdependent organisms and abiotic components of the environment matter and energy flow through these systems (VCSSU121)	• exploring interactions between organisms, for example, predator/prey, parasites, competitors, pollinators and disease vectors • using modelling to examine factors that affect population sizes, for example, seasonal changes, destruction of habitats, introduced species
10.2 Matter and energy in food webs	BS: Ecosystems consist of communities of interdependent organisms and abiotic components of the environment matter and energy flow through these systems (VCSSU121)	• exploring interactions between organisms, for example, predator/prey, parasites, competitors, pollinators and disease vectors • using modelling to examine factors that affect population sizes, for example, seasonal changes, destruction of habitats, introduced species
10.3 Human impact on ecosystems	BS: Ecosystems consist of communities of interdependent organisms and abiotic components of the environment matter and energy flow through these systems (VCSSU121)	• using modelling to examine factors that affect population sizes, for example, seasonal changes, destruction of habitats, introduced species
11 Digital technology		
11.1 Communication	SHE: Advances in scientific understanding often rely on developments in technology and technological advances are often linked to scientific discoveries (VCSSU115)	• considering how the properties of electromagnetic radiation relate to its uses, for example, radar, medical diagnosis and treatment, mobile phone communications and microwave cooking
11.2 Electronics	PS: Electric circuits can be designed for diverse purposes using different components; the operation of circuits can be explained by the concepts of voltage and current (VCSSU130)	• investigating the properties of components such as LEDs, and temperature and light sensors
11.3 Robotics and control	PS: Electric circuits can be designed for diverse purposes using different components; the operation of circuits can be explained by the concepts of voltage and current (VCSSU130)	• investigating the properties of components such as LEDs, and temperature and light sensors • exploring the use of sensors in robotics and control device

ISBN 978 1 4202 3736 8

Online resources

Throughout this book you will find links to activities and video or audio files. The activities are for students to practise key skills, or to reinforce learning on key concepts. Activities vary in type and include crosswords, matching, drag and drop, labelling, multiple choice, true and false, and sequencing activities. Students can repeat these activities as revision, and practise them at any time.

Each activity is scored and the teacher can review student progress in the digital mark book. In *ScienceWorld 9* there are approximately 85 activities.

When one or more activities are available, you will find an icon on the page of the book where it is most relevant to learning.

 **Digital activity**

 **Audio or video material**

 **Supporting documentation**

We hope you enjoy using these activities to improve learning outcomes.

An agricultural scientist is working on ensuring sustainable food supplies for the future. In this greenhouse researchers are developing disease-resistant varieties of crops. You can find out more about disease and the micro-organisms that cause disease in chapter 5, 'Microbes'. In chapter 10, 'Energy in ecosystems', you will also learn about how populations of pests are controlled in the farming of crops, and how farming and sustainable ecosystems can exist side by side while maintaining ecosystem diversity.

ISBN 978 1 4202 3736 8

IN THIS CHAPTER

Science Inquiry Skills

- solve a problem by planning, conducting and evaluating an experiment
- evaluate an experiment, making sure the results are reliable and valid
- assess whether experiments with people and animals are valid and ethical
- learn how to write a science magazine article
- use a line of best fit to analyse experimental data and make predictions (interpolations and extrapolations)
- develop skills for collecting data in the field
- evaluate experiments and suggest ways of obtaining more reliable results

CH•1 Science is investigating

ISBN 978 1 4202 3736 8

GET STARTED: *EXPLORE*

Science skills can be used to solve everyday problems. For example, suppose Emily's bicycle has a flat tyre and she wants to know why. Emily and her friend Nick investigate this.

Emily: Hey Nick, my back tyre is flat! There must be a leak somewhere. We'll have to find out where the air is getting out before we can fix it.

Nick: Perhaps there's a nail in it.

Emily: I can't see one.

Nick: There might be a cut in the tyre.

Emily: No, it seems OK.

Nick: What about the valve? Someone told me you can test it by putting some spit on it. If air is getting out, a bubble will form in the spit.

Emily: Let's try that.

Nick: I'll just pump some air in.

Emily: Hey look, the bubble is slowly getting bigger.

Nick: Then the valve must be leaking.

Emily: Well, let's put a new tube and valve in.

In their investigation Emily and Nick used several science skills. Try to identify the following in their conversation:

- observations
- inferences
- an experiment
- a prediction.

ISBN 978 1 4202 3736 8

1.1 Steps in investigating

Planning an experiment

There are four main steps in a scientific experiment, as shown.

1 *Planning the experiment*

- Identify the problem.
- Identify the variables.
- Write a research question or a hypothesis that can be tested.
- Work out which variable you will change, which you will measure and which you will control.
- Work out the method and select the equipment you will use.

2 *Conducting the experiment*

- Carry out the experiment.
- Observe, measure and record data.

3 *Processing data*

- Organise the data, draw graphs and do calculations.
- Identify patterns in the data and relationships between the variables.
- Use scientific knowledge to explain the patterns and relationships.

4 *Evaluating the experiment*

- Evaluate the design of the experiment and the methods used.
- Discuss the results. Are they reliable?
- Evaluate the findings in relation to the original problem, question or hypothesis.
- Write a conclusion.
  Make sure it is valid.

ISBN 978 1 4202 3736 8

Planning an experiment

Imagine you work for a motoring organisation. You have read an overseas report that says that the brand of tyres used on a car makes little difference to the car's stopping distance when braking in an emergency. You decide to investigate this claim under Australian conditions, using the steps in investigating on the previous page.

1 In your own words, write down the problem to be investigated.
2 Rewrite the problem as a hypothesis—a generalisation that can be tested by an experiment.
3 What are the variables involved; that is, what factors could affect the results of the experiment?
4 What method will you use to test your hypothesis?
5 Which variable will you purposely change in your experiment? This is the independent variable.
6 Which variable will you measure? This is the dependent variable.
7 Which variables will you need to control?
8 What equipment will you need?
9 What data will you collect and how will you record it?
10 How will you know whether your hypothesis is correct or not?

Evaluating an experiment

When you have finished an experiment, you should think carefully about how successful it was and whether you could improve it. This is called *evaluating an experiment*. For example, were you able to make accurate measurements? Did you repeat your measurements and calculate an average? The more measurements you make, the more *reliable* the average will be, but three measurements are usually enough.

After evaluating the experiment, you may need to repeat it with some modifications. You also need to be able to evaluate other people's experiments. Scientists do this often, and they sometimes do the experiments themselves to see if they obtain the same results. They may be able to suggest ways to improve the experiment.

It is also important to check any conclusions or generalisations made from the data collected in an experiment to make sure they are logical or *valid*. Sometimes poor thinking or reasoning can lead to incorrect or invalid conclusions. Also, not everyone will reach the same conclusions after analysing the same data.

In the next activity you can practise evaluating an experiment and a conclusion.

ISBN 978 1 4202 3736 8

Part A: Evaluating an experiment

The manufacturer of a brand of paintbrush has made the following claim:

Scientific tests show that Super Soaker has greater paint pick-up than any other brand.

Five brands of paintbrush were tested as follows.

1. Paint was added to the tray until the reading on the electronic balance was 500 g exactly.
2. The first brush was attached to the lever. It was lowered into the paint, then lifted out.
3. The new mass of the tray plus paint was recorded, and the mass of paint picked up was calculated by subtraction.
4. The same procedure was followed for all five brushes.
5. The test was repeated four times for each brush and the masses were averaged. The results in the data table on the right show the average masses.

What variables would need to be controlled in this experiment?

Are the results reliable? Give a reason for your answer.

Do you consider the manufacturer's claim to be correct? Explain.

How could you improve the experiment?

lever

tray of paint

balance

SUPER SOAKER

477:00

Brand of paintbrush	Final mass of tray plus paint (g)	Mass of paint picked up (g)
Bettabrush	478	22
Easy Paint	491	9
Slurp	485	15
Super Soaker	**477**	**23**
Thickbrush	483	17

Part B: Evaluating a conclusion

James and Tjanda wanted to know which was the best all-purpose pesticide. To do this they recorded the death rate for flies, mosquitoes and spiders using four different pesticides.

James concluded that Bingo was the best all-purpose spray, but Tjanda said that No More Flies was the best.

Who do you agree with? Explain your choice clearly.

Pesticide	Percentage death rate		
	Flies	Mosquitoes	Spiders
Bingo	80	60	60
Bugaway	30	20	90
No More Flies	95	100	15
Zap	40	40	40

ISBN 978 1 4202 3736 8

Investigating velcro

In the experiment on the next page you will investigate the strength of a velcro strip. Before you do this, however, you need to know something about velcro.

ACTIVITY

Your teacher will give you a small piece of velcro (both hook and loop strips). Examine both strips using a hand lens or stereomicroscope.

- Sketch the appearance of the surface of both strips.
- Explain how the two strips link together.
- Can you make a join with two pieces of tape of the same type? Explain.

Figure 1.1 Velcro weed sticking to socks

Velcro is a trademark name from the French words *velours* (velvet) and *crochet* (hook). Swiss engineer Georges de Mestral had the idea for velcro after getting burrs caught in his clothing and in his dog's fur while walking in the forest. When he examined the burrs under a microscope, he found tiny hooks that could attach themselves to anything with loops in it, such as hair or clothes. Velcro (or nylon press tape) is made in two parts, one with hooks and one with loops. It is now a universal fastener, for everything from disposable nappies to sandals.

A 5 mm square piece of velcro may contain 3000 hooks and loops, although they will not all be hooked together. Two 5 cm squares pressed together can support a person weighing 80 kg! You may have seen velcro jumping, where a person leaps off a trampoline and sticks to a velcro wall.

As the diagram shows, less force is required to detach velcro when it is pulled off at an angle than when it is pulled at right angles or parallel to the velcro. This is because you are pulling a single row rather than all the hooks and loops together. This smaller force is sufficient to disconnect one row after another, producing the familiar ripping sound.

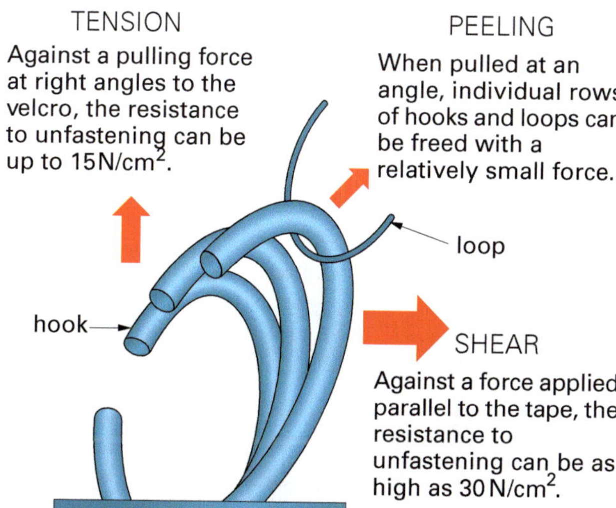

ISBN 978 1 4202 3736 8

EXPERIMENT 1.1 Testing velcro

Research questions

Rachel is thinking of buying a new pair of training shoes that use a velcro strap instead of laces. She likes the idea because it is so easy to change into and out of her shoes.

Rachel is fascinated with the velcro idea and wants to check whether the information on velcro on the previous page is correct.

These are some of the questions she would like answered.

1 What peeling force is needed to unfasten the velcro strip?

2 What shear force will be necessary to undo the strap off her shoe?

3 Will the strap keep its strength if it is unfastened and fastened many times?

4 Will the strength of the strap be affected by grit and material fluff that gets caught in the velcro?

Designing your experiment

1 Work in a small group and discuss which test or tests you would like to do.

2 Write a hypothesis for your experiment.

3 Make a list of the equipment you will need.

4 Write a draft of your plan, including the variables you will be controlling and the data you are going to record.

5 Discuss the draft design with your teacher, then write your final design.

Hints and tips

1 You can buy hook and loop strips in cheap variety stores or in fabric stores.

2 You should record the results of the tests as the force used (in newtons) per area of velcro (for example, per cm^2).

3 You will have to design a clever way to attach the force measurer (usually a spring balance) to the velcro. Stitching, using a small clamp or gluing are three possible methods.

Risk assessment and planning

- Do a risk assessment to identify any safety hazards and decide on necessary precautions.
- Prepare a data table for your results. Remember, your results will be more reliable if you take at least three measurements and find the average.

Writing your report

Write a report of your experiment using the seven headings Title, Aim, Materials, Method, Results, Discussion and Conclusion.

Your description of what you did needs to be good enough so that if someone else follows your method they will get very similar results. A diagram will help.

In the discussion, say how well your method worked and suggest how you might be able to get more reliable results.

In your conclusion, you need to answer the research questions you investigated.

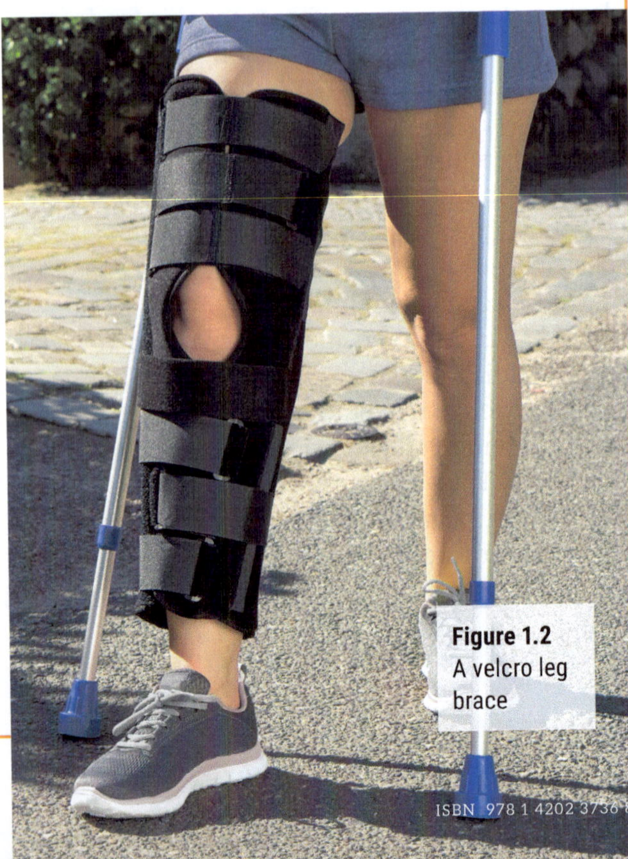

Figure 1.2 A velcro leg brace

ISBN 978 1 4202 3736 8

SCIENCE AS A HUMAN ENDEAVOUR

Experiments using animals

No new drug can be put on the market until extensive information has been obtained on the effects it is likely to have on humans. One tragic example where this was not done properly was with the drug thalidomide. It was used in the 1950s to stop morning sickness and as a sleeping pill by pregnant women, but was later identified as a cause of deformities in newborn babies.

New drugs are usually tested first on laboratory animals, mainly rats and mice. Sometimes animals are also used to test the safety of food additives and household cleaning products. However, many people feel that this testing is unethical, and for this reason very few cosmetics are now tested this way.

Experiments involving living things require special methods. This is because no two individuals are the same. Also, it is not possible to control the behaviour of live subjects, or to control attitudes if people are used. However, scientists take care to control as many variables as possible. For example, if they were using mice, they would control the following variables:

- genetic differences—all mice would be descended from the same stock
- age—all mice would be the same age
- environment—all mice would be kept in similar cages and be given the same food and water
- no diseases—the mice would be kept in the best of health.

When conducting such experiments, scientists normally use a test group and a control group. The test group is given the drug and the control group is not. Any differences in response can then be said to be caused by the drug.

Experimenting on people

When experimenting with people, it is important that the subjects do not know whether they are in the test group or the control group. Suppose a drug company wants to test a new drug that they claim can help smokers give up smoking. A test group and a control group are given tablets—real ones for the test group and fake ones for the control group. However, the volunteers do not know which tablets they have been given. The fake tablets are called **placebos (pla-SEE-bows)** and appear to be exactly the same as the real tablets. After several months, the smoking behaviour of the volunteers is checked, and conclusions can then be drawn. This procedure is called a **blind experiment** because the subjects are unaware of (or blind to) whether they are in the test group or the control group.

A blind experimental design helps to overcome differences in the attitudes of the people involved in the trial. Some people may want to give up smoking more than others, and some may think that no treatment will work for them. Despite this special experimental design, however, the results may still be inconclusive. For example, suppose 20% of the test group give up smoking and 10% of the control group give up smoking. Before you can draw a conclusion from this, you need to analyse the data to decide whether the differences could have arisen by chance alone or whether there are real differences.

In some experiments, the scientists are 'blind' as well as the subjects. This design is called a **double-blind experiment**. Suppose a scientist wanted to test a new ingredient X, which is supposed to reduce acne (pimples). She could

The Venetian Bros Laboratory specialises in double blind experiments.

ISBN 978 1 4202 3736 8

arrange for a large number of bottles of lotion to be made, half with X in them and half without. The bottles could then be numbered and given to volunteers to use. With this design, however, neither the volunteers nor the scientist would know which volunteers were using ingredient X and which were not. The scientist could then judge the effect on the pimples of each volunteer without prejudice. Only after the experiment would the scientist find out who had been given ingredient X.

Ethical or unethical experiments?

Some people say it is unethical for researchers to give sick people placebos, or no treatment, if effective treatments are already available.

A needle-exchange study with heroin addicts was conducted in 1997 in Anchorage, Alaska. Half of the addicts were given needles and the other half were not. The study was to see how many in each group got hepatitis B, even though there is an effective hepatitis B vaccine. The vaccine was offered to all participants after the study, but critics of the study claim it was designed to prove that needle-exchange programs work, rather than to help the addicts.

Questions

1 Explain the differences between a blind experimental design and a double-blind experimental design.

2 Suppose a drug company has developed a new drug called Nodec that they claim will reduce tooth decay. They arrange to test Nodec at your school using this method:
- Company representatives visit the school to explain the experiment and call for volunteers.
- They select 100 students and each student is examined to record the number of fillings.
- Each student is given a jar of tablets—either Nodec or a placebo. Students are to take one tablet each day. The drug company claims that their representatives do not know who is given Nodec and who is given the placebo.
- After six months the students are examined again and the data recorded. When the trial is complete, the drug company sends the following summary to the school.

	Total number of fillings	
	Before	After
Placebo (50 students)	56	73
Nodec (50 students)	47	59

a Evaluate the design of the experiment and the results obtained.

b On the basis of this experiment, would you use Nodec? Explain.

3 In a group discuss whether animals should be used to test drugs, cosmetics and other products intended for use by humans. You could research this topic on the internet or have a class debate.

4 Consider the Anchorage needle-exchange program described on this page.

a Do you think this study was ethical? Explain.

b Two of the principles of the Declaration of Helsinki (October 2000) are:
- Medical progress is based on research that ultimately must rest in part on experimentation involving humans.
- In medical research on humans, considerations relating to the well-being of the human should take precedence over the interests of science and society.

Were these principles used in the Anchorage needle-exchange study? Explain.

ISBN 978 1 4202 3736 8

1 Match these four words with the four statements below:

inference observation

hypothesis prediction

a My pulse rate is 56 beats per minute.
b My pulse rate will increase when I run.
c The more active you are, the higher your pulse rate.
d I think my pulse rate is caused by my heart beating.

2 What is a variable? Why is it so important to control variables in an experiment?

3 Write down in the correct order the four steps in an investigation.

4 a A magnet moving in and out of a coil of wire generates an electric current. What variables could be changed to produce a larger electric current?
b Milk left open out of a refrigerator turns sour much more quickly than unopened milk kept in a refrigerator. What variables can affect the rate at which milk turns sour?
c When a hot concentrated solution of copper sulfate was poured into a watch glass, small crystals started to grow around the edge of the solution. What variables could influence the growth of these crystals?

5 Jessica set up five pots, each containing 10 small cabbage plants. Each plant was 4–5 cm tall, and each pot had the same amount of soil in it. On the day after the cabbages were planted, Jessica added different amounts of liquid fertiliser to each pot. From then on, she watered the plants the same amount each day. She observed the growth of the plants over 10 days, and her results are shown below.

Jessica seemed unaware of the plants' attempts at telepathic communication.

a What problem was Jessica investigating?
b What variables did she control in her test?
c What conclusions can you draw from her results?

Pot	Amount of liquid fertiliser added (mL)	Observations after 10 days	
		Colour of leaves	Average height (cm)
1	none	pale green	8
2	5	green	8
3	10	green	15
4	15	green	16
5	20	yellow	8

6 Dominic is a keen tennis player and has played on several different surfaces. He wants to know which surface causes balls to bounce highest. Design an experiment to answer this question. Make sure you list all the variables Dominic will have to control.

7 Work in a group and discuss how you would investigate these research questions.
a Which coloured flowers do bees prefer?
b Do the phases of the moon affect the weather?

ISBN 978 1 4202 3736 8

8 Tom wanted to find out which type of nut contained the most stored energy. For each nut, he followed the steps in the box below.

a Do you think Tom's conclusions would be valid?

b How could he improve his experiment?

1 Put some water in a test tube and clamp it in place as shown.
2 Measure the temperature of the water.
3 Pick up the nut using a metal skewer and light it in a burner.
4 Heat the water in the tube using the flame from the nut.
5 Note the increase in temperature of the water.
6 Repeat steps 1 to 5 for the other nuts.

CHALLENGE

1 Suggest why velcro loses strength when it collects thread or fluff (called lint) during washing.

2 a For each of the following hypotheses, write down the independent variable and the dependent variable.

A Punch brand batteries last longer than GoGo batteries.

B Small marble chips dissolve more quickly in acid than large chips do.

C Light-coloured clothing is cooler to wear than dark-coloured clothing.

D Iron rusts faster in sea water than in fresh water.

E The chirp rate of crickets increases in warmer weather.

b Design an experiment to test one of the hypotheses in a.

3 Four pairs of students carry out an experiment into the effects of exercise on pulse rate. Their methods are as follows.

A Kiri and Monique run on the spot for 2 minutes, then take each other's pulse.

B Drew runs on the spot for 2 minutes. Felicity then measures his pulse.

C Samara takes Mimaki's pulse while Mimaki is seated. Mimaki then runs on the spot for 2 minutes and Samara takes her pulse again.

D Adam runs on the spot for 2 minutes, then takes his own pulse. Bradley sits and takes his pulse.

Evaluate the method used by each pair of students. Which students are most likely to be able to make a valid conclusion about the effect of exercise on pulse rate? How could their experiment be improved?

4 When planning an experiment, it is a good idea to use your knowledge of science to change the question you are investigating into a hypothesis. For example:

Question: Which part of your skin is most sensitive to touch?

Hypothesis: Fingertips are the part of your skin most sensitive to touch.

Use your knowledge of science to change the following questions into testable hypotheses.

a Which objects are attracted to a magnet?

b Do plants grow better under green plastic or clear plastic?

c What is steam?

d What causes silver to tarnish?

ISBN 978 1 4202 3736 8

1.2 Collecting data

Some science investigations require you to collect data in the field. For example, you might be investigating the condition of the water in a creek. You will have to take water samples and measure the pH, temperature and the amount of dissolved nutrients such as nitrates and phosphates, and test the clarity of the water. You might also want to find out the types and numbers of organisms that live in the creek.

Let's look at some techniques used to collect data in the field.

Estimating numbers in a population

Some organisms, such as barnacles, are fixed in place in their habitat, while most other organisms are mobile and move from place to place. Different techniques are needed to study and count fixed and mobile organisms.

In field studies you can usually never count every organism in a population. You have to use methods to *sample* the population, and then *estimate* the total number.

The quadrat method

This method is used to study populations that are fixed in position. A **quadrat** is a square frame made of plastic, wire, wood or even string, which can be used to sample the organisms in a particular area.

Suppose you need to estimate the population of barnacles and molluscs in an area of a rocky shore. Quadrats vary in size but are usually 1 m × 1 m square; however, this is far too large to count small rocky shore organisms. A suitable quadrat might be 200 mm × 200 mm.

You can use the quadrat to count the various organisms in a number of different places chosen at random in the area. For a reliable estimate you should sample at least 10 places. The data can then be used to estimate the various populations in the selected habitat, or the population density per square metre.

Suppose you were studying a rocky shore and wanted to estimate the number of barnacles in a 10 m × 5 m area. You drop your 200 mm × 200 mm square quadrat at random over the selected area. You do this 10 times and each time you count the number of barnacles inside the quadrat.

Number of barnacles in each of the 10 quadrats:

5, 9, 5, 8, 11, 14, 7, 5, 10, 6 Total = 80 barnacles

Area of 1 quadrat = 200 mm × 200 mm
= 0.2 m × 0.2 m
= $0.04\ m^2$

Area of 10 quadrats = 10 × 0.04
= $0.4\ m^2$

$$\text{Population size} = \frac{\text{no. of barnacles in 10 quadrats}}{\text{area of 10 quadrats}} \times \text{total area}$$

$$= \frac{80}{0.4} \times 50$$

= 10 000 barnacles

So, you can estimate there are 10 000 barnacles on the shore.

The capture–recapture method

This sampling method is used to estimate mobile populations of organisms. In this method a sample of the population is caught, counted and tagged. The organisms are then released back into the habitat. After some time when they have dispersed throughout the population, a second sample is taken. Some organisms will be tagged, others will be untagged. Both are counted and an estimate is calculated as shown below.

Suppose 200 fish were caught in a lake. They were tagged and released. One month later, 100 fish were caught. Among these were 25 tagged fish that had been caught previously.

The capture–recapture method works on the principle that the proportion of tagged fish in the second sample is the same as the proportion of tagged fish in the total population.

$$\text{Proportion of tagged fish in 2nd sample} = \frac{25}{100}$$

$$\text{Proportion of tagged fish in whole population} = \frac{200}{\text{total}}$$

$$\frac{25}{100} = \frac{200}{\text{total}}$$

$$\text{Therefore, total} = \frac{100}{25} \times 200$$

= 800 fish

So, you can estimate there are 800 fish in the lake.

ISBN 978 1 4202 3736 8

Estimating populations

Aim

To estimate the size of a population using the quadrat and capture–recapture methods.

Materials

- a large container of plastic-coated, coloured paperclips
- 1 m of heavy wire (fencing or coathanger wire)
- at least 5 m of string
- bar magnet (optional)

Risk assessment and planning

- It is best to work in groups of three or four.
- Carefully read through the Methods for Part A and B and decide which part you will do first.
- Prepare data tables for your results in both parts.

PART A Quadrat method

Method

1 Bend the wire into a square 200 mm × 200 mm frame. This is your quadrat.

2 Count the paperclips. Then scatter them over an area of at least 2 m × 2 m in the room or outside.

Instead of dropping the quadrat at random, you will use a *transect*. This is a line across your area along which you place your quadrats. Take your 10 samples along this line.

3 Have two people hold the ends of the string and *without looking* lay it across the area containing the scattered paperclips.

4 Use the quadrat to take 10 samples along the transect.

Count and record the number of paperclips.

Discussion

1 Find the total number of paperclips in the 10 quadrats.

2 The area of the 200 mm × 200 mm quadrat is 0.04 m^2. Find the total area of the 10 quadrats.

3 Use the equation below to estimate the population of paperclips.

$$\text{Population size} = \frac{\text{no. of paperclips in 10 quadrats}}{\text{area of 10 quadrats}} \times \text{total area}$$

4 How does the estimated paperclip population compare with the known count of paperclips?

5 Calculate the population density in numbers per square metre. (Use the estimated population.)

6 Suggest ways to improve this investigation so that you obtain more accurate results.

7 What is the advantage in taking samples along a transect? Can you think of another way to sample the population that will give you reliable results?

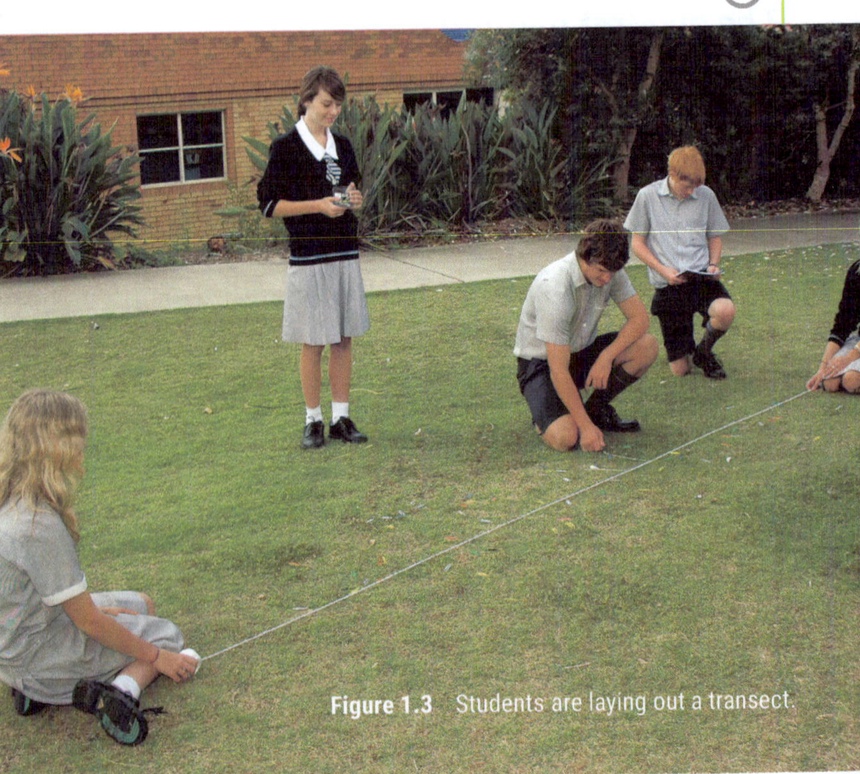

Figure 1.3 Students are laying out a transect.

ISBN 978 1 4202 3736 8

PART B Capture–recapture method

Method

1 Empty the container of paperclips on the desk. Select a colour and count all the paperclips of this colour. These represent your *tagged* paperclips in the total population.

Record this number.

2 Return the paperclips to the container and mix them up well.

3 Use a small container about the size of an eggcup or a kitchen measuring spoon to scoop out some paperclips. Alternatively you can dip a bar magnet into the paperclips.

4 Count the number of paperclips in the sample and also the number of the selected colour (tagged) paperclips.

Record your results.

5 Return the sample to the container, mix well and repeat steps 3 and 4 for a total of 10 samples.

Discussion

1 Use the ratio formula below to calculate the estimated population size for each of the 10 samples.

$$\frac{\text{total no. tagged}}{\text{population size}} = \frac{\text{no. tagged recaptured}}{\text{no. in sample}}$$

2 Find the average population size for the 10 samples. Compare this with the known size of the population of paperclips.

3 Comment on the reliability of your results. Could you improve your method?

4 This method assumes that the number of individuals in a population remains the same throughout the sampling. Would this be true of a population of fish in a lake? What factors might affect this assumption?

Sampling in the field

Collecting data on the types and numbers of organisms is one part of a field study; obtaining data on the physical factors in the environment is the other part.

You know from previous studies in science that physical or abiotic factors such as temperature, availability of water, soil types and soil nutrients, and the pH of water and soil (see pages 216–217) play a large part in determining the abundance and distribution of organisms in a particular habitat.

Figure 1.4 Sampling the distribution of seaweed on a rocky shore using a quadrat

ISBN 978 1 4202 3736 8

Sampling physical factors

Aim

To practise various techniques for sampling physical factors in the environment, such as temperature, pH, dissolved salts and dissolved oxygen.

This investigation is best done in the field. If this is not possible, you will be supplied with four buckets of creek or pond water. Assume that each bucketful of water has been collected from one of four sites in a creek as shown in the diagram.

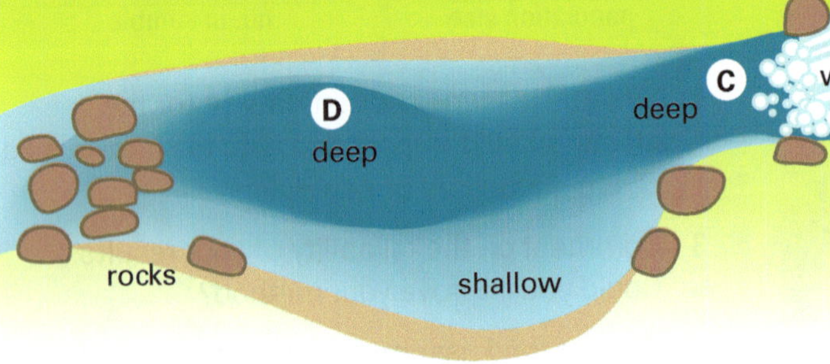

Note: You may be able to use a datalogger with a temperature probe and a pH probe instead of the thermometer and pH paper.

Risk assessment and planning

- Work in groups of three or four.
- Carefully read through each part. Make sure you know what to do. Decide which part your group will do first, then discuss this with your teacher.
- Prepare data tables for your results in each part.
- Make a list of the safety issues in each part. Your teacher will discuss these as a class before you start.

PART A Temperature and pH

Materials

- four buckets of pond water or specially prepared water
- 100 mL beaker or glass jar
- thermometer
- pH paper, universal indicator solution and colour card, or swimming pool pH kit
- distilled water

Method

1 Make sure your beaker is clean and rinsed in distilled water.

2 Number the buckets. Take about half a beakerful of water from the first bucket.

3 Use the equipment to find the temperature and pH of the water.

Record the results in the data table.

4 Tip out the water and rinse the beaker in distilled water.

5 Repeat steps 2 and 3 for the other three buckets of water and record your results.

Discussion

1 Why is it necessary to rinse the beaker in distilled water after each test?

2 Reliable results are obtained when you take a number of samples and average the results. Would you average the data from all four sites in the creek or just some of the sites, or use individual readings? Use the map of the creek to justify your answer.

ISBN 978 1 4202 3736 8

PART B Conductivity

Conductivity is a measure of the amount of dissolved salts and nutrients in the water. Water from a salt water swimming pool is tested for conductivity to determine how much dissolved salt is in the water.

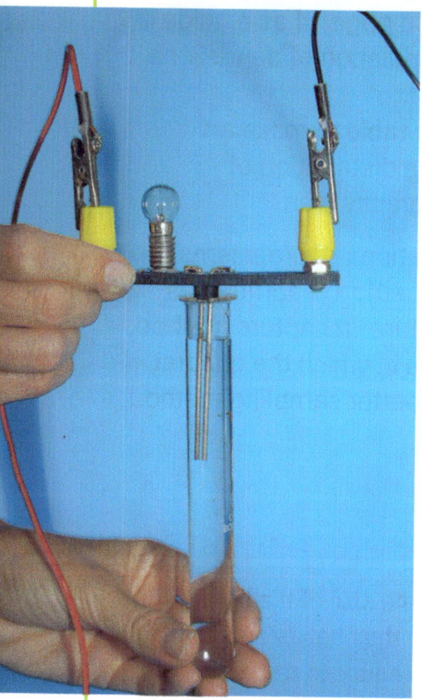

A conductivity probe contains two metal electrodes. When the battery is switched on and the electrodes are dipped into the water, the ions in the water carry the current between the electrodes. A meter reads how much current flows. This reading is proportional to the concentration of the dissolved salts.

Materials

- 100 mL beaker or glass jar
- conductivity kit or datalogger with conductivity probe

Method

1 Clean the beaker and rinse it in distilled water.

2 Take a sample of about 50 mL from one of the buckets of water. Record the number of the bucket.

3 Use the equipment to find the conductivity of the water.

Record your results in the data table.

Discussion

1 Calculate the average conductivity of all four samples.

2 What conditions would change the conductivity of the water in a creek?

PART C Dissolved oxygen

Dissolved oxygen (DO) is a very important factor in determining the distribution and abundance of aquatic organisms. Some organisms can survive only in water with high levels of dissolved oxygen, while others can tolerate very low levels.

Materials

- 100 mL beaker and glass jar with screw lid
- oxygen meter with probe, or DO test kit

Note: Your teacher will show you how to use the oxygen meter and probe if your school has one. Alternatively you will be shown how to use the dissolved oxygen (DO) test kit.

Method

1 Clean the beaker and rinse it in distilled water.

2 Without disturbing the surface of the water too much, slowly dip the beaker into a bucket of water and collect about 70 mL of water.

3 Use the oxygen meter or the DO test kit to find the level of DO in the water.

Record your results in the data table.

4 Repeat steps 1 to 3 for the other buckets of water.

Record your results.

5 Take another water sample from any bucket and pour it into the glass jar. Screw the lid on and shake it vigorously. Then test for DO.

Record your results.

Discussion

1 Compare the DO in the shaken jar with the water in each of the buckets. Account for the differences.

2 Why was it necessary to avoid disturbing the water when you took your samples from the buckets?

3 What biotic and abiotic factors might change the level of dissolved oxygen in a creek?

ISBN 978 1 4202 3736 8

1 Match these words with their descriptions:

sample quadrat conductivity
transect population abiotic

a A measure of the concentration of ions in water
b A square frame used to count organisms in a particular area
c A line across a selected area, which is used as a guide to sample organisms
d A number of organisms of the same kind in a particular area
e A small group of organisms selected from the total population
f The physical or non-living factors in the environment

2 The table (top right) shows the number of dandelion plants in a grassy area in 1 m × 1 m quadrats taken along a transect. The grassy area measured 10 m × 25 m.

Quadrat	1	2	3	4	5	6	7	8	9	10
Number of dandelion plants	4	4	9	9	11	6	8	9	7	4

a Find the total number of dandelions in the 10 quadrats.
b Use the equation on page 14 as a guide to find the total population of dandelions in the grassy area.
c Use the data in the table to make an inference about the distribution of dandelions in the grassy area.

3 When sampling populations of organisms in the field, the quadrat method is sometimes preferred over the capture–recapture method. Describe the situations in which the quadrat method would be the better sampling method to use.

1 The grid below shows the number of feral horses in a particular area. The horses were photographed from an aircraft and their positions (●) were placed on the grid shown.

1	2	3	4	5
6	7	8	9	10
11	12	13	14	15
16	17	18	19	20
21	22	23	24	25

Biologists want to estimate the size of the horse population so they can study their habits and try to reduce the damage that they cause to native wildlife.

The biologists used the quadrat method to sample the horses. They selected five squares at random (shown in blue) and also five squares along a transect (shown in yellow).

a Use the random squares (blue) to estimate the size of the horse population.
b Now use the transect squares (yellow) to estimate the horse population.
c Do your answers for a and b indicate that one method gives a more accurate estimate of the total horse population than the other? Explain.
d Select another five squares to show that results can vary when using the quadrat method. How did you select the quadrats?
e Why was the quadrat method used by the biologists? Could the biologists have used the capture–recapture method instead? Give reasons for your answer.

ISBN 978 1 4202 3736 8

1.3 Processing data

Once you have done an experiment and collected your data, you need to organise and display it. This makes it easier to identify any patterns or trends in the data. It also makes it easier to discover any cause-and-effect relationships or links between the variables. That is, does increasing (or decreasing) one variable have any effect on another variable?

Over 300 years ago an English schoolteacher called Robert Hooke found a relationship between the amount a spring stretches and the force used to stretch the spring. You can repeat Hooke's experiment yourself on the next page.

SKILLBUILDER

Drawing lines of best fit

In the next investigation, you are going to use your data to draw a graph to show the relationship between two variables.

You will find that the points you plot on the graph will lie close to, but not exactly on, a straight line. You need to draw a **line of best fit**, rather than joining all the points. A line of best fit averages out any errors you made in your measurements in the investigation.

SKILLBUILDER

Writing a science magazine article

Robert Hooke 1635–1703

Robert Hooke has been described as the greatest experimental scientist of the 17th century. Yet he is not nearly as famous as Isaac Newton.

Your task is to research information about Robert Hooke and write an interesting science magazine article about him (maximum 500 words).

Structure of the science article

Here are some hints and tips on writing an article for a science magazine.

- Write more of a human interest story than a science story.
- The *introduction* is very important. You should entice your reader with emotion, drama, descriptions and quotations.
- The *body* of the article needs to expand the ideas from the introduction.
- The *conclusion* should be short and punchy and remind the reader of the key points of the story.
- Write in the active voice, e.g. 'Robert Hooke used his artistic talents to draw the organisms he saw with his newly invented microscope'.
- Avoid lengthy paragraphs. Two or three sentences will do for each paragraph.

Suggestions

- Use the websites below or search for *Robert Hooke* in your browser.
- Write your article electronically. You can download images from websites. Make sure you reference any sources and check copyright requirements.
- Make sure your article is scientifically and historically accurate. Don't make up information!

EXPLORE ONLINE

Check out the links to the websites below.

Robert Hooke (1635–1703)

This website contains useful information and links to other sites.

Robert Hooke

This website is dedicated to Robert Hooke and has useful information and pictures that can be downloaded.

Robert Hooke—natural philosopher, inventor …

This is an interesting website with a large amount of information about his discoveries and achievements.

ISBN 978 1 4202 3736 8

EXPERIMENT 1.2 Hooke's spring

Research question

What is the relationship between the force (load) on a spring and the extension of the spring?

Materials

- helical spring
- 50 g mass hanger and standard masses
- stand and clamp
- metre ruler
- brick or other heavy mass
- graph paper

Risk assessment and planning

- Use the research question above to write a hypothesis linking the load and the extension of the spring.
- List the steps you will take in your experiment. Use the photo as a guide to setting up your apparatus.
- Draw up a suitable data table in which to record the load (mass added) and the extension (amount of stretch) of the spring.
- List any safety issues.

Planning hints

1 To find the load in newtons, divide the mass (in grams) by 100.

2 After adding the first mass, remove it and check that the spring returns to the zero mark. If it does not, you may not be able to form a valid conclusion from your results. Continue in this way by adding extra masses and recording the extensions. (If the spring does not return to the zero mark between measurements, it is best to stop the experiment and try another spring.)

Note: You could enter your data into a computer spreadsheet such as Excel.

Discussion

1 Look closely at your data. Do they support your hypothesis?

2 Which is the independent variable (the one you purposely changed in the experiment)?

3 Which is the dependent variable (the one you measured)?

4 Use graph paper to draw a line of best fit as shown below.

5 Compare your data with the data collected by other students. Explain any differences.

Spring extension vs load

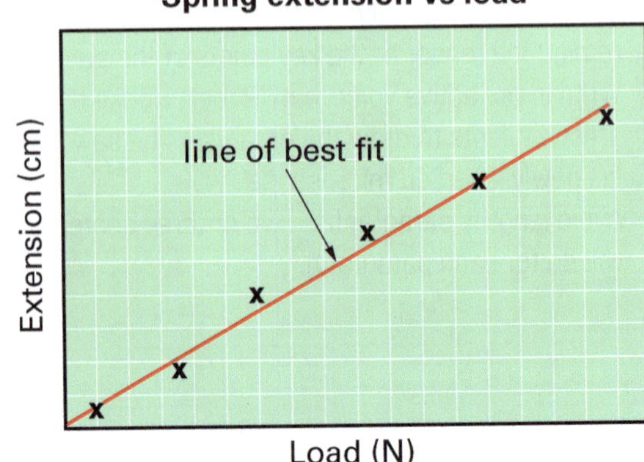

ISBN 978 1 4202 3736 8

Interpreting graphs

Graphs are a very useful way of displaying patterns or trends in data. For example, the graph in the previous experiment shows that the extension of the spring and the load on the spring are directly related to each other. An increase in one variable causes an increase in the other. Similarly, a decrease in one causes a decrease in the other. The fact that the graph is a straight line means that the increases or decreases are proportionally equal. For example, if you double the load, you double the extension, and if you triple the load, you triple the extension.

Sometimes an increase in one variable causes a decrease in the other. For example, the number of hits on an archery target, as shown in Figure 1.5, decreases as the distance from the target increases. In this case the variables are inversely related.

Note that a line of best fit does not go through all the points, but it does go close to them. It tends to 'average' the points and reduce any inaccuracies due to the experimental method used. It is also possible to draw a curve of best fit, as shown in Figure 1.6.

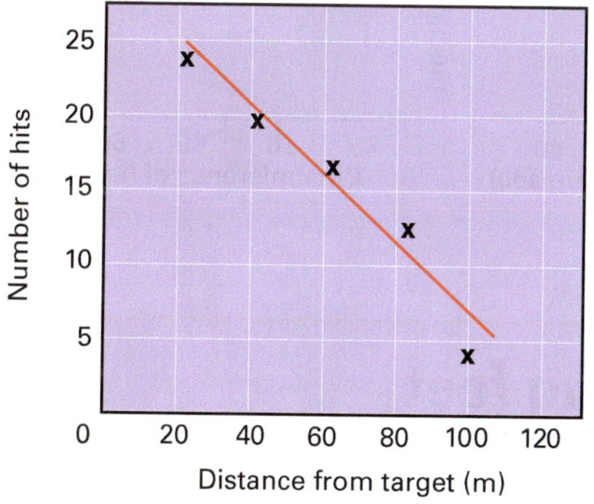

Figure 1.5 An increase in one variable causes a decrease in the other.

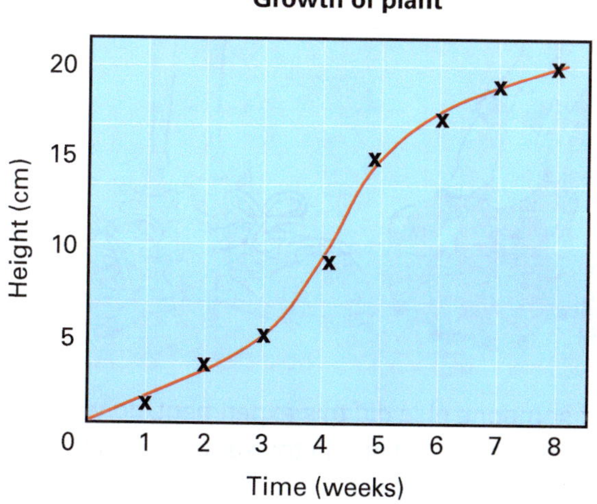

Figure 1.6 A curve of best fit

Predicting from graphs

Graphs not only show patterns but can also be used to make predictions. If the prediction is *between* two measurements, the process is called **interpolating** (in-TERP-oh-late-ing). On the graph in Figure 1.7, for a load of 3.5 N you can predict a spring extension of about 20 cm.

It is also possible to make predictions for values *beyond* the measured values. This process is called **extrapolating** (ex-STRAP-oh-late-ing). For example, for a load of 7 N you can extend the straight line and predict an extension of about 40 cm. However, the graph may not be a straight line at that point—for example, it may curve upwards. (This is what happens if the spring does not return to its original length when the load is removed.) If this is the case your prediction of a 40 cm extension for a load of 7 N will be far too small. This is why you often see widely different predictions for such things as world population or global warming.

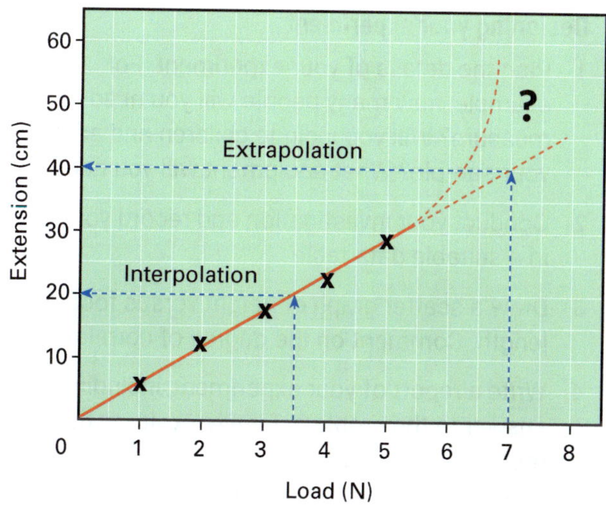

Figure 1.7 Interpolating and extrapolating on a graph

ISBN 978 1 4202 3736 8

Scatter graphs

Suppose you want to see if there is a relationship between the mass and the height of students. You simply plot all the points and look for any *pattern* in the scatter of the points. This type of graph is called a **scatter graph.**

In graph 1, there is an obvious trend. The taller the student, the heavier they are likely to be. There is a direct relationship between the two variables. We say there is a *high* **correlation** between them. In fact, you could draw a line of best fit through the points.

In graph 2, there is no direct relationship, but there is *some correlation*. Most plants tend to grow in soils with water content between 10 and 25 g/100 g water.

In graph 3, there is no relationship between the size of a person's head and their intelligence. There is *no correlation*.

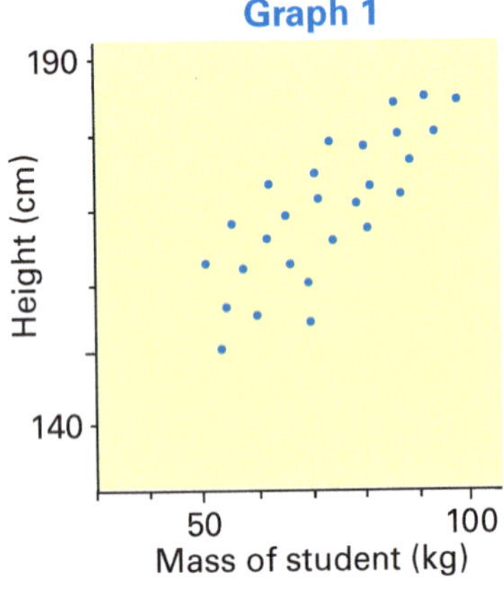

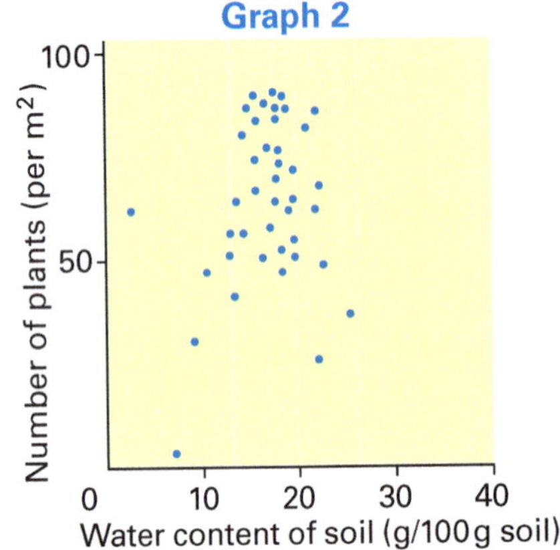

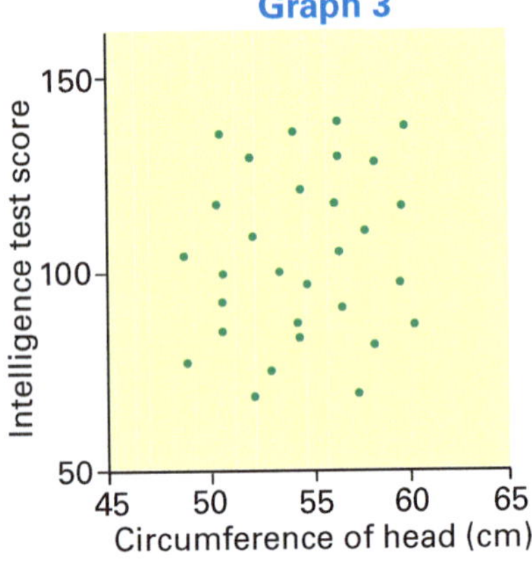

Figure 1.8 Scatter graphs

EXPERIMENT 1.3 Measuring feet

The problem to be solved

Is there any correlation between the length of a person's foot and their height?

Designing your experiment

1. Plan the details of your experiment. For example, how many people will you need to measure? Will you include children and adults in your sample? What equipment will you need?
2. Conduct your investigation and record your data in a suitable data table.
3. Draw a scatter graph of height versus foot length. Comment on the degree of correlation.
4. Write a report of your experiment, including the answer to the problem. Finally, evaluate the method you used. Are there things you could do to make your results more reliable?

If someone else did this experiment, do you think they would obtain the same results? Explain your answer.

ISBN 978 1 4202 3736 8

Solving problems

Ryan and Brittany set up the equipment below and recorded their results in a data table.

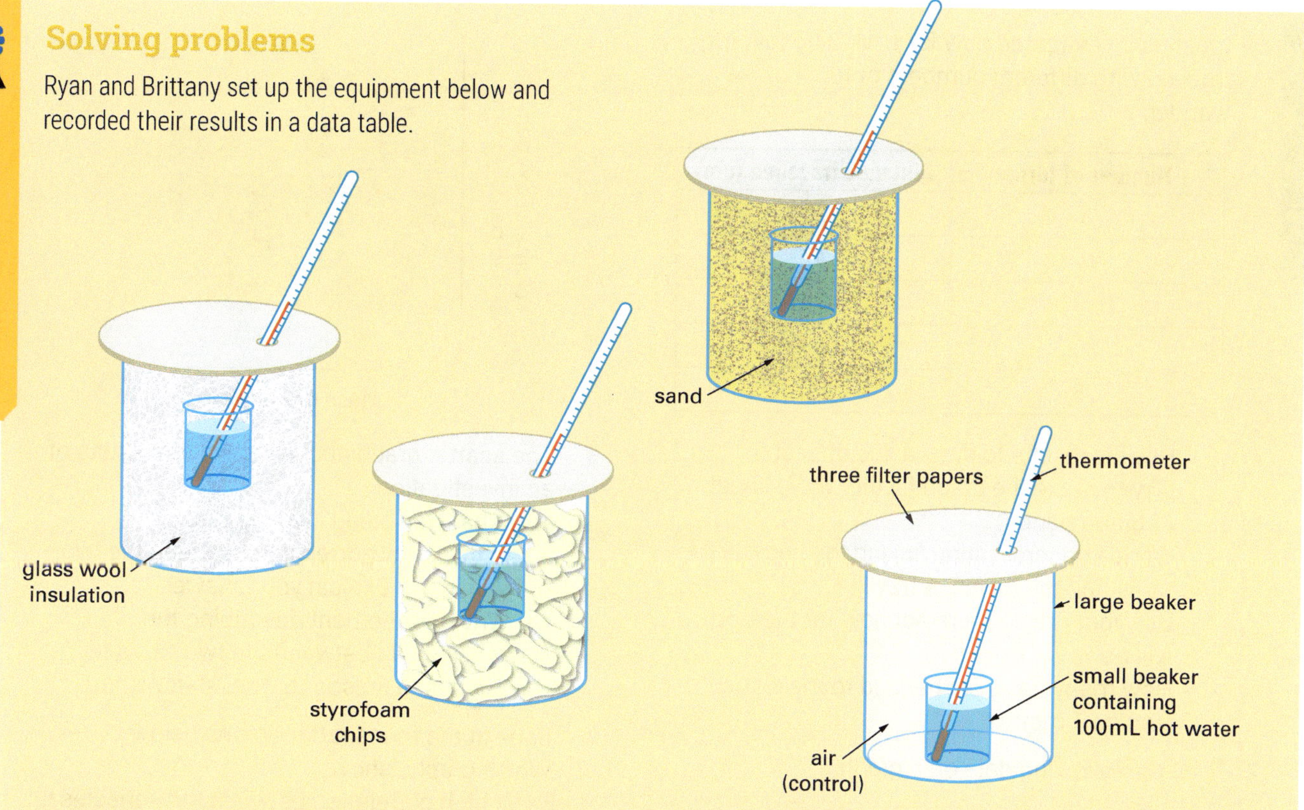

Insulating material	Temperature (°C) after ... min									
	1	2	3	4	5	6	7	8	9	10
glass wool	61.5	59	57.5	56	55	54	53	52	51	50.5
styrofoam chips	61	59	57	55.5	54	53	52	51	50	49.5
sand	61.5	59.5	58	57	55.5	54.5	53.5	52.5	52	51
air (control)	61	58.5	56.5	55	53	52	50.5	49.5	48.5	48

Questions

1 What problem were Ryan and Brittany trying to solve by doing an experiment? Write your answer as a question.

2 Suggest why they put filter papers on top of each beaker.

3 Ryan and Brittany were careful to change only one variable and keep all the others the same.
   a Which variable did they change?
   b Which variables did they keep the same? (There are at least four.)

4 What was the purpose of the control beaker that contained only air?

5 Plot the results on graph paper. Use a different coloured pencil for each material and label the lines.

6 Summarise the results, making sure you answer the question Ryan and Brittany were trying to solve.

7 The initial temperature of the hot water was the same in all four beakers. Use your graph to extrapolate what this temperature was.

8 Could Ryan and Brittany improve their experiment? How?

9 What are the scientific skills that Ryan and Brittany used in solving their problem?

10 Do you think they will use these skills when they leave school and get jobs? Explain your answer.

ISBN 978 1 4202 3736 8

CHECK

1 Joshua investigated how far a wind-up toy frog moved with different numbers of turns of the winder.

Number of turns	Distance travelled (cm)
5	23
10	47
15	70
20	90
25	117

a Use his results to draw a line of best fit. (Try to make the graph fill the whole sheet of graph paper.)
b Write a generalisation linking the number of turns and the distance travelled.
c Use your graph to predict how far the frog will go with 12 turns.
d How many turns are needed to make the frog go 1 metre?

2 Plot the following data on a graph.

Air temperature (°C)	Distance hiked in 1 hour (km)
9	8.6
15	6.4
22	4.3
25	3.2
30	2.1

a Draw a line of best fit.
b Write a statement describing the relationship between the two variables.
c Use the graph to predict how far you would expect to be able to hike at 20 °C and at 35 °C.
d Which of these two predictions do you think is more accurate? Why?

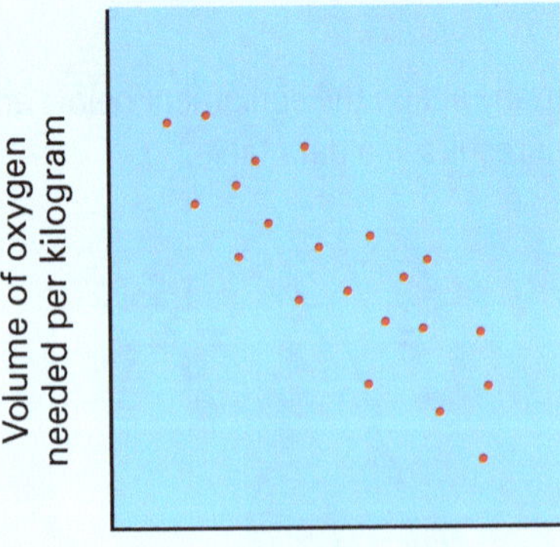

3 The scatter graph above shows the results of an investigation.
a What was being investigated?
b Is there any correlation between the two variables? Explain your answer.
c Write a statement describing the relationship between the two variables.
d Suggest a reason for the relationship.

4 Look at the four scatter graphs below.
Which graphs show:
- a high correlation between the variables?
- a low correlation between the variables?
- no correlation between the variables?

Give a reason for each choice.

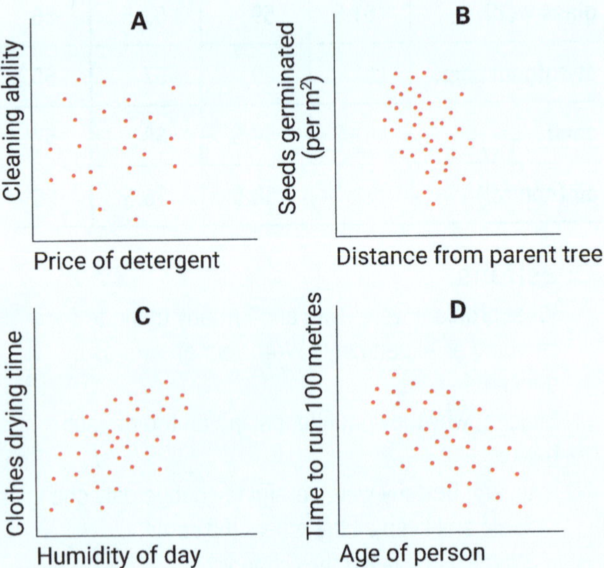

ISBN 978 1 4202 3736 8

1 The table shows the average stopping distance of a car on dry and wet roads.

a Which is the independent variable and which is the dependent variable?

b Plot both sets of data on the one graph and draw curves of best fit. Label one curve 'dry road' and the other 'wet road'.

c What conclusions can you make from the graph?

d What variables do you think would have been controlled in this investigation?

Speed (km/h)	Stopping distance (m)	
	Dry road	Wet road
0	0	0
20	8	8.5
40	20.5	22
60	38	43
80	60.5	71
100	87.5	106
120	119	149

2 Use the scatter graph below to answer the following questions.

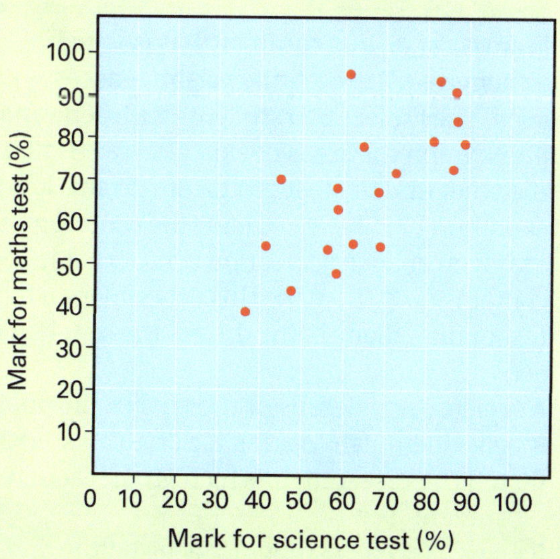

a Which two variables are plotted on the graph?

b What was the highest mark on the science test?

c What was the lowest mark on the maths test?

d What correlation is there between the two sets of marks?

e Draw a line of best fit through the points.

f If a student obtains a score of 70% on the science test, predict their score on the maths test.

g Compare your prediction with those made by others. Explain any differences.

3 Write a sentence to describe the relationship between the variables in each of the six graphs below.

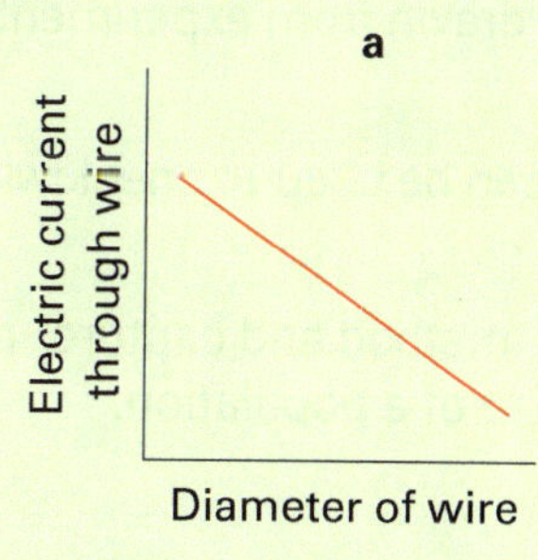

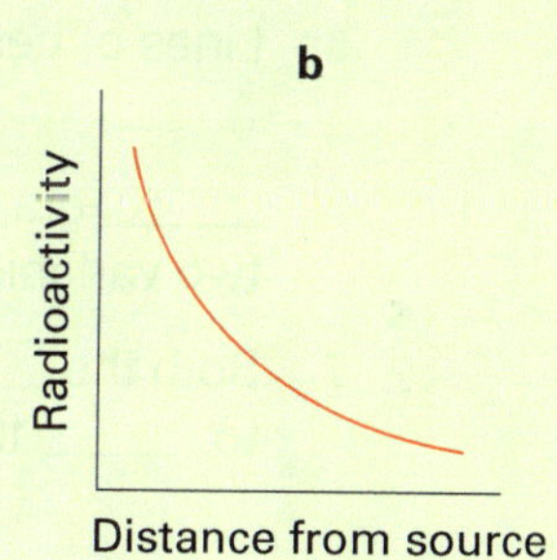

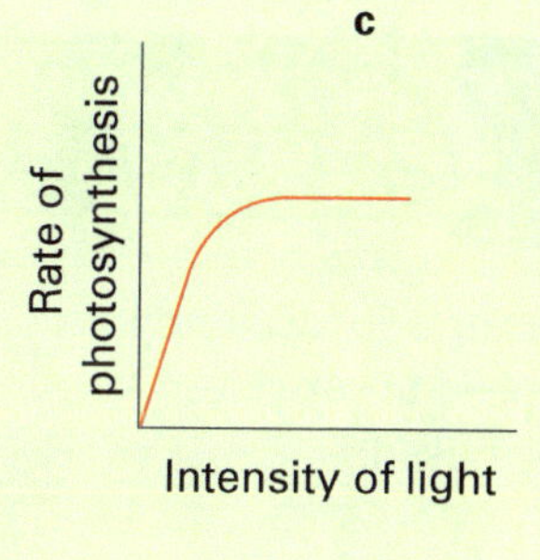

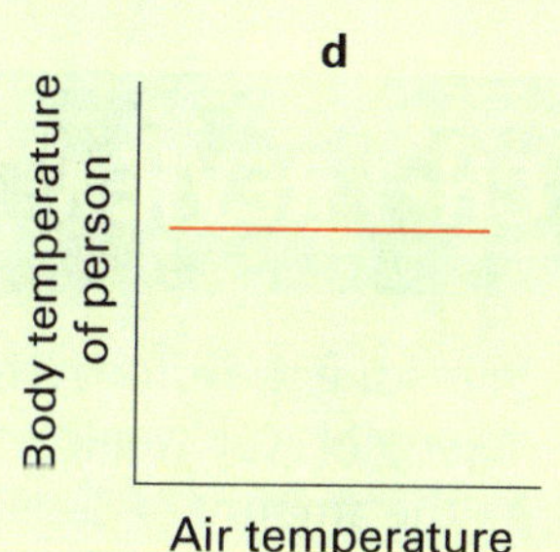

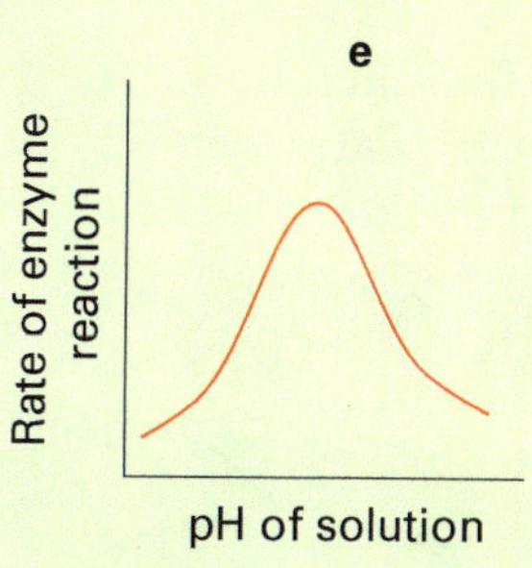

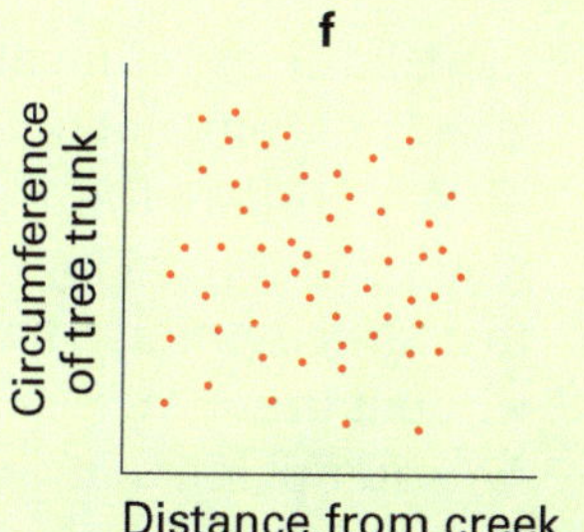

ISBN 978 1 4202 3736 8

MAIN IDEAS

Copy and complete these statements to make a summary of this chapter. The missing words are on the right.

1 The four main steps in a scientific investigation are ______, doing the experiment, processing the data and ______ the experiment.

2 To obtain ______ results in an experiment, you usually need to take repeated measurements and calculate an ______.

3 To evaluate an investigation, you think about how you could improve the experiment and whether your conclusions are ______.

4 Processing data involves looking for ______ or trends showing relationships between the ______ being investigated.

5 Lines of best fit drawn from experimental data can be used to make ______.

6 ______ graphs can be used to check what correlation there is between two variables.

7 Both the ______ method and capture–recapture method can be used to ______ the size of a population.

quadrat
estimate
patterns
valid
average
planning
variables
predictions
evaluating
scatter
reliable

CH•1 REVIEW

1 In testing the effectiveness of a dishwashing detergent, you would not need to consider:
A the amount of detergent used.
B the time of day.
C the temperature of the water.
D how the dishes were washed.

2 Glen did an experiment to find out if fertiliser affects the amount of oxygen a water plant makes. He used the apparatus shown below. A suitable control for this experiment would be to use the same apparatus but without the:
A water plant.
B test tube.
C fertiliser.
D water.

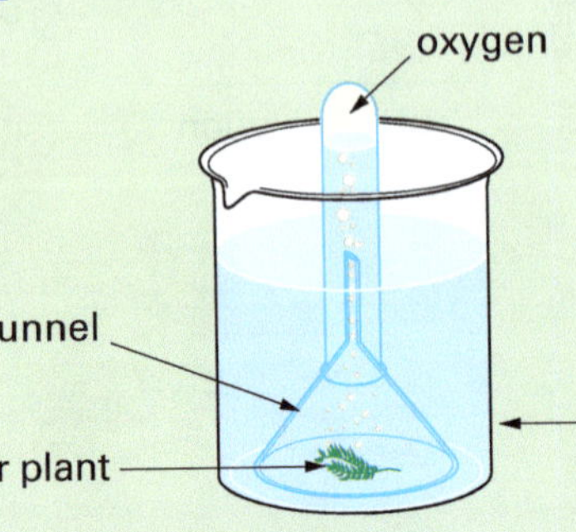

3 From coral, a drug company has isolated a chemical (Z), which they claim reduces acne. They select 200 students with acne and photograph the areas of skin affected. Half of the students (the test group) are given a lotion containing Z. The other half (the control group) are given an identical lotion except that it contains no Z. To make this experiment a fair test of ingredient Z, the drug company should *not*:
A release any details of the trial to the public.
B give identical-looking lotion to all students.
C tell the researchers which group each student is in.
D compare each student's acne before and after the experiment.
Justify your answer.

ISBN 978 1 4202 3736 8

4 A biologist found that if you eat a meal containing a lot of carbohydrates, your blood sugar level rises rapidly then drops off almost as rapidly. If you eat a meal containing a lot of protein, your blood sugar level rises more slowly to a lower peak. It also drops more slowly, but it does not fall as far as with a high carbohydrate meal.

Which graph correctly shows these findings?

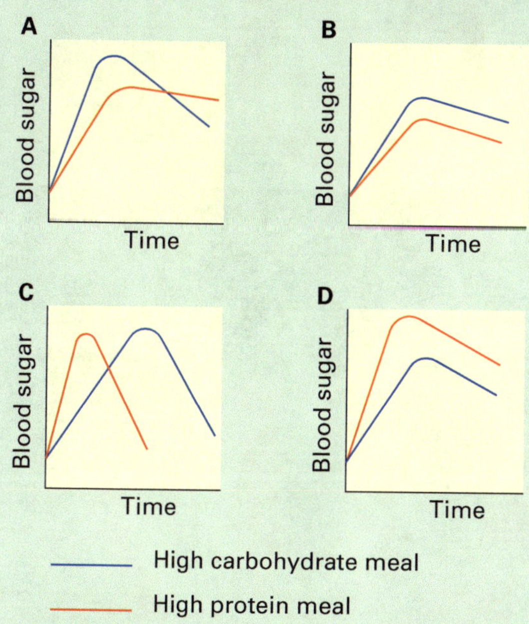

5 Describe how you would use the capture–recapture method to estimate the population of mullet in a section of a river.

6 Two students used the quadrat method to estimate the population of periwinkles on a rocky platform close to the water's edge. The rocky platform measured 20 m × 5 m, and ten 1 m × 1 m quadrats were sampled along a transect.

Quadrat	1	2	3	4	5	6	7	8	9	10
Number of periwinkles	10	12	15	13	12	9	16	14	10	9

a Find the total number of periwinkles in the 10 quadrats.

b Estimate the periwinkle population on the rocky shore platform.

c Suggest why the quadrat method was used by the students instead of the capture–recapture method.

7 Nancy and Daniel were both given a new bicycle for Christmas. Nancy's was a mountain bike with 22 gears and Daniel's was a BMX bike with 10 gears. Nancy argued that her bike was safer because its larger wheels meant it would stop more quickly than Daniel's BMX bike with smaller wheels.

To settle the argument, Nancy and Daniel rode their bikes down a hill and braked when they reached a particular spot on the road. Nancy stopped in 22 m and Daniel stopped in 14 m. Daniel claimed that Nancy was wrong—small wheels stop you more quickly than large wheels.

a What Nancy and Daniel did was not a fair test of wheel size and braking ability. List at least three uncontrolled variables that could have affected the results.

b Suggest ways in which Nancy and Daniel could improve their test.

8 Chung investigated the relationship between the diameter of a rope and its breaking strain.

Diameter of rope (cm)	Breaking strain (kg)
1	400
2	500
3	750
4	950
5	1100

a Use Chung's results to draw a line of best fit.

b Write a statement of the relationship between the two variables.

Check your answers on page 296.

ISBN 978 1 4202 3736 8

IN THIS CHAPTER

Science Understanding

- describe the energy changes that occur in a simple electric circuit
- investigate factors such as conductivity that affect the transfer of energy through an electric circuit
- investigate the differences between series and parallel circuits
- describe how the discovery of conducting plastics has led to the development of new technologies

Science Inquiry Skills

- write a generalisation about the types of materials that conduct and do not conduct electricity
- connect up and investigate simple electric circuits
- use your knowledge of electric circuits to invent a useful electrical device

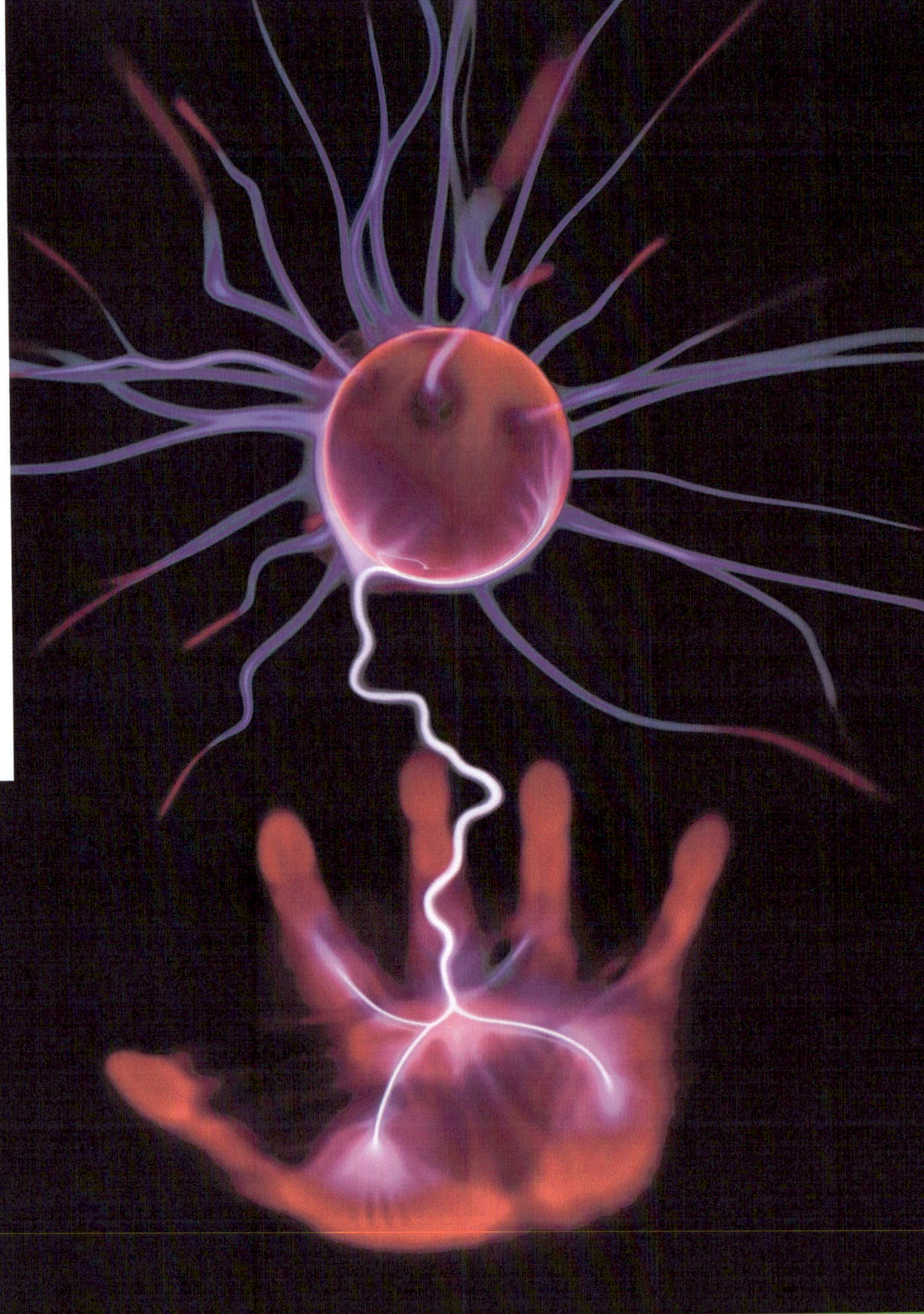

CH•2 Introducing electric circuits

ISBN 978 1 4202 3736 8

GET STARTED: *EXPLORE*

In this photo at the Franklin Institute in Philadelphia, USA, a person sits in a cage which is being struck by a massive electrical discharge. The giant Van de Graaff generator (the columns and spheres behind the cage) builds up electrical charge which jumps through the air and hits the cage. The voltage produced can reach as much as 2.5 million volts—your home electricity is only 240 volts.

> The cage is designed so that no electrical discharges will reach inside. What do you think the cage would be made of? Explain why.
> The lightning hits the cage, but why doesn't the lightning reach the person inside? Where does the electricity go after hitting the cage?
> Do you think a cage like this could protect sensitive electrical equipment from lightning strikes? Explain.
> Can you think of any other uses for a cage like this?
> This is an example of static electricity. Do you know any other examples?

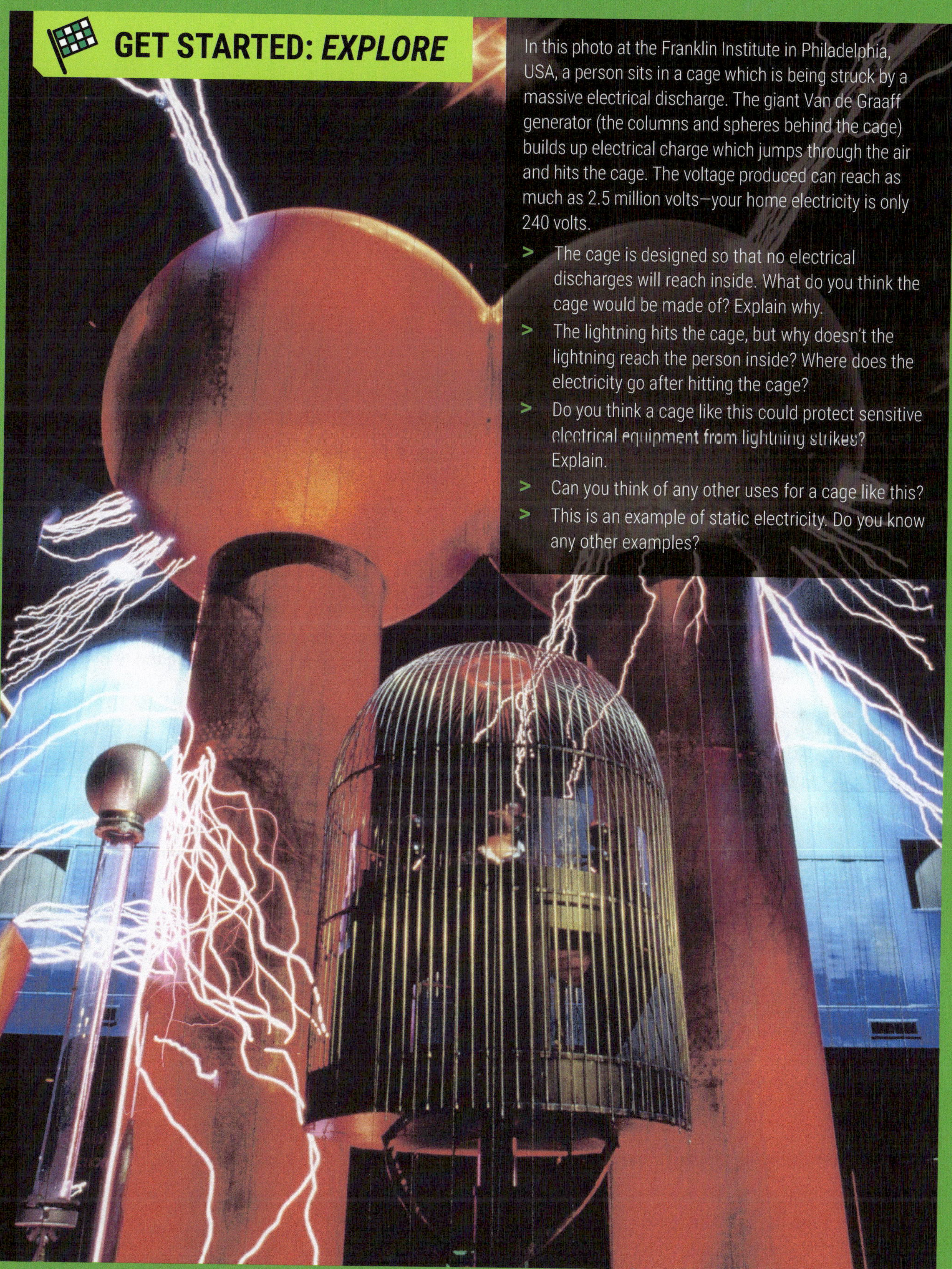

ISBN 978 1 4202 3736 8

2.1 Electric charges

You have probably experienced static electricity. Maybe by rubbing your feet on the carpet, jumping on a trampoline, brushing your hair or going down a slide. Maybe you have even seen sparks when pulling a woollen jumper over your head. If so, you probably noticed that your hair began to stand up, you felt a tingling sensation and when you touched something or somebody you got a little electric shock.

Electric charges can build up on objects that are rubbed together, due to the friction between them. This build-up of electric charges on objects is called *static electricity*, because the charges stay on the object. They are stationary.

Figure 2.1 Static electricity can be caused by rubbing a balloon on your hair.

INVESTIGATION 2.1 Electric charges

Aim

To make and investigate electric charges.

Risk assessment and planning

This investigation can be done only on dry days. Also, make sure everything (including your hands) is grease-free. Wash the equipment in soapy water and dry it thoroughly. You may need to warm some equipment in an oven. Read through the four parts. Note that Part B is a teacher demonstration.

PART A

Materials

2 balloons and string

Method

1 Blow up a balloon and tie it.

2 Rub the balloon on a jumper or woollen cloth. Hold the balloon against a wall, then let it go. What happens?

3 Charge a second balloon in the same way. What happens when you hang the two charged balloons close together?

PART B

If your school has a Van de Graaff generator, your teacher may demonstrate how it is able to generate a static electric charge on its dome. You may even be able to make your hair stand on end.

Figure 2.2 A Van de Graaff generator can produce static electricity.

ISBN 978 1 4202 3736 8

PART C

Materials

- piece of fur or silk
- plastic rod

Method

Rub the plastic rod vigorously with fur or silk and bring it near (but not touching) a trickle of water.

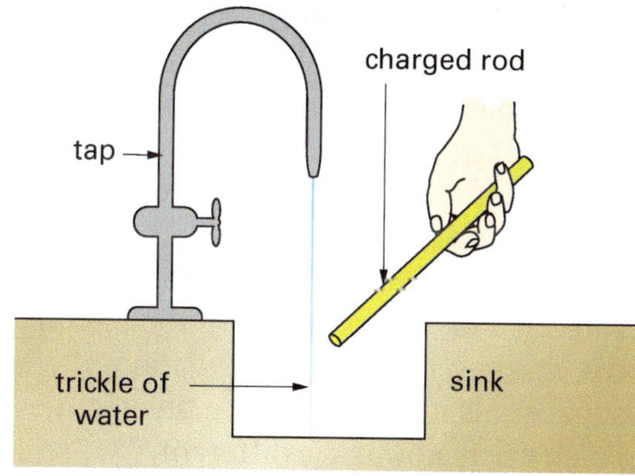

Describe what happens.

Predict what will happen if you *do* touch the water with the rod. Give a reason for your prediction. Now try it.

PART D

Materials

- 2 perspex rods
- 2 ebonite rods
- piece of wool or fur
- piece of silk
- watch glass
- Blu-Tack
- cooking oil
- tile

Method

1 Put a watch glass on top of a drop of oil on a tile. Place a small amount of Blu-Tack on either side of the watch glass, as shown.

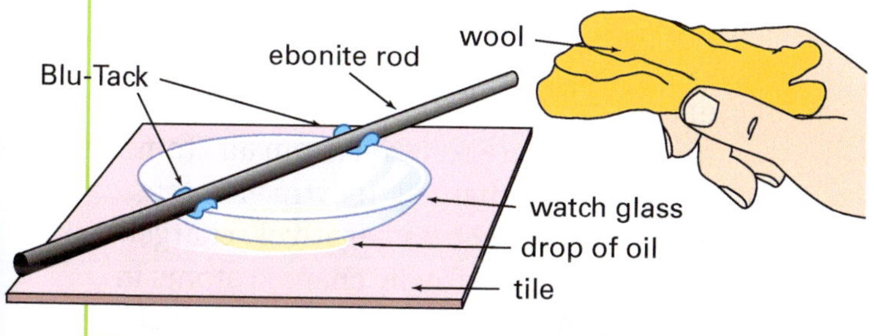

2 Rub an ebonite rod (the black one) with wool and place it on the Blu-Tack. Bring the wool near one end of the rod. Try the other end as well.

Record your observations.

3 Take the ebonite rod off the watch glass. Rub a perspex rod (the clear one) with silk, place it on the watch glass, and bring the silk near one end.

Record your observations.

4 Rub an ebonite rod with wool and place it on the watch glass. Rub a second ebonite rod with wool and bring it near one end of the rod on the watch glass. Repeat the test, but this time use two perspex rods rubbed with silk.

5 Repeat step 4 but this time bring a charged perspex rod up to a charged ebonite rod, and vice versa.

Record the results for steps 4 and 5 in a data table as shown below.

Rod 1	Rod 2	What happened
ebonite with wool	ebonite with wool	
perspex with silk	perspex with silk	
ebonite with wool	perspex with silk	
perspex with silk	ebonite with wool	

Discussion

1 Shannon tried to do the tests by placing the rods on the desk top instead of on a watch glass. She saw nothing happen. Suggest a reason for this.

2 Do both the charged rods behave in the same way? Explain your answer.

Conclusion

Write a generalisation to explain the results of your tests with charged rods.

ISBN 978 1 4202 3736 8

Attraction and repulsion

You have seen that rods rubbed with different types of cloth can move one another by non-contact forces. But why do electric charges sometimes attract and sometimes repel?

Let's hypothesise that an electric force is something like a magnetic force—another type of non-contact force. With magnets, two like poles repel each other, while two unlike poles attract.

So if two perspex rods rubbed with silk repel each other, they have the same electric charge on them. Similarly, two ebonite rods rubbed with wool repel each other, so they also have the same charge. However, a perspex rod rubbed with silk and an ebonite rod rubbed with wool attract each other, so they have opposite charges.

A magnet can attract some unmagnetised metals, so you might also expect that a charged rod can attract some uncharged objects.

There are three laws that describe electric forces.

1. Charged objects attract uncharged objects. For example, a charged plastic rod will attract small pieces of paper or a stream of water (as in Investigation 2.1 Part C).
2. Like charges repel each other. It does not matter whether they are both positive or both negative.

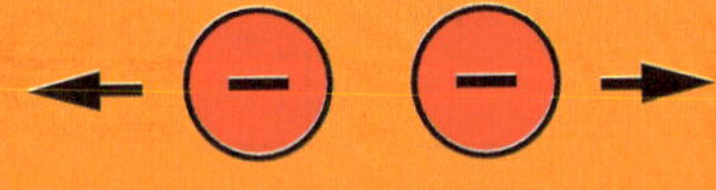

force of repulsion

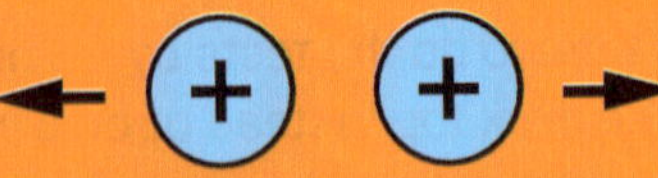

3. Unlike charges attract each other.

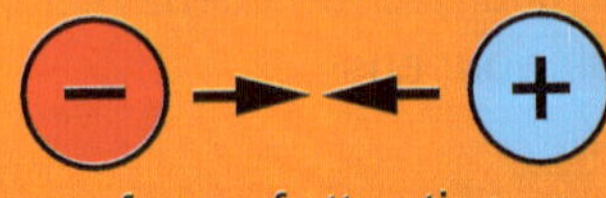

force of attraction

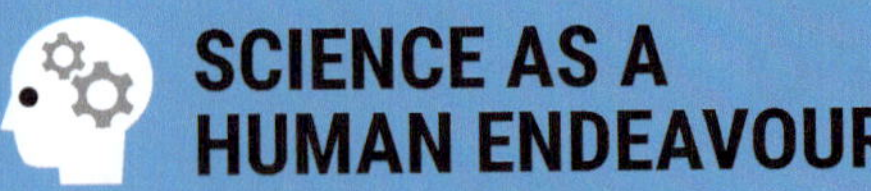

SCIENCE AS A HUMAN ENDEAVOUR

Benjamin Franklin

The great American scientist Benjamin Franklin was the first person to explain successfully the charging of an object by rubbing. He suggested that the two types of charge could be called *positive* and *negative*.

He inferred that there was an 'electric fluid' that could be moved from one object, to another. If this electric fluid was added to an object, then it gained a positive charge. If electric fluid was removed, then the object developed a negative charge.

Franklin's ideas were useful for explaining electric charges, but other observations do not support his inferences about an electric fluid. Scientists now use a different explanation (see the next page).

Benjamin Franklin is famous for flying a kite in a thunderstorm—an extremely dangerous thing to do. When lightning struck the kite, electricity flowed down the string to a key. Luckily he survived.

Find out more online.

EXPLORE ONLINE

Inside atoms

Inside atoms there is a small central nucleus, which is positively charged. It contains protons, which are positively charged, and neutrons, which are neutral (no charge). Moving around the nucleus are electrons, which are negatively charged. If there are equal numbers of protons and electrons, the charges balance each other and the whole atom is uncharged (neutral).

If some electrons are removed from an atom, it becomes positively charged. If extra electrons are added, the atom becomes negatively charged. When the number of positively charged atoms in an object just balances the number of negatively

ISBN 978 1 4202 3736 8

charged atoms, the whole object is uncharged. But if the numbers become unequal, then the object has an electric charge.

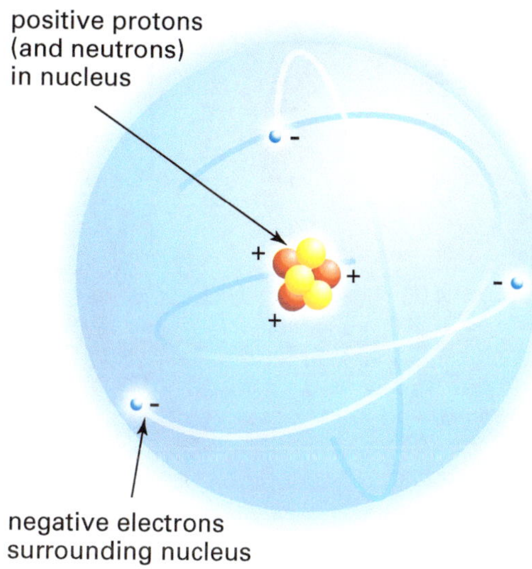

Figure 2.3 A picture of an atom. It is neutral, with no overall charge.

Explaining electric charges

What happens when you rub a perspex rod with a silk cloth? The frictional forces of the rubbing cause electrons to be removed from atoms on the surface of the rod and to become attached to atoms on the silk. This leaves the rod with a positive charge and the silk with a negative charge.

A different type of cloth may give electrons to the rod and make it negatively charged. This cloth will, of course, then have a positive charge.

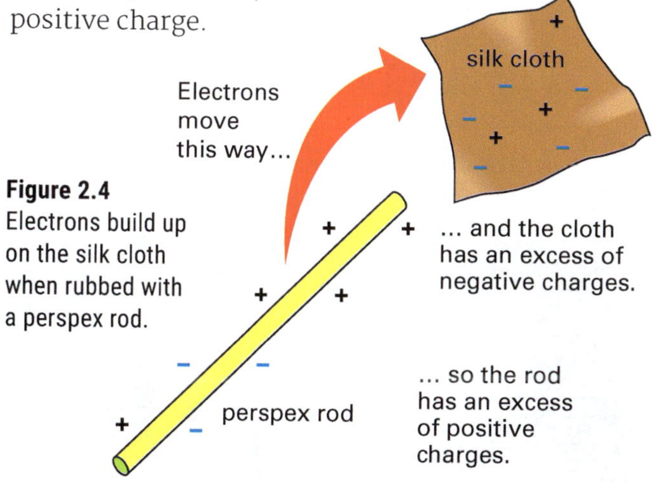

Figure 2.4 Electrons build up on the silk cloth when rubbed with a perspex rod.

In the case of rubbing a balloon on a woollen jumper, electrons rub off the jumper onto the balloon. When the balloon is held to the wall it attracts positive charges, which make it stick to the wall as shown in Figure 2.5.

Figure 2.5 A balloon can stick to the wall when negatively charged.

ISBN 978 1 4202 3736 8

Everyday static electricity

The tingle you get when you walk across a synthetic carpet and then touch something metallic is due to static electricity. The friction between your shoes and the carpet causes your body to become charged. When you touch a metal object, the static electricity is discharged (allowed to escape). As the electricity flows across your skin, you feel a slight electric shock.

Flying an airplane causes friction between particles in the air and the metal plane. This causes charge to build up on the plane. During World War I, pilots landing small rubber-tyred aircraft often received a powerful shock when they stepped onto the ground. Today aircraft have special tyres that have metal in them. This lets the static electricity pass harmlessly to the ground when they land and prevents shocks and electrical problems.

The rapid movement of drops of water in thunderclouds can cause a separation of positive and negative charges. The tops of the clouds normally become positive, and the bottoms negative. If these charges become big enough, electrons can jump from one part of the cloud to another, causing a spark. The air is heated so much it glows, producing lightning. The intense heat also makes the air expand suddenly, causing the loud noise of thunder. Lightning can also spark to the ground, or to other clouds.

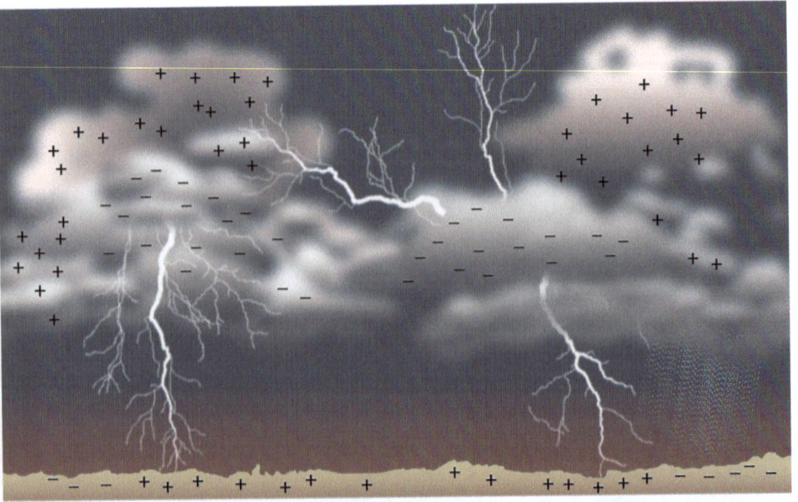

Figure 2.6 Lightning can spark within a large cloud, from a cloud to the ground, or from cloud to cloud.

What to do in a thunderstorm

Each year in Australia lightning claims up to 10 lives and causes over 100 injuries. Many of these injuries happen when people use telephones during thunderstorms.

If you are caught outdoors in a thunderstorm:

- Seek shelter in a hard-top vehicle or solid building.
- If swimming or surfing, leave the water immediately.
- If in a boat, go ashore to shelter as soon as possible.
- Never shelter under trees.
- Don't use a mobile phone.
- Don't handle fishing rods, umbrellas or golf clubs.
- Stay away from metal poles, wire fences, sheet metal, clothes lines etc.
- Don't ride a horse or bike, or drive an open vehicle.
- If you are in a car, park away from trees and power lines. Close the windows and avoid touching metal parts of the car.
- If caught in the open, crouch down with your feet together.

Figure 2.7 Lightning strikes the lightning conductor on the Q1 tower on the Gold Coast.

Applications of static electricity

Operating theatres

In operating theatres the sudden movement of blankets, clothes or equipment can produce electrostatic sparks. (*Electrostatic* means 'relating to static electricity'.) These sparks are very dangerous because of the large amount of oxygen in the air and other flammable gases used to anaesthetise the patient. Many precautions are therefore taken to make sure static charges do not build up anywhere.

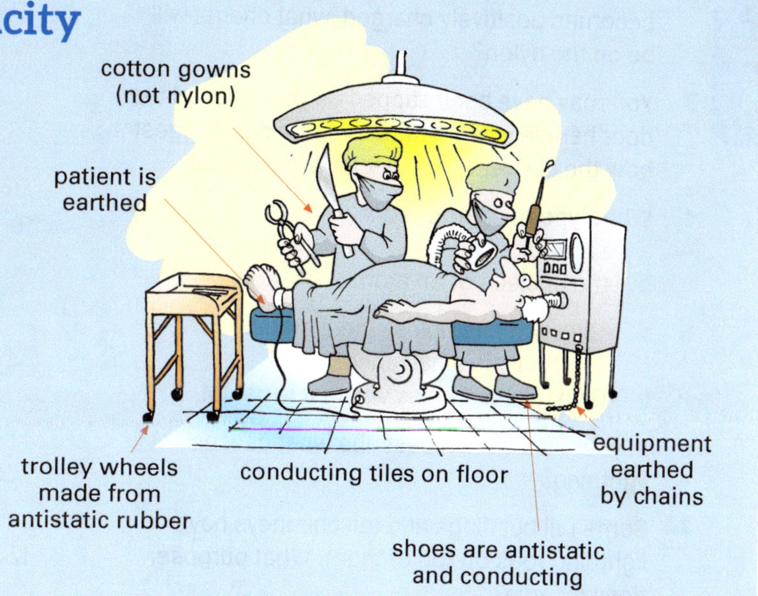

Photocopiers

Photocopiers work by an electrostatic process. The main part of the machine is a rotating light-sensitive drum onto which the image of the document is projected. The positively charged paper attracts the negatively charged toner from the drum, forming an image. The paper then passes between heated rollers that fuse (melt) the toner onto the surface of the paper.

image to be copied (face down)
lamp
lens
drum
cartridge containing toner powder (–)
+
printed image
heated rollers

Powder coating

When objects are powder-coated, they are charged so they will attract the powder. This gives a much more even coating than other methods of spraying, and the powder reaches all parts of the object's surface. However, great care has to be taken to keep dust particles out of the air, or they too will be attracted onto the object's charged surface.

ISBN 978 1 4202 3736 8

CHECK

1 Why do you sometimes notice a crackling noise when you take off your clothes?

2 If a rod is rubbed with nylon cloth and the rod becomes positively charged, what charge will be on the nylon?

3 You may have been zapped as you touched the door handle when getting out of a car. Suggest how the car becomes electrically charged.

4 What type of charge is on:
   a an electron?
   b the nucleus of an atom?

5 a Give two examples where static electricity is a nuisance.
   b Give two examples where it is useful.

6 In your own words, describe what causes lightning.

7 Some tall buildings and tall chimneys have lightning rods on top of them. What purpose does it serve?

8 A piece of plastic held in your hand can be electrified by rubbing it with a cloth, but it is impossible to electrify a piece of metal in the same way. Why?

9 A suspended, positively charged rod has a second rod brought near to it. What is the charge—positive, negative or no charge—on the second rod if it:
   a repels the suspended rod?
   b attracts the suspended rod (two answers)?

10 Look at the labels on the cartoon of the operating theatre on the previous page.
   a The equipment and the patient are *earthed*. What does this mean?
   b What does the word *conducting* mean?
   c What does *antistatic* mean?

11 Have you noticed that computer and TV screens become dustier than the things around them? Suggest a reason for this.

12 Explain ways in which static electricity and magnetism are similar. In your answer use the terms *non-contact force* and *force field*.

CHALLENGE

1 Five different rods (A, B, C, D, E) were given an electric charge by rubbing them with two different cloths. The rods were then tested in pairs to see whether they repelled or attracted. A attracted C and C attracted E. A repelled D and B repelled E. Predict what will happen if you bring D and E together and B and C together.

2 The photo below shows a light plane being refuelled. Suggest why there is a wire between the fuel hose and the plane.

EXPLORE

1 You have been asked to solve the problem of the two sides of a plastic bag sticking together.
   a Why do you think this problem occurs?
   b Design an experiment to show how the sides stick together.
   c Suggest experiments you could try to overcome the problem.

2 Which type of carpet is most likely to give you an electric shock when you walk about on it? Design and carry out an experiment to find out.

3 In a very dark room, rub a spare fluorescent tube with wool, fur or clear plastic wrap. Can you see it glow?

4 Bring a charged rod near the smoke from a burning mosquito coil. What happens?

ISBN 978 1 4202 3736 8

2.2 Electric currents

Static electricity is electricity that is stationary. If, somehow, this electricity can be made to move, you have *current electricity* or an **electric current**. A torch battery provides the energy to drive the current. When the battery is connected by wires to a bulb, electrons flow to light up the bulb.

INVESTIGATION 2.2 Simple electric circuits

Aim

To investigate different ways of connecting a torch battery and bulb.

PART A Lighting a bulb

Materials

- 1.5 volt torch battery without holder
- torch bulb (2.5 volt) without holder
- 2 connecting wires

Risk assessment and planning

Read through Part A and describe to your partner what you have to do. Your partner can then describe Part B to you.

Method

1 Use the battery and one connecting wire to make the bulb light.

Draw a diagram of how you connected the battery and bulb.

2 See if you can find at least one other way of making the bulb light.

Draw diagrams of any ways that you discover.

What special places must be touched on the bulb for it to light?

What special places must be touched on the battery?

3 Can you make the bulb light using *two* connecting wires?

Draw diagrams of your set-ups.

PART B Using a switch

Materials

- 1.5 volt torch battery with holder (or power pack)
- torch bulb (2.5 volt) with holder
- 3 connecting wires with alligator clips
- switch

Method

1 Use the holders and the three connecting wires to connect the battery and bulb as shown.

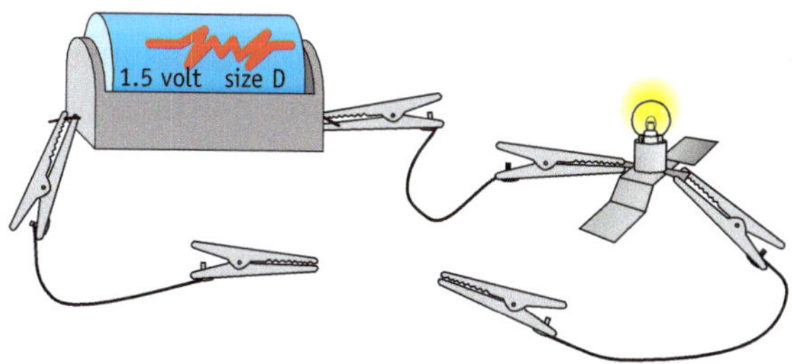

2 Make the bulb go on and off by touching the alligator clips together.

3 Now connect the switch into the circuit as shown. Switch the bulb on and off.

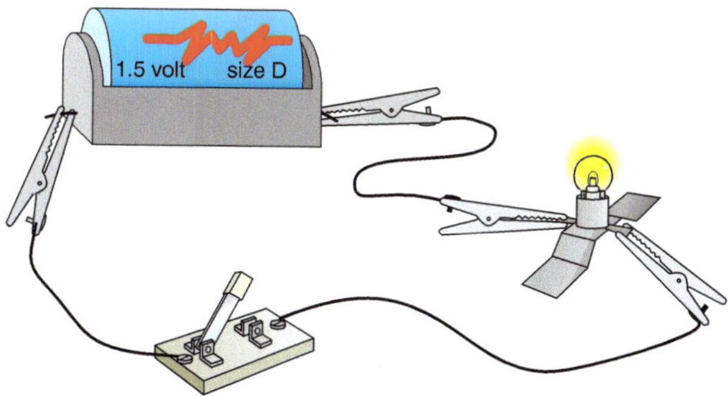

Does it make any difference if you reverse the connections to the battery?

ISBN 978 1 4202 3736 8

What is a circuit?

In Investigation 2.2 you should have noticed these things:

1. Both ends of the battery must be connected to the bulb before it will light. These metal connection points are called *terminals*. The top of the battery is positive ⊕ and the bottom is negative ⊖.
2. The bulb has to be connected in two special places. The metal side of the bulb is one terminal, and the bottom is the other. They are both the same. There is no positive or negative.
3. For the bulb to light, there has to be a closed path (or circuit) joining the battery and the bulb. This is called an **electric circuit**. When there is a gap in the circuit, the light is off. A switch lets you open and close the circuit.

Current

An electric current can be compared to water flowing through pipes. The battery is like a water pump—it gives energy to the electrons just as the pump forces the water through the pipes (see Figure 2.9).

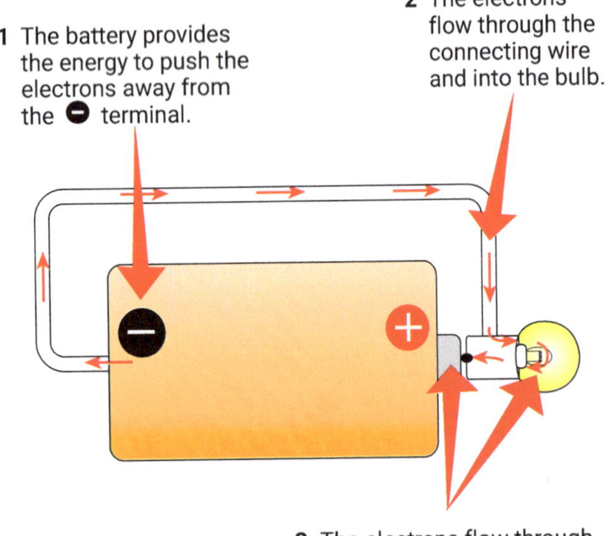

Figure 2.8 This shows how a battery lights a bulb.

A water meter measures how many litres of water are flowing through a pipe each second. In an electric circuit, the electric current or number of electrons passing each second is measured using an **ammeter** (AM-eat-er). An ammeter measures electric current in *amperes* (abbreviation amps, symbol A) or milliamps (1000 mA = 1 A).

Voltage

Voltage is a bit like the pressure in the pipes. It is a measure of how much energy can be given to the moving electrons in a circuit. It is measured in *volts* (V) using a *voltmeter*. A torch battery has 1.5 volts. A 6 volt battery can push a larger current around the same circuit.

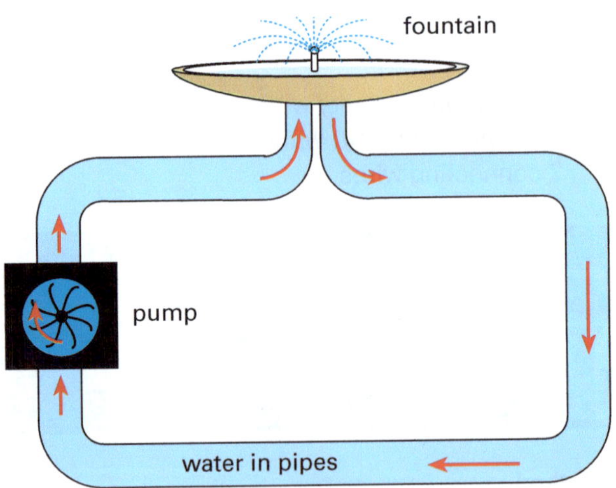

Figure 2.9 An electric current flowing from a battery through a bulb can be compared to water flowing in pipes.

Conductors and insulators

If one of the connecting wires in Investigation 2.2 was replaced by a piece of string, the light bulb obviously would not glow. String does not let electricity through and is called an **insulator**. A substance such as wire that does let electricity through is called a **conductor**.

ISBN 978 1 4202 3736 8

INVESTIGATION 2.3 Does it conduct?

Aim

To test various substances to see how well they conduct electricity.

Materials

- 1.5 volt battery and holder (or power pack)
- torch bulb and holder
- ammeter or multimeter
- 4 connecting wires
- variety of objects, e.g. paperclip, plastic and glass rods, nail, coin, carbon rod, copper rod, matchstick, rubber band, aluminium foil, strip of paper, piece of string

Risk assessment and planning

Discuss the investigation with your teacher. You may use a 6 volt battery or a power pack instead of the 1.5 volt battery.
Look at the ammeter. The red or + terminal must be connected to the + terminal of the battery.

- Suggest why you use an ammeter in this investigation.
- Draw up a data table like the one shown. List at least 10 objects in the left-hand column. Write down what material each object is made of.

Object	Material	Does bulb grow?	Ammeter reading (mA)
paperclip stirring rod	steel glass		

Method

1 Set up a circuit as shown. Ask your teacher to check your circuit before you go on to step 2.

2 Touch the two alligator clips together. Observe what happens to the bulb.
Record the electric current reading on the ammeter.

3 Connect one of the objects between the alligator clips.

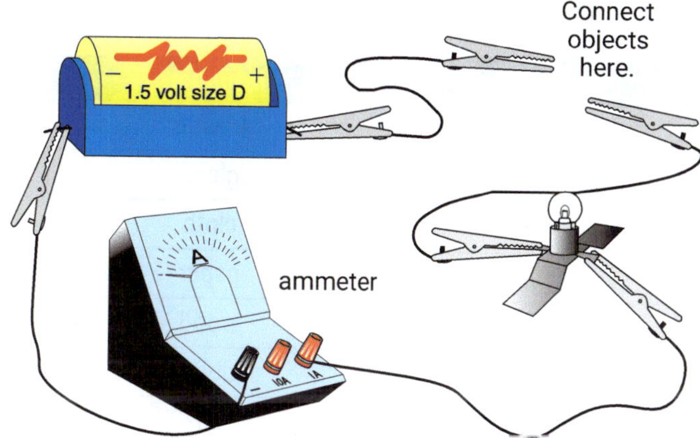

Record whether the bulb glows.
Record the ammeter reading. (This tells you how much current passes through the object.)

4 Test each of the other objects.
Record the results in your data table.

5 Is your skin a conductor or an insulator? Does it make any difference if your skin is wet or dry?

Discussion

1 Which materials are good conductors of electricity? How do you know?

2 Which materials are poor conductors (insulators)?

3 Use the ammeter readings to decide which one of the materials is the best conductor.

4 Why is it that some materials did not cause the bulb to glow, yet gave a reading on the ammeter?

5 Is air a conductor or an insulator? How do you know?

6 How could you test whether water is a conductor or an insulator?

Conclusion

How are the materials that conduct electricity similar? Write a generalisation about the types of materials that conduct and do not conduct electricity.

ISBN 978 1 4202 3736 8

More on conductors and insulators

All metals are conductors, while most non-metals are insulators.

Conductors	Insulators
carbon	plastic
salt water	glass
acids	cloth
silver	paper
copper	wood
gold	rubber
aluminium	air

Insulators are very important in the supply and use of electricity. The poles that carry electricity from power stations to cities need insulators to stop electricity from escaping to the ground (Figure 2.11). The handles of screwdrivers and pliers are often coated with plastic insulation. The casings of electric plugs, sockets and switches are all made from plastic.

An electric current is a flow of electrons. So a conductor is a material through which electrons can flow.

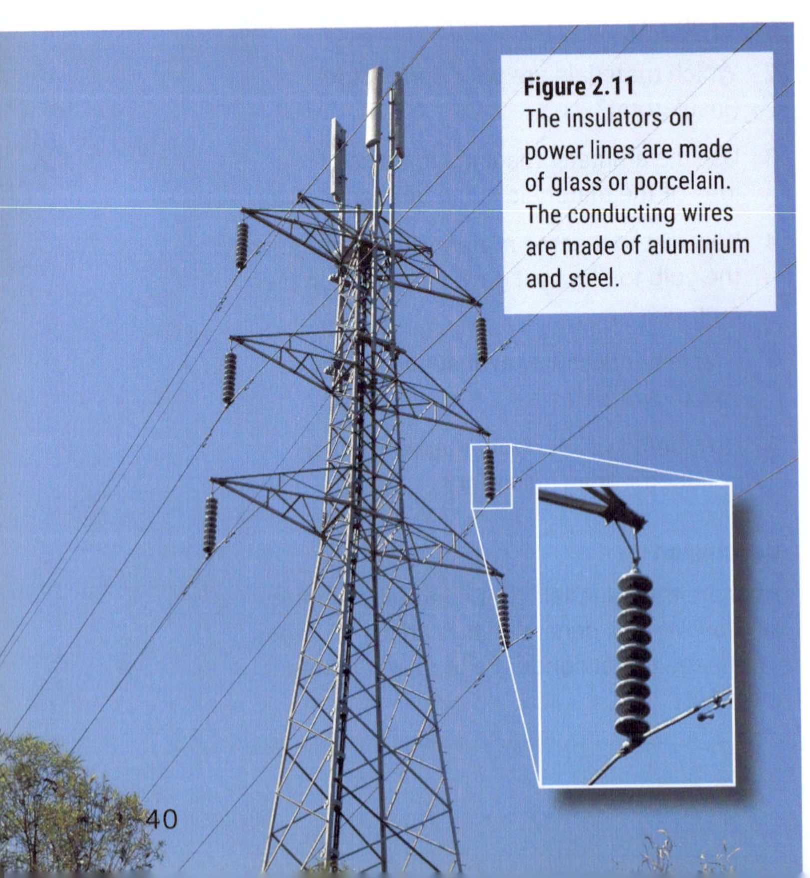

Figure 2.11 The insulators on power lines are made of glass or porcelain. The conducting wires are made of aluminium and steel.

Conductor

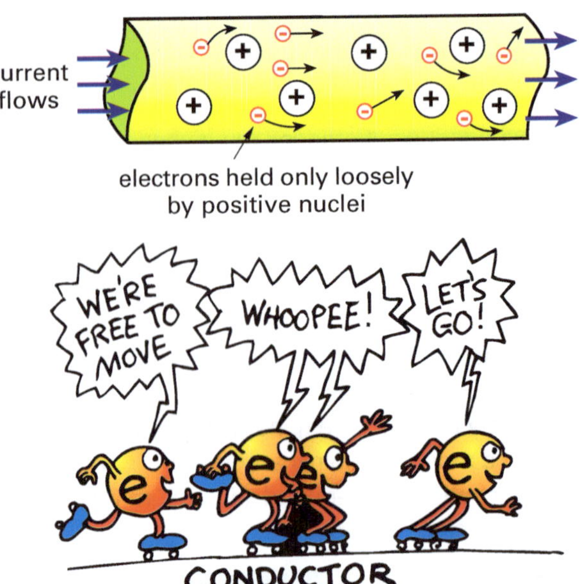

Insulator

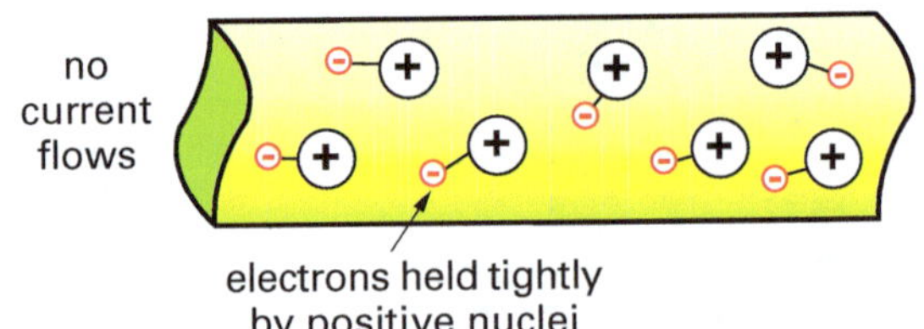

Figure 2.10 Electrons in a conductor are free to move, but not in an insulator.

A metal consists of an arrangement of positive nuclei in a 'sea' of electrons. These electrons are not strongly attracted to any one nucleus. So, when the metal is connected to a battery, the electrons can move easily through the metal to produce a current.

ISBN 978 1 4202 3736 8

In an insulator, the electrons are held tightly by the positive charges. Because of this, the electrons cannot move, and no electric current can flow when the insulator is connected to a battery.

If you charge an insulator such as a plastic rod by rubbing, the charge stays on the surface of the insulator. But the insulator slowly loses its charge to the air. The charge is also lost quickly if you touch the insulator with your hand. This process allows the charge to flow to the ground, and is called *earthing*. You cannot charge a conductor by rubbing. Any charge you produce flows through the conductor to the ground immediately.

Resistance

When an electric current moves through a conductor, there is always some **electrical resistance** to the current. This is because of the level of attraction of the electrons to the positive nuclei of the atoms in the *conductor*.

High resistance	Low resistance
Strong attraction of electrons to positive nuclei.	Weak attraction of electrons to positive nuclei.
Electrons do not move as easily.	Electrons move more easily.
More energy is lost as electrons are pushed through the conductor.	Less energy is lost as electrons are pushed through the conductor.
More heat is produced.	Less heat is produced.

As the electrons are pushed through a conductor, they lose some of their energy as heat. This waste heat can be a nuisance; for example, computers get hot when used. However, the waste heat is sometimes useful. For example, because nichrome wire has a fairly high resistance, it is used to make the heating elements in many electrical appliances used around the home. It is usually coiled to take up less space.

In electric power lines, there is always loss of energy due to the resistance of the metal in the wires. For this reason, scientists are trying to make cheap *superconductors*, which offer ***no*** resistance to the flow of electricity. The use of such materials would save billions of dollars.

Superconductors are also used in maglev trains. These trains float above the tracks supported by the non-contact forces between large electromagnets.

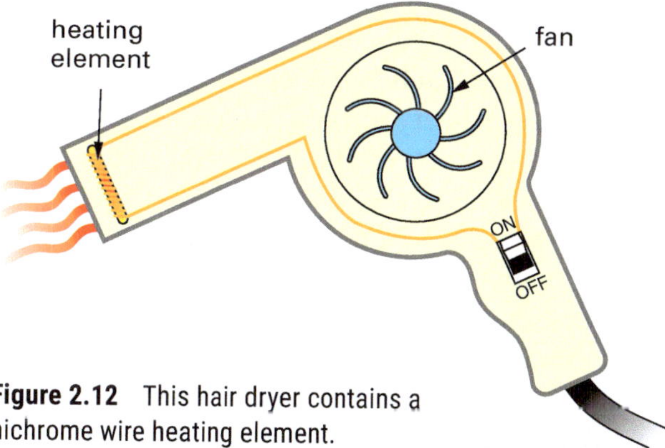

Figure 2.12 This hair dryer contains a nichrome wire heating element.

Figure 2.13 The chilled superconductor (bottom) is acting like a magnet. It repels the magnet (top), making it hover in the air above the superconductor.

ISBN 978 1 4202 3736 8

SCIENCE AS A HUMAN ENDEAVOUR

Conducting plastics

Plastics are insulators—they don't usually conduct electricity. However, in the mid-1970s three scientists discovered a plastic that was somewhere between an insulator and a conductor.

A Japanese scientist, Hideki Shirakawa, was trying to make a plastic called polyacetylene. By accident he added 100 times as much catalyst as he intended, and a shiny metallic-looking film appeared on the inside of his reaction vessel. At about the same time two other scientists, Alan MacDiarmid and Alan Heeger, were experimenting with metallic films at the University of Pennsylvania in the United States. MacDiarmid and Shirakawa met by chance during a coffee-break at a seminar in Tokyo. When MacDiarmid heard about Shirakawa's accidental discovery, he invited him to work with him in his laboratory in the United States.

MacDiarmid, Heeger and Shirakawa did many experiments and found that if they exposed the polyacetylene to bromine vapour, its electrical conductivity increased by a factor of 10 million! They immediately published their discovery of a conducting plastic, and in 2000 they were jointly awarded the Nobel Prize in Chemistry.

In 2011 a team led by Paul Meredith and Ben Powell from the University of New South Wales discovered a technique for producing a conducting plastic that is completely flexible. It is easy to produce and still acts like a plastic, but has a high conductivity just like metals. Until this point conducting plastics were expensive, hard to produce and had many problems. This new technology has already led to the production of low-cost flexible thermometers, and could lead to flexible touch screens and e-paper. Conductive polymers are already used in fuel cells, computer displays and microsurgical tools, and are now being used in the field of biomaterials and biosensors.

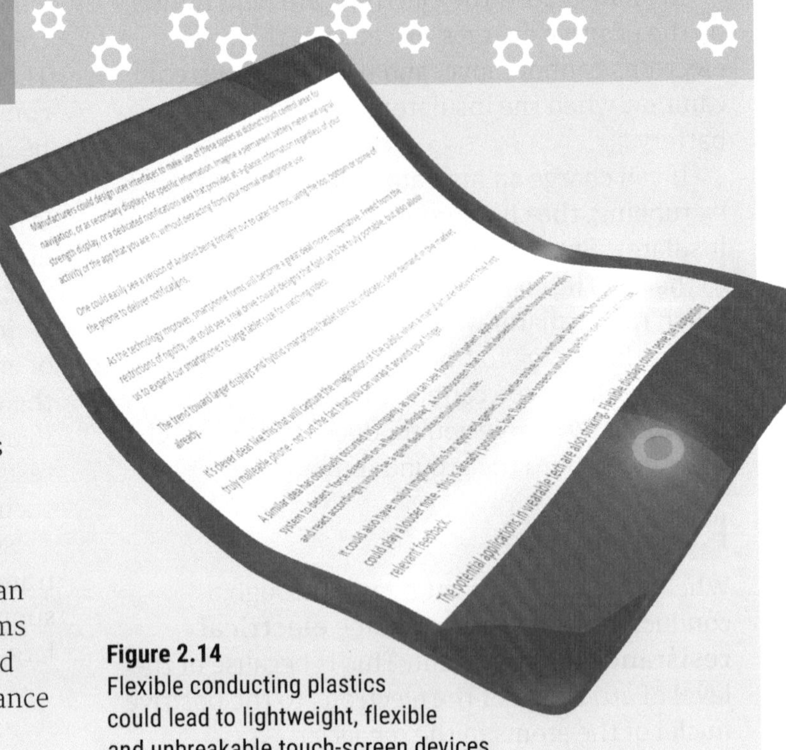

Figure 2.14
Flexible conducting plastics could lead to lightweight, flexible and unbreakable touch-screen devices.

Conducting plastics could also be used to make solar cells in a continuous roll. These are cheaper and more versatile than the present silicon-based solar cells. The solar cell plastic can also be made into fabric to make clothes that can convert light into electricity.

Other applications of conducting plastics that are available are:

- rechargeable plastic batteries for use in portable electronic equipment
- windows that you can darken during the day by passing a small electric current through them
- antistatic material for use in offices and operating theatres, where it is important to avoid a build-up of static electricity.

Questions

1 What is the important development in science described on this page?
2 What new technologies have been developed as a result of this development in science?
3 Which of these technologies do you think has the most potential for the future? Explain your answer.

ISBN 978 1 4202 3736 8

1 Copy and complete the following sentences.
- a A path for electricity is called a _____.
- b A _____ lets you open and close a circuit.
- c Moving electrons in a wire are called an electric _____.
- d An _____ is an instrument used to measure electric current.
- e The unit for electric current is the _____.
- f _____ is a measure of the energy given to the electrons in a circuit.
- g Substances that do not allow an electric current to flow through them are called _____.
- h Metals are _____ because they allow an electric current to pass through them.
- i Opposition to the flow of current in a circuit is called _____.
- j If the resistance in a circuit is increased, the current _____.

2 In which of these circuits will the bulb glow? For the other circuits, explain why the bulb *won't* glow.

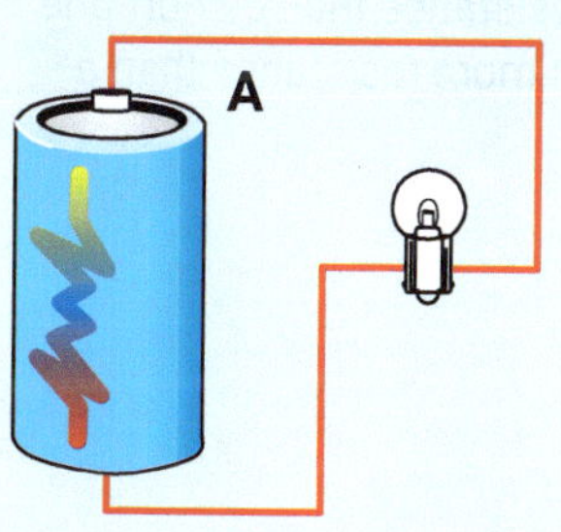

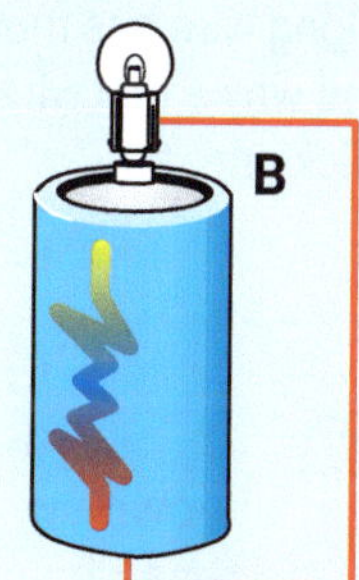

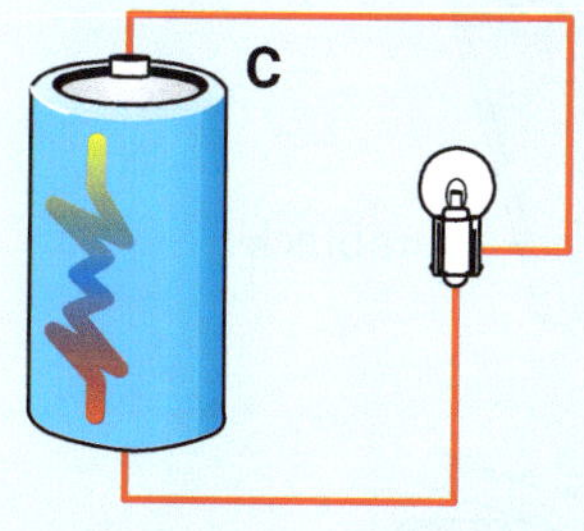

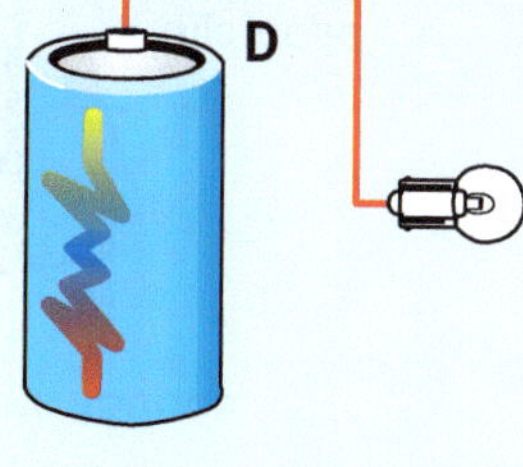

3 Into what two forms of energy is electrical energy changed in a light bulb?

4 Which battery can supply the most energy to electrons in a circuit: 1.5 volt, 6 volt or 9 volt? Why?

5 Lisa connected a bulb to a battery. The wires were connected properly, but the bulb did not glow. What could be wrong (two possibilities)?

6 Explain in your own words the difference between an insulator and a conductor of electricity.

7 Why are electrical connecting wires covered with plastic?

8 This ammeter measures current in two different ranges: 0 to 1 A and 0 to 10 A.

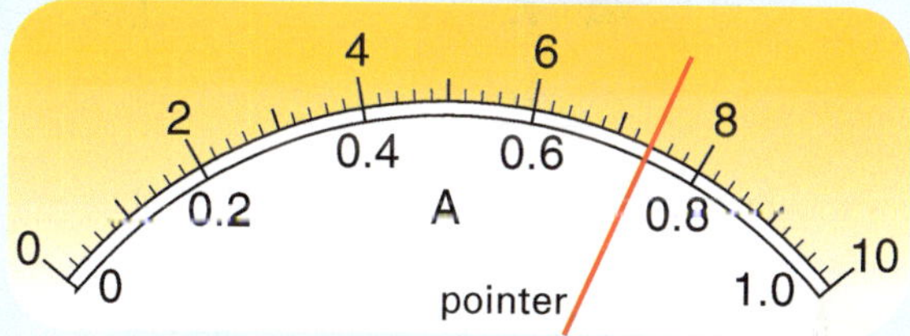

- a What is the reading if the 0–10 A range (top) is used?
- b What is the reading if the 0–1 A range (bottom) is used?

9 Ngoc tested how well different types of pencil 'lead' of the same length and thickness conduct electricity. His results are shown in the table below:

Type of 'lead'	Electric current (amperes)
4H	0.03
HB	0.10
3B	0.70

- a Which type of 'lead' has the greatest resistance?
- b Pencil 'leads' contain graphite, which is a conductor. Which type of pencil 'lead' would you infer contains the most graphite?

10 Why is it safer to wear shoes than to go barefoot in an electrical storm?

ISBN 978 1 4202 3736 8

1 Explain why the battery in a torch eventually goes flat.

2 When you push down the switch, the torch produces a beam of light. Explain in detail how this happens.

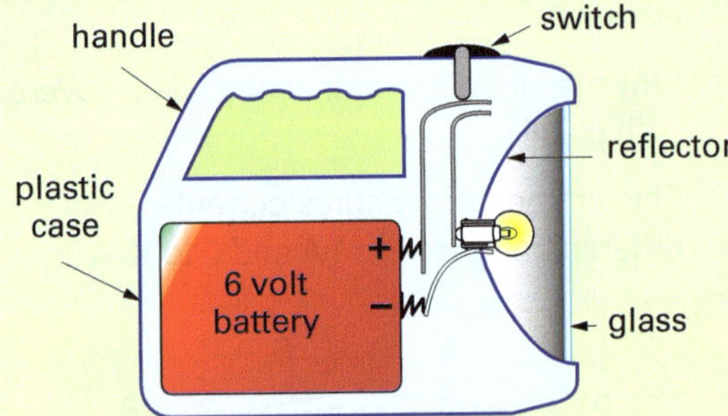

3 What do you think is the most likely cause of the following?

a Your radio starts to get quieter and quieter. Turning up the volume doesn't seem to help much.

b Your torch is very bright but suddenly goes out.

c Your CD player stops working, but when you tap on the case it works again.

4 A company produces an all-metal electric kettle, but the government bans its sale. Suggest why it was banned.

5 Explain why the element in a toaster becomes red-hot, while the wires connecting the toaster to the mains power supply remain cool.

6 One of the things that 'lie-detectors' measure is skin resistance. Lying is supposed to make you sweat. How do you think this lie-detector works?

7 Why don't the materials that conduct current electricity hold static electricity?

8 Using what you know about resistance, explain why a long wire has more resistance than a short one, and why a thin wire has more resistance than a thick one.

EXPLORE

1 Find out whether tap water will conduct an electric current. Set up the circuit shown, using a conductivity kit. You could also test rainwater, distilled water and salt water.

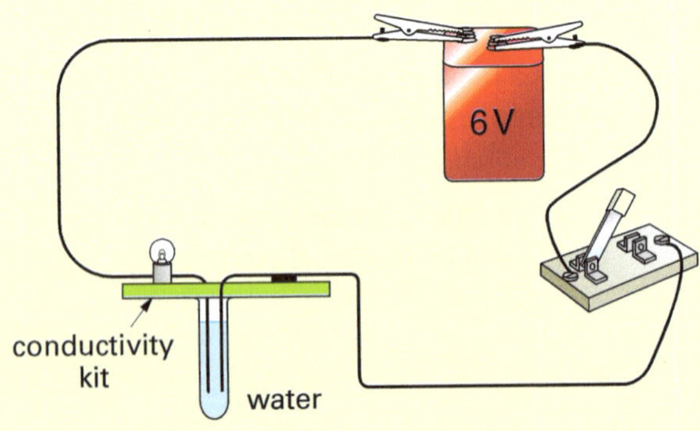

2 Make your own switch. Here are some designs. Try out your switch in a circuit.

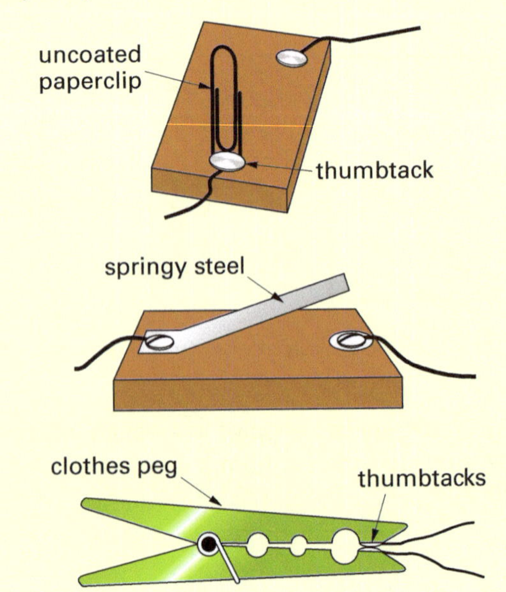

ISBN 978 1 4202 3736 8

2.3 Electric circuits

Circuit diagrams

Look at the two circuits in Figure 2.15. They look different, but they are actually the same.

If you wanted to tell someone how to set up this circuit, you might confuse them if you drew these sketches. Also, drawing diagrams like these takes time. So electricians have decided on a simple way to draw electric circuits with a symbol for each component (part). These symbols are listed below.

The wires in a circuit are drawn straight and at right angles. For example, the circuit on the right can be drawn as shown. This is called a **circuit diagram**.

Electrical symbols

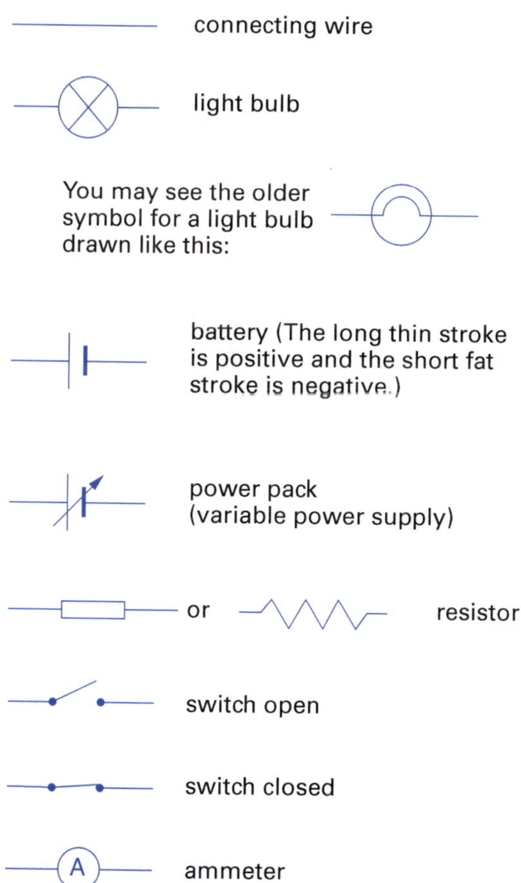

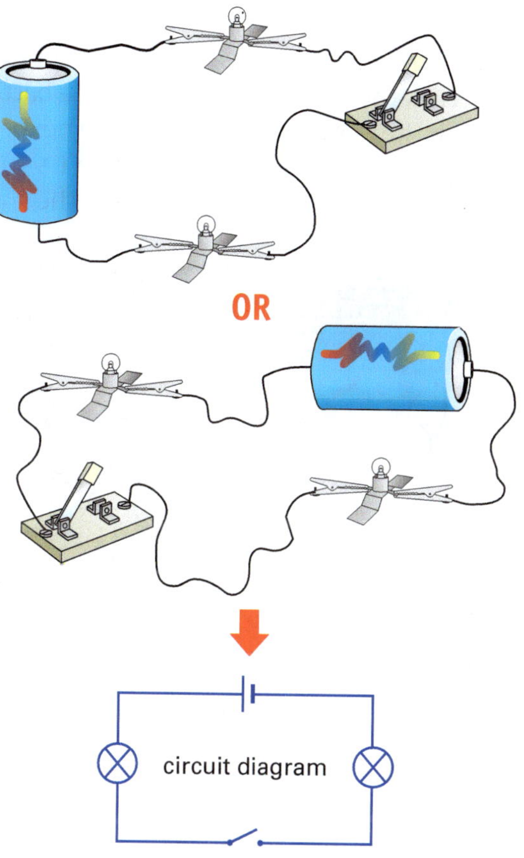

Figure 2.15 How to draw a circuit diagram

Series and parallel circuits

The parts of a circuit can be arranged in two different ways. Take, for example, two torch bulbs. They can be connected one after the other as shown in Figure 2.16. This is called a **series connection**. Note that there is only one path for the electric current to flow, and the current is the same everywhere in the circuit. As you connect more bulbs in series, the current decreases, and the bulbs don't glow as brightly.

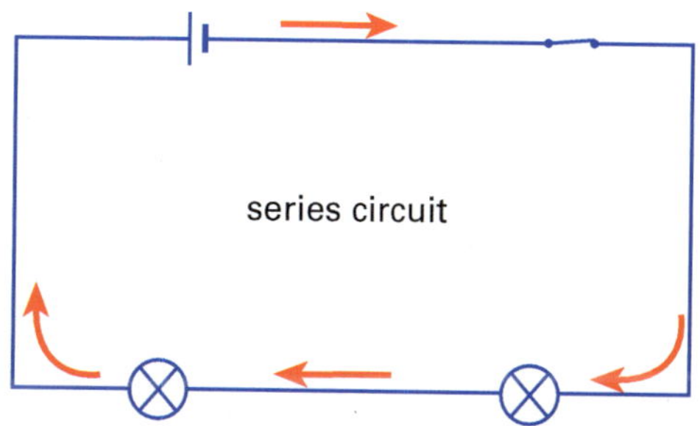

Figure 2.16 A series circuit

ISBN 978 1 4202 3736 8

Many electrical appliances use several batteries connected in series. When you put in the batteries, the positive terminal of one battery must touch the negative terminal of the next. For example, a 3 volt toy usually has two 1.5 volt batteries arranged in series as shown in the cartoon.

Two bulbs can also be connected side by side. This is called a **parallel connection**. Look at Figure 2.17. At A, the electric current splits and follows two different paths. The electrons flowing through each bulb get the full push from the cell—they don't share it as in a series circuit. As a result, each bulb glows as brightly as if it was the only bulb in the circuit. A master switch can be used to turn off both bulbs together, or separate switches can be used to turn each bulb off independently.

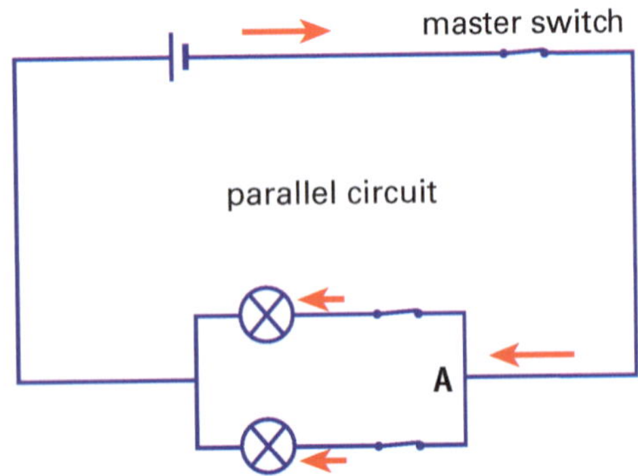

Figure 2.17 A parallel circuit

Sometimes it is not easy to tell whether the components of a circuit are connected in series or in parallel. However, if you can trace the complete circuit using one finger, then the components are connected in series. Those parts of a circuit that branch and where you have to use more than one finger are connected in parallel. Note that a circuit may contain a mixture of series and parallel connections (Figure 2.18).

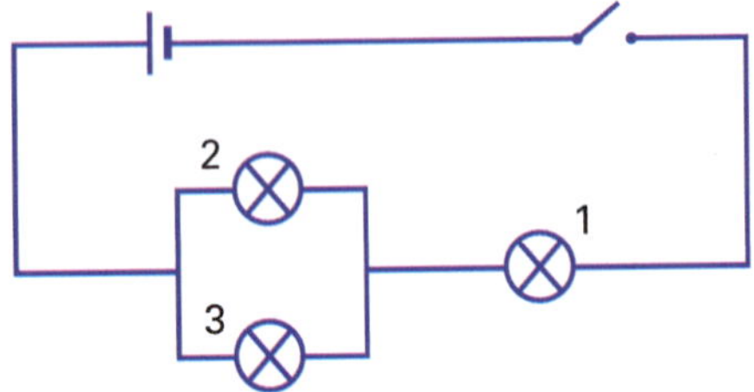

Figure 2.18 Bulbs 2 and 3 are in parallel, but they are in series with bulb 1, the switch and the battery

In Investigation 2.4, you can investigate series and parallel circuits for yourself.

ISBN 978 1 4202 3736 8

INVESTIGATION 2.4 Series and parallel circuits

Aim

To investigate series and parallel circuits.

Materials

- two 1.5 V batteries and holders (or power pack)
- 3 torch bulbs and holders
- 6 connecting wires
- ammeter or multimeter

Risk assessment and planning

- Carefully read through the instructions for the three parts.
- To which terminal of the battery do you connect the positive (+) terminal of the ammeter?

PART A Lighting a bulb

Method

1 Connect up a circuit with a battery, a switch and one bulb. Close the switch and observe the brightness of the glow of the bulb.

2 Connect a second bulb in series with the first bulb, as shown below.

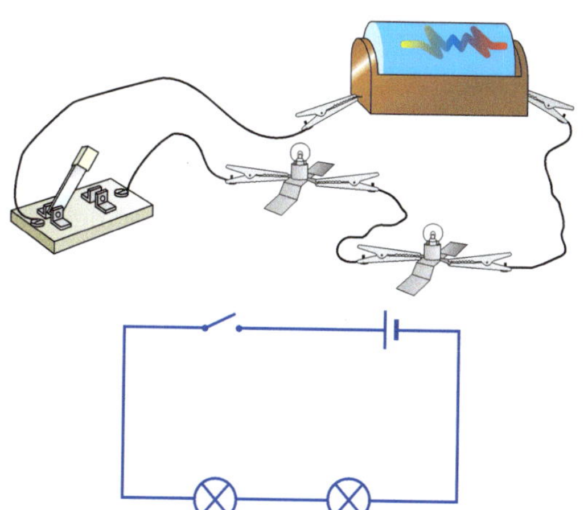

Does each bulb glow as brightly as the single bulb in step 1?

Unscrew one of the bulbs from its socket.

Record what happens.

3 Repeat step 2 with *three* bulbs.

4 Connect up a second circuit with the two bulbs in parallel, as shown in the next column.

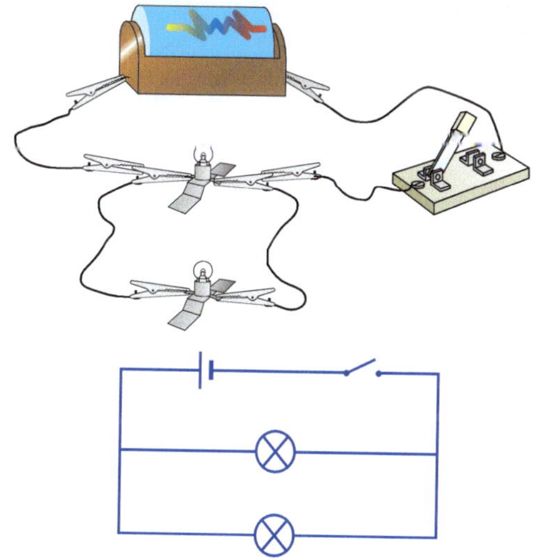

In which two-bulb circuit do the bulbs glow more brightly? Suggest a reason for this.

What happens if you unscrew one of the bulbs in the parallel circuit?

5 Add a third bulb in parallel with the other two. What happens?

Discussion

1 What is the effect of increasing the number of bulbs in series in a circuit?

2 If one bulb in a series circuit blows, the others also go out. Why?

3 Describe the effect of adding more bulbs in parallel in a circuit.

4 When one bulb in a parallel circuit fails, the others continue to operate. Why?

5 Parallel circuits are used in the electrical wiring of a house. Suggest reasons for this.

ISBN 978 1 4202 3736 8

PART B Battery problem

Inquiry: *Can you make the bulb glow more brightly by adding a second battery?*

Experiment to find out whether you should add the second battery in series or in parallel.

Write a brief report of your findings.

Notes for Part B

1 When connecting batteries in series, you must connect the positive of one to the negative of the other, as shown.

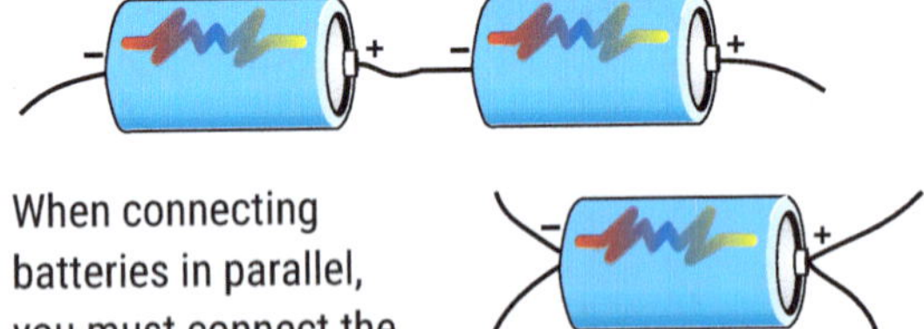

2 When connecting batteries in parallel, you must connect the positive of one to the positive of the other.

PART C Using an ammeter

Inquiry: *How can you use an ammeter to find out whether the current is the same in all parts of a series and a parallel circuit?*

Discuss the research question in a group and design an experiment. Check it with your teacher before you start.

Don't forget to connect the positive terminal of the ammeter to the positive terminal of the battery or power pack, as in the circuit below.

Write a report of what you find.

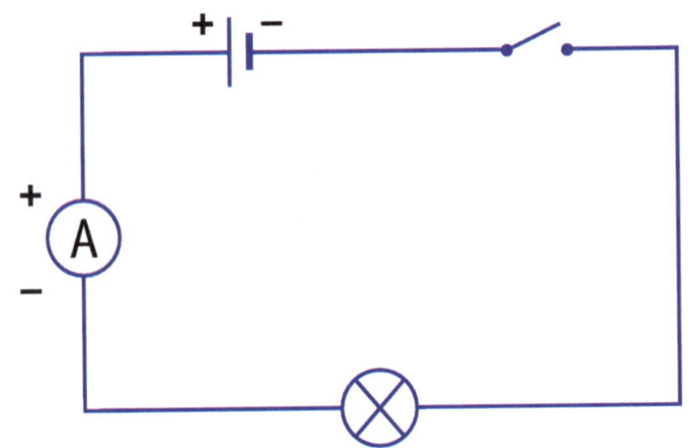

1 Copy and complete the sentences by selecting the correct words to describe the circuit below.

- a In this circuit the electricity has _____ (one / two) paths to follow.
- b This circuit is _____ (open / closed).
- c If bulb A went out while the switch was closed, bulb B would (stay on / go out).
- d If more bulbs were added to the circuit, each bulb would glow _____ (more / less) brightly.
- e If the circuit had only one bulb, it would glow _____ (more / less) brightly.
- f The bulbs are connected in _____ (series / parallel).

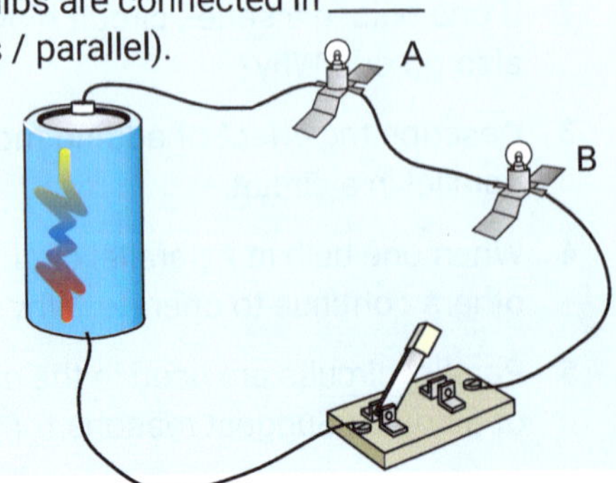

2 Answer these questions about the circuit below.

- a How many paths can the electric current follow?
- b Does the current have to pass through bulb A for bulb B to glow?
- c If bulb B blew would bulb A continue to glow?
- d What would happen if you added a third bulb in parallel?

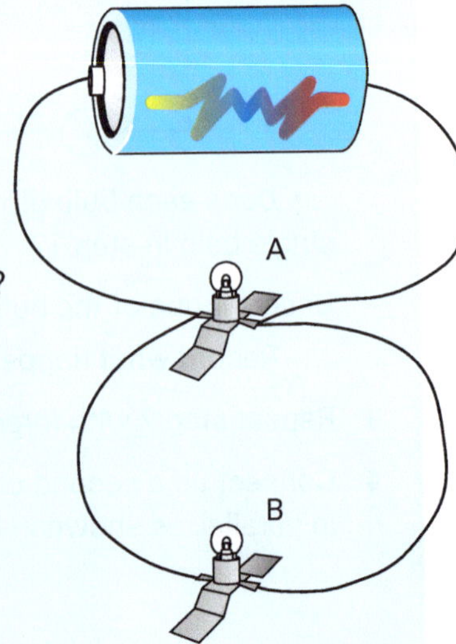

ISBN 978 1 4202 3736 8

3 Write out a list of the equipment needed to set up circuit A. Do the same for circuit B.

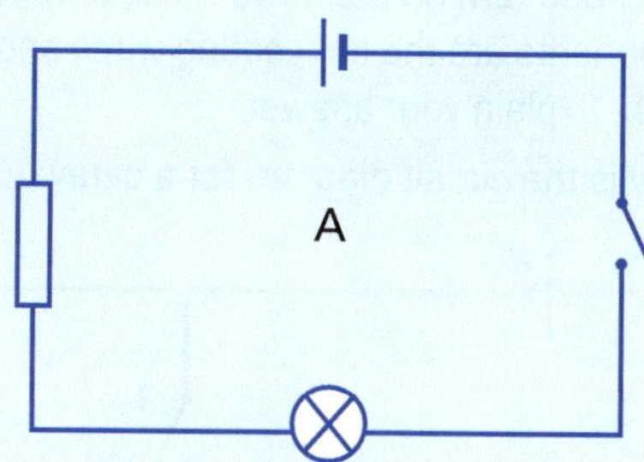

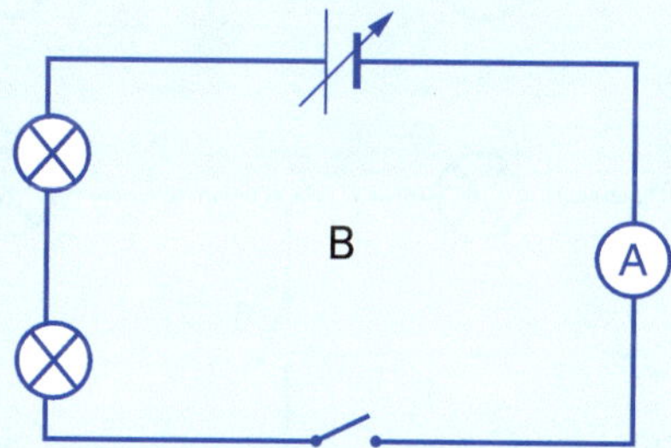

4 Draw a circuit diagram for each of these circuits.

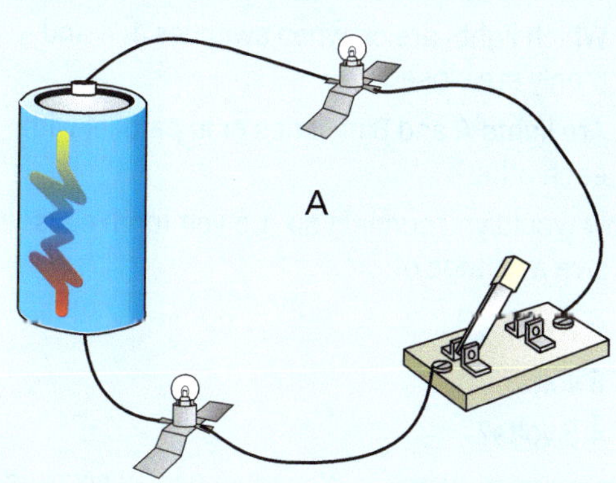

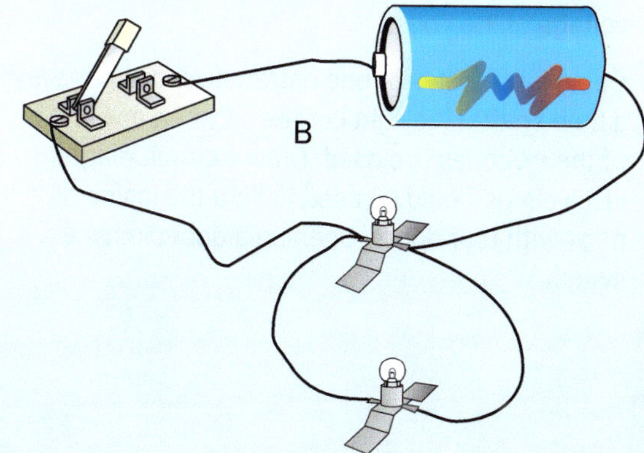

5 What voltages are being used in these two electrical appliances?

Torch

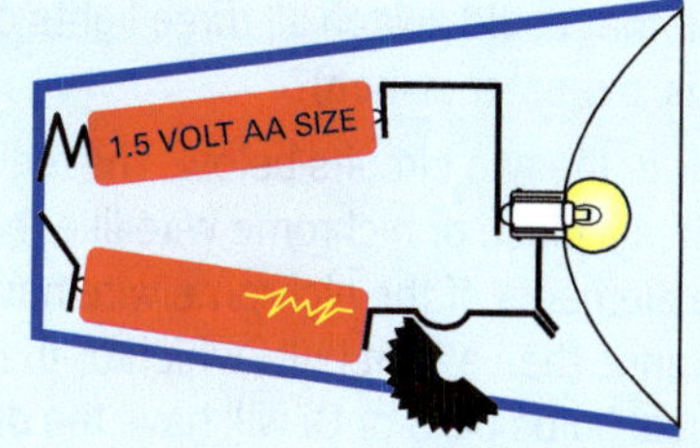

Battery compartment of radio

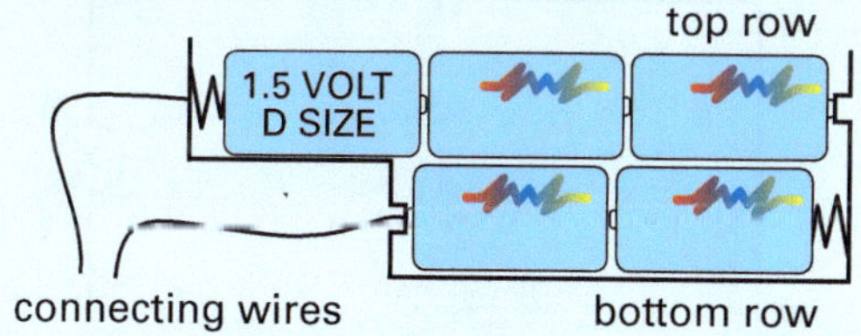

6 Draw a circuit diagram that has:

a two batteries and a bulb in series
b one battery and two bulbs in series
c two batteries in parallel and a bulb in series
d two batteries in parallel and two bulbs in series
e a power pack and a string of eight decorative bulbs in parallel.

7 Draw a circuit using two batteries and two bulbs that makes the bulbs glow most brightly.

8 In the circuit diagram below, what happens to each of the bulbs A, B and C when you:

a close switch 1?
b then close switch 2?
c then open switch 3?

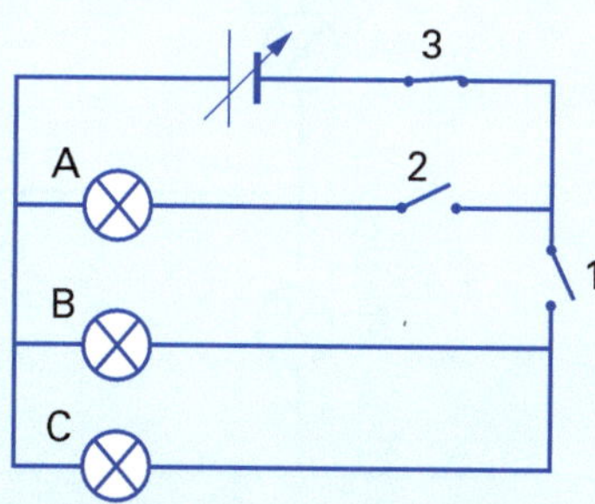

9 Give two reasons why lights in parallel are better than lights in series.

ISBN 978 1 4202 3736 8

CHALLENGE

1 Draw a circuit diagram with a battery, three lights and three switches so that each switch turns on only one light. Where would you place a fourth switch that could switch all three lights on and off (that is, a master switch)?

2 Consider the two circuits below. The resistor in the circuit is a piece of nichrome wire like that used in kettle elements. If the nichrome wire has a greater resistance than a light bulb, which of the three identical bulbs (A, B or C) will have the dimmest glow? Explain your answer.

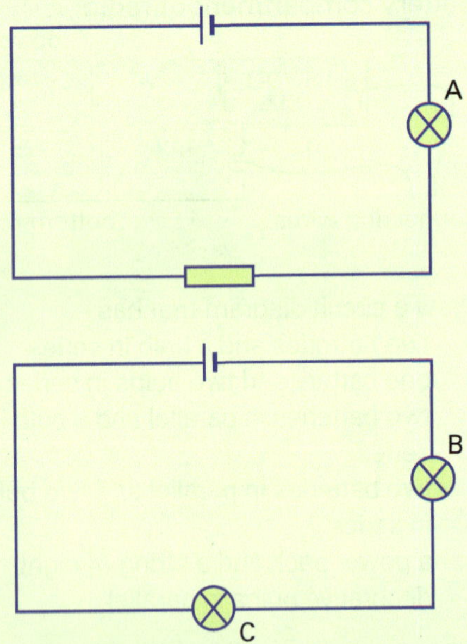

3 The bulbs in this circuit are both dimly lit when the switches are open. Predict what will happen when:

a switch 1 is closed (two things)

b switch 2 is closed as well.

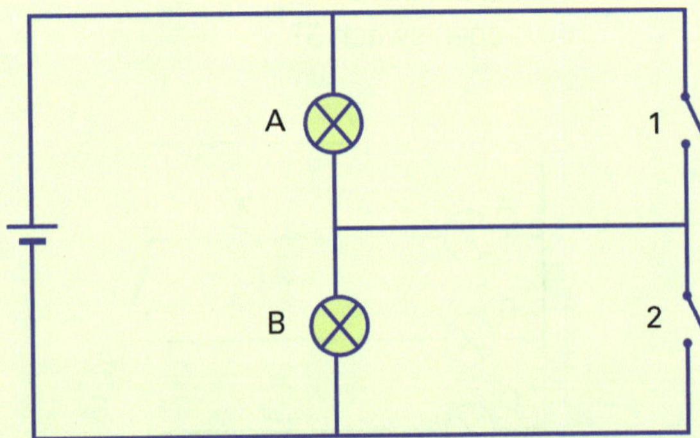

4 How is adding an ammeter (very low resistance) to a circuit different from adding a light bulb or electric motor?

5 Predict what will happen if you connect a 1.5 volt torch bulb in a circuit that is powered by a 12 volt battery and turn on the switch. The only other circuit components are the connecting wires and the switch. Explain your answer.

6 Below is the circuit diagram for a caravan.

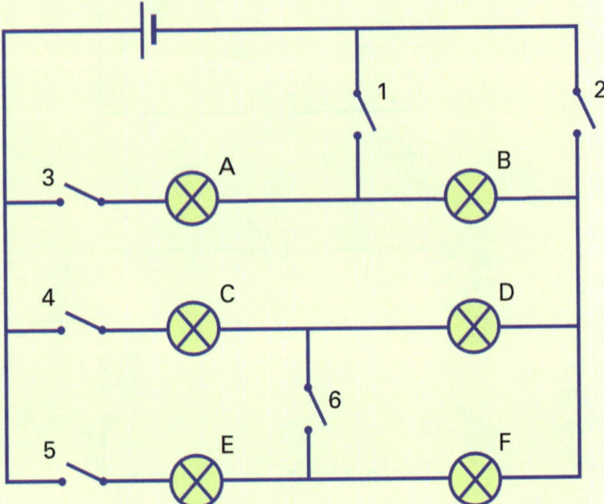

a Which switches do you need to close so that only one light stays on?

b Which lights are on when switches 1, 4 and 6 only are closed?

c Are lights A and B in series or in parallel with each other?

7 How would you connect six 1.5 volt torch cells to give a voltage of:

a 9 volts?

b 6 volts?

c 4.5 volts?

Draw circuit diagrams. You must use all six cells. (Hint: Two 1.5 volt cells in parallel have a total voltage of 1.5 volts.)

8 Design a circuit with one battery, four switches and a bulb so that the light comes on when any one of the switches is closed. Draw a circuit diagram. (This circuit could be used to light the inside of a car with four doors. Opening a door *closes* a switch.)

ISBN 978 1 4202 3736 8

EXPERIMENT 2.1 Your invention

Aim

To use what you have learnt in this chapter to invent a useful electrical device.

Method

1 Study the two inventions on the right. Explain to another student how one of them works. Your partner will explain to you how the other one works.

2 Use your imagination to design your own invention, or use the ideas below.

- a battery tester
- a circuit in which you can switch a light on in one place and turn it off somewhere else
- a burglar alarm in which a bell rings, a light flashes or a trapdoor opens to catch the burglar
- a model house in which you can turn the lights on and off independently
- an alarm to warn you of strong wind
- a device to warn you when a water tank is about to overflow
- an alarm clock using a candle
- an electric maze
- a way of dimming a light (Hint: A long wire has more resistance than a short one.)
- a pinball machine (Hint: A rolling metal ball could be used to close a switch.)

3 Draw a sketch of your design before you start. Try to draw a circuit diagram too.

4 Make a list of the things you will need to make your invention.

5 Check your design with your teacher, and then go ahead and make it. (You may be able to work on your invention at home.)

6 Prepare a report of your invention for the rest of the class. Make sure you report any problems, as well as your successes. Other students may be able to suggest ways of improving your design. (If your invention is good enough, you may be able to enter it in a science contest.)

Traffic lights

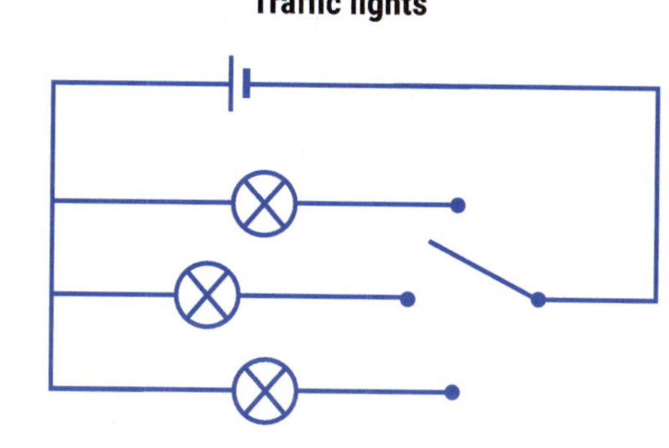

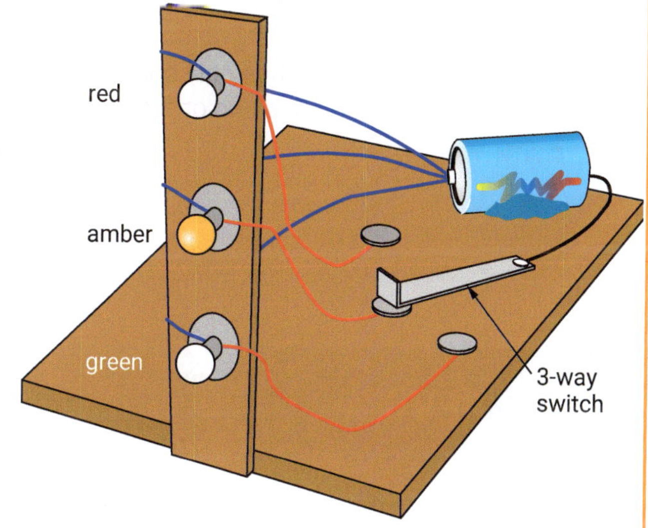

Wiper glasses

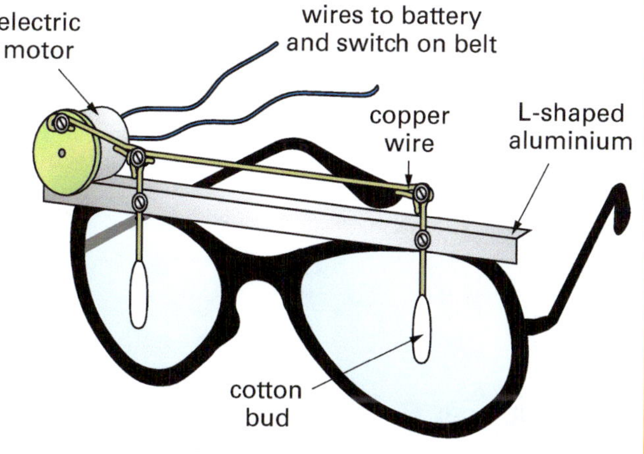

ISBN 978 1 4202 3736 8

MAIN IDEAS

Copy and complete these statements to make a summary of this chapter. The missing words are on the right.

1 Objects can be given an electric ______ by rubbing. Gaining electrons makes an object negatively charged, and losing electrons makes it positively charged.

2 Like charges ______ each other, while unlike charges ______ each other.

3 Electric current will flow only if it has a continuous path or ______.

4 Electric current is a flow of ______. It is measured in amperes, using an ______.

5 Batteries supply the ______ to push electrons around a circuit. ______ is a measure of how much energy can be given to the moving electrons in a circuit. It is measured in volts.

6 ______ offer little resistance to the flow of electricity. ______ offer a great deal of resistance.

7 A series circuit has only one conducting path for electrons, whereas a ______ circuit has two or more alternative paths.

energy
ammeter
voltage
insulators
circuit
parallel
electrons
repel
conductors
attract
charge

CH•2 REVIEW

1 What happens to two charged rods held near each other if they have:
 a the same charge?
 b opposite charges?

2 What charge is left on a material if it has been rubbed and:
 a loses electrons?
 b gains electrons?

3 Which of the following are conductors and which are insulators?
 a copper
 b plastic
 c steel
 d air
 e wood
 f salt water

4 Look at the diagrams below.
 a Which is the correct way to put two batteries in a torch?
 b Are the batteries connected in series or in parallel?

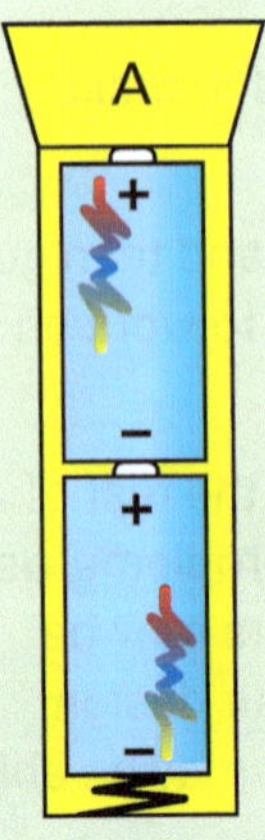

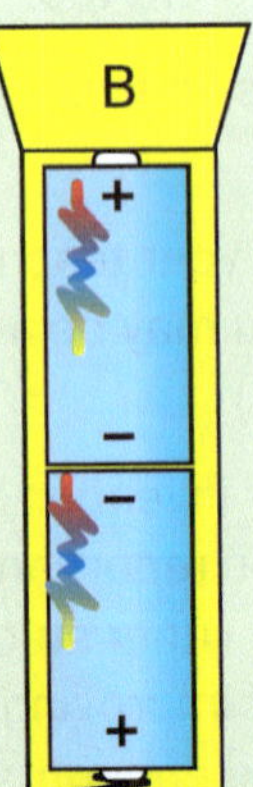

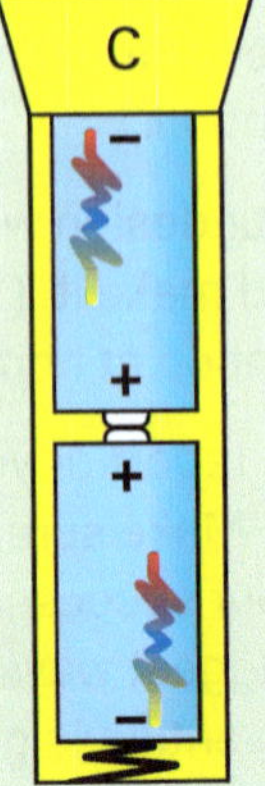

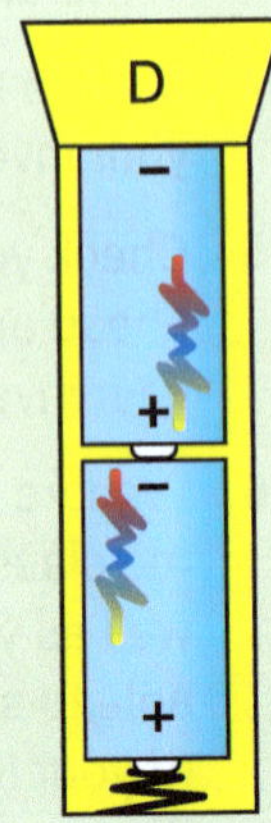

ISBN 978 1 4202 3736 8

5 Consider the circuits below.

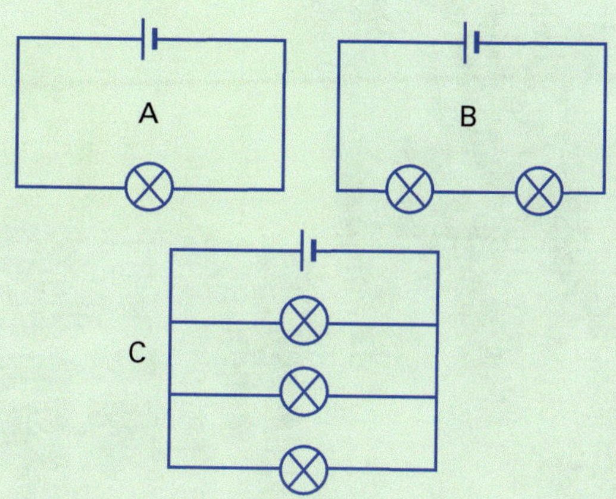

a How will the brightness of the bulbs in circuit B compare with the bulb in circuit A? Why?

b How bright will the bulbs in circuit C be compared with the bulb in circuit A? Why?

c Without changing the number of bulbs, how could you make the brightness of the bulbs in circuit B the same as the bulb in circuit A? Draw a diagram of the new circuit.

6 Consider the circuit below.

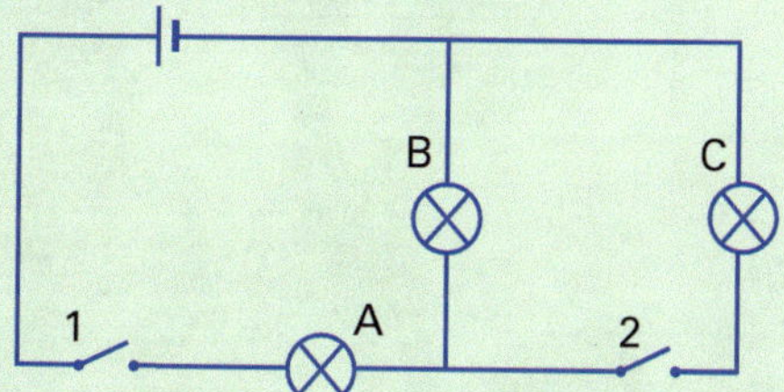

When switch 1 is closed and switch 2 is open:

A none of the bulbs lights up.

B only bulb A lights up.

C bulb A and bulb B light up.

D all the bulbs light up.

7 Explain why you sometimes get an electric shock when you walk on a nylon carpet and then touch something made of metal.

8 Design a circuit with a cell, a switch and a bulb, so that the light goes *off* when the switch is closed.

LAB REVIEW

Work with a partner. From the two circuits below, select the one you are going to set up. Your partner (or your teacher) will mark you on your performance.

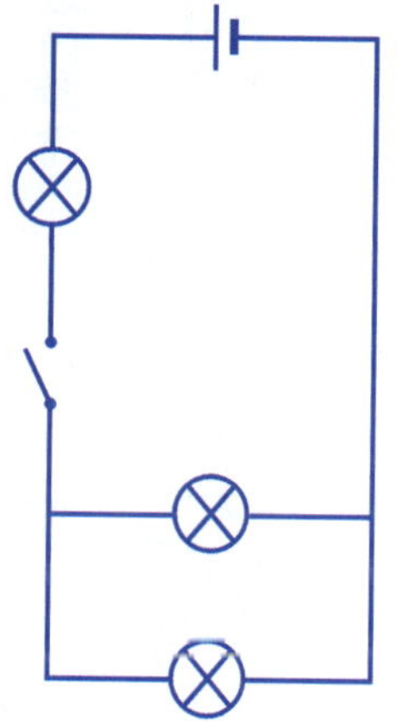

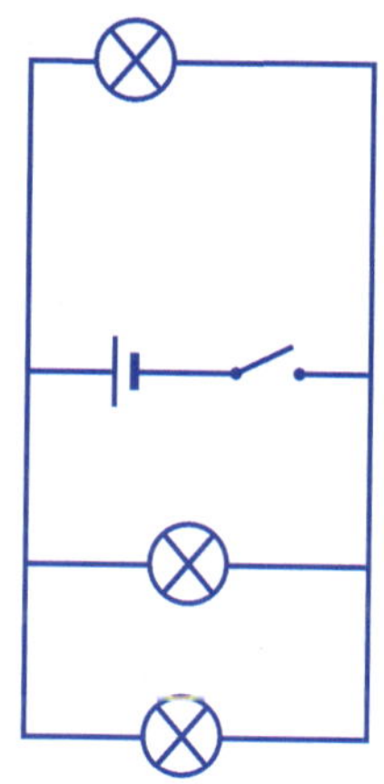

First write down a list of the equipment that you will need to set up the circuit. Then set up the equipment correctly and promptly.

How to score

List of equipment:

A chose the equipment perfectly

B left out a small item, such as a connecting wire

C left out a major item, such as a bulb or a battery

D was not sure of the equipment needed

Setting up the equipment:

A set up the circuit correctly and promptly

B set up the circuit correctly, but took quite a while to do it

C set up the circuit promptly, but with a slight error in it

D set up the circuit slowly, but with a slight error in it

E was not sure how to set up the circuit

Dismantle the circuit. Now swap roles, so that this time you mark the performance of your partner setting up the other circuit. (Don't forget to return all equipment.)

Check your answers on page 297.

ISBN 978 1 4202 3736 8

IN THIS CHAPTER

Science Understanding

- describe and model the structure of atoms in terms of the nucleus, protons, neutrons and electrons
- describe the differences between protons, neutrons and electrons
- describe the three types of radiation released by radioactive material—alpha and beta particles and gamma radiation
- describe and distinguish between chemical and nuclear reactions in terms of conservation of mass

Science Inquiry Skills

- investigate the historical development of models of the structure of matter
- reflect on how knowledge about nuclear fission was used to develop the atomic bomb

CH•3 Inside the atom

ISBN 978 1 4202 3736 8

GET STARTED: *QUESTION*

Answer the questions below to find out how much you know about the structure of the atom.

Record your answers and check them when you have finished the chapter. Correct your answers and rewrite them once you know more about the atom.

1. What are the three particles inside the atom?
2. What is a quark?
3. What is the difference between a chemical reaction and a nuclear reaction?
4. What is the significance of Einstein's equation $E = mc^2$ for nuclear reactions?
5. What is a chain reaction?
6. How does a nuclear power station produce electricity?
7. What is the difference between nuclear fission and nuclear fusion?
8. What is radioactivity?
9. How are radioactive substances used in medicine?
10. When was the first atomic bomb made?

ISBN 978 1 4202 3736 8

3.1 Atomic structure

Until about the end of the 19th century, scientists thought that atoms were like tiny balls that could not be split into anything smaller. But discoveries were being made that suggested that there might be even smaller particles *inside* the atom. This idea was difficult to test because nobody could see atoms, and they certainly couldn't see inside them. However, from the results of their experiments, scientists were able to form models to represent what is inside the atom.

The plum pudding model

In 1897, an Englishman called JJ Thomson (or JJ as he was usually called) was experimenting with electricity in gases. He found tiny, negatively charged particles that were much, much smaller than atoms. These new particles were called **electrons**, and scientists thought they would probably be found inside atoms. Thomson suggested a model for the atom that was like a plum pudding of positive charge with negatively charged electrons scattered through it like raisins.

Rutherford's experiment

The next big discovery was made eight years later by a New Zealander, Ernest Rutherford. He and two of his students were doing an experiment at Cambridge University in England. They were firing positively charged particles from radioactive radium at an extremely thin piece of gold foil in a vacuum, as shown below. Circling the foil was a photographic film to record any particles that hit it. What they found was that most of the particles went straight through the foil and ended up at A. Some passed through but changed direction slightly, striking the film at other points, for example B. But occasionally a particle bounced straight back (like C).

Rutherford described his experiment like this: 'It was quite the most incredible event that has ever happened to me in my life. It was almost as incredible as if you fired a 15 inch shell [40 cm in diameter] at a piece of tissue paper and it came back and hit you.' Thomson's plum pudding model didn't fit Rutherford's observations, because it didn't predict that the large positively charged particles would bounce straight back. So Rutherford proposed a new model in which the positive charges in the atom were concentrated in a small central core or **nucleus**. This nucleus would have a big enough charge to repel the positively charged particles, causing them to bounce straight back. He inferred that most of the particles did not bounce back because they went through the empty space inside the gold atoms.

Figure 3.1 Rutherford's gold foil experiment

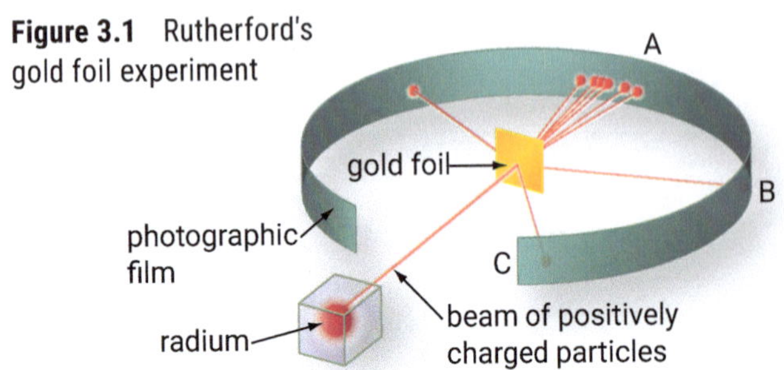

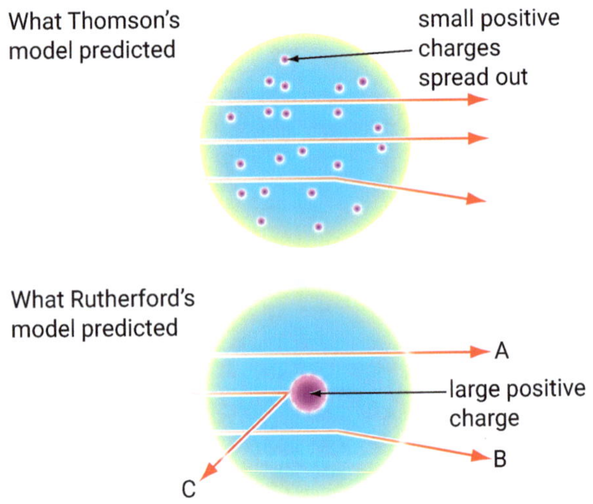

Figure 3.2 Comparing Thomson's and Rutherford's models of the atom

A changing model

Rutherford's model of the atom explained where the positive charges are, but not where the negative electrons are. Niels Bohr, a young Danish scientist, worked with Thomson and Rutherford. In 1913, he did some calculations that suggested that the electrons move rapidly around the nucleus in fixed orbits, like planets orbiting the sun.

ISBN 978 1 4202 3736 8

In the years that followed, scientists began to think that the nucleus itself could be made up of even smaller particles. They fired particles at the nucleus and found that smaller particles were occasionally knocked out. One of these was the positively charged **proton**. The other type of particle wasn't affected by the charged particles fired at it. So scientists inferred that it had no charge, and called it a **neutron**. Protons and neutrons have about the same mass, and are almost 2000 times heavier than an electron.

Atomic structure

From all these discoveries, scientists have put together the following picture (model) of what is inside the atom. There are three kinds of particles—protons, neutrons and electrons. These are often called *subatomic* particles (*sub* means 'under' or 'smaller'). The protons and neutrons are packed together tightly in the nucleus of the atom. The electrons are attracted to the positively charged nucleus, and move rapidly in the area around it.

If the atom were enlarged to the size of a football stadium, the nucleus would be the size of a pinhead in the centre of the field. The electrons would be moving rapidly around the stadium, but even at this scale they would be far too small to see. Most of the atom is empty space.

The electrons are so small and move so rapidly that it is impossible to say exactly where they are at any particular time. This is why they are sometimes shown as an electron cloud—a sort of fuzzy area around the nucleus where the electrons are most likely to be.

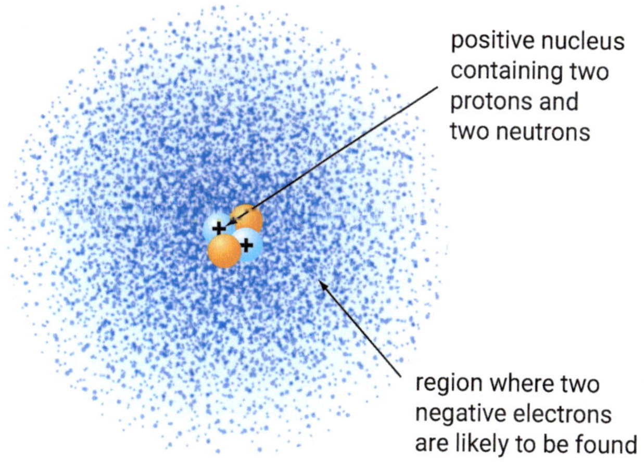

Figure 3.3 A model of a helium atom, with 2 protons, 2 neutrons and 2 electrons

ACTIVITY

Mystery boxes

1 Your teacher will give you a numbered mystery box. Each mystery box contains an unknown object. You are to find out as much as you can about the object, *without opening the box*. You can tilt the box, shake it, or anything else, but you must not open it or damage it in any way.

Each time you do something with the box, record your observations.

The idea is not so much to guess what the object is, but to describe the size, shape and any other properties of the object. You can then draw it. Your drawing is a model of the object.

How is what you have done in this activity similar to the way scientists found out what is inside atoms?

If you have time, try another mystery box. Your teacher *may* tell you what is in each box. If so, you can compare your model with the real thing.

2 The diagram below shows a novel way to obtain information about an unknown object in a box by examining the paint left on the moveable cardboard strip.

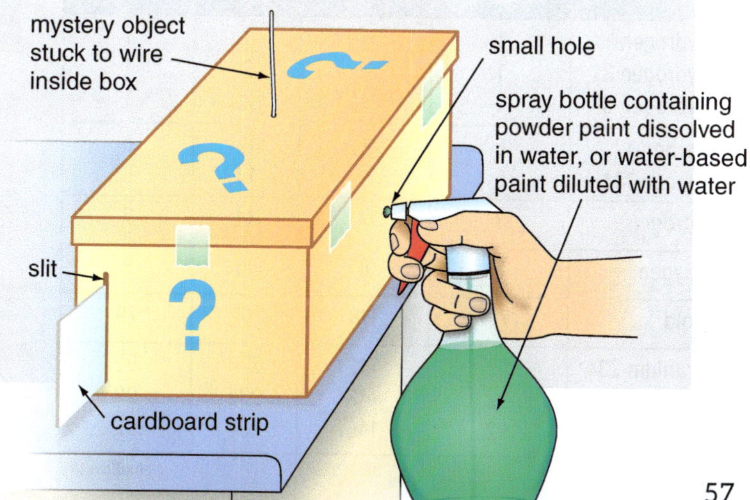

ISBN 978 1 4202 3736 8

Atomic number

Different atoms have different numbers of protons in their nucleus. For example, hydrogen atoms have only one proton, but uranium atoms have 92 protons. The number of protons in an atom is called its **atomic number**. The number of electrons is the same as the number of protons. This means that the positive and negative charges balance each other, and the atom has no overall charge—it is neutral.

The only thing that makes one atom different from another is the number of particles it contains. For example, the difference between a nitrogen atom and an oxygen atom is that the oxygen atom has one more proton and one more electron. You can see each atom's atomic number in the periodic table on the next page.

Isotopes

Early in the 20th century, it was found that most naturally occurring elements contain atoms that are not all exactly the same. For example, there are three different forms of the element hydrogen. These are called hydrogen-1, hydrogen-2 (deuterium) and hydrogen-3 (tritium). The nucleus of a hydrogen-1 atom contains only one proton, but hydrogen-2 has a proton and a neutron in its nucleus. Hydrogen-3 has a proton and *two* neutrons. The table below shows the numbers of subatomic particles in a few common isotopes. Note that the number of protons is always the same as the number of electrons.

Isotope	Number of protons (atomic number)	Number of neutrons	Mass number	Number of electrons
hydrogen-1	1	0	1	1
hydrogen-2	1	1	2	1
hydrogen-3*	1	2	3	1
carbon-12	6	6	12	6
carbon-14*	6	8	14	6
nitrogen	7	7	14	7
oxygen	8	8	16	8
gold	79	118	197	79
uranium-234*	92	142	234	92
uranium-235*	92	143	235	92
uranium-238*	92	146	238	92

***Radioactive**

The three different forms of hydrogen are called **isotopes** (EYE-so-topes). *Iso* means 'equal' or 'the same'. These isotopes have the same chemical properties, because they all have the same number of protons. However, they have different masses because their nuclei are different. The number 1, 2 or 3 after the name is used to tell them apart. This number is the total number of protons and neutrons in the nucleus. It is called the **mass number**.

Figure 3.4 Like these characters, isotopes are forms of the same element. They have the same chemical properties but different masses (different nuclei).

Radioisotopes

If the number of neutrons is much greater than the number of protons, the nucleus may be unstable and break up. When this happens, a nuclear reaction occurs and the isotope is said to be radioactive. An isotope that is radioactive is called a **radioisotope**.

ACTIVITY

Isotope models

1. Make small balls from two different colours of plasticine to represent protons and neutrons.
2. Make a model of hydrogen-1 by hanging a proton inside a round balloon. Put a small dot on the balloon to represent the electron.
3. Also make models of hydrogen-2 and hydrogen-3.
   What is the difference between the three isotopes of hydrogen?
4. Using the information in the table, make models of the two isotopes of carbon, and the atoms of nitrogen and oxygen. Remember that as the atomic number increases, so does the size of the atom.

ISBN 978 1 4202 3736 8

Periodic table of the elements

Elements in the same vertical column have similar properties

Key

The colour of the name for each element indicates its state at room temperature:
black—solid
blue—liquid
pink—gas

Elements in red are synthetic.

Chemical families

- Alkali metals
- Alkaline earth metals
- Transition metals
- Rare earth metals
- Other metals
- Metalloids
- Other non-metals
- Halogens
- Noble gases

1	2	3	4	5	6	7	8	9	10	11	12	13	14	15	16	17	18
1 H Hydrogen																	2 He Helium
3 Li Lithium	4 Be Beryllium											5 B Boron	6 C Carbon	7 N Nitrogen	8 O Oxygen	9 F Fluorine	10 Ne Neon
11 Na Sodium	12 Mg Magnesium											13 Al Aluminium	14 Si Silicon	15 P Phosphorus	16 S Sulfur	17 Cl Chlorine	18 Ar Argon
19 K Potassium	20 Ca Calcium	21 Sc Scandium	22 Ti Titanium	23 V Vanadium	24 Cr Chromium	25 Mn Manganese	26 Fe Iron	27 Co Cobalt	28 Ni Nickel	29 Cu Copper	30 Zn Zinc	31 Ga Gallium	32 Ge Germanium	33 As Arsenic	34 Se Selenium	35 Br Bromine	36 Kr Krypton
37 Rb Rubidium	38 Sr Strontium	39 Y Yttrium	40 Zr Zirconium	41 Nb Niobium	42 Mo Molybdenum	43 Tc Technetium	44 Ru Ruthenium	45 Rh Rhodium	46 Pd Palladium	47 Ag Silver	48 Cd Cadmium	49 In Indium	50 Sn Tin	51 Sb Antimony	52 Te Tellurium	53 I Iodine	54 Xe Xenon
55 Cs Caesium	56 Ba Barium		72 Hf Hafnium	73 Ta Tantalum	74 W Tungsten	75 Re Rhenium	76 Os Osmium	77 Ir Iridium	78 Pt Platinum	79 Au Gold	80 Hg Mercury	81 Tl Thallium	82 Pb Lead	83 Bi Bismuth	84 Po Polonium	85 At Astatine	86 Rn Radon
87 Fr Francium	88 Ra Radium		104 Rf Rutherfordium	105 Db Dubnium	106 Sg Seaborgium	107 Bh Bohrium	108 Hs Hassium	109 Mt Meitnerium	110 Ds Darmstadtium	111 Rg Roentgenium	112 Cn Copernicium	113 Nh Nihonium	114 Fl Flerovium	115 Mc Moscovium	116 Lv Livermorium	117 Ts Tennessine	118 Og Oganesson

57 La Lanthanum	58 Ce Cerium	59 Pr Praseodymium	60 Nd Neodymium	61 Pm Promethium	62 Sm Samarium	63 Eu Europium	64 Gd Gadolinium	65 Tb Terbium	66 Dy Dysprosium	67 Ho Holmium	68 Er Erbium	69 Tm Thulium	70 Yb Ytterbium	71 Lu Lutetium
89 Ac Actinium	90 Th Thorium	91 Pa Protactinium	92 U Uranium	93 Np Neptunium	94 Pu Plutonium	95 Am Americium	96 Cm Curium	97 Bk Berkelium	98 Cf Californium	99 Es Einsteinium	100 Fm Fermium	101 Md Mendelevium	102 No Nobelium	103 Lr Lawrencium

ISBN 978 1 4202 3736 8

Use the periodic table on page 59 then answer these questions. You may need to research the answers to some questions.

1 a How many elements are there in the periodic table?
  b Which element is the smallest and which is the largest? Justify your answer.

2 Find the elements with symbols C, Ar, Br and No. For each one write down its name, atomic number and two other characteristics.

3 Do you know what the element used in these signs is called?

4 Find the information on the periodic table that tells you how to check if an element is a solid, liquid or gas at room temperature. Use this information to name two solids, two gases and two liquids.

3 Find five other elements that you have heard of and write down their names, atomic number, and what you know about them.

Use the key to identify the chemical families of the periodic table.

Note that each column of the periodic table is called a group and has a number from 1 to 18. Use this information to answer questions 4–6.

4 a What is the name of the chemical family in column 2?
  b Identify the column number and name of the group made up of gases only.
  c How many groups are made up of metals?

5 Examine the elements in the halogens chemical family.
  a Which element is the smallest and which is the largest?
  b Which of the elements are solids? What are the states of the other elements in the family?

6 The elements B and Si are in the metalloids family. What characteristics might these elements have?

Silicon

7 Create a mnemonic to help you remember the first 20 elements. To do this, take the symbols of each element and make up a sentence or sentences that start with the symbols as the start of each word. Here is an example to get you started:

Happy **He**nry **Li**kes **Be**ing **B**eside **C**razy **N**erds …

8 You have learnt about how elements join together to form molecules. Look at the periodic table and think about which of the elements might join together, and write down the chemical formulas of some molecules that you know, for example carbon dioxide, CO_2.

ISBN 978 1 4202 3736 8

SCIENCE AS A HUMAN ENDEAVOUR

Searching for the God particle

Physicists now think that protons and neutrons are made of even smaller particles. Using special machines called synchrotrons (SINK-row-trons), they have accelerated protons to enormously high speeds, almost the speed of light (300 000 km/s). When these speeding protons hit neutrons and other protons, they are shattered into even smaller particles.

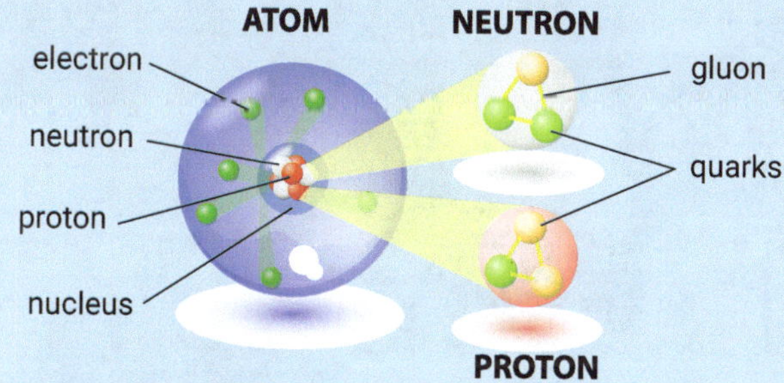

To explain the results of these experiments, scientists have proposed that protons and neutrons are made of particles called **quarks**. There are six different types of quarks—up, down, charm, strange, top and bottom. These quarks are held together by particles called gluons. A proton is made up of two up quarks and one down quark. A neutron is made of two down quarks and one up quark. As well as quarks, there are also leptons and guage bosons. See Figure 3.6.

In 2012 physicists detected and measured the mass of a new particle that had long been searched for. This particle, called the Higgs boson (or God particle), is about 126 times the mass of a proton, and was discovered using the Large Hadron Collider (see next page). The existence of the Higgs boson is considered very important as it supports the Standard Model of particle physics, and confirms the existence of the Higgs field. The Higgs field is an invisible energy field that exists throughout the universe, and gives all other particles their mass. In the Standard Model of particle physics, the Higgs particle is a boson with no spin or electric charge. It is very unstable and decays into other particles immediately, and therefore was very hard to create and measure. It is also not yet confirmed whether there is more than one type of Higgs boson.

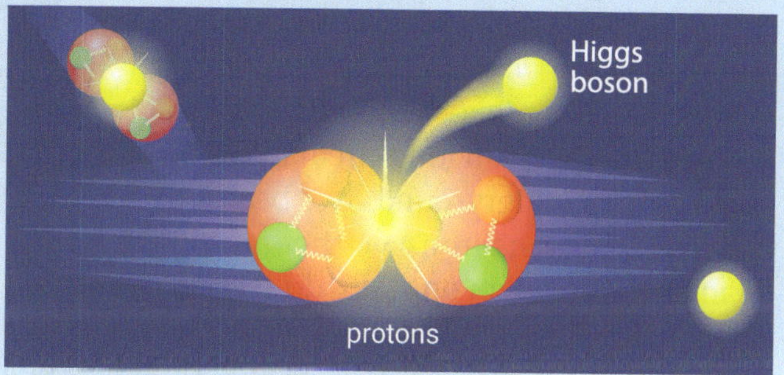

Figure 3.5 The Higgs boson is formed by colliding protons.

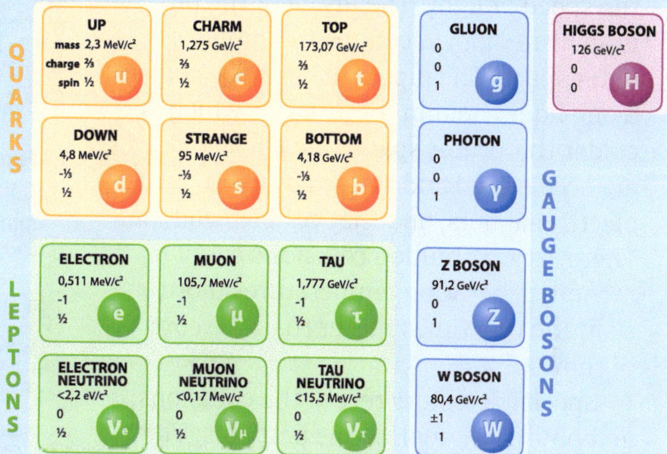

Figure 3.6 Elementary particles of matter

Figure 3.7 This is a representation of particle tracks from a proton–proton collision. A large number of particles (orange lines) were created in the collision. Among the particles produced was a Higgs boson. The particle was not seen, but is shown through its decay products: two electrons and two positrons (green lines).

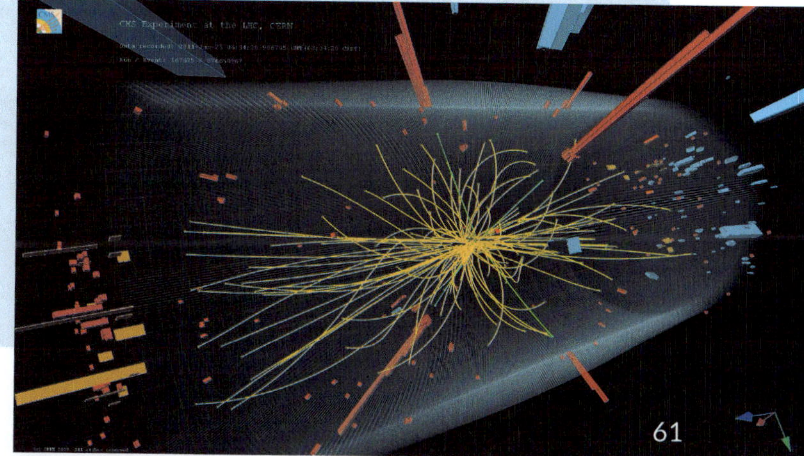

ISBN 978 1 4202 3736 8

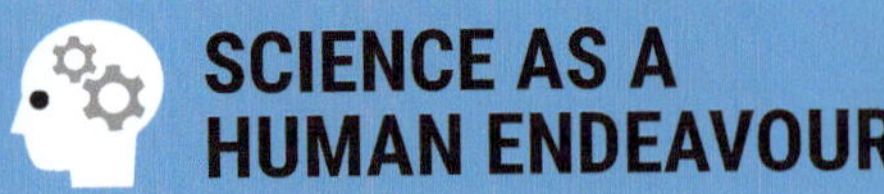

SCIENCE AS A HUMAN ENDEAVOUR

The Large Hadron Collider

There are many unanswered questions in particle physics, and this is why the Large Hadron Collider has been built. This giant synchrotron is near Geneva, where it spans the border between Switzerland and France. It is a 27 km long circular tunnel about 100 m underground. Two beams of particles called hadrons—either protons or lead ions—travel in opposite directions in a pipe inside the tunnel, gaining energy every lap. Trillions of particles race around this pipe, 11 245 times a second, at 99.9999991% the speed of light. The inside of the pipe is an ultra-high vacuum—as empty as outer space. The pipe also has to be cooled using liquid helium to −271.3°C, which is colder than outer space. The particle beams are accelerated and steered by 9300 giant electromagnets, and they collide head-on. There are 600 million collisions every second, generating temperatures more than 100 000 times hotter than the core of the sun!

Special detectors record the collisions, and physicists analyse the particles. They recreate conditions similar to those in the very early universe, just after the Big Bang. These conditions are too hot and energetic for the gluons to hold the quarks together, and physicists use this to observe the Higgs boson (God particle).

There is a much smaller synchrotron in Melbourne that is used to study the molecular structure of materials.

Figure 3.8 The Large Hadron Collider is a 27 km long tunnel of superconducting magnets.

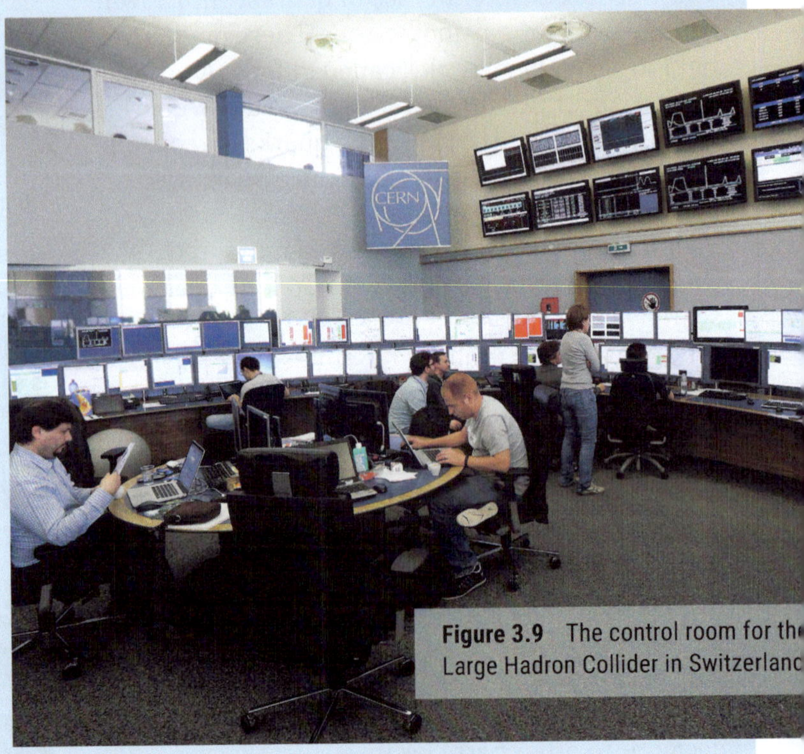

Figure 3.9 The control room for the Large Hadron Collider in Switzerland

Question

The Large Hadron Collider cost $9 billion to build. Do you think this huge expenditure can be justified? Explain.

ISBN 978 1 4202 3736 8

CHECK

1 What makes one atom different from another?

2 Name the three subatomic particles.

3 Copy the diagram of a tritium atom below and add labels for A, B, C and D.

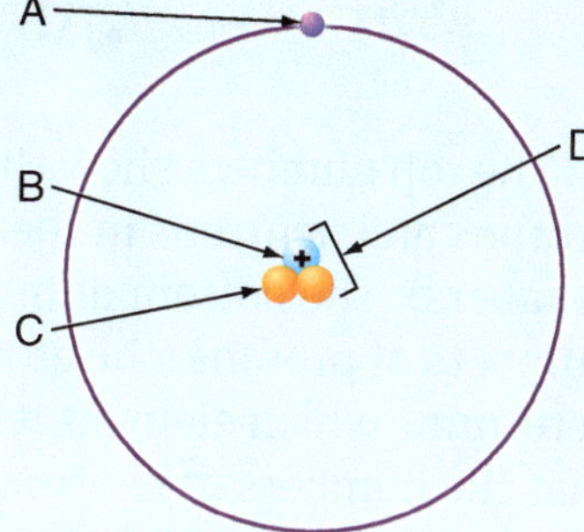

4 Which of the three subatomic particles are:
- a in the nucleus?
- b moving rapidly around the nucleus?
- c the smallest?
- d positively charged?
- e negatively charged?
- f neutral?

5 An atom has six protons in its nucleus. How many electrons does it have? Explain your answer.

6 A lithium atom has three protons, four neutrons and three electrons.
- a What is its atomic number?
- b What is its mass number?
- c Draw a picture of it.

7 Copy the diagram of a boron atom below.

- a Label the nucleus and the electrons.
- b How many protons, neutrons and electrons does a boron atom have?
- c Compare the boron atom with the helium atom on page 57. How are the two atoms different?

8
- a What do the isotopes of a particular element have in common?
- b How do isotopes of a particular element differ from each other?

9 Two atoms both have 10 neutrons in their nuclei. One has 8 protons, the other has 9. Are they isotopes? Why or why not?

10 What are radioisotopes?

CHALLENGE

1 Imagine that you have shrunk to a size smaller than that of an atom. At this size, you can wander around inside a gold atom. It has an atomic number of 79 and a mass number of 197. Describe what you see.

2 Why did Rutherford change Thomson's model of the atom?

3 Explain the meaning of the term *isotope* using the terms *atomic number* and *mass number*.

4 Different isotopes of the same element react in the same way during chemical reactions. Why?

5 Scientists can't see protons, neutrons and electrons—they only infer that they exist from the results of their experiments. What does this statement mean?

6 Sir Isaac Newton once said, 'If I have been able to see further than others, it is because I stood on the shoulders of giants'. Write a paragraph using the history of the atom to explain what he meant.

7 The neutron was not discovered until more than 30 years after the discovery of the proton and the electron. Why was the neutron so difficult to detect?

8 When a giant star explodes, its core collapses to form a neutron star. A matchbox full of material from a neutron star would have a mass of millions of tonnes. Use what you learnt about atoms to explain this incredibly high density.

ISBN 978 1 4202 3736 8

3.2 Nuclear reactions

A chemical reaction

Figure 3.10 is a representation of the chemical reaction that occurs when hydrogen gas burns in air. Hydrogen molecules (H_2) react with oxygen molecules (O_2) from the air to form water molecules (H_2O). Count the number of hydrogen atoms and oxygen atoms before and after the reaction. What do you notice?

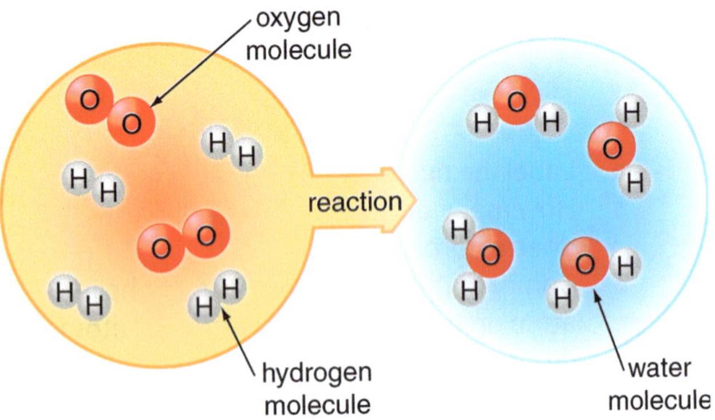

Figure 3.10 Hydrogen and oxygen react to form water. Mass is conserved.

Conservation of mass

All the atoms present before the chemical reaction are still there after the reaction. No atoms are destroyed and no new atoms are formed. In other words, the atoms are *conserved*. This is called the **law of conservation of mass**. The molecules break apart and rearrange themselves to form new molecules. The hydrogen and oxygen atoms form chemical bonds with each other by sharing their electrons, but the nuclei of the atoms are not affected. A new substance (water) is formed, but no new elements—that is, no new atoms are formed.

A nuclear reaction

In this section you will learn about **nuclear reactions**. For example, an atom of uranium-238 has 92 protons in its nucleus and 146 neutrons, giving it a mass number of 238. Uranium-238 is radioactive and its nucleus is very unstable. As a result, two of the protons and two of the neutrons break off to form a helium nucleus, which has 2 protons and 2 neutrons. What remains is no longer uranium-238, because it has only 90 protons. It is the element thorium. This nuclear reaction can be represented as follows:

$$^{238}_{92}\text{U} \rightarrow {}^{234}_{90}\text{Th} + {}^{4}_{2}\text{He}$$

The top numbers show the total numbers of protons and neutrons in the nuclei (the mass numbers). The bottom numbers show the numbers of protons (the atomic numbers), which determine which element it is. You will notice that the numbers of neutrons and protons are the same on both sides of the equation. Despite this, scientists have found that the combined mass of the two atoms on the right is very slightly less than the mass of the original uranium atom. So mass has mysteriously been lost. This never happens in a chemical reaction. It only occurs in nuclear reactions where the nuclei are involved in the reaction. Scientists have found that the lost mass (m) in a nuclear reaction is converted directly into energy (E), according to Albert Einstein's famous equation $E = mc^2$.

Figure 3.11 Albert Einstein proposed that mass can be converted to energy.

In this equation **c** is the velocity of light—a very large number—so a tiny loss of mass in a nuclear reaction produces a huge amount of energy. For example, a loss of 1 gram produces 90 million megajoules—enough energy to meet the needs of a city of about 140 000 people for a day!

ISBN 978 1 4202 3736 8

Nuclear fission

In 1919 Ernest Rutherford fired nuclei of helium atoms at nitrogen atoms, and occasionally oxygen atoms were produced. This was a nuclear reaction because it involved changes to the nuclei of the atoms. We now know that the nucleus of the nitrogen atom had gained a proton (and a neutron) to become an oxygen atom. This discovery caused much excitement among scientists. It seemed that the dream of the alchemists of converting common metals into gold might come true.

When neutrons were discovered in 1932, scientists used them to bombard other atoms in the hope of causing more nuclear reactions. At the time, uranium was the largest atom known, and scientists were trying to make larger atoms by bombarding it with neutrons. Eventually they were successful and made completely new elements such as plutonium and americium. However, they also discovered that the nuclei of some atoms could be split into two smaller nuclei. This process was called **nuclear fission**. Fission simply means splitting into parts.

When a neutron collides with a uranium-235 atom, the uranium atom becomes extremely unstable. It then splits into two smaller atoms and three or four neutrons, and releases a huge amount of energy.

The neutrons released collide with other uranium nuclei, which release more neutrons. These collide with even more nuclei, and so on. This is called a **chain reaction** (see Figure 3.13).

Huge amounts of energy can be released in a fraction of a second. If the chain reaction is not controlled, there is a nuclear explosion. If the chain reaction is controlled, the heat released can be harnessed to generate electricity in a nuclear power station.

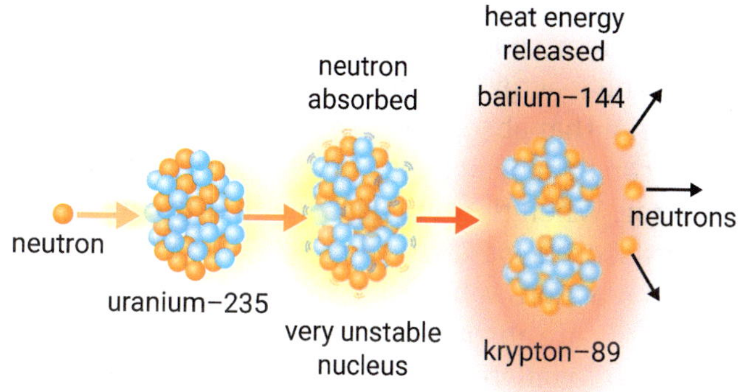

Figure 3.12 Nuclear fission of a uranium-235 atom

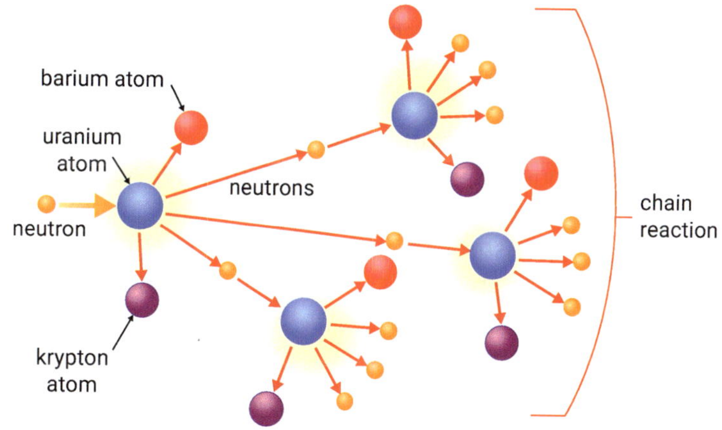

Figure 3.13 Nuclear fission of a uranium atom, leading to a chain reaction

ACTIVITY

Chain reaction

1. Arrange a set of dominoes so that each domino that falls will knock over two others. The more dominoes you use, the more dramatic the effect will be.

   Would you say this was a controlled chain reaction or an uncontrolled one? Why?

2. Set up the dominoes again. This time arrange them so that when they fall, some knock over two dominoes, some knock over one, and some don't knock any over.

   What happens this time? Is this a controlled chain reaction or an uncontrolled one? Why?

   Of the two set-ups, which is most like what happens when a nuclear bomb explodes? Which is like what happens in a nuclear power station?

3. Another way to model a chain reaction is to use a box full of mousetraps, with two wads of paper on each trap as shown. Then drop a paper wad into the box.

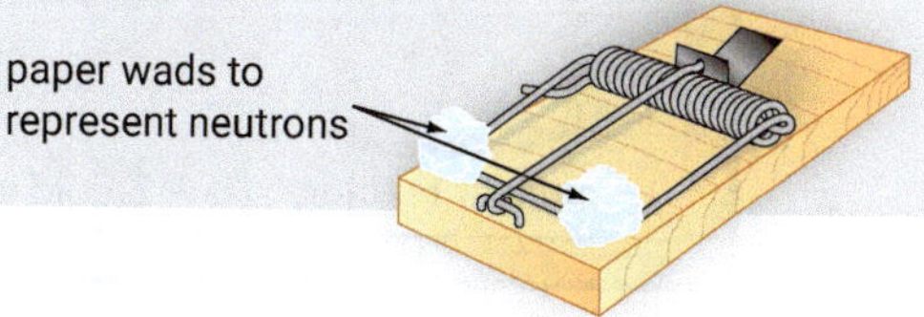

ISBN 978 1 4202 3736 8

Nuclear power stations

In a nuclear bomb huge amounts of energy are released in a fraction of a second. In a power station the chain reaction is carefully controlled, so that the energy is released at a steady rate.

The central part of a nuclear power station is the **nuclear reactor**, which contains the uranium fuel. The reactor has *control rods*, made from a material that absorbs neutrons. These control rods can be moved in and out of the *core* of the reactor. When they are pushed into the core, they absorb a lot of neutrons, and this stops the chain reaction. When they are pulled out, fewer neutrons are absorbed. As a result, there are more neutrons available, and the chain reaction speeds up. So, by moving the control rods in and out, the chain reaction can be controlled.

The heat produced in the reactor is taken away by the coolant, which is usually pressurised water. This heat is used to boil water and produce steam. The steam then spins turbines that are connected to electric generators. During the fission reaction, dangerously radioactive materials are produced. The thick concrete containment building around the reactor stops the radiation from escaping.

There are two main isotopes of uranium—uranium-235 and uranium-238. Only uranium-235 atoms will split, but uranium ore contains less than 1% of this isotope. Scientists have solved this problem by removing some of the U-238 using a gas centrifuge to produce enriched uranium (about 3% U-235) for use in nuclear reactors. The uranium used in nuclear bombs must be enriched even further (at least 20% U-235).

Some ships and submarines are nuclear-powered and work in much the same way as a nuclear power station. The ships don't need to carry fuel oil as normal ships do, and can keep operating for several months without refuelling.

Nuclear fusion

There is a second type of nuclear reaction called **nuclear fusion**, which is the reaction that powers the sun and other stars. It occurs when two nuclei of deuterium or tritium (isotopes of hydrogen) come together or *fuse*. This process releases huge amounts of energy, even more than nuclear fission does.

However, fusion needs enormous temperatures to get it going—around 100 000 000 °C. At this temperature, matter is in the plasma state—a gas consisting of positive ions and fast-moving electrons. The major problem is how to contain the plasma, since no material known can withstand these temperatures. So scientists have been experimenting with 'magnetic bottles', in which the plasma is contained inside invisible magnetic fields.

So far scientists have only been able to produce fusion for a few minutes. Much more research and development will be necessary before fusion power stations are possible. The promising thing about nuclear fusion is that it produces less radioactive waste than nuclear fission. Also, the fuel needed is in almost unlimited supply since deuterium (heavy hydrogen) can be extracted from seawater.

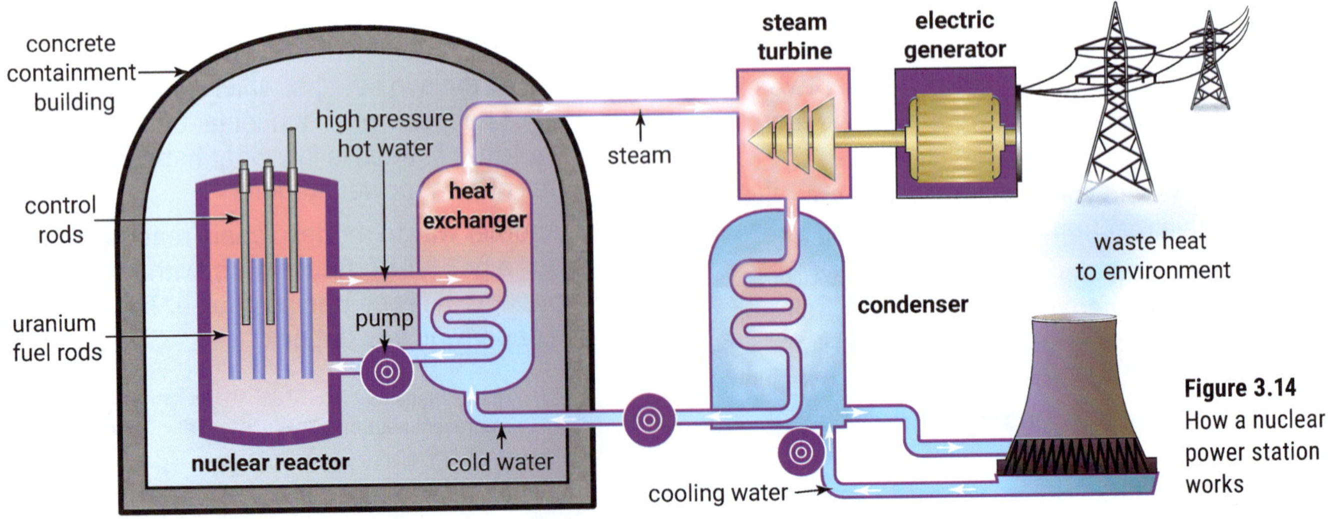

Figure 3.14 How a nuclear power station works

ISBN 978 1 4202 3736 8

SCIENCE AS A HUMAN ENDEAVOUR

Lise Meitner and nuclear fission

Lise Meitner (LEEZ-a MITE-ner) was born in Vienna, Austria, in 1878. She was one of eight children and her family was Jewish. She was a shy girl. At that time girls didn't go to high school, so her parents paid for a personal tutor. She did well at maths and science and loved music. She went on to university, even though women at the time didn't do this. Lectures bored her but the laboratory fascinated her, so she decided to study physics.

Meitner started experimenting with radioactive elements and for 30 years worked with Otto Hahn in Berlin in Germany. In 1917, she and Hahn discovered a new element called protactinium. They were also hoping to discover an element heavier than uranium.

In 1938, Hitler invaded Austria, and Meitner was forced to flee from Germany into Sweden. This cut her off from her laboratory and the scientists she had been working with, but she kept in touch by mail. Hahn and Fritz Strassman continued the experiments they had begun earlier with Meitner—bombarding uranium with neutrons. To their surprise, they found that the products of the nuclear reactions were *lighter* than uranium, not heavier.

Hahn wrote to Meitner in Sweden saying 'it can't really break up into barium ... try to think of some other possible explanation'. While in Denmark for the Christmas holidays, she and her nephew Otto Frisch proved that the splitting of the uranium atom was theoretically possible. They published their findings and called the process 'fission'.

Hahn and Strassman then showed by experiment that Meitner was correct. Hahn published his findings without listing Meitner as a co-author, even though he had worked with her on fission for 30 years. In 1944, he was awarded the Nobel Prize in Chemistry, and Meitner missed out. Hahn said he couldn't give credit to Meitner because of the political situation. Even in later years, Hahn never fully acknowledged the part that Meitner played in the discovery.

Meitner never complained about not getting the Nobel Prize. She refused to work on the project to develop an atomic bomb and did not like being called 'the mother of the atom bomb'. When Hollywood wanted to make a movie about her, she said that she 'would rather walk the length of Broadway in the nude than see herself in a movie'.

In 1997, Lise Meitner was rewarded with a permanent place in the periodic table. The synthetic element with atomic number 109 was called meitnerium (mite-NEAR-ee-um) after her.

Figure 3.15 Lise Meitner, a pioneer of nuclear chemistry

Questions

1. Why did Lise Meitner have to flee from Germany in 1938?
2. Why do you think Meitner wasn't awarded the Nobel Prize with Otto Hahn?
3. Do you think Meitner was treated fairly? Explain.

ISBN 978 1 4202 3736 8

CHECK

1 Explain the difference between a chemical reaction and a nuclear reaction.

2 Explain in your own words the law of conservation of mass.

3 What is used for fuel in a nuclear reactor?

4 Which subatomic particles produced during nuclear fission are capable of causing a chain reaction?

5 Use a diagram to explain the term *nuclear fission*.

6 In a chain reaction, huge quantities of energy are released. How does this happen?

7 Use the diagram on page 66 to explain in your own words how a nuclear power station works.

8 a What is the function of the control rods in a nuclear reactor?

b What would happen to a nuclear reactor if the control rods were pulled out and then for some reason could not be pushed back in?

9 Why is a nuclear reactor enclosed in concrete?

10 What is the difference between fission and fusion? Which produces more energy?

11 What would be the advantages and disadvantages of using a nuclear reactor to power a spaceship?

CHALLENGE

1 The photo at right shows sulfur burning in a jar of air (oxygen). The circles below the photo show the atoms and molecules involved in this chemical reaction. Does the reaction obey the law of conservation of mass? Explain.

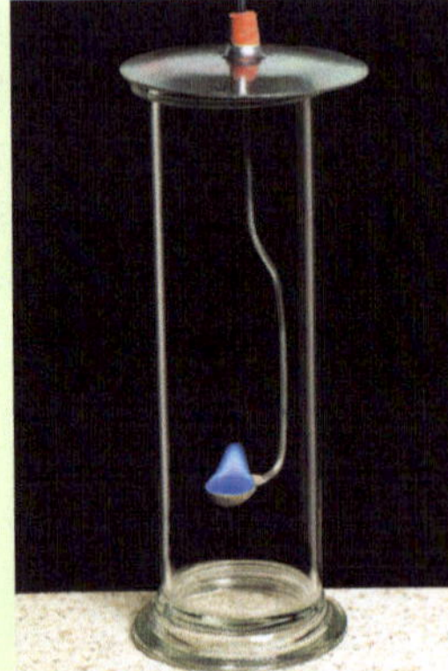

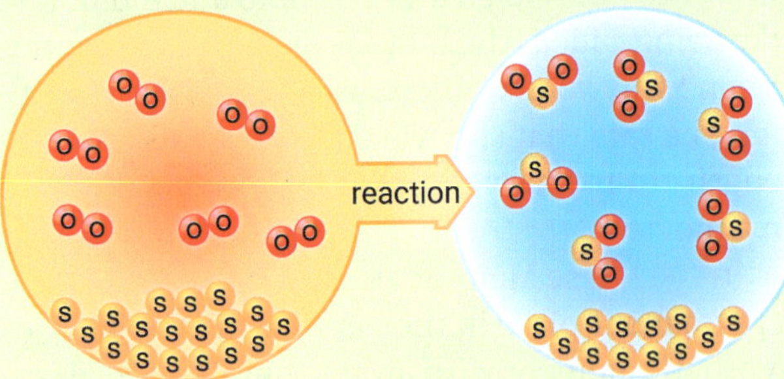

2 The equation for the reaction between hydrogen and oxygen (see page 64) is as follows:

$$2H_2 + O_2 \rightarrow 2H_2O$$

How does this equation illustrate the law of conservation of mass?

3 In 1772, the chemist Lavoisier concentrated heat on a diamond using a large magnifying glass. The diamond burnt away completely. Does this chemical reaction obey the law of conservation of mass? Explain.

4 Explain how both of the following statements can be true:

- Nuclear fusion has not been used as an energy source on Earth.
- Nuclear fusion is the most important energy source on Earth.

5 How could the use of nuclear power help avoid global warming?

6 Given the choice, would you prefer to live near a coal-burning power station or a nuclear power station? Explain your answer.

7 Australia's only nuclear reactor is at Lucas Heights in Sydney. Research to find out:

a what it is used for

b whether there are any safety concerns.

8 Research the details of how a nuclear reactor works. Use materials such as cardboard, wire and papier-mâché to build a model that incorporates the information you have found. Label all the parts of your model.

ISBN 978 1 4202 3736 8

3.3 Radioactivity

You probably know that radioactivity is dangerous, but what exactly is radioactivity?

Uranium was discovered in 1896, and was named after the planet Uranus. Henri Becquerel, a French scientist, placed some crystals of a uranium compound in a dark drawer in which he had some photographic plates wrapped in black paper. Sometime later when he wanted to use the plates, he discovered they were blackened where the crystals had been sitting on them. To explain this, he inferred that the uranium must have given off a type of radiation that could pass through the black paper and affect the photographic plates. He had discovered radioactivity.

Becquerel knew Marie Curie and her husband Pierre, and he suggested that they follow up his findings. The Curies found that all uranium compounds were radioactive. They also discovered two more radioactive elements—radium and polonium.

Ernest Rutherford investigated radioactivity further. He put some radium in a hole in a block of lead to form a beam of radiation. He then put magnets near the beam as shown below. He found that part of the beam was not bent, while other parts of the beam were bent in opposite directions. From this he inferred that the two bent beams had opposite charges. Also, the negative beam was bent more than the positive beam, so he inferred that the particles in it were lighter than those in the positive beam. He inferred that the beam that was not affected by the magnetic field had no charge.

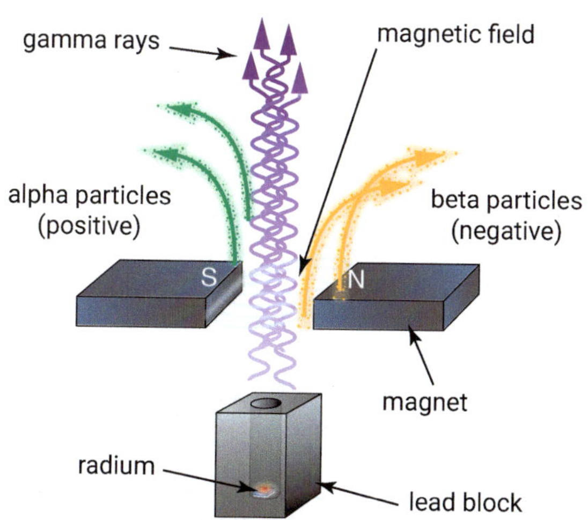

Figure 3.16 Rutherford's experiment detected three types of radiation.

Types of radiation

From Rutherford's experiment, we know that radioactivity is made up of three different types of radiation. **Alpha particles** are helium nuclei made up of two protons and two neutrons. They have a positive charge and move at speeds of up to one-tenth the speed of light. They travel in air for only a few centimetres and can be stopped by a sheet or two of paper.

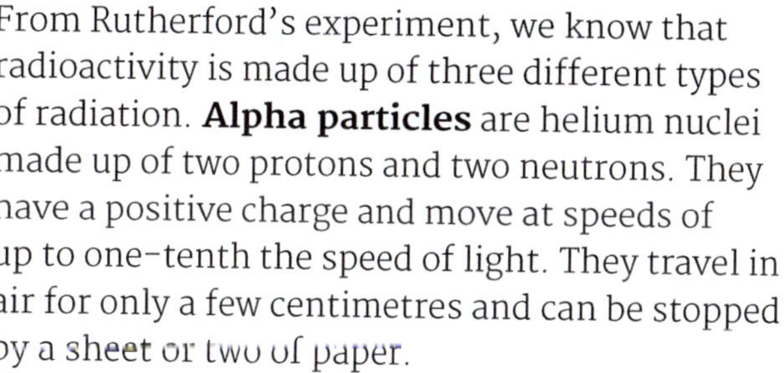

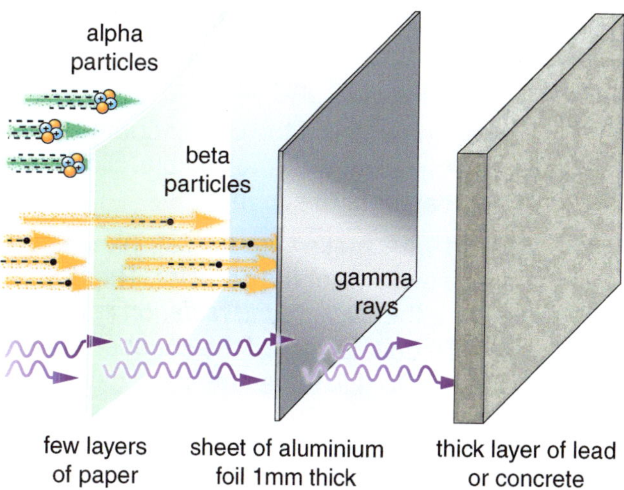

Figure 3.17 Alpha, beta and gamma radiation

Beta particles are high-speed negatively charged electrons. They can travel a few metres through air, but are stopped by objects such as a sheet of aluminium or a centimetre thickness of wood. They move at speeds up to nine-tenths the speed of light.

Gamma rays are not particles but a type of electromagnetic radiation, like light, UV and X-rays. They have very high energy and can travel kilometres through air. They can be stopped by objects such as thick concrete blocks or 2–3 cm of lead. Gamma rays are often released with alpha or beta particles. Like all electromagnetic radiation, gamma rays move at the speed of light.

ISBN 978 1 4202 3736 8

Detecting radiation

To detect radioactivity you use a *Geiger counter*. It consists of a glass tube containing a gas at low pressure. When radiation enters the tube, it causes *ionisation*, splitting gas molecules into positive ions and negative electrons. The electrons produce a pulse of electricity, causing the counter to click. The faster the clicking, the higher the level of radiation. A Geiger counter is shown on page 54.

Using a Geiger counter (teacher demonstration)

In this activity, your teacher will show you how to use a Geiger counter to measure radiation levels. You will need a Geiger counter and a range of samples to test, e.g. white sand, black mineral sand, piece of granite.

1. Find out how to use the Geiger counter. If possible, switch on the audible 'click' so you can hear the count.
2. To start with, make sure the samples are well away from the detector. Then measure the *background radiation* in the room. To do this, record the count after 1 minute. Do this three times and calculate the *average* count.
3. Now place a test sample 1 cm away from the detector. Measure and record the radiation as before.
4. Repeat for the other samples.
   What is the background radiation in your classroom?
   What is the radiation count for each of the samples? Is it greater than the background radiation?

Radioactive decay

If an isotope has an unstable nucleus, it will break down or *decay* to a more stable nucleus. When this happens it will emit radiation in the form of alpha particles, beta particles or gamma rays.

It is impossible to predict when a particular nucleus will decay. It could be in the next second or the next 500 years. The decay process is random. However, one aspect of radioactive decay is predictable and it is called the **half-life** of an isotope. This is the time taken for half of the nuclei in a sample of the isotope to decay. No matter how much of the isotope there is, it always takes the same time for half of its nuclei to decay.

Half-lives of isotopes can range from more than a billion years to less than a billionth of a second. For example, the half-life of radioactive carbon-14 is 5730 years. After one half-life, only half the radioactivity remains. After two half-lives (11 460 years), only a quarter of its radioactivity remains. After three half-lives (17 190 years), only an eighth of its radioactivity remains.

As a radioisotope decays, the amount of radiation emitted decreases, and fewer unstable nuclei remain. Isotopes with short half-lives emit more radiation, but their radioactivity dies away more quickly.

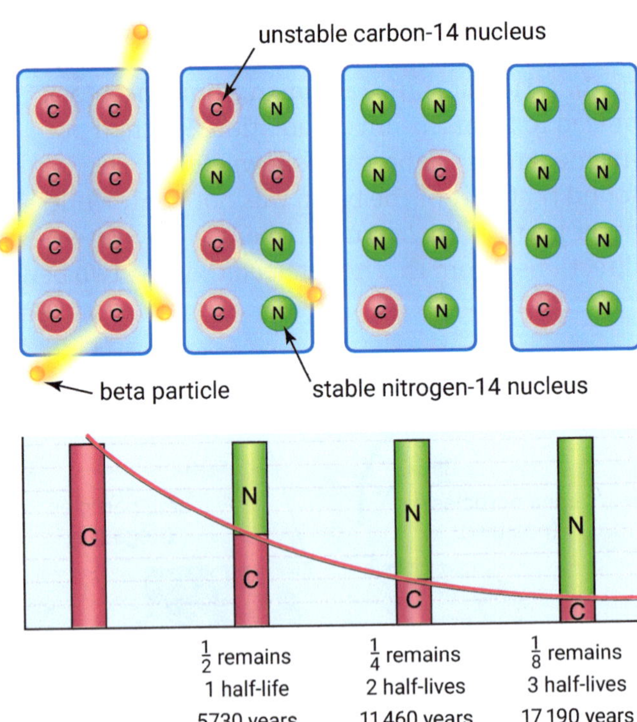

Figure 3.18 Radioactive carbon changes to a stable nitrogen isotope when a neutron decays to form a proton and a beta particle is released.

ISBN 978 1 4202 3736 8

ACTIVITY

Radioactive decay model

We don't know when a particular atom in a radioisotope will decay, but we do know that half of the atoms will decay in a certain time. This is like tossing a coin. You don't know whether it will be heads or tails, but if you toss it 100 times, you would expect about 50 heads and 50 tails.

You will need 100 coins or plastic counters with a dot on one side, and a container with a lid.

1 Put 100 coins or counters in the container and mix them thoroughly.
2 Tip the coins or counters onto the bench and remove the heads or the counters with dots facing up. These represent the nuclei that have decayed after one half-life.
3 Count and collect the remaining coins. These represent the radioactive atoms remaining.

Record this number in a table.

Number of half-lives	Radioactive nuclei remaining
0	100
1	
2	
3	

4 Repeat steps 1–3 six times, recording the number of tails or counters with no dots left each time.
5 Plot your results with time in half-lives on the *x*-axis and radioactive nuclei remaining on the *y*-axis. Draw a curved line of best fit through the plotted points.

Describe the shape of the graph and what it tells you.

Radioactive dating

Cosmic radiation from space is constantly bombarding our atmosphere. When it hits nitrogen atoms, it can change them into radioactive carbon-14, called radiocarbon. These carbon-14 atoms combine with oxygen to form radioactive carbon dioxide ($C + O_2 \rightarrow CO_2$).

This radioactive CO_2 is taken in by green plants such as trees, along with normal CO_2 containing carbon-12 atoms. During photosynthesis the CO_2 is converted into compounds in the tree's tissues. When the tree dies, the radiocarbon gradually decays to nitrogen-14. So the amount of carbon-14 in the wood slowly decreases.

After about 5730 years, there are only half as many carbon-14 atoms as there were when the plant died. So, by measuring how much carbon-14 there is in an object, you can estimate how long ago the plant from which it was made was alive. The less carbon-14 compared with carbon-12 there is, the older the object.

Radiocarbon dating is useful for finding the age of such things as wood, bones, shells and objects left by early humans, for example the Dead Sea scrolls (about 2000 years old). In 1988 radiocarbon dating was used to date the shroud of Turin. This ancient piece of cloth shows marks that some people believe were made by Christ's body after the crucifixion. However, the cloth was found to be only 600–700, not 2000, years old.

After about 50 000 years the amount of radiocarbon becomes too small to measure accurately, so other radioisotopes are used. For example, all rocks contain small amounts of radioactive elements such as uranium and potassium. These have longer half-lives than carbon-14 and can be used to estimate the ages of rocks and the fossils they contain.

Figure 3.19 The Dead Sea scrolls

ISBN 978 1 4202 3736 8

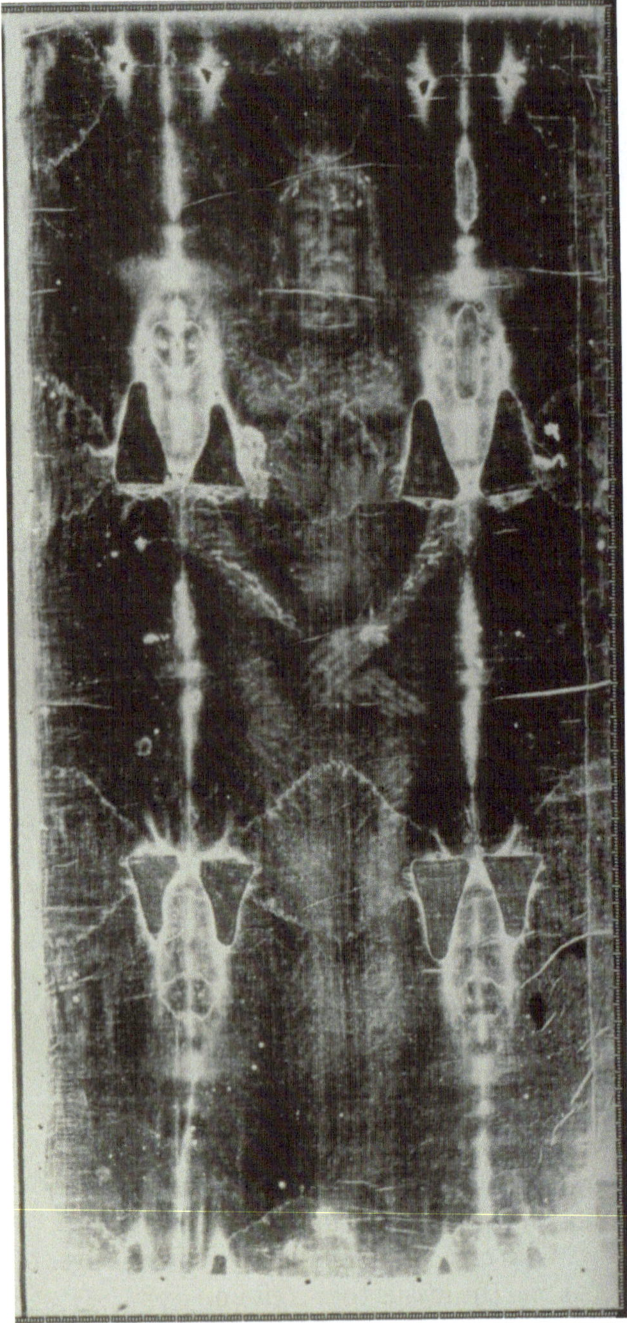

Figure 3.20 The shroud of Turin

Most of the harmful radiation comes from the natural background radioactivity in the Earth, mainly from the radioactive uranium and thorium in rocks. These elements decay to release a radioactive gas called radon into the air you breathe. You also receive radiation from cosmic rays from space and from the natural radiation of your food and drink.

You cannot do anything about the natural radiation you receive, but it is obviously important to keep the radiation from medical and other artificial sources to a minimum. We each receive a dose of about 2 millisieverts every year from natural radiation. The *sievert* is a unit used for measuring radiation. At present there is an international guideline that no member of the public should receive more than 1 millisievert of artificial radiation per year. A single chest X-ray results in a dose of about 0.1 millisieverts. Researchers, hospital staff and workers in nuclear power stations wear special detectors called dosimeters. These measure the amount of radioactivity they receive.

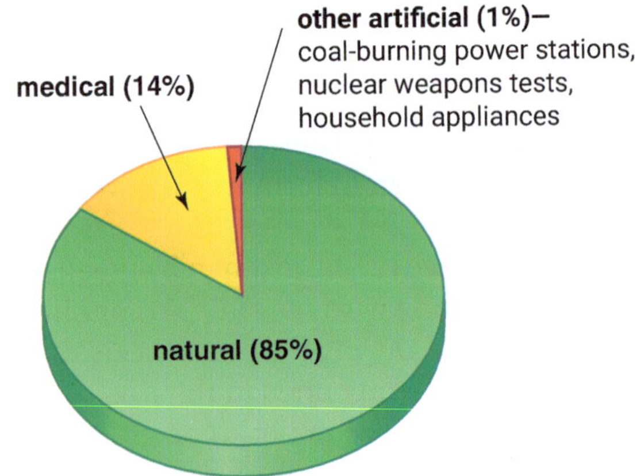

Figure 3.21 Sources of radiation in Australia

Is radiation harmful?

There are many different types of radiation around you, but it is only the high-energy radiation that is harmful. Examples are UV radiation, X-rays, gamma rays and cosmic rays (from space). All these types of radiation can cause cancer. The effects of radiation on your body depend on how much radiation you receive.

Uses of radioisotopes

Radioisotopes are commonly made by placing stable isotopes into a nuclear reactor, where they are bombarded by neutrons. In Australia, radioisotopes are made at the Lucas Heights nuclear reactor in Sydney. Radioisotopes can also be made in a *cyclotron*. This device accelerates charged particles to very high speeds. These

ISBN 978 1 4202 3736 8

particles hit a target of specially prepared material in which nuclear reactions occur and some atoms become radioactive.

Nuclear medicine

Alpha, beta and gamma radiation is sometimes called *ionising radiation* because it can ionise atoms and molecules; that is, give them an electric charge. These ions are more likely to become involved in chemical reactions. In the body, these ions may cause chemical reactions that destroy cells, or they may cause uncontrolled cell growth, which causes cancer.

Cancer cells are more sensitive to radiation than normal cells, so radiation is often used to treat cancer. In *radiotherapy*, a radioisotope is placed in a shielded box with an opening through which a beam of gamma rays emerges. The beam is aimed at the area where the cancer is. This ensures that the cancer cells receive a large dose, killing most of them, while the surrounding healthy cells receive only a small dose.

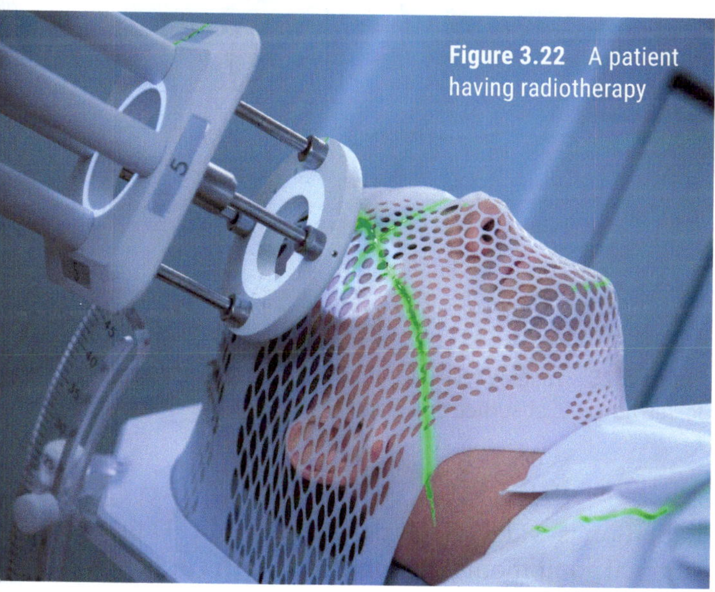

Figure 3.22 A patient having radiotherapy

Radioisotopes with short half-lives can also be used to diagnose the condition of internal organs, blood vessels and bones. Radioactive tracers that emit gamma rays are swallowed or injected, and these tend to collect in particular parts of the body. They can then be detected by a gamma camera outside the body, which converts the gamma rays to an image.

A recent technique called PET (positron emission tomography) can show doctors how your body tissues and organs are working. Radioisotopes are injected or inhaled into the body. These radioisotopes emit *positrons*, which are like electrons except that they have a positive charge. When a positron meets an electron in an atom in the body tissue, the particles destroy each other and produce gamma rays. Detectors around the patient detect these gamma rays and feed the information into a computer. Here the exact position of the nucleus that emitted the positron can be calculated. By combining the information from many separate emissions, the computer builds up a picture of a slice of the patient's body.

Industry

Many industrial problems can be solved using radioisotopes. For example, small leaks in complicated pipe systems can be traced. You simply add a radioisotope to the liquid in the pipe and follow its movement with a radiation detector.

When paper is manufactured it is important to make it the correct thickness. Beta radiation is passed through the paper and detected on the other side. The thicker the sheet, the less radiation passes through it. The detector can be linked to the rollers to automatically control the thickness of the paper. If the amount of radiation passing through the paper decreases, the rollers move closer together to make a thinner sheet. A beta source such as strontium-90 is usually used. For plastic or metal sheets, a gamma source is used because its radiation can pass through these materials.

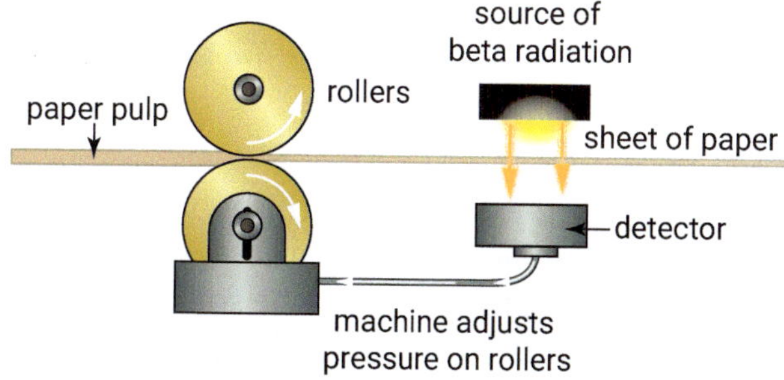

Figure 3.23 Beta radiation is used to monitor the thickness of paper.

ISBN 978 1 4202 3736 8

Smoke detectors

Most homes have smoke detectors installed. Older types contain a small amount of americium-241, a synthetic radioisotope made in nuclear reactors. Alpha particles emitted by the americium ionise the air and create a small electric current that stops the alarm from sounding. If smoke enters the detector, it absorbs the alpha particles and cuts off the current, setting off the alarm. Newer smoke delectors are photoelectric.

Figure 3.24 A smoke alarm

smoke
americium-241
alpha particles absorbed by smoke
electric circuit containing battery and alarm

Agriculture

To investigate the effect of fertiliser on plants, the fertiliser is 'labelled' with a beta-emitting radioisotope and injected into the soil around the plant. The fertiliser is absorbed by the plant and its path in the plant can be traced by a detector. In this way, you can find out how plants use fertilisers and which fertiliser is the best to use.

Another use of radioisotopes is the control of insects without using chemical insecticides. Male insects are sterilised with radiation before they hatch. They are then released into an infected area. When they mate with females, no offspring are produced. In this way, the insect pests in the area can be greatly reduced.

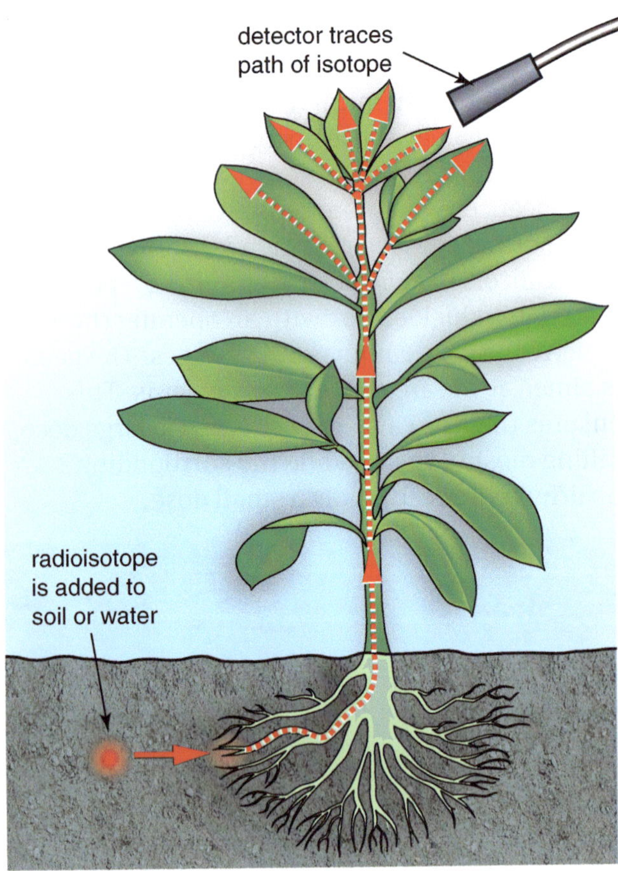

Figure 3.25 Tagging fertiliser with a radioisotope allows us to see how it is absorbed in a plant.

Irradiation

Radiation is widely used to sterilise medical and surgical equipment by killing all bacteria and other micro-organisms so that they cannot infect patients. For example, disposable syringes are sealed in plastic and the package is then exposed to gamma rays from a cobalt-60 source. This process is called *irradiation*.

Irradiation can also be used to preserve food because it kills the bacteria that cause the food to spoil. Food that has been irradiated keeps longer than food that hasn't.

ISBN 978 1 4202 3736 8

When food is irradiated, it does not become radioactive. Astronauts eat food preserved by irradiation. However, the radiation may destroy vitamins in the food and produce unwanted chemicals. For this reason, many people are at present not convinced that food irradiation is a good idea.

Environmental monitoring

Radioisotopes can be used to monitor the movement of materials in the environment. For example, radioactive gold-198 is used to monitor the flow of sewage from pipes off the coast near Sydney. The movement of the radioisotope is easily traced by using a radiation detector, as shown in Figure 3.26.

Radioisotopes can also be used to track the movement of sand and silt. Suppose the sand along a beach is disappearing and you want to know where it is going. You simply add a small amount of radioactive tracer to the sand and follow its path with a detector.

Figure 3.26 This food irradiation machine uses gamma rays to kill bacteria and pathogens in oysters.

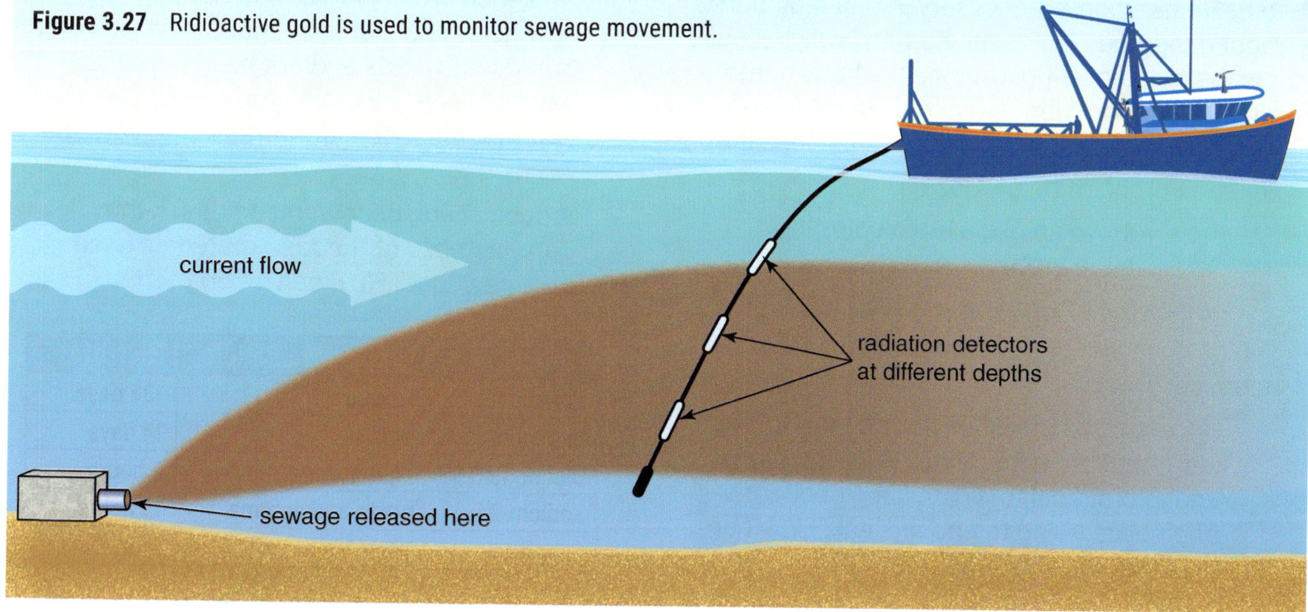

Figure 3.27 Ridioactive gold is used to monitor sewage movement.

ISBN 978 1 4202 3736 8

Forensic analysis

By bombarding a sample with neutrons in a nuclear reactor, the atoms in the sample can be made radioactive and then identified. This technique is used in forensic analysis to analyse material gathered at crime scenes. For example, residue from the skin of a person believed to have fired a gun can be analysed for the presence of metals such as lead.

In one famous case, forensic scientists used this technique to investigate a few strands of Napoleon Bonaparte's hair, 140 years after he died. They found arsenic in the last few centimetres of each hair. From this they inferred that he may have been poisoned. However, he may have taken arsenic as a medicine, or he may have absorbed arsenic from the wallpaper in his house.

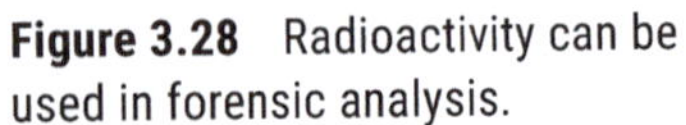

Figure 3.28 Radioactivity can be used in forensic analysis.

Problem solving

When radioisotopes are used, it is important to choose the right one. If the half-life is too short, the radioisotope may cease its activity before the job is complete. But because the radiation could be dangerous, you don't want the half-life to be longer than is necessary. Alpha radiation is most dangerous inside the human body because it is the most ionising. Outside the body gamma radiation is most dangerous because it is so penetrating, even though it is the least ionising.

For each problem below decide:

- how you would use a radioisotope to solve the problem
- which radioisotope you would use, and why.

You can use any of the isotopes in the table opposite.

1 Underground pipelines sometimes spring leaks that are hard to find. It is expensive and impractical to dig up large sections of pipeline trying to find the leak. How can these leaks be located?

2 You are a hospital consultant specialising in the treatment of breathing problems. You suspect that one of your patients has a blockage in an air passage in one of his lungs. How can you check? (X-rays are unsuitable.)

3 An agricultural chemist has developed a new herbicide that is intended to attack only broad-leaf weeds and not be absorbed at harmful levels by grain crops. How can she test that the herbicide works in the way she wants it to?

4 A council engineer wants to check the compactness of the road materials used on a new road. How can he do this?

Isotope	Type of radiation	Half-life
polonium-210	alpha	138 days
phosphorus-32	beta	14 days
technetium-99	gamma	6 hours
sodium-24	beta and gamma	15 hours
cobalt-60	gamma	5 years

ISBN 978 1 4202 3736 8

SCIENCE AS A HUMAN ENDEAVOUR

The story of the atomic bomb

In 1938, a year before the start of World War II, many people feared that Hitler would build an atomic bomb. However, because Hitler was persecuting Jewish scientists, many of them fled to the United States. One such scientist was Albert Einstein. He gave up his opposition to war and urged then president Franklin Roosevelt to develop an atomic bomb before Hitler did. Eventually Roosevelt agreed and the Manhattan Project was set up to build an atomic bomb.

The Manhattan Project was carried out in extreme secrecy, even though by 1945 the project had nearly 40 laboratories that employed about 20 000 people. Among these people were some of the greatest scientists who have ever lived. First the scientists had to enrich the uranium to make it suitable for bomb making. They also had to build nuclear reactors to make plutonium.

To make a fission bomb, there must be enough uranium or plutonium to keep the chain reaction going. This is called the *critical mass*. There must also be enough fast-moving neutrons to start the reaction. There are two explosions that occur. The first explosion is that of a conventional explosive, which forces two masses of uranium or plutonium together to make a critical mass. This explosion then triggers the chain reaction that causes the nuclear explosion.

On 16 July 1945, after six years of research and development, the first atomic bomb was tested in the desert of New Mexico. It vaporised the metal tower supporting it, and all the desert sand within a distance of 700 m was turned into glass. Some of the scientists, including Einstein, were worried about the power of the bomb and wanted the project stopped. However, on 6 August 1945 a four-tonne uranium bomb nicknamed *Little Boy* was detonated over the city of Hiroshima in Japan, resulting in 66 000 instantaneous deaths. Total vaporisation from the blast measured 800 m in diameter, and total destruction was over 1.6 km. Three days later, a plutonium bomb named *Fat Man* was detonated over Nagasaki, killing another 39 000 people. On 15 August the Japanese surrendered, and World War II was over.

Figure 3.29 The atomic bomb explodes over Nagasaki, 1945.

Rain that follows the explosion of an atomic bomb is heavily contaminated with radioactive particles, and many survivors of the initial blasts eventually died due to radioactive poisoning. Those who didn't die suffered severe burns, nausea, vomiting, fatigue, diarrhoea and hair loss. Others passed on leukaemia to their children.

Question

In 1954, a year before his death, Einstein said to his old friend Linus Pauling, 'I made one great mistake in my life ... when I signed the letter to President Roosevelt recommending that atom bombs be made; but there was some justification —the danger that the Germans would make them'. Do you agree that Einstein made a mistake? Justify your answer.

ISBN 978 1 4202 3736 8

1 Copy and complete this table.

Type of radiation	What is it?	Distance travelled in air	Can be stopped by ...
alpha	helium nuclei		
beta			
gamma			

2 Technetium-99 decays by emitting gamma rays, and has a half-life of 6 hours. If you have 100 g of technetium-99 now, how much will be left in:

a 6 hours?

b 12 hours?

c 24 hours?

3 You have 64 g of a radioisotope with a half-life of 5 days. How much will be left after 5 days? After 10 days? How many days will it take for the mass to decrease to 1 g? Show this information on a graph.

4 If the half-life of carbon-14 is 5730 years, about what fraction of the original radioactive carbon-14 would you expect to find in:

a an Aboriginal boomerang 11 000 years old?

b a human skull 23 000 years old?

5 The graph below shows the radioactive decay curve for a radioisotope.

a How long did it take for the radioactivity to fall below 3 counts/second?

b What is the half-life of the radioisotope?

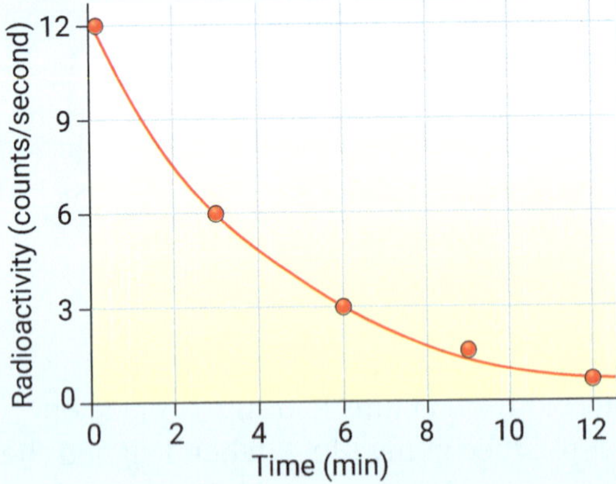

6 Would an alpha particle emitter be suitable for measuring the thickness of cardboard in a factory manufacturing cardboard boxes? Explain.

7 a What is the danger of using radioisotopes to kill cancer cells?

b Why do you think gamma radiation is used to treat cancers inside the body, but beta radiation is used to treat skin cancers?

8 To treat a patient with a cancerous tumour, a doctor chooses strontium-90, a beta source with a half-life of 28 years. Should he inject the patient with this or should he use it externally? Explain your answer.

9 Why is gamma radiation rather than beta radiation used for sterilising dressings?

10 There is a mining operation upstream from a town. The residents want to know if any of the waste products from the mine are reaching the town. How could they use a radioisotope to check this?

11 The manager of a breakfast cereal factory wants to make sure the packets are filled to the correct level. Draw a labelled diagram showing how this could be done using a radioisotope.

12 Figure 3.30 shows how radioisotopes for use in medicine and industry are packaged for transport by air from Sydney to other parts of Australia. Describe the safety features.

13 A beta particle emitter is not suitable for use in a smoke alarm (page 74). Why not?

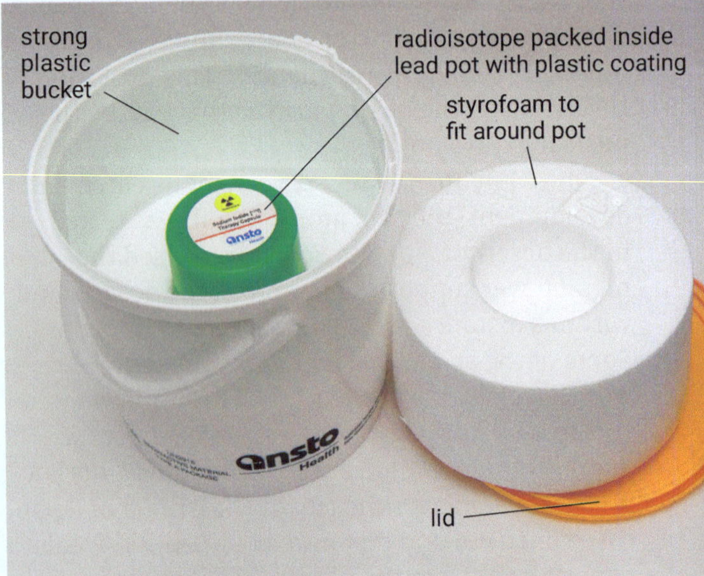

Figure 3.30 Packaging for radioactive isotope transport

ISBN 978 1 4202 3736 8

1 An electric field is set up between two charged plates as shown, and the alpha, beta and gamma radiation from a radioactive radium source passes into the electric field. Some of the radiation is attracted to the positive plate, some to the negative plate and some is unaffected and passes straight through.

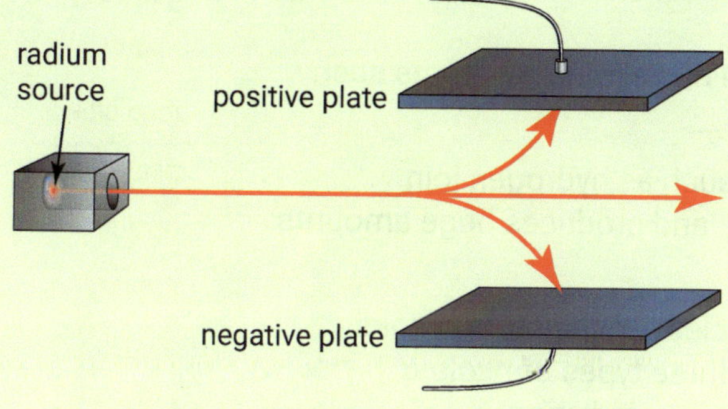

a Which type of radiation is attracted to the positive plate? Explain your answer.

b Which radiation is attracted to the negative plate? Why?

c Which radiation passes through the electric field unaffected? Why?

2 Sources of alpha particles are not usually regarded as dangerous unless they are breathed in or taken in with food. Why is this?

3 Suggest a way of storing a radioisotope that emits only beta particles.

4 You have designed a building that is radiation-proof. How could you test that it works?

5 The radioactivity count rate for a sample of a radioisotope fell from 800 to 100 counts/min in 15 hours. What is the half-life of the radioisotope? Explain your working.

6 The table below shows the radiation count from a radioisotope over a period of 40 minutes.

Time (min)	0	5	10	15	20	25	30	35	40
Count	152	115	87	66	50	38	29	22	17

a Plot a line graph to display the data.

b How long does it take for the count to drop from 100 to 50 counts?

c How long does it take for the count to drop from 50 to 25 counts?

d What is the half-life of the radioisotope?

e Predict how long it would take for the count to reach 10.

7 How could an agricultural researcher use a radioisotope to find out whether most of the fertiliser spread on a paddock is ending up in the food the cattle eat or whether it is being washed into a creek and wasted.

8 Tom used a Geiger counter to measure the radiation 1 cm away from three different radioactive sources. He also measured the radiation when he placed sheets of various materials between the source and the detector of the Geiger counter. He then repeated his measurements with two unknown sources.

All his results are counts per minute, and he found that the background radiation at 1 cm was 20 counts/min.

Source	Po-210	Sr-90	Co-60	A	B
Radiation type	alpha	beta	gamma		
No shielding	198	5152	2119	523	2728
2 sheets of paper	28	4796	2092	461	2492
1 sheet of plastic	26	2704	2046	210	969
1 sheet of aluminium	23	2137	1993	156	635
1 sheet of lead	20	22	1773	80	22

a Which source was the most radioactive?

b Which type of shielding was the most effective at stopping radiation?

c Did the sheet of lead stop all the radiation from the Po-210 source? Explain.

d Suggest why the sheet of lead didn't stop all the gamma radiation.

e Use Tom's data to infer which type of radiation was released by the two unknowns—A and B. Explain your answers fully.

ISBN 978 1 4202 3736 8

MAIN IDEAS

Copy and complete these statements to make a summary of this chapter. The missing words are on the right.

1 Atoms are composed of a positively charged ______ surrounded by negatively charged ______. Inside the nucleus are ______ (positive charge) and neutrons (no charge).

2 Isotopes are atoms that have the same number of protons and electrons, but different numbers of ______.

3 Nuclear energy is produced by splitting the nuclei of large atoms such as uranium-235. This is called nuclear ______.

4 Nuclear ______ occurs when small atoms such as hydrogen join together. This process occurs in the ______ and produces huge amounts of energy.

5 ______ materials are those in which the nucleus of the atom breaks down and radiation is given off. There are three types of nuclear radiation: alpha and ______ particles and ______ radiation.

6 The rate at which the radiation is given off is measured by the ______. This is the time taken for half of the ______ in a sample to decay.

7 Isotopes that are radioactive are called ______. They have many uses, especially in ______, industry and radioactive dating.

radioisotopes
beta
fusion
electrons
gamma
nucleus
half-life
medicine
neutrons
fission
protons
nuclei
radioactive
sun

CH•3 REVIEW

1 If an atom has 9 protons and 11 neutrons, then the number of electrons around the nucleus is:
- **A** 2.
- **B** 9.
- **C** 11.
- **D** 20.

2 What happens when an unstable nucleus breaks down?
- **A** Electrons are forced into the nucleus.
- **B** Particles and energy leave the nucleus.
- **C** Two nuclei join together.
- **D** Huge amounts of energy are needed.

3 The half-life of a radioisotope is:
- **A** the time it takes for half of it to decay.
- **B** the time before it starts to give off radiation.
- **C** when it has only half of its life left.
- **D** the same for all radioisotopes.

4 A common use for the carbon-14 radioisotope is:
- **A** cancer therapy.
- **B** detecting thyroid disease.
- **C** as a fuel in nuclear power stations.
- **D** finding the age of ancient objects.

5 Name the particles that are found in an atom. State what their charge is and where they are found in the atom.

ISBN 978 1 4202 3736 8

6 Name the three types of radiation shown in the diagram below.

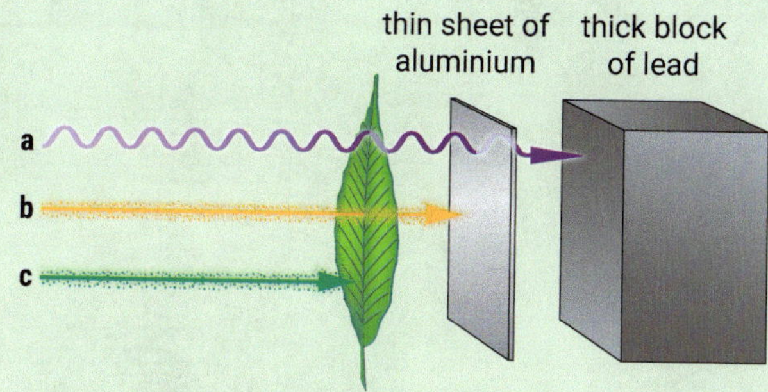

Figure 3.31 How far different types of radiation can penetrate

7 a What is the difference between nuclear fission and nuclear fusion?
b Why has it been possible to use both fission and fusion in bombs, but only fission in nuclear reactors?

8 Three different isotopes of helium are shown below.

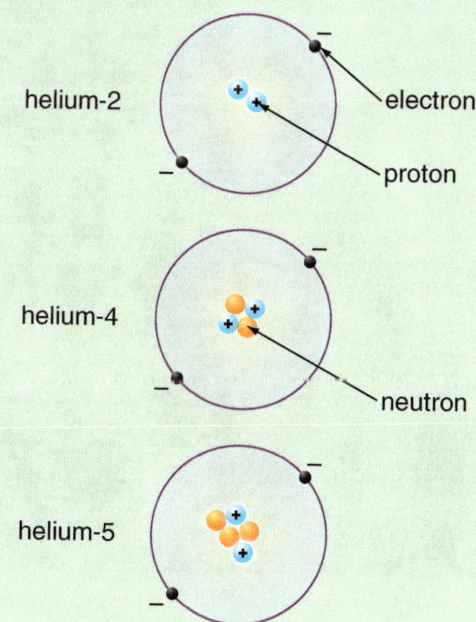

a What do the numbers 2, 4 and 5 indicate?
b How are the three isotopes similar?
c How are they different?
d Which isotope would you expect to be radioactive? Why?

9 When using a radioisotope to measure the thickness of cardboard, why is it important to use a source of beta radiation rather than alpha radiation?

10 A barium isotope has a half-life of 6 minutes. If there are 800 g of barium now, how much will there be in half an hour from now?

11 There is a leak in a nuclear reactor and strontium-90 is released into the surrounding countryside, which is used for farming. Two months later, breastfed babies are found to have small amounts of strontium-90 in their bodies. Explain how this could have happened.

12 Radiation is more destructive to cells that are dividing and growing rapidly than it is to normal cells.
a How can this be an advantage?
b How can it be a problem?

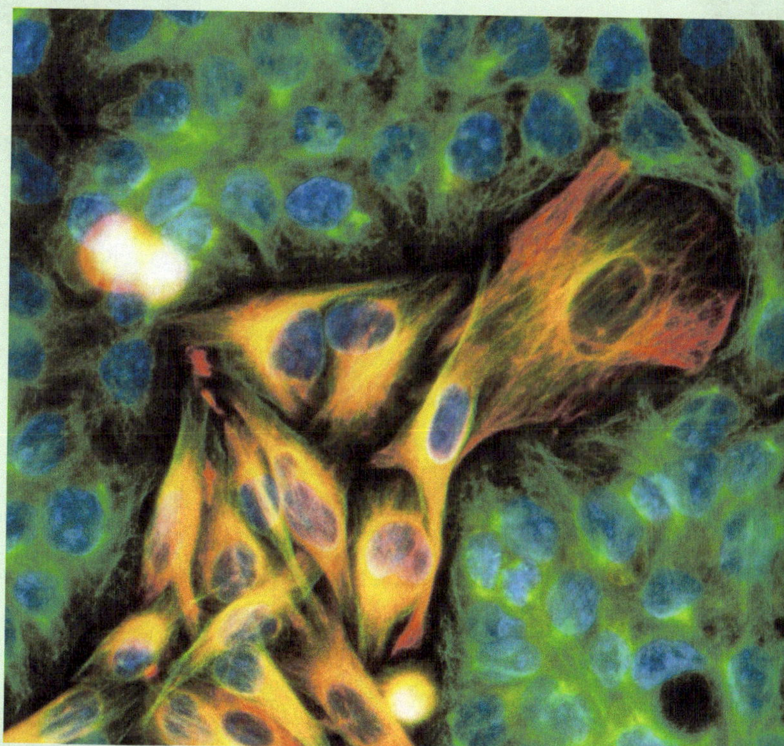

Figure 3.32 Cancer cells (orange) invading normal skin cells (green).

Check your answers on page 298.

ISBN 978 1 4202 3736 8

IN THIS CHAPTER

Science Understanding

- investigate the effects of magnetic forces on familiar objects
- investigate magnetic fields around magnets of different shapes
- give examples of how magnets and electromagnets are used in everyday life
- investigate how electric generators and electric motors work
- follow the historical development of our knowledge of magnetism
- use Dr Cathy Foley's research with superconductors to investigate the link between science and technology

Science Inquiry Skills

- make an electromagnet and investigate ways of making it stronger
- map the magnetic fields around magnets

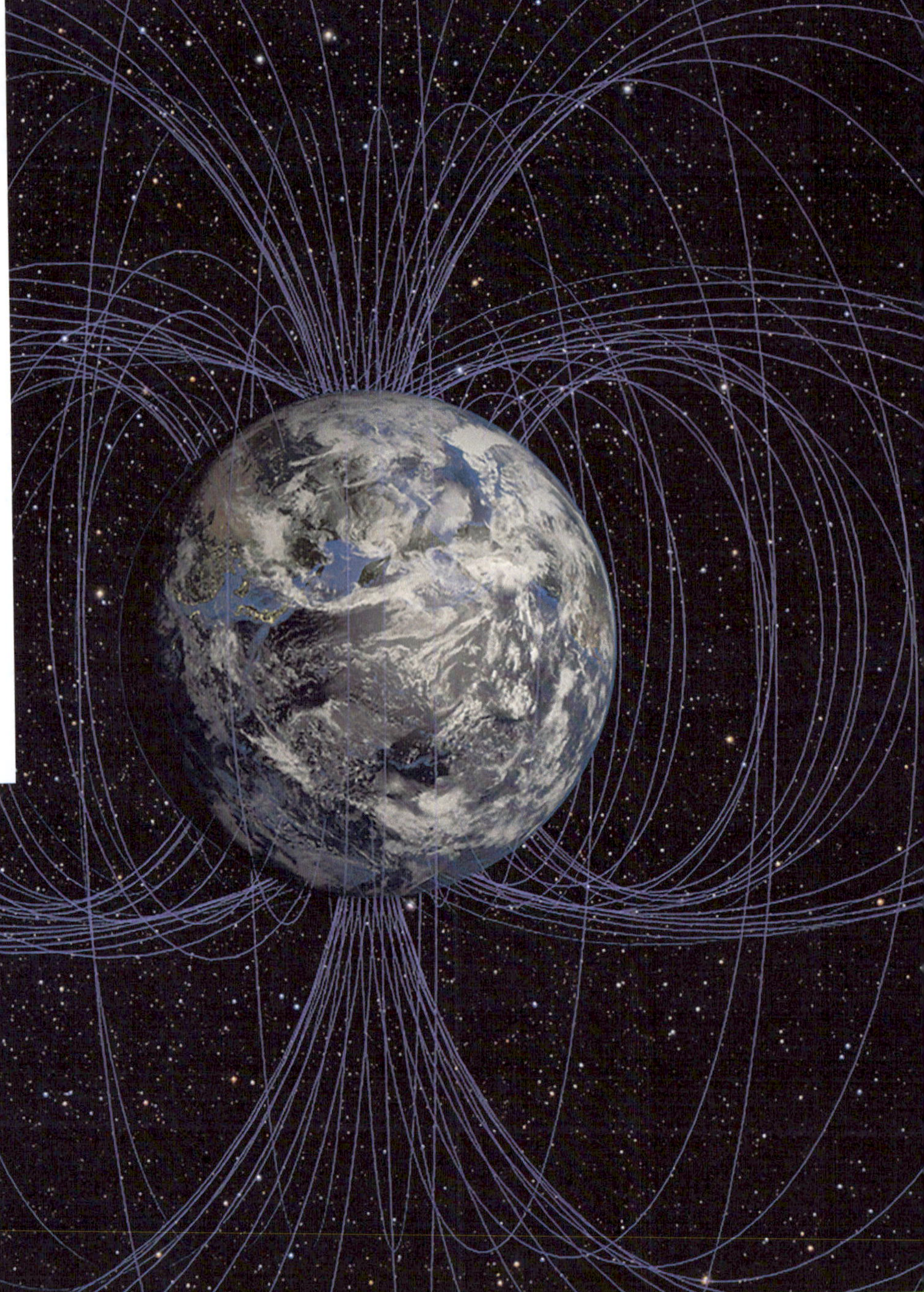

CH•4 Magnetism and electricity

ISBN 978 1 4202 3736 8

GET STARTED: *QUESTION*

In a small group, discuss each of the three questions below.

> Some animals (for example, honeybees, homing pigeons, salmon, whales and turtles) have magnetic material in their bodies. Can you suggest a purpose for this?

> This machine can be used to separate objects made of iron and steel from other rubbish. How does it work?

> This magnetic metal disc is floating in the air. How can you explain this?

ISBN 978 1 4202 3736 8

4.1 Investigating magnets

INVESTIGATION 4.1 Properties of magnets

Risk assessment and planning

Read through the investigation and discuss with your teacher which parts you will do. Note that in Part E and in Inquiry in Part B you will need to work out what to do for yourself.

PART A

Materials

- bar magnet and horseshoe magnet
- box of paperclips, tacks or pins

Method

1 Hold the bar magnet as shown and bring it down slowly onto a pile of paperclips.

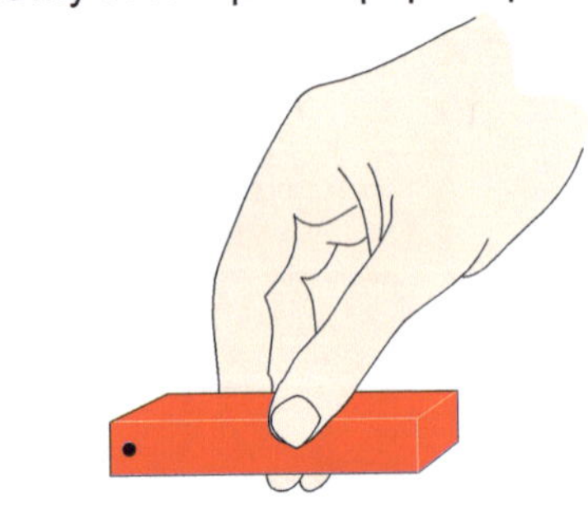

2 Now lift up the magnet, and observe which parts of the magnet hold the most paperclips and which part holds the least.
Record your observations using a simple sketch.
Write an inference to explain your observations.

3 If possible, repeat the test with a horseshoe magnet.

PART B

Materials

- 2 bar magnets with poles marked (most magnets have a dot or mark on the north pole end).

Method

1 Place two bar magnets on a table near each other as shown in **a** below.
Record whether the magnets attract each other or repel (push apart) each other.

2 Repeat with the magnets in positions **b**, **c** and **d**.

3 Look carefully at your results. What patterns can you see?
In one sentence, write a generalisation about these patterns.

a

b

c N S S N

d S N N S

Inquiry

Using what you have learnt in Part B, design a way of making one magnet float on top of another.

ISBN 978 1 4202 3736 8

PART C

Materials

- bar magnet
- Petri dish or small round take-away container
- ice-cream container or bowl
- small magnetic compass

Method

1 Fill the ice-cream container with water.
2 Place the magnet in the middle of the Petri dish and float the dish on the water in the container.
Record what happens to the magnet and Petri dish.
3 Turn the Petri dish in another direction and then release it.
Which way is it pointing now?
4 Use a compass to find out where north, south, east and west are in your classroom. Make sure you keep the compass away from other pieces of iron or steel and other magnets. In which direction does the magnet in the Petri dish point?

5 Copy and complete these sentences.
  a One end of a floating magnet always points ______, and the other end points ______.
  b The floating magnet is acting like a ______.

PART D

Which substances do magnets attract?
Your teacher will give you a bar magnet and a box containing objects made from different metals and non-metals.
Use the objects, and any other objects you would like to try, and classify them into two groups—those attracted to a magnet and those that aren't.

PART E

In this part you have to design a way of measuring the strengths of magnets. Two ideas are shown, but try to design your own method.
In a small group, discuss your design. Show your plan and list of required materials to your teacher, then go ahead and try it.
Write a report of your experiment.

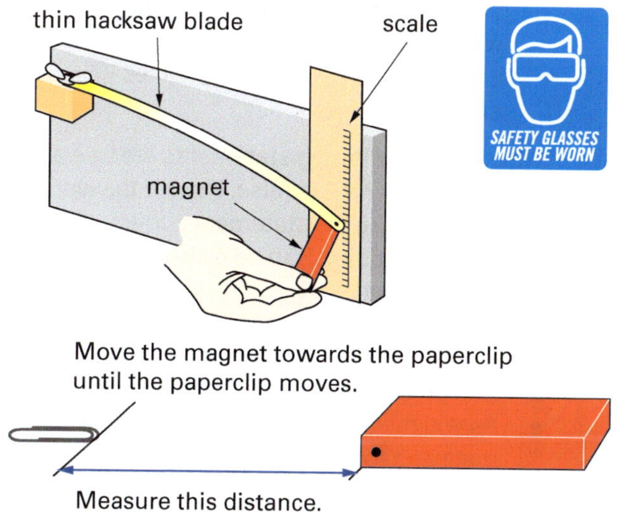

ISBN 978 1 4202 3736 8

Facts about magnets

Fact 1

From the last investigation, you can see that a magnet can push or pull something, without even touching it. A magnetic force is a non-contact force.

Fact 2

The magnetic force is strongest near a magnet, and becomes weaker as you get further from it.

Fact 3

Magnets attract certain metals only—iron, nickel or cobalt. These materials are said to be *magnetic*. Steel, which contains iron, is also magnetic. Only magnetic materials can be magnetised (made into magnets).

Fact 4

The ends of a magnet are more magnetic than the middle. These ends are the **poles** of the magnet. If you suspend a magnet or float it on water it always points north–south. The end that points north is called the north pole. The other end is called the south pole. A north or a south pole will attract magnetic material that has not been magnetised.

Fact 5

If you place the north poles of two magnets together, they will *repel* each other. Two south poles also repel each other. However, the north pole of one magnet will *attract* the south pole of another magnet. You can write this as a generalisation: *like poles repel*, and *unlike poles attract*.

1 Copy and complete the sentences below.
   - a A substance that is attracted to a magnet is said to be _____.
   - b A magnet can exert a _____ on a piece of iron without touching it.
   - c Magnets can be made of _____, _____ or _____.
   - d All magnets have two _____.
   - e One end of a magnet is called the _____ _____. The other end is called the _____ _____.
   - f Two south poles or two north poles are called _____ poles.
   - g When two magnets push each other apart we say they _____.
   - h Like poles _____, and _____ poles attract.

2 Which of the following statements are true and which are false? If it is false, change the words in italics to make the statement correct.
   - a Any *metal* object can be picked up by a magnet.
   - b A magnet is strongest at its *middle*.
   - c A north pole will *attract* a south pole.
   - d Like poles *attract* each other.
   - e As you move away from a magnet the magnetic force *decreases*.

3 An electrician has a magnetised screwdriver to help her hold screws. Which of the following will the screwdriver attract: steel screws, brass screws, aluminium washers, steel washers, copper wire?

4 For each situation below, say whether the magnets will attract or repel each other.

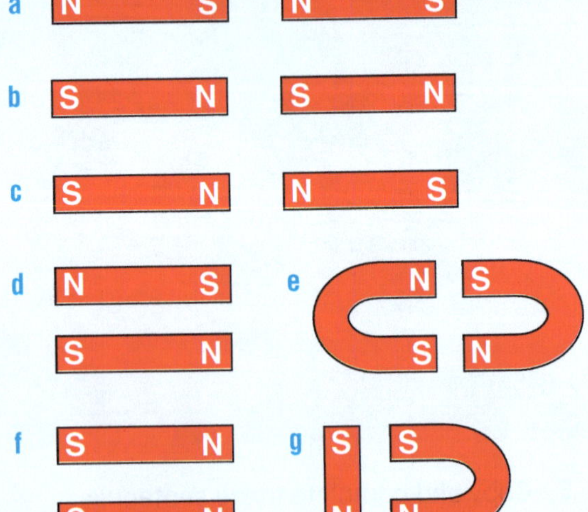

5 When you use a bar magnet to pick up small iron or steel objects, you use the *end* of the magnet. Why?

ISBN 978 1 4202 3736 8

6 Use the words *magnetic* and *non-magnetic* to explain how you can separate a mixture of iron filings and copper turnings.

7 Make a list of everyday uses of magnets.

8 The magnetic couplings in the Lego train on the right are either blue or red. They will link up only if you join a red to a blue. Suggest a reason for this.

9 Suppose you were given a bar magnet whose poles were unmarked. How could you use a compass to find which was the north pole and which was the south pole?

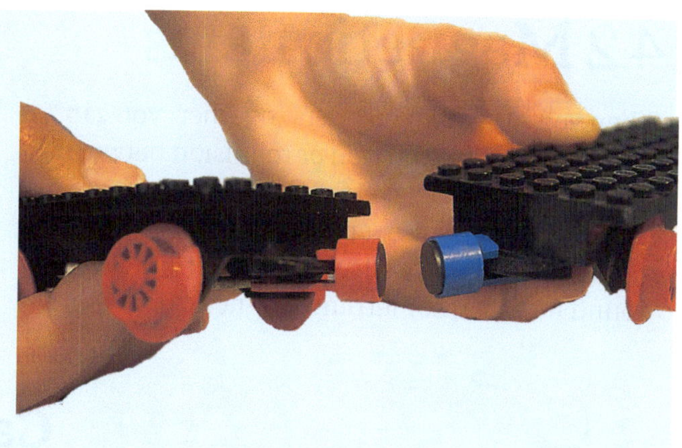

CHALLENGE

1 How could you arrange three magnets in a triangle so that they all attract each other? Sketch your arrangement, showing the poles, then try it.

2 These two magnets repel when placed like this:

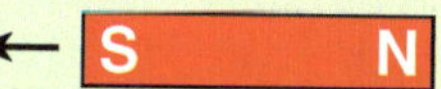

Will they attract or repel when placed like this? Explain.

3 A magnet attracts a metal bar X no matter which pole is used to do it. The same magnet attracts and repels a metal bar Y depending on which pole is used.

a Which metal bar is a magnet and which is a piece of iron?

b Why is repulsion the only sure test that the metal bar is a magnet?

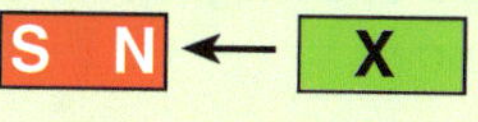

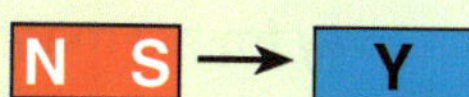

4 Explain why two needles, hung from the ends of a bar magnet, lean towards each other as shown.

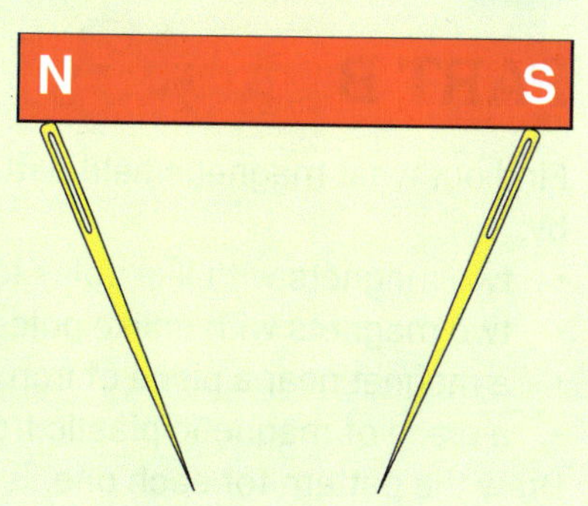

EXPLORE

Can you do the trick shown in the photo with two magnets?

EXPLORE ONLINE

1 To find out more about magnets and to do some experiments with magnets go to **Magnet man**.

2 Use the library to find out when magnetism was discovered. Who found out about it?

3 Most magnets are made of a substance called alnico. Find out what this is.

ISBN 978 1 4202 3736 8

4.2 Magnetic fields

When you bring two magnets together, you can feel the force of attraction or repulsion between them. The magnets exert a non-contact force on each other, i.e. they exert a force without touching.

A magnet is said to have a **magnetic field** around it. When something magnetic is in this field, it will experience a force. The closer it is to a magnet, the stronger the magnetic field. So, as the magnetic substance gets closer to a magnet, the attractive force increases. Non-magnetic substances like your finger are not affected.

Magnetic fields are invisible. However, you can see where they are by using a magnetic compass or iron filings.

Can you see a magnetic field?

Aim

To map invisible magnetic fields using iron filings.

Risk assessment and planning

Before you start, wrap the magnet in some cling film. This stops the iron filings sticking to the magnet. Iron filings are very difficult to get off a magnet!

PART A

Materials

- 2 bar magnets and a horseshoe magnet
- sheet of transparent plastic film (OHP film)
- iron filings (in a salt shaker)
- small magnetic compass
- piece of cling film

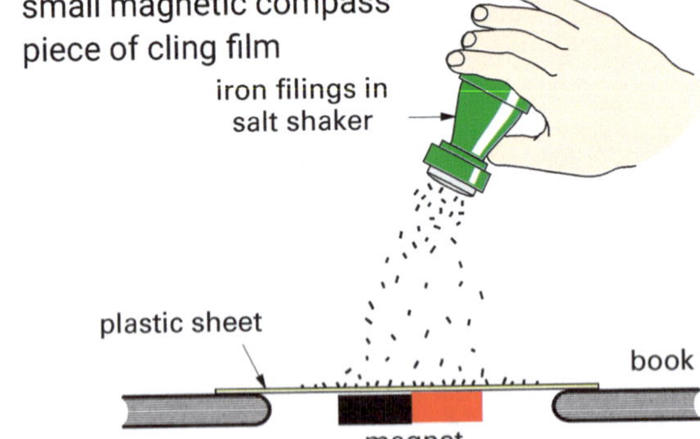

Method

1 Put a bar magnet on the desk. Lay a sheet of plastic film over it. Hold up the ends of the sheet by putting books or wooden blocks under it, as shown.

2 Put a compass on top of the magnet. Notice that the compass is affected by the magnet, and no longer points north–south. Move the compass about. The direction it points is the direction of the magnet's invisible field.

3 Gently sprinkle iron filings onto the sheet, over the magnet. Lightly tap the sheet to spread the iron filings.

4 Observe the pattern that appears around the magnet. Compare it with the patterns produced by other students.
Draw a diagram of the iron filings pattern.

5 Repeat the activity with the horseshoe magnet.

You can preserve your iron filings patterns by spraying them carefully with hair spray.

Discussion

1 Where is the magnetic field of a magnet strongest? How do you know?

2 What happens to the magnetic field as you move away from a magnet?

PART B Inquiry

Find out what magnetic field pattern is produced by:

- two magnets with like poles together
- two magnets with unlike poles together
- a magnet near a piece of iron or steel
- a piece of magnetic plastic from a fridge.

Draw the pattern for each one.

ISBN 978 1 4202 3736 8

Magnetic field shapes

In Investigation 4.2 you investgated magnetic fields around magnets. The following figures show the shape of the magnetic fields in two different situations. How do your results compare to these?

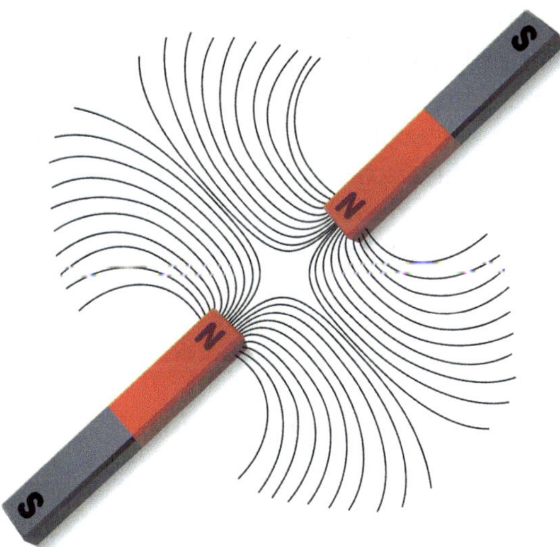

Figure 4.1 The field shape around repelling bar magnets

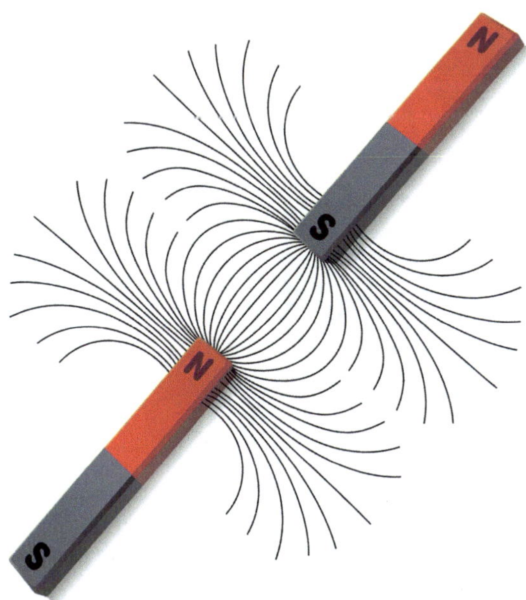

Figure 4.2 The field shape around attracting bar magnets

What causes magnetism?

ACTIVITY

1. Half fill a test tube with iron filings and stopper it.
2. Bring the test tube close to a magnetic compass. It should attract either end of the needle, like any other magnetic substance.
3. Now stroke the test tube along its length about 10 times with one pole of a bar magnet. Make each stroke in the same direction, as shown.

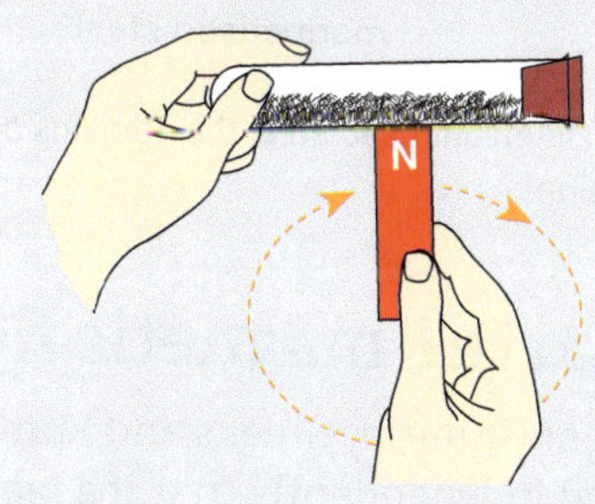

4. Bring the test tube near the compass again. Is it a magnet now? That is, does it attract one end of the needle and repel the other?
5. Shake the test tube a few times. Is it still a magnet?

On the basis of activities like the one above, scientists have worked out a *model* to explain what causes magnetism.

Imagine a piece of steel to be made up of a very large number of tiny patches of magnetism. These tiny patches are called *domains*. When the steel is unmagnetised, the domains are all mixed up. This is rather like what you would get if you threw a pile of small bar magnets into a box. The poles of the magnets would point in different directions. As a result, the magnetic forces tend to cancel each other out.

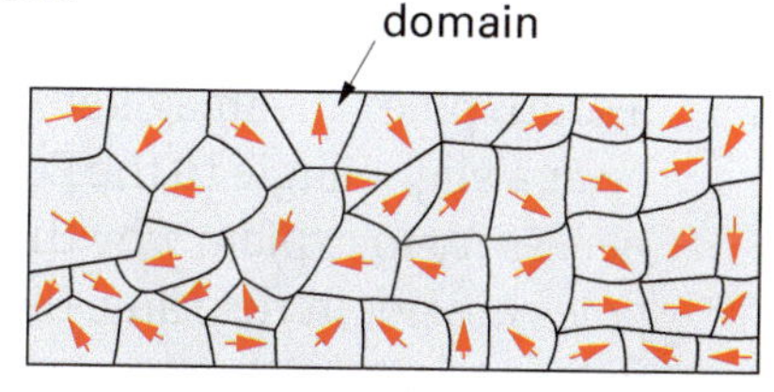

Figure 4.3 In unmagnetised steel the domains all point in different directions.

ISBN 978 1 4202 3736 8

When the piece of steel is stroked by one end of a magnet, the domains are all forced to point in the same direction. As a result, the substance becomes magnetised. The more domains that are turned in the same direction, the stronger the magnet will be. Notice that there are north poles at one end of the piece of steel, and the same number of south poles at the other end.

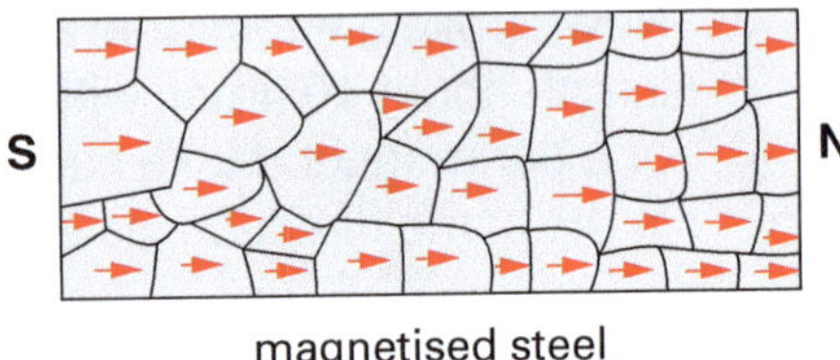

Figure 4.4 In magnetised steel the domains all point in the same direction.

The Earth's magnetic field

The ancient Chinese, Greeks and Romans discovered independently that if a piece of magnetite (a magnetic rock) is free to rotate, it will line up in a north–south direction (just like the magnet in Investigation 4.1 Part C). The first compasses used for navigation were simply magnetised needles on pivots. However, it was hundreds of years before anyone knew *why* a compass points north–south.

In the year 1600, Sir William Gilbert proposed an inference to explain why a compass points north–south. He suggested that the Earth behaves as if there is a gigantic bar magnet at its centre. It therefore has its own magnetic field. Gilbert was the private doctor of Queen Elizabeth I of England, and explained his ideas about magnetism to her.

Gilbert said that the south pole of his imaginary magnet was near the Earth's North Pole. In this way he could explain why the north pole of a suspended magnet always points to the Earth's North Pole, since a north pole is attracted to a south pole. The imaginary magnet also needs to be at a slight angle to the Earth's axis. This is necessary to explain why the Earth's magnetic poles are not in the same place as the geographic poles.

So what causes the Earth's magnetic field? Scientists are still not sure but they infer that it is caused mainly by electric currents in the molten core of the Earth.

The Earth's magnetism has other effects as well. It helps to trap a layer of invisible charged particles in the atmosphere around the planet. This layer is called the ionosphere (eye-ON-os-fear). Radio signals can be sent right around the world by bouncing them off this layer.

Beautifully coloured lights (called auroras) can sometimes be seen in the night sky near the North Pole and South Pole (see Figure 4.6 on the next page). The Aurora Australis can sometimes be seen from southern Australia. It is formed when charged particles from the sun hit the top layer of the Earth's atmosphere. These particles move towards the poles because they are deflected there by the Earth's magnetic field.

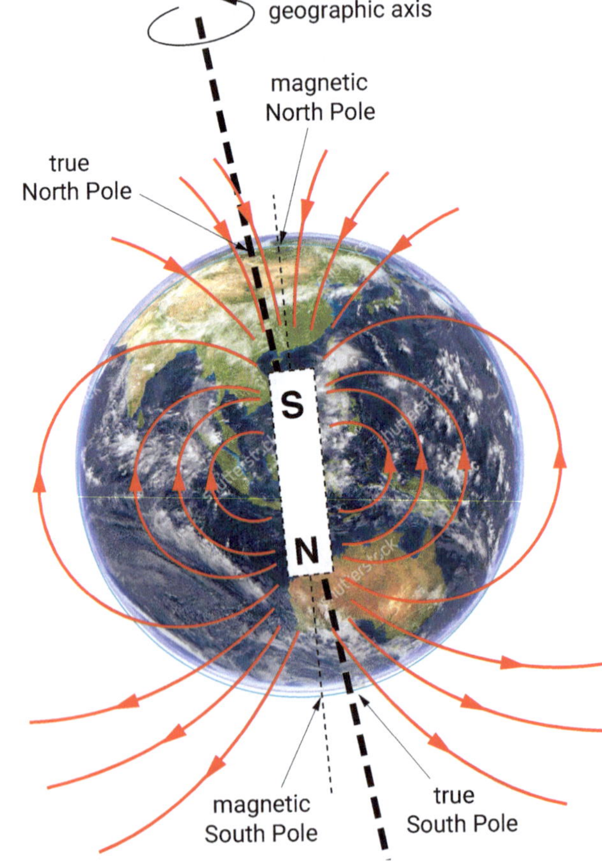

Figure 4.5 An explanation of the Earth's magnetic field: the arrows show the direction in which the north end of a compass needle points.

ISBN 978 1 4202 3736 8

Figure 4.6 The beautiful Aurora Australis (or Southern Aurora) as seen from the International Space Station. The aurora forms over the South Pole as a result of the Earth's magnetism.

Magnetic shielding

Magnetic fields pass through all substances except magnetic ones. Magnetic materials gather in, or concentrate, magnetic fields. These materials therefore act as shields, stopping the magnetic field from passing through them.

Look at Figure 4.7, which shows a piece of plastic and a piece of iron near a bar magnet. Notice how the magnetic field near the south pole is not affected by the piece of plastic—it passes right through it. At the north pole, however, the field is gathered in by the piece of iron and does not go into the area above it. Anything above the piece of iron is *shielded* from the magnetic field.

Many sensitive scientific instruments need to be shielded from stray magnetic fields. This can be done by enclosing the object in a steel box.

In Investigation 4.3 on the next page, you can investigate magnetic shielding for yourself.

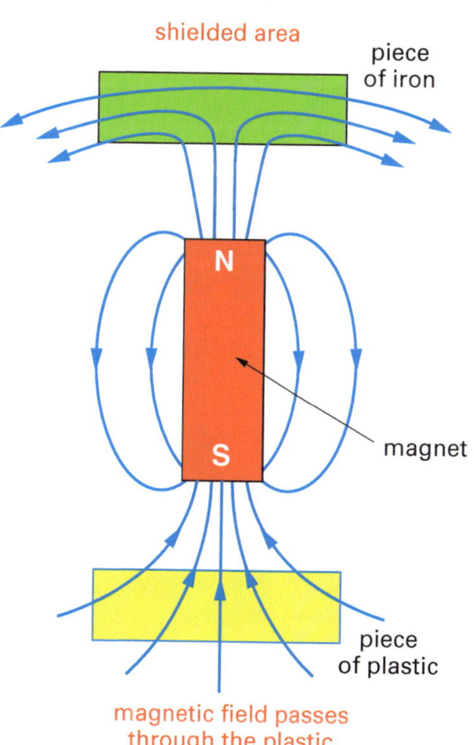

Figure 4.7 How a piece of iron can act as a magnetic shield.

ISBN 978 1 4202 3736 8

Magnetic shielding

Aim

To find out which materials act as magnetic shields.

Risk assessment and planning

Read the investigation and study the diagram below carefully.

Which materials will you be testing as magnetic shields?

PART A

Materials

- bar magnet
- paperclip
- piece of cotton thread
- piece of plasticine
- adhesive tape
- metal stand and clamp
- sheets of various materials, e.g. paper, glass, plastic, tin, iron, wood, copper, aluminium foil

Method

1 Set up the magnet and paperclip as shown.

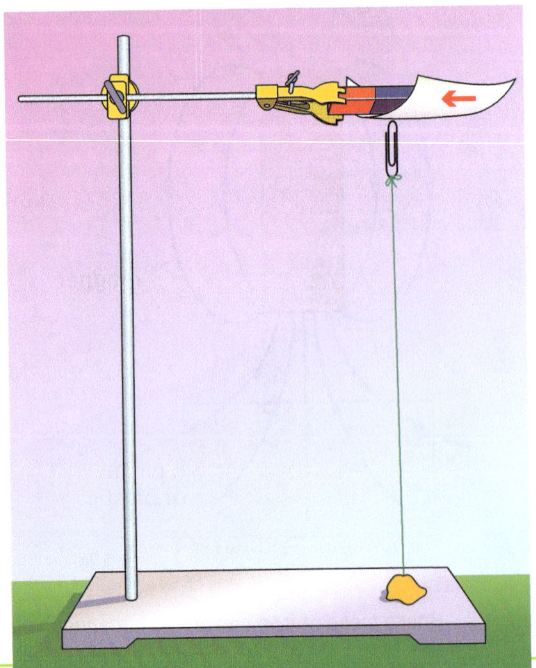

The paperclip is attracted to the magnet because it is in its magnetic field, but it is held away from the magnet by the thread. What is the largest gap you can leave between the magnet and the paperclip?

2 Place a sheet of paper between the magnet and the paperclip. Did the paperclip fall? That is, did the sheet of paper stop the magnetic field and act as a magnetic shield?

3 Use sheets of various materials to see if they can act as magnetic shields.
Record your results.

Discussion

1 Write a generalisation about the types of materials that can act as magnetic shields.

2 Consider the paperclip suspended below the magnet. What two forces are acting on the paperclip? Are they balanced? What happens if the forces become unbalanced?

3 Predict what will happen if you burn the thread with a match. Give a reason for your prediction. Then check your prediction.

PART B Inquiry

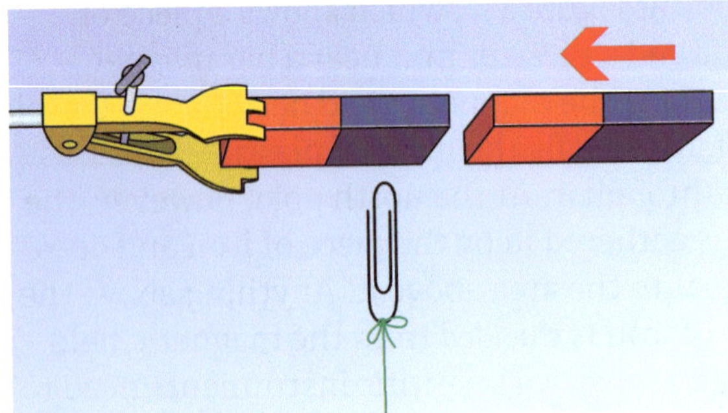

Predict what will happen to the paperclip if you bring a second magnet near the first one. Does it matter whether it is a south or a north pole?

Record your observations, and try to explain them in terms of magnetic fields.

ISBN 978 1 4202 3736 8

CHECK

1 What is a magnetic field? Can you see one? Explain.

2 The diagram shows a magnet and its magnetic field. **A**, **B** and **C** are steel ball bearings.

a Which ball bearing does the magnet attract most strongly? Which does it attract the least?

b In which position is the magnetic field strongest? Where is it weakest?

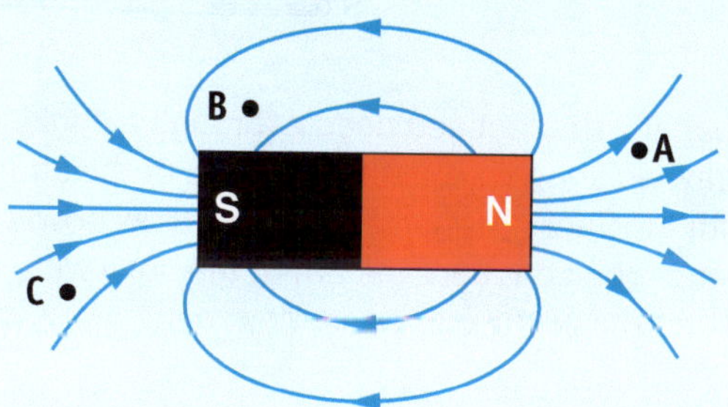

3 What makes a compass needle point to the north?

4 How is the magnetic south pole different from the geographic South Pole?

5 Sketch the magnetic field around a horseshoe magnet.

6 In which of the following situations would a compass probably give an incorrect reading?

A in the bush

B in an aluminium dinghy

C in a car

D near an iron ore mine

7 Why is a magnetic compass of little value inside a submarine?

8 Heating and rough handling of a magnet will reduce its strength. How can you explain this in terms of the domain model?

9 Look at the diagram in the Activity on page 89. When you stroke the test tube this way, predict which end will be a north pole. Check your prediction.

CHALLENGE

1 Using what you know about magnetic fields, describe a gravitational field.

2 It has been shown that the Earth's magnetic field has reversed several times in the past. The north and south magnetic poles have swapped over. Which one of the following competing inferences explains this swapping better? Why?

A The Earth's magnetic field is due to a large magnet inside it.

B The Earth's magnetic field is due mainly to electric currents in its molten core.

3 a An iron tube is placed near the end of a bar magnet as shown. Draw the magnetic field that you would expect. Is there an area where there would be no magnetic field; that is, is there a shielded area?

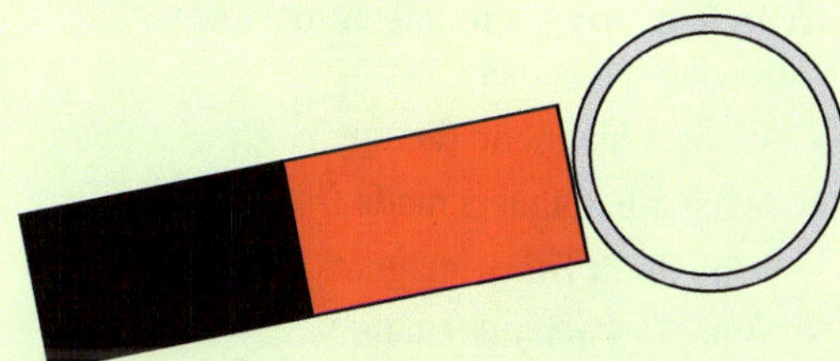

b Trina finds that a steel ball bearing inside the tube does not move when she brings up the magnet. How can you explain this?

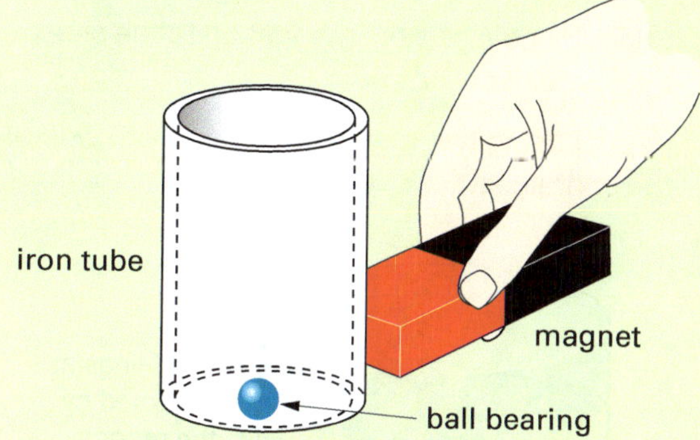

4 Use the domain model discussed on page 89 to predict what will happen if you break a magnet in half. (To test your prediction, magnetise a paperclip or easily breakable piece of metal, by stroking it as in step 3 of the Activity on page 89, then cut or break it in half.)

5 Do you think there is a limit to the amount of magnetism that can be produced in a piece of steel by stroking it with a magnet? Explain in terms of the domain model.

ISBN 978 1 4202 3736 8

EXPLORE

1 Make a model of the magnetic Earth by moulding a ball of plasticine around a bar magnet. Roll the ball in iron filings to show up the magnetic poles. Use a small compass to show that the needle points to the North Pole wherever the compass is.

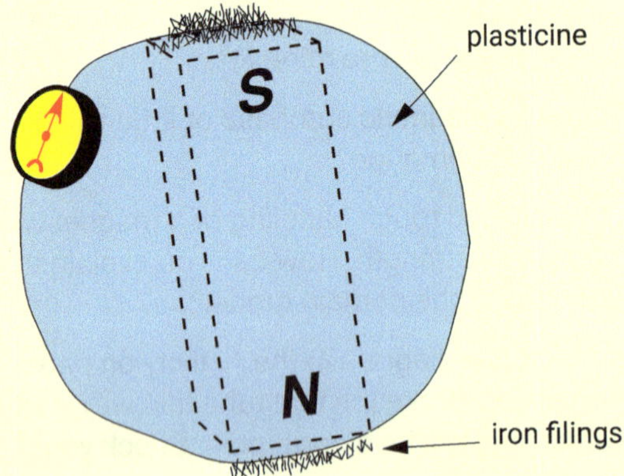

2 Here is a way to observe magnetic fields in three dimensions.

Put some iron filings in a small, clear plastic or glass container, then fill the container with cooking oil. Put on the lid and shake the container vigorously to mix the iron filings with the oil.

Bring the pole of a bar magnet up to the side of the container as shown.

Repeat using two magnets, one on each side of the container.

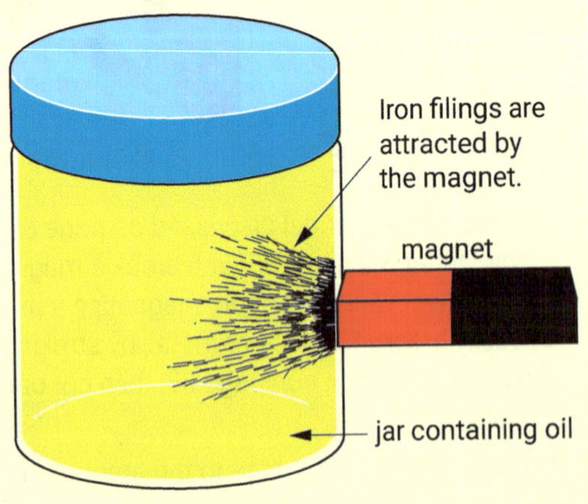

3 Here is a way to make your own compass.

Magnetise a large sewing needle by stroking it with one pole of a bar magnet, as shown. You will need to do this at least 10 times.

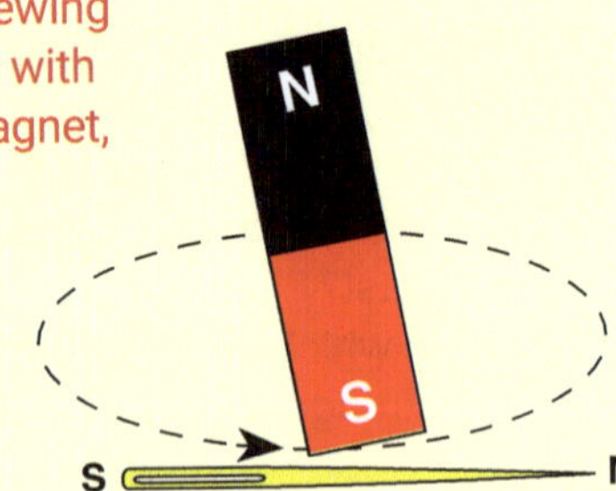

Push the magnetised needle through a drinking straw as shown and plug each end of the straw with Blu-Tack. Then float it in a shallow bowl of water. You can mark the rim of the bowl with the points of the compass.

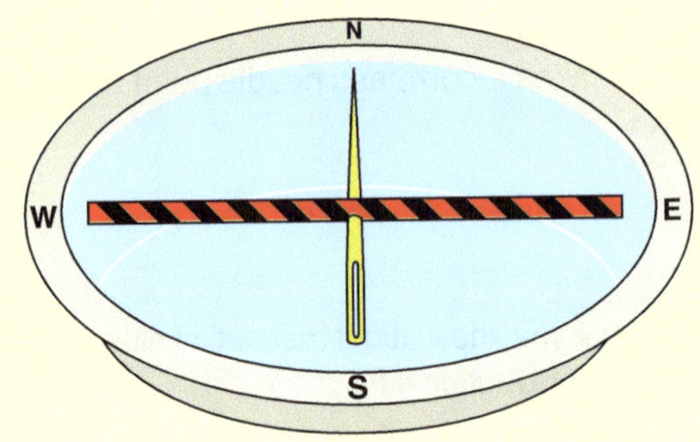

EXPLORE ONLINE

1 Credit cards have a magnetic strip that can be read by a scanner in an ATM. How are magnetic strips made?

Go to **Mad Scientist Network**.

Type *magnetic strip* in the search box and many sites will come up.

You can use Ask-A-Scientist to ask questions or browse through questions that other people have asked (with their answers).

2 Use the Mad Scientist Network website to try to find answers to one or more of the following questions.

- What is magnetic therapy?
- What are magnets made from?
- Why don't fridge magnets have poles?
- What is a magnetometer?
- What is magnetic levitation?

ISBN 978 1 4202 3736 8

4.3 Electricity and magnets

In 1820, a Danish scientist called Hans Oersted was giving a lecture in Copenhagen in which he did an experiment to show (so he thought) that there was no connection between electricity and magnetism. He put a compass needle and a wire side by side and then he passed an electric current through the wire. Imagine his surprise, and embarrassment, when the compass needle swung around. Oersted had found, by accident, that when an electric current flows through a wire a magnetic field forms around it. An Italian scientist had made this discovery 18 years earlier, but nobody realised how important it was.

With a single wire, the magnetic field is weak. But if the wire is made into a coil, the field is much stronger. A coil of wire like this is called a *solenoid*. Its magnetic field is similar to that of a bar magnet.

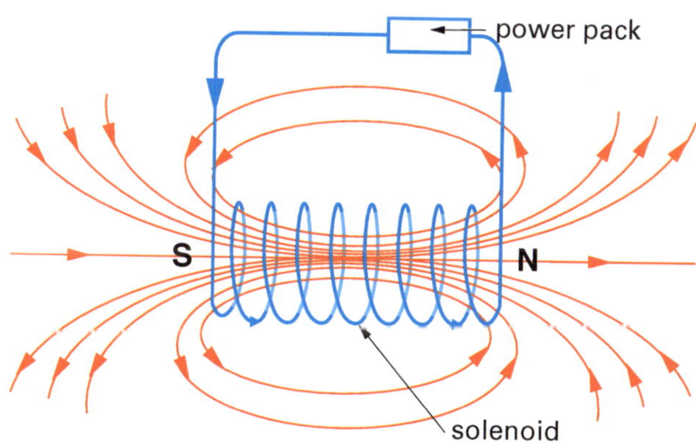

Figure 4.8 A solenoid can produce a magnetic field.

You can make the magnetic field of a solenoid even stronger by placing a piece of iron (called a core) inside it. This makes an **electromagnet**. This was discovered in 1825 by a British shoemaker called William Sturgeon when he was working on his hobby of experimenting with electricity.

An electromagnet is a temporary magnet because the magnetic field disappears when the electricity is turned off. A bar magnet, on the other hand, is called a permanent magnet because you cannot switch off its magnetic field.

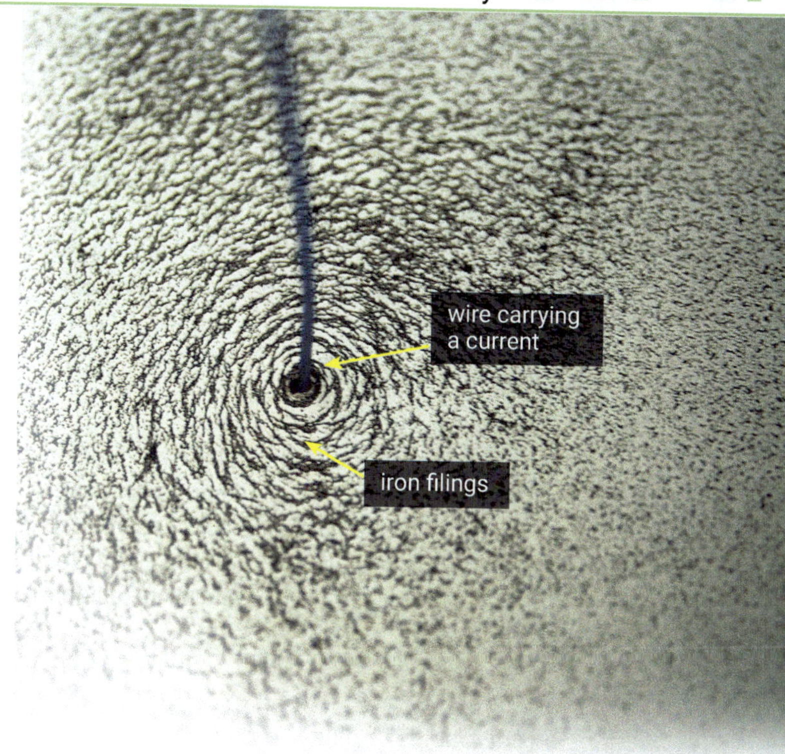

Figure 4.9 Iron filings can be used to show the circular magnetic field around a wire carrying an electric current.

Check out Oersted's discovery for yourself.

1. Put a small magnetic compass in the tray of a matchbox. Wind about a metre of connecting wire around the box as shown below.
2. Rotate the matchbox so that the north-pointing compass needle is parallel to the wire.
3. Connect one of the wires to a battery. Then *briefly* touch the other end to the other terminal. Observe what happens.

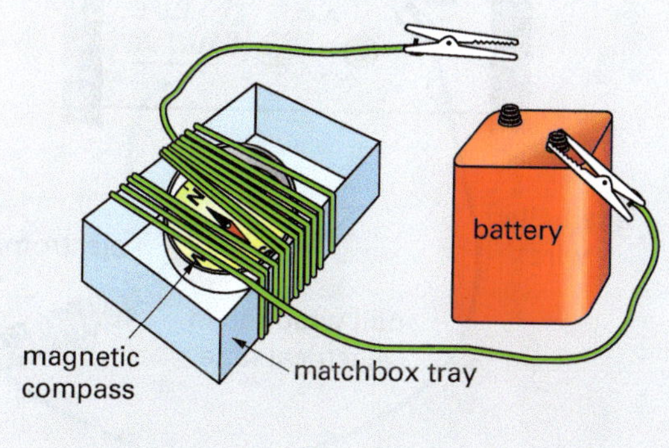

ISBN 978 1 4202 3736 8

INVESTIGATION 4.4 Making an electromagnet

Aim

To make an electromagnet and investigate how it can be made stronger.

Materials

- small magnetic compass
- large steel nail or bolt (at least 75 mm long)
- about 2 metres of enamelled copper wire (0.8 mm diameter) with alligator clips at each end. (Note: Scrape the enamel coating off the ends of the wire before you attach the alligator clips.)
- electrical tape or duct tape
- power pack
- box of staples or pins

Risk assessment and planning

Read the investigation carefully. Note that you will use 25, 50 and 100 turns of wire around the nail.

- What precautions are necessary when using a power pack?
- Suggest why you need to wrap electrical tape around the nail or bolt.

Method

1 Wrap some electrical tape around the large nail or bolt to insulate it.

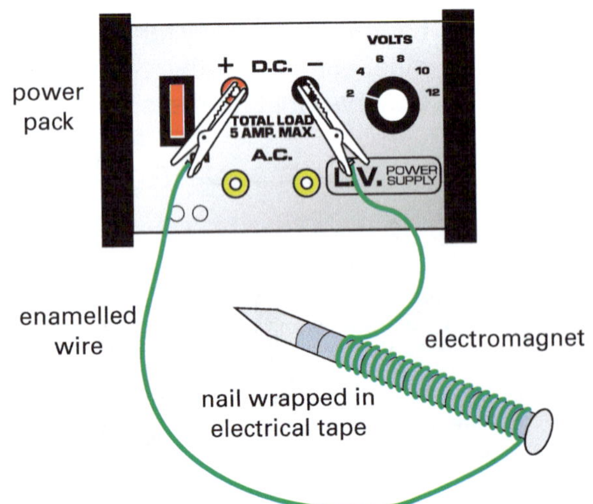

2 Wrap about 25 turns of the enamelled wire around the nail.

3 Connect the alligator clips to a power pack. (It doesn't matter which alligator clip you connect to the positive or negative.)

4 Set the power pack to 2 volts DC. Turn it on.
Has the nail been magnetised? Will it pick up staples? How many?
What happens when you turn off the power pack?

5 Turn the electromagnet on again, and bring a compass near the sharp end of the nail. Decide whether the sharp end of the electromagnet is a north pole or a south pole. Also test the other end of the electromagnet.

6 Reverse the connections to the power pack.
Have the poles of the electromagnet changed?

7 Wrap 50 turns of wire around the nail.
How many staples will it pick up this time?
How many staples will it pick up with 100 turns?

8 Investigate the effect of increasing the voltage from the power pack.

Discussion

1 Why is an electromagnet called a 'temporary magnet'? What advantage does it have over a bar magnet?

2 How can the poles of the electromagnet be reversed?

3 List two ways of increasing the strength of an electromagnet.

4 Sketch the magnetic field around your electromagnet. Show the directions of the field lines.

ISBN 978 1 4202 3736 8

Uses of electromagnets

Huge electromagnets are used in scrapyard cranes. They are powerful enough to lift cars. In a hospital, a surgeon can use an electromagnet to remove a piece of iron or steel from the eye of a patient.

Telephones

In the mouthpiece of a telephone, sound waves are changed into electric currents. In the earpiece of another telephone, these electric currents pass through an electromagnet. When the current is on, the metal disc is attracted to the magnet, and when it is off, it springs back. This causes the disc to vibrate, producing sound waves, which you can hear.

Metal detectors

The metal detectors at airports contain solenoids. Anything metallic you are carrying when you walk through these will affect the magnetic field of the solenoid. This immediately alerts the security officer.

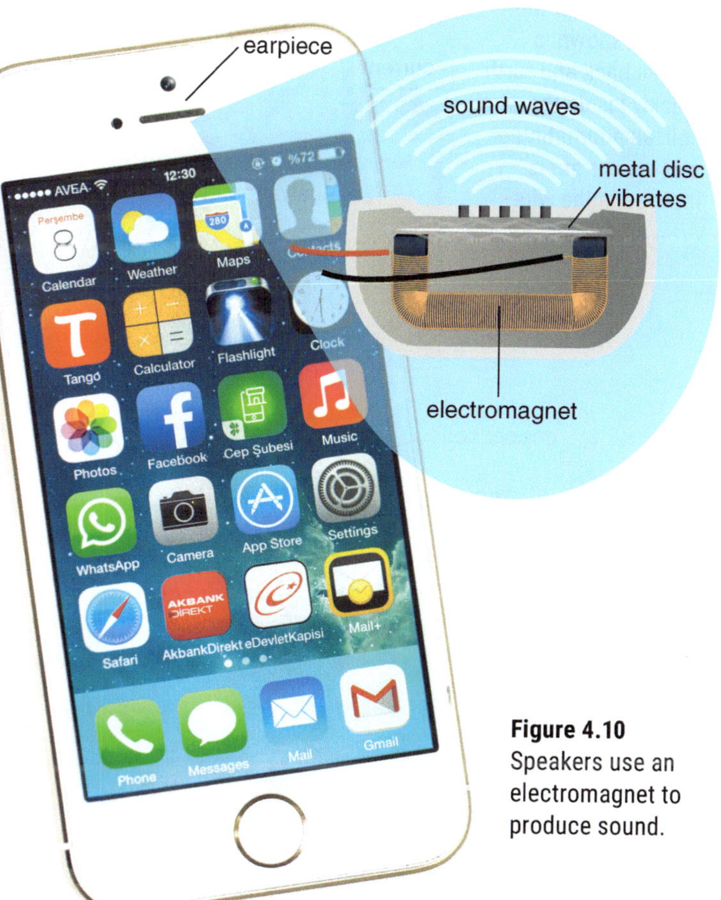

Figure 4.10 Speakers use an electromagnet to produce sound.

Figure 4.11 An electromagnet can be very powerful.

Figure 4.12 Metal detectors use a solenoid to detect metal when there is a change in the magnetic field.

ISBN 978 1 4202 3736 8

Maglev train

Electromagnets are used in a high-speed train called a *maglev train*, which is short for *magnetic levitation train*. There are electromagnets on the track (called a guideway) and underneath the train. The electromagnets keep the maglev train 'floating' about 10 mm above the guideway so there is no contact with the ground and, therefore, no friction. There is also no need for wheels.

How does the maglev train move?

The maglev train is propelled by the electromagnets along the guideway rather than by an on-board engine. By changing the magnetic fields in the electromagnets, you can make the maglev train start, stop or change its speed.

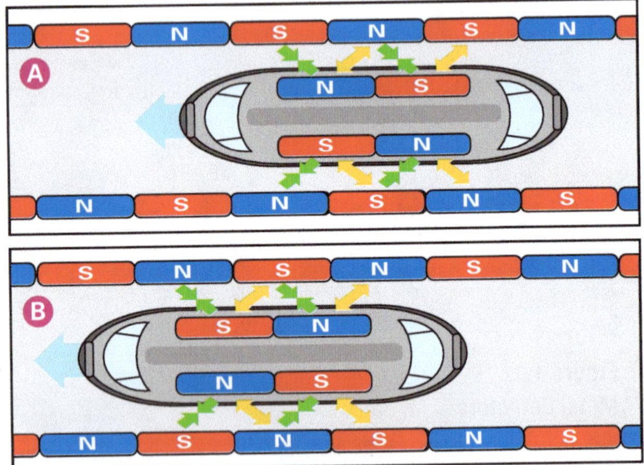

Figure 4.13 The forces of attraction and repulsion between a series of electromagnets propel the maglev train. As the train moves forward from A, the electromagnets in the train reverse poles at B.

Figure 4.14 A maglev train.

Search the internet to make a list of the advantages and disadvantages of maglev technology. Go to **How a maglev train works**, or you might want to try **How to build a maglev train**.

Generating electricity

Magnets can also be used to generate electricity. When a magnet is moved inside a coil of wire, it causes electrons in the atoms inside the wire to move. The moving electrons in the wire create an **electric current**. In the simple experiment shown in Figure 4.15, this electricity can be detected with a galvanometer—an instrument designed to detect very small electric currents. As the magnet shown is moved back and forth quickly inside the wire coil, an electric current is *induced* in the wire. The galvanometer needle moves back and forth as the magnet moves and the electric current changes direction. This is called an **alternating current** (AC), meaning it changes or alternates from one direction to the other.

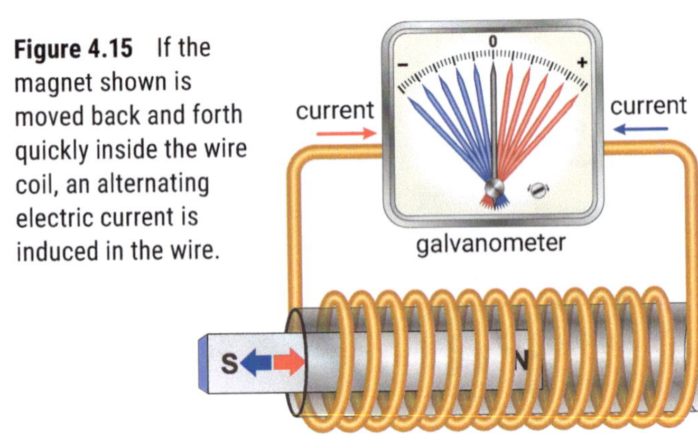

Figure 4.15 If the magnet shown is moved back and forth quickly inside the wire coil, an alternating electric current is induced in the wire.

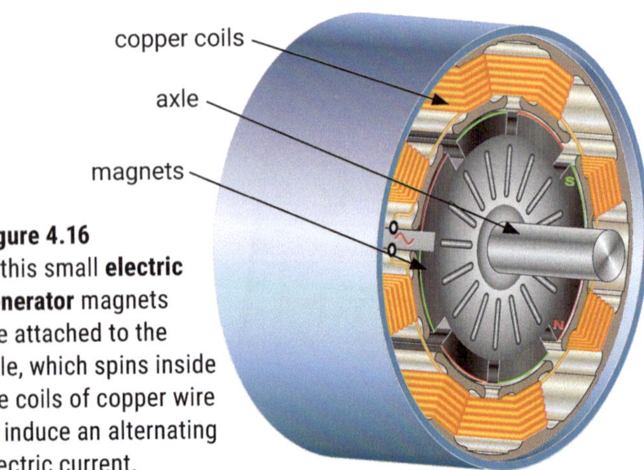

Figure 4.16 In this small **electric generator** magnets are attached to the axle, which spins inside the coils of copper wire to induce an alternating electric current.

ISBN 978 1 4202 3736 8

In larger generators in coal-fired power stations, hydro-electric plants and wind turbines, the spin axle has large electromagnets attached to it.

Figure 4.17 The inside of a large electric generator at a power station

Electric motors

Electric motors consist of a coil of wire between magnets. When an electric current flows through the coil it interacts with the magnetic field and the coil moves, or spins. This is the opposite of how a generator works.

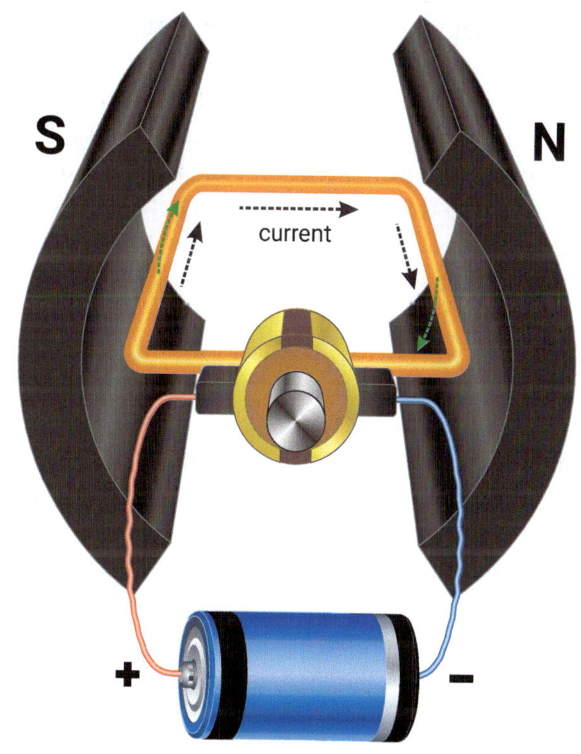

Figure 4.18 When the battery is connected, electricity flowing through the coil interacts with the magnetic field and the coil spins.

ACTIVITY

Make an electric motor

Construct a simple electric motor like the one shown here, using the following materials:

- AA battery
- thin enamelled copper wire
- two large safety pins or needles
- small circular magnet
- sticky tape or electrical tape
- modelling clay

Your motor should spin until the battery goes flat!

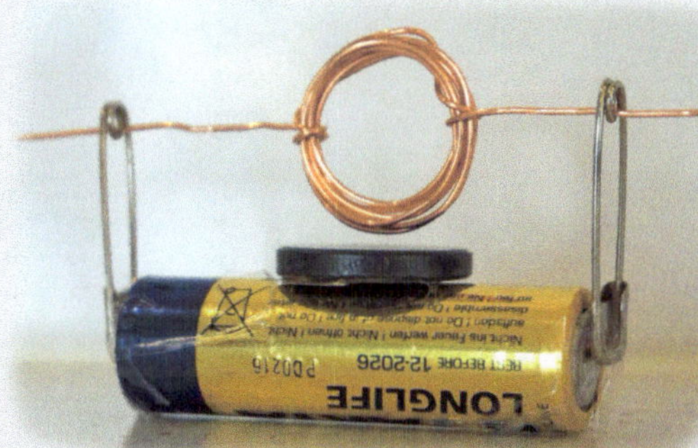

EXPLORE ONLINE

Go online to find instructions on how to build this model. There are many variations available to build a simple motor like this. Start with the one at Education.com and then see if you can build other variations.

ISBN 978 1 4202 3736 8

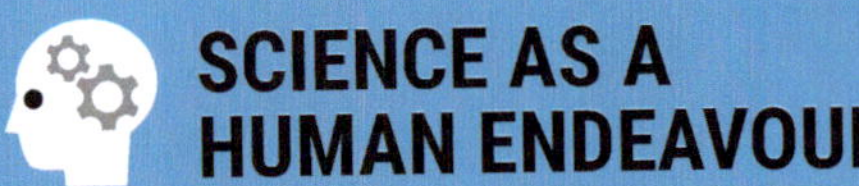

SCIENCE AS A HUMAN ENDEAVOUR

Superconductors

You have something wrong with you and a doctor sends you for an MRI scan (photo below). The MRI scanner uses a powerful magnetic field made possible by the use of superconductors. The maglev trains you read about on page 98 also use superconductors. But what are superconductors?

Superconductors are materials that lose all resistance to the flow of electricity at very low temperatures. The first superconductors had to be cooled to −270 °C. This could only be done using liquid helium, which is very expensive and difficult to work with. Then in 1986 two scientists in Switzerland discovered a ceramic compound that became a superconductor at only −200 °C. This meant that liquid nitrogen could be used for cooling. Liquid nitrogen is much cheaper and easier to use than liquid helium.

In Australia Dr Cathy Foley heads a team of 140 researchers at CSIRO in Sydney studying so-called high-temperature superconductors. Her team has developed a device called a SQUID (Superconducting Quantum Interference Device) for detecting extremely weak magnetic fields. These devices have been used to explore for minerals and have found $6 billion worth of mines in Australia and around the world. Dr Foley's team is also researching other uses for this technology. For example, it can be used to detect defects in the heartbeat of a foetus, or to monitor adult hearts without using electrodes. It can also be used to check for unwanted metal fragments in food, and can detect submarines.

Scientists are hoping that they will discover how to make superconductors that work at normal temperatures. In electric power lines there is always some loss of energy as heat, due to the resistance of the metal in the wires. If these wires could be made cheaply from superconductors, no energy would be lost and billions of dollars could be saved. Many scientists also think that it won't be long before a completely new type of superfast quantum computer is invented that uses superconductors.

Writers of science fiction have imagined all sorts of uses for superconductors and SQUIDs—building a space elevator to carry people and materials into space, devices to read people's minds and teleportation devices. If scientists make new discoveries, some of these things may come true in your lifetime.

Questions

Use the information on this page to explain:

- the difference between science and technology
- how developments in science have led to the development of new technologies.

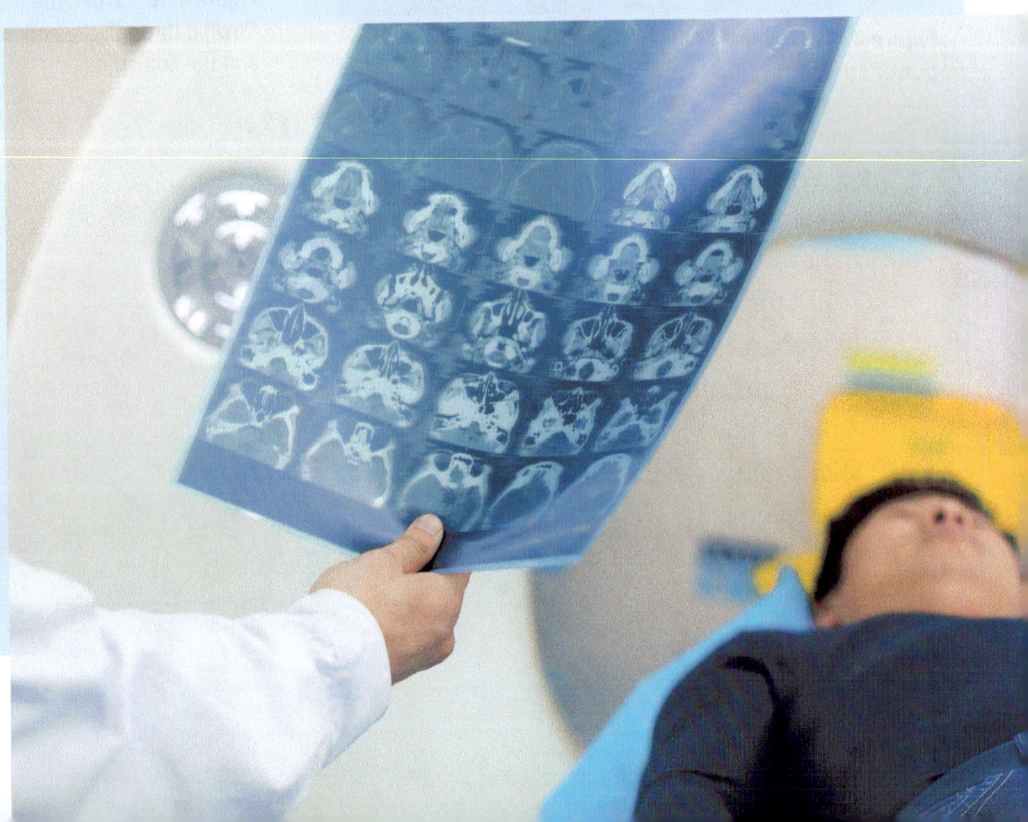

1 How could you show that electricity and magnetism are related?

2 Leanne wants to find out whether her electromagnet is as strong as Scott's. To do this she counts how many paperclips each electromagnet will pick up. She repeats her measurement three times for each electromagnet. Her results are as follows:

	Test 1	Test 2	Test 3
Leanne's electromagnet	16 clips	18 clips	16 clips
Scott's electromagnet	16 clips	14 clips	14 clips

a Which electromagnet is stronger—Leanne's or Scott's?

b Would the result have been the same if Leanne had tested each electromagnet only once? Explain.

3 The diagram on the right shows a musical doorbell.

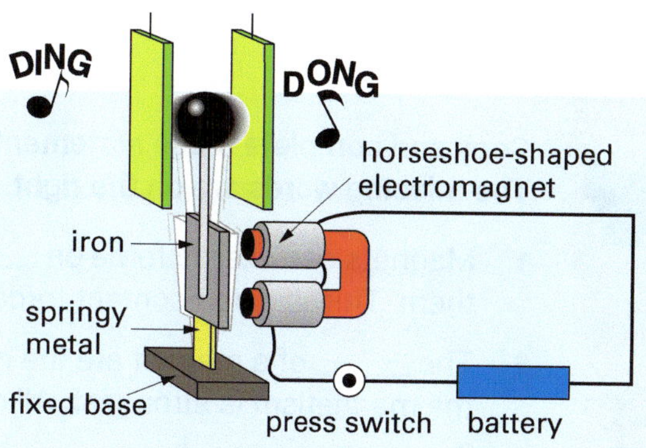

a Explain how the doorbell works.

b When the switch is pressed and then released, would the bell go 'ding-dong' or 'dong-ding'?

c How could the design be adapted to make it go 'ding-dong' rather than 'dong-ding'?

4 Explain what electric current is.

5 Compare an electric generator and an electric motor to show how they are similar and how they are different.

6 Draw a diagram to show the main parts of an electric motor.

1 Rory did an experiment using an electromagnet. He was able to vary the number of turns in the coil, and he could use either an iron core or a steel core. He recorded his results in a data table. (He noticed that when the electricity was switched off, the paperclips dropped from the iron-cored electromagnet but not from the steel-cored one.)

Electro-magnet	Core material	Number of turns in coil	Voltage (volts)	Number of paperclips lifted
1	iron	10	2	2
2	iron	20	2	4
3	iron	20	4	17
4	iron	20	6	26
5	iron	20	8	28
6	iron	30	2	6
7	steel	20	3	7
8	steel	30	2	5

a How can you tell how strong each electromagnet is?

b Write down three things that affect the strength of an electromagnet.

c Which electromagnets do you need to work out how the strength depends on the voltage?

d Copy and complete the bar graph shown here for iron-cored electromagnets with 20 turns in the coil. Is it true that the higher the voltage, the stronger the electromagnet?

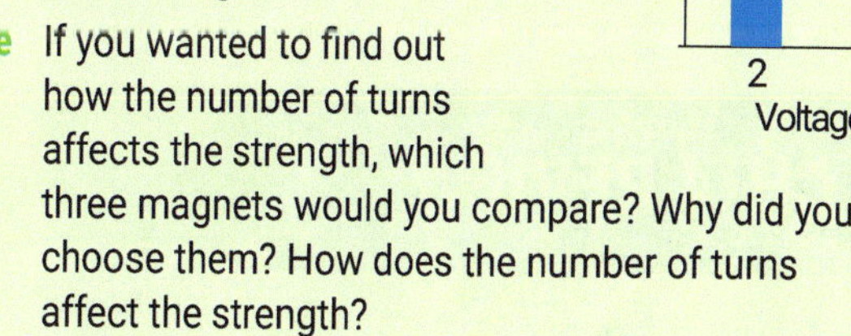

e If you wanted to find out how the number of turns affects the strength, which three magnets would you compare? Why did you choose them? How does the number of turns affect the strength?

f Iron is a better core material than steel. Suggest two reasons for this.

2 Suppose your school has some old bar magnets that have lost their strength. Suggest how you could renew their magnetism.

3 Why would hospitals use an electromagnet rather than a permanent magnet for removing a metal splinter from a patient's eye?

4 Can someone with a stainless steel part in their body walk through a metal detector without the alarm sounding? Use your knowledge of magnetic substances to make a prediction.

ISBN 978 1 4202 3736 8

MAIN IDEAS

Copy and complete these statements to make a summary of this chapter. The missing words are on the right.

1. Magnets can exert a force on ______ materials, without even touching them. This is a non-contact force.
2. The ______ of a magnet are the places at the ends of a magnet where the magnetism is strongest. All magnets have two poles, a north and a ______.
3. ______ magnetic poles repel each other, and unlike poles ______ each other.
4. A magnetic ______ is the area around a magnet in which a force is exerted on a magnetic substance.
5. The Earth has a magnetic field. This is why a ______ and a floating or suspended magnet point north–south.
6. When an electric ______ flows through a wire, a magnetic field forms around it.
7. An ______ is a temporary magnet made from a coil of wire (______) wound round a piece of iron. The electromagnet creates a magnetic field when electricity passes through the coil.
8. An electric current is the flow of ______ in a wire. An electric current can be induced by ______ a magnet in an electric field.
9. An electric motor works by sending electric current through a coil of wire between magnets, making the coil move or ______.

south
compass
field
magnetic
permanent
poles
solenoid
like
current
moving
electromagnet
attract
electrons
spin

CH•4 REVIEW

1. Magnets attract:
   - **A** other magnets only.
   - **B** all other materials.
   - **C** metals only.
   - **D** some metals only.
2. You want to build a box to shield a scientific instrument from stray magnetic fields. Which one of these materials would you use?
   - **A** Aluminium
   - **B** Lead
   - **C** Steel
   - **D** Wood
3. Electricity flows through the coil of an electromagnet, producing a north and a south pole. Which one of the following would reverse the poles of the electromagnet?
   - **A** Increasing the number of turns of wire
   - **B** Changing the direction of the electric current
   - **C** Increasing the electric current
   - **D** Turning off the current
4. **a** What are the unknown poles X and Y?

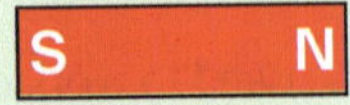

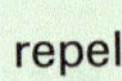

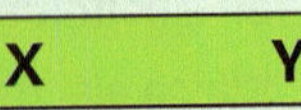

ISBN 978 1 4202 3736 8

b Predict what will happen in each of the two situations below.

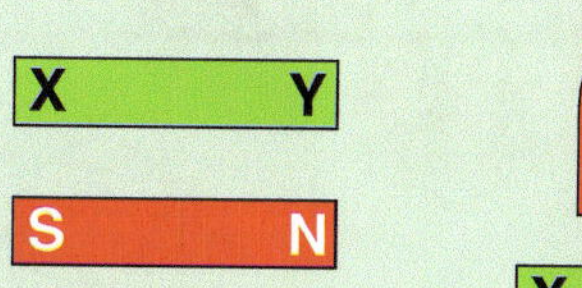
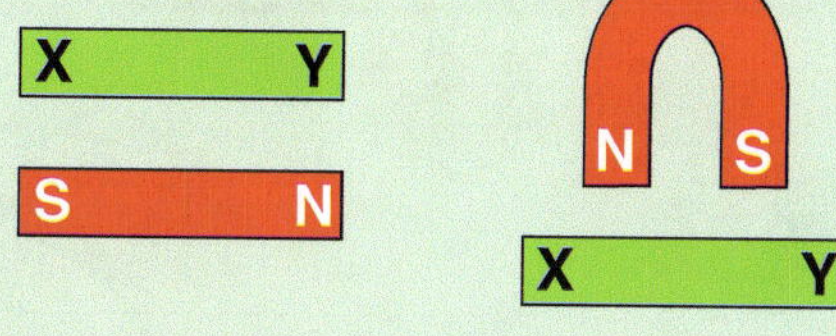

5 How could you show that there is a magnetic field around a bar magnet? Give at least two different ways.

6 a Sketch what the Earth's magnetic field would be like if it was reversed.
b How would a compass behave?

7 Two metal rods, A and B, are placed side by side, about 3 cm apart, on a smooth table.

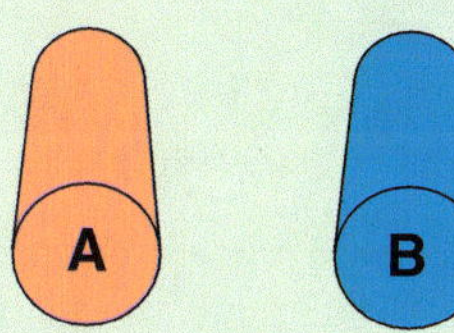

Predict what will happen if:
a A and B are both magnets with like poles next to each other
b A is a magnet and B is a piece of iron
c A is a magnet and B is a piece of brass
d A and B are both unmagnetised pieces of iron.

8 A disc magnet has poles as shown.
a Draw what you think would happen if you dipped this magnet in iron filings

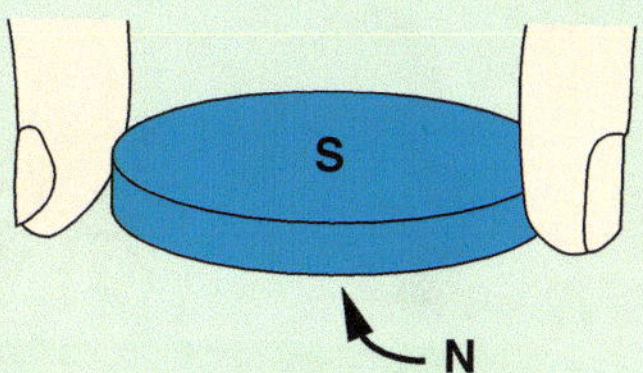

b Draw the magnetic field you would expect around the magnet.
c Explain how you could use this magnet as a compass.

9 Two pieces of iron were each suspended from a length of string, as shown at the top of the next column. The north pole of a bar magnet was brought close to end A of the first piece of metal. End A moved away from the north pole.

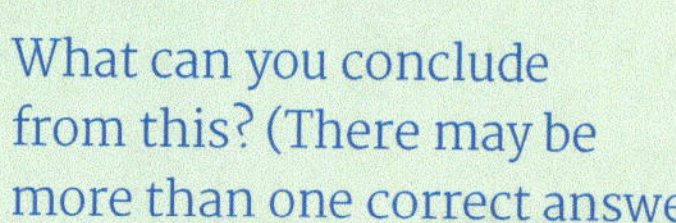

The north pole of the magnet was then brought close to end B of the other piece of metal. End B moved towards the north pole of the magnet.

What can you conclude from this? (There may be more than one correct answer.)
A End A could be the south pole of a magnet.
B End B could be the north pole of a magnet.
C End B could be the south pole of a magnet.
D End B might not be the pole of a magnet.
E End A might not be the pole of a magnet.

Explain your reasoning.

10 A steel ball bearing is placed on a smooth, flat table at equal distances from four very strong electromagnets, as shown below. The electromagnets are identical, except for the number of turns in their coils, and the number of batteries.
a In which direction will the ball bearing move when all the electromagnets are turned on?
b What will happen if electromagnet 4 is now turned off?

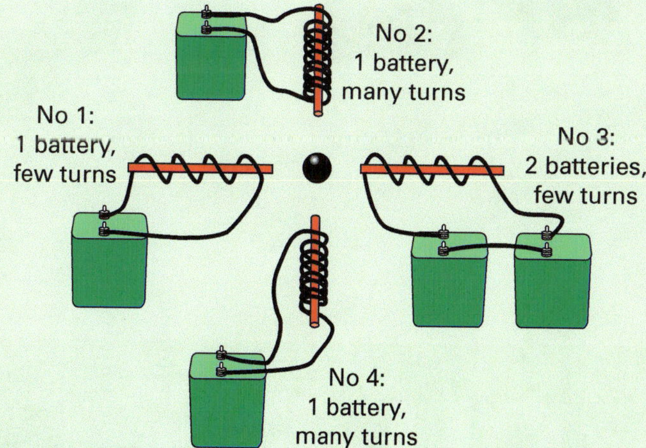

11 a Explain how electricity can be generated using a piece of wire and a magnet.
b What would happen if the magnet was moved back and forth in Figure 4.15 but the wires were disconnected from the galvanometer?

Check your answers on page 299.

ISBN 978 1 4202 3736 8

IN THIS CHAPTER

Science Understanding

- understand the role of decomposers in recycling matter in the environment
- investigate the responses of the body to the presence of micro-organisms
- describe the technologies involved in stopping the decay of food
- appreciate the work of Professor Ian Frazer in developing a vaccine for cervical cancer

Science Inquiry Skills

- suggest testable hypotheses about sick chickens
- critically analyse the validity of information on internet sites
- design and carry out investigations and review their effectiveness

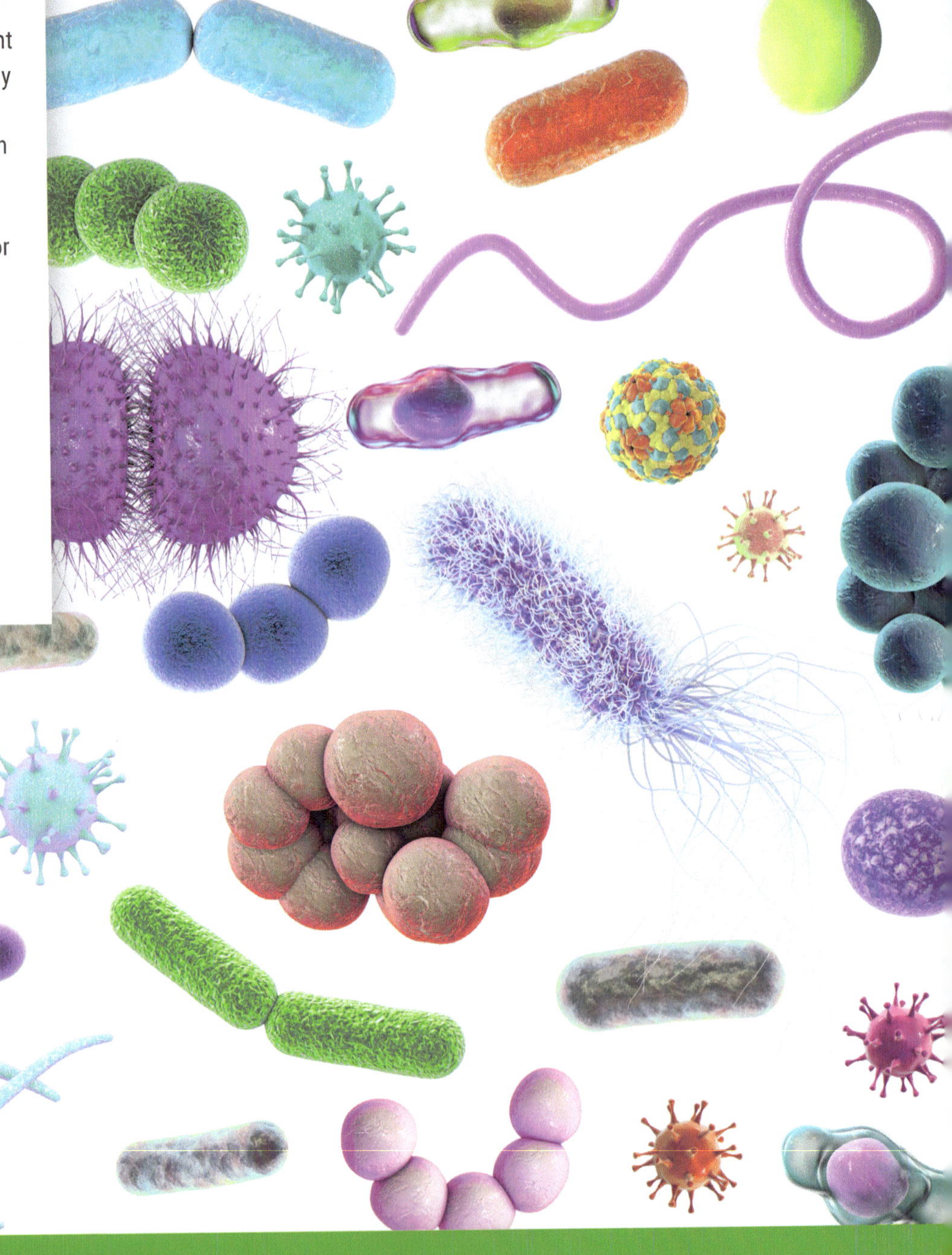

CH•5 Microbes

ISBN 978 1 4202 3736 8

GET STARTED: *IMAGINE*

Imagine you have just contracted influenza, usually known as the flu. You are feeling very unwell, and your body is starting to respond to the invasion of the virus that causes this illness.

1 Describe the symptoms you have.
2 What is a virus? Are viruses different from bacteria?
3 Look at the picture of the influenza virus on this page. Is a virus dead or alive? Imagine ways in which the virus can get into your body and reproduce itself.
4 You notice that your friend has spent a lot of time around you, but does not have the flu. She says she had the flu vaccine. What is a vaccine or vaccination, and how does it protect you?
5 Why is it that if you have a flu vaccine this year do you need a new one again next year?
6 You think, 'Hey, maybe I should visit the doctor and get some antibiotics!' Will antibiotics work against the flu? Explain why or why not?
7 How does your body respond to a disease and how do these responses help you to fight off the illness?
8 What are some other diseases you know of? Try to think about whether each is caused by bacteria or a virus, or some other pathogen (disease -producing organism).
9 What are some of the things you can do to ensure you do not contract an illness in future?

Influenza virus

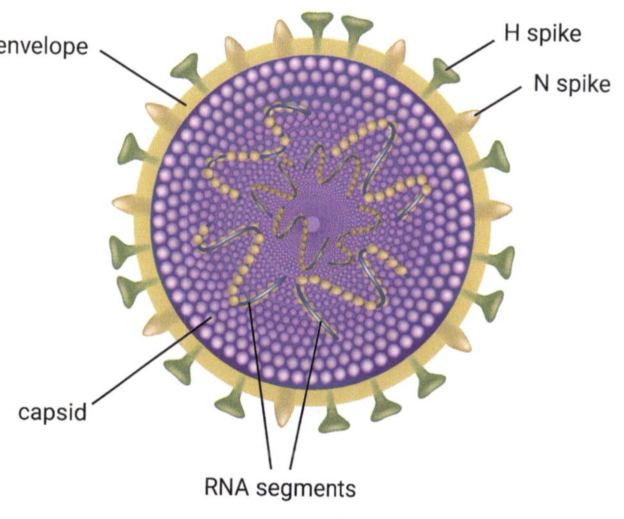

ISBN 978 1 4202 3736 8

5.1 Microscopic life

Microscopic organisms were first observed just over 300 years ago. Yet today most people have never seen one. We call microscopic organisms micro-organisms or **microbes**, and to observe them you need a microscope.

Most microbes belong to the Protist and Monera kingdoms. But some microscopic fungi such as yeasts and moulds are also classified as microbes.

Protists

Protists live in oceans, rivers, lakes and ponds or in the water in moist soil. Some protists are plant-like because they contain chlorophyll and can make their own food by photosynthesis. These protists are called *algae* (singular alga).

Other protists do not contain chlorophyll. They catch and eat food from the water around them. These protists are more animal-like and are called *protozoans*.

Many protists have structures on the outside of the cell to help them move through the water. For example, some have tiny hair-like *cilia* (SILL-ee-a) on the outside of the cell. These beat rhythmically and propel the organism through the water. Others have large whip-like *flagella* (fla-JELL-a), which act like paddles to move the organism.

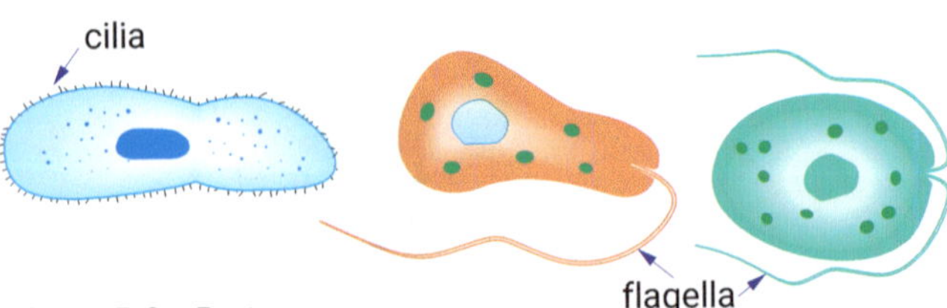

Figure 5.2 Protozoans are able to move.

Large protists

Not all protists are microscopic. Some protists, such as seaweeds or marine algae, are multicellular and can grow very large.

The brown alga called kelp, shown in the photo below, grows in the cooler waters off the southern Australian coast, and can grow up to 60 metres in length.

The kelp forests in the Southern Ocean and Atlantic Ocean provide food and protection for a large diversity of animals.

Figure 5.1 This green freshwater alga is made up of cells joined end to end like a necklace.

Figure 5.3 Kelp can grow up to 60 metres in length.

ISBN 978 1 4202 3736 8

Microscopic life in pond water

Aim

To observe and identify some of the microscopic organisms found in pond water.

Materials

- pond water
- small glass bowl (crystallising basin)
- microscope
- 3 cavity slides and cover-slips
- dropper
- methyl cellulose or gelatin solution (see Teacher Book for preparation)
- neutral red or methylene blue stain (see Teacher Book for preparation)

Risk assessment and planning

- Carefully read through the Method.
- Explain to your partner how to prepare a slide without getting air bubbles underneath the cover-slip.
- You will be given a small bottle of pond water with some sludge on the bottom. The water just above the sludge contains a lot of micro-organisms.
- What is the purpose of the methyl cellulose and neutral red in this investigation?

Note: To help you identify the micro-organisms in the pond water, your teacher may set up a camcorder on a microscope and send the images to a TV or computer.

Method

1 Add a drop of the neutral red or methylene blue stain to each of two microscope slides. Put the slides in a warm place and allow the stain to dry. You will use these in step 5.

2 Use the dropper to add one or two drops of the pond water to a slide. Then add a cover-slip. (Avoid air bubbles.)

3 Look at the slide under a microscope. Notice how quickly some of the organisms move.

4 To slow the organisms down, lift the cover-slip and add one drop of methyl cellulose or gelatin solution. Then replace the cover-slip.

How many different types of pond organisms can you observe? Use the diagrams on the next page to try to identify some of them.

5 Place a drop of the pond water on top of the dried stain. Then add a drop of methyl cellulose. Place a cover-slip on the slide.

The stain on the slide dissolves in the water and diffuses into the micro-organisms. The stain makes it easier to see cell structures such as nuclei, cilia and flagella.

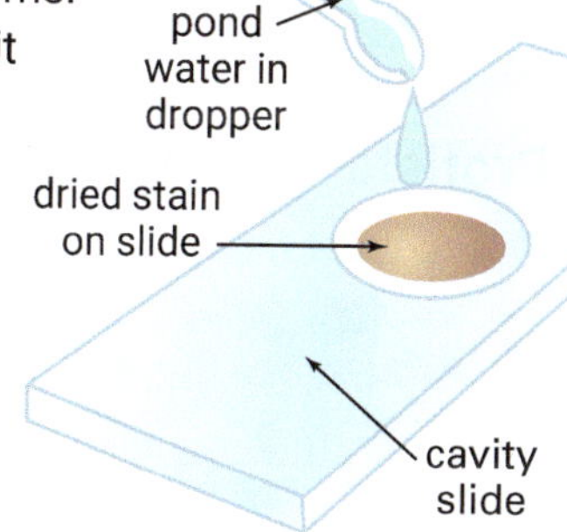

Discussion

1 Most of the micro-organisms that you observed are able to move. List the various methods of locomotion used by micro-organisms.

2 Using your observations, the internet and reference materials from the library, make a list of the ways micro-organisms capture and eat their food.

3 Explain how you could measure the size of the protists and microscopic animals that you observed in this investigation.

Check out **a virtual pond dip.**
Click on the images in the jar of pond water and read the factfiles about the micro-organisms.

ISBN 978 1 4202 3736 8

Micro-organisms in pond water

Protists—algae

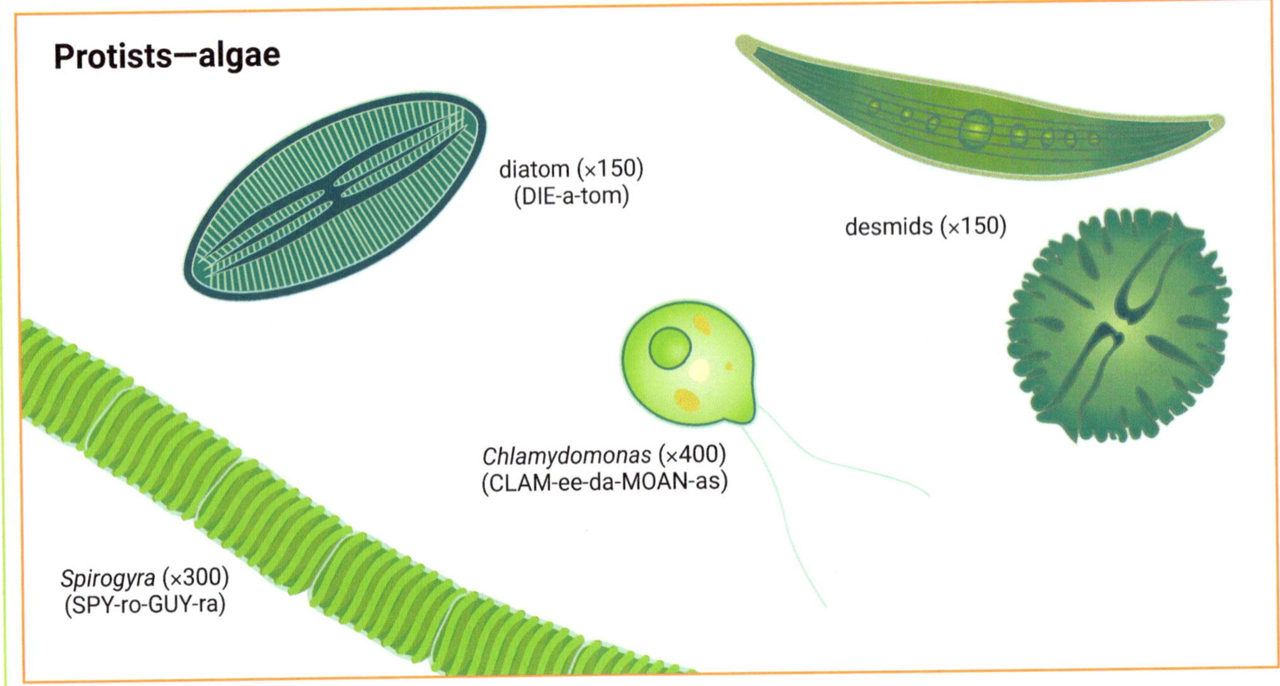

Protists—protozoans

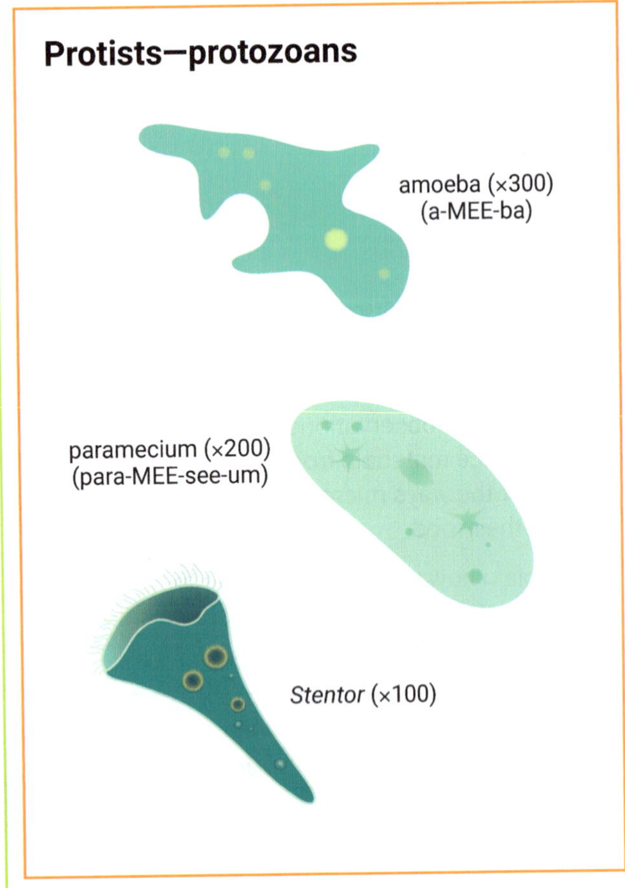

Microscopic animals

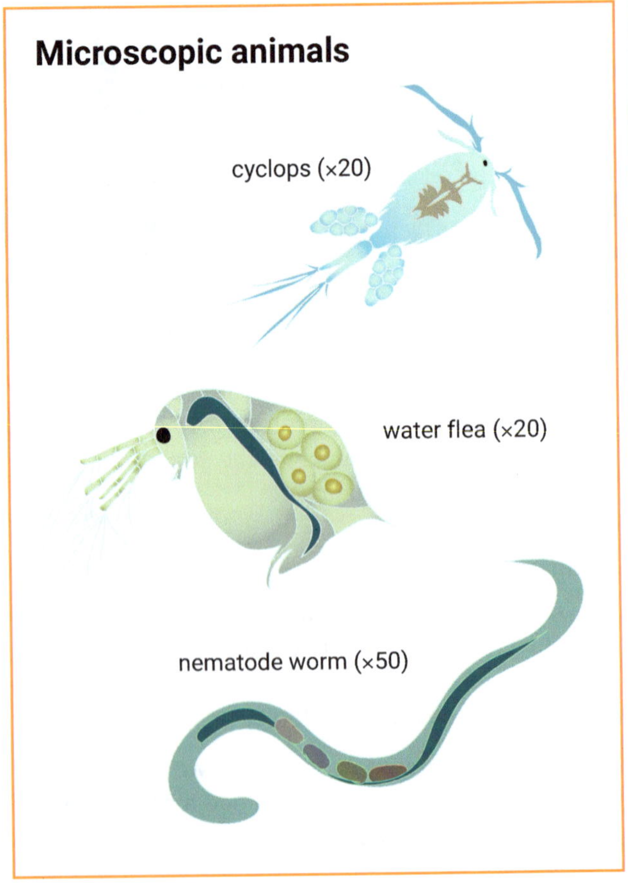

ISBN 978 1 4202 3736 8

Bacteria

Most monerans are unicellular and include **bacteria** and blue-green algae. Unlike other organisms, they have no distinct nucleus, and they have a very simple cell structure.

Bacteria are very small. Most are too small to be seen with an ordinary microscope. Very powerful electron microscopes are needed to observe their structure.

Bacteria can be classified according to their shape. There are rod-shaped, spherical and spiral types (Figure 5.4). Some of the rod-shaped and spiral bacteria have flagella that they use for movement. Some bacteria are joined together in chains, while others live in groups. They have a coating on the outside of their cell wall, which helps them to stick together and also to stick to other objects.

Bacteria live in all sorts of places. Many prefer places that are warm and wet. They need water in order to absorb oxygen and for reproduction. Most bacteria die if they dry out. This is why fewer bacteria are found in the dry air of deserts. Some types of bacteria are found in swampy areas, where they live on the decaying vegetation and produce methane gas (swamp gas).

Most bacteria rely on other organisms for food. They live on dead organisms that are gradually decomposed into simpler substances. For this reason, bacteria (as well as fungi) are called *decomposers*. Certain types of bacteria can make their own food using sunlight, others use chemicals such as ammonia or hydrogen sulfide as their energy source.

Some bacteria live inside other living organisms, and sometimes cause diseases. For example, in humans, bacteria cause diseases such as tetanus, food poisoning and cholera.

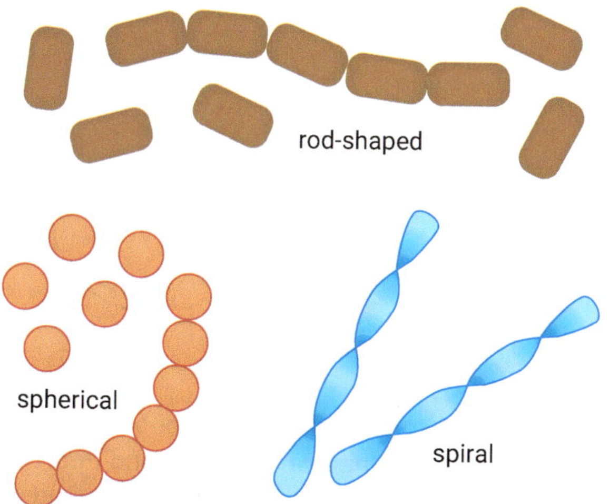

Figure 5.4 The shapes of the three types of bacteria

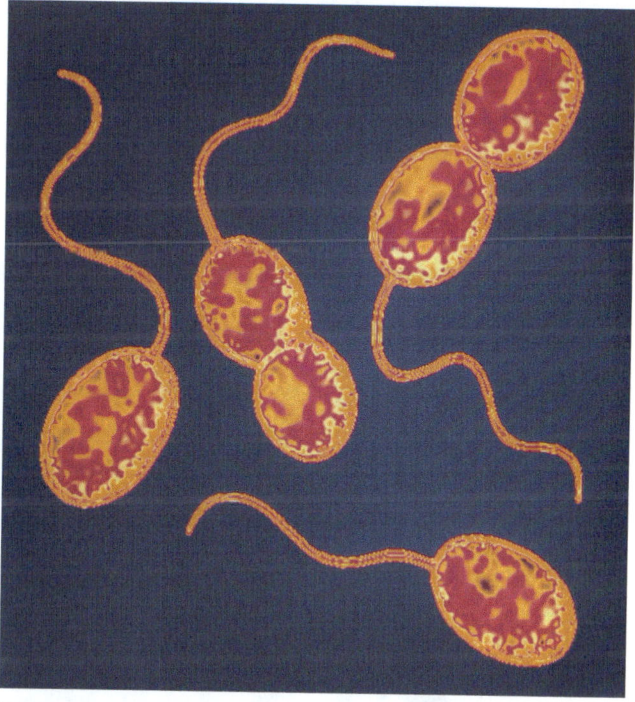

Figure 5.5 Some bacteria live in hostile places. These sulfur bacteria live in hot springs where the water temperature can reach 60 °C.

Figure 5.6 This rotting fruit is being attacked by bacteria and moulds (fungi). The fruit will gradually decompose and the broken-down material will be returned to the soil.

ISBN 978 1 4202 3736 8

INVESTIGATION 5.2 Growing microbes

Aim

To observe the growth of microbes on agar plates.

Materials

- 4 sterile Petri dishes containing nutrient agar
- sterile cotton bud
- adhesive tape
- sterile forceps
- marking pen
- soap for washing hands
- paper towel
- disposable gloves (optional)

Risk assessment and planning

- There are some very important safety issues to be aware of in this investigation. Carefully read through the Method and the other points in this Risk assessment and planning before you start.
- You should always wash your hands with soap and water before handling agar plates, as well as after the investigation.
- Because harmful bacteria could grow in the Petri dishes, you must not lift the lids of the Petri dishes until you are ready to use them. After use, they must be sealed with adhesive tape.
- The dishes will be put in a warm dark place below about 30 °C. If an incubator is available, do not set it above 30 °C. This will minimise the growth of bacteria, such as *E. coli*, which are harmful to humans.
- When you have finished the investigation, put the Petri dishes in the container provided by your teacher. Do not throw them in the waste bin.
- The jelly-like agar in the Petri dishes contains moisture and nutrients (similar to beef broth) which the micro-organisms use as food.
- Prepare a full-page data table for your observations in Method step 9.

Method

1 Take a cotton bud out of its wrapping and rub it over a desk or the floor.

2 Raise the lid of one Petri dish and rub the cotton bud over the surface of the agar as shown below.

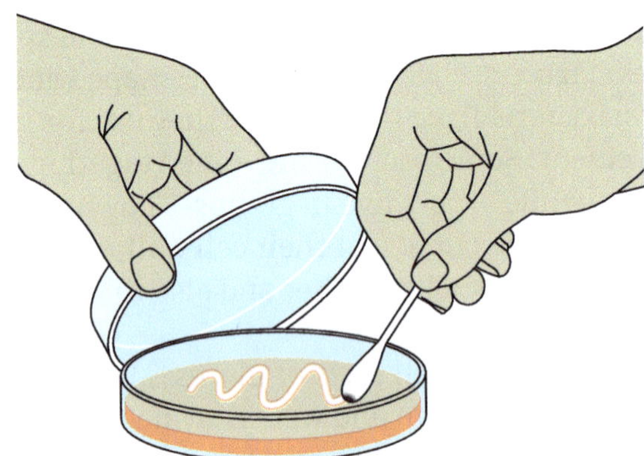

3 Immediately put the lid back on the Petri dish. Tape the lid on with adhesive tape.

4 Turn the Petri dish upside down. (This stops water drops from forming under the lid and falling on the agar.) Label it number 1.

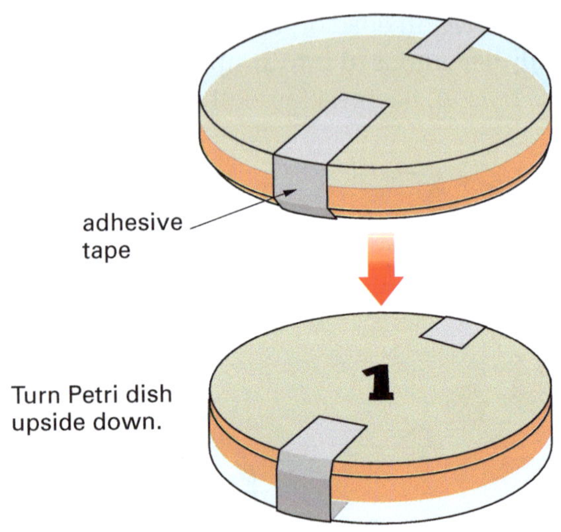

5 Use forceps to rub a dead insect or some dead grass over the agar in the second Petri dish. Tape the lid on, turn it upside down, and label it number 2.

ISBN 978 1 4202 3736 8

6 Open the lid of the third Petri dish and lightly stroke your fingers across the surface of the agar. Tape the lid on, turn it upside down, and label it number 3.

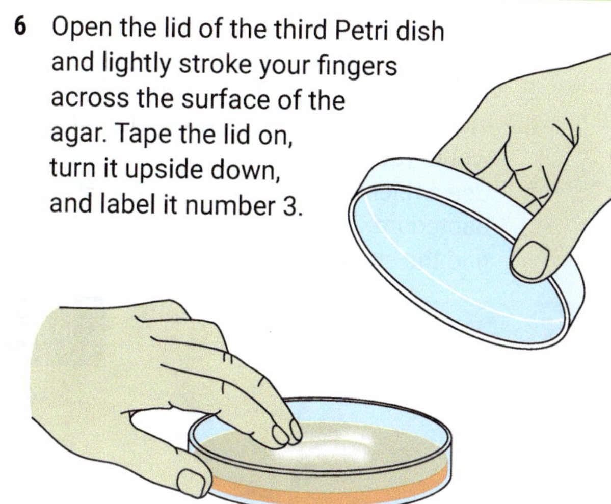

Lightly stroke the surface of the agar.

7 The fourth Petri dish is the **control**. Don't open it. Tape the lid on, turn it upside down and label it number 4.

> **Note:** A control is used to compare any changes that you observe in the other dishes with those that occur in your control dish. A control is a comparison.

8 Leave the dishes turned upside down in a warm, dark cupboard. Check them each day for the next five days. DO NOT REMOVE THE LIDS.

9 Read the information on the growth of microbes in the box below. Then prepare a full-page data table for your results. The table should include space for a sketch of each dish and a description of the colonies.

Record the growth in the Petri dishes each day. Sketch the four dishes and show the number, size, colour and texture of the colonies.

Discussion

1 Which Petri dish contained the fastest growing or largest colonies? Suggest a reason for this.

2 In Petri dish 1, do the microbes grow only on the places that were rubbed with the cotton bud? Why do you think this happens?

3 What was the reason for including Petri dish 4 in the experiment?

4 Which type of microbe colony became visible first—bacterial or fungal? Did this happen in each dish?

5 Do you think that the bacterial and fungal colonies would keep on growing in the dishes? Give a reason for your answer.

6 Predict how the results for Petri dish 3 would be different if you had not washed your hands before touching the agar.

Growth of microbes

The microbes that grow on the agar are bacteria and fungi. As they reproduce and grow in number, they form tiny dots or strips called colonies on the agar. These colonies become larger as the numbers of bacteria or fungi increase.

Bacterial colonies are usually shiny, smooth and sometimes coloured. Fungal colonies are fuzzy or furry. Yeast (a type of fungus) grows in a colony that looks like a spider's web.

The growth of colonies of bacteria and fungi increases as the temperature increases, but the Petri dishes in your investigation were kept at a temperature lower than human body temperature to reduce the risk of growth of harmful micro-organisms. The agar kept at this temperature also encourages a wider variety of micro-organisms.

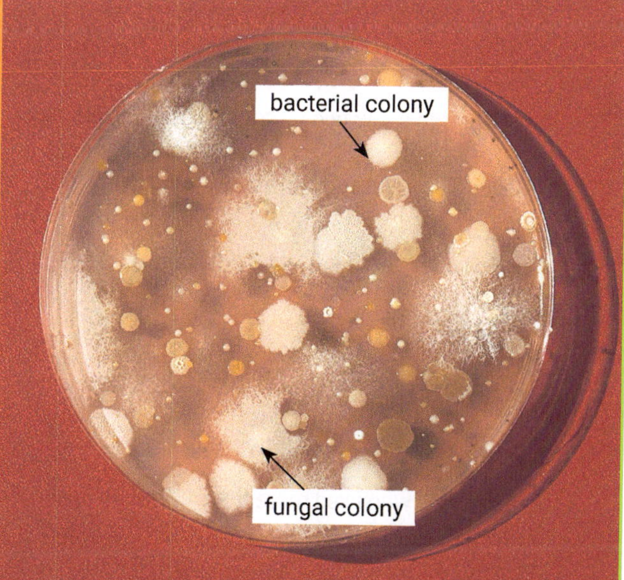

Figure 5.7 Bacterial and fungal colonies on an agar plate

ISBN 978 1 4202 3736 8

SW•9 Victorian Curriculum

CHECK

1 What does the word *microbe* mean? Which groups of organisms are usually referred to as microbes?

2 Use the diagrams to answer the questions below.

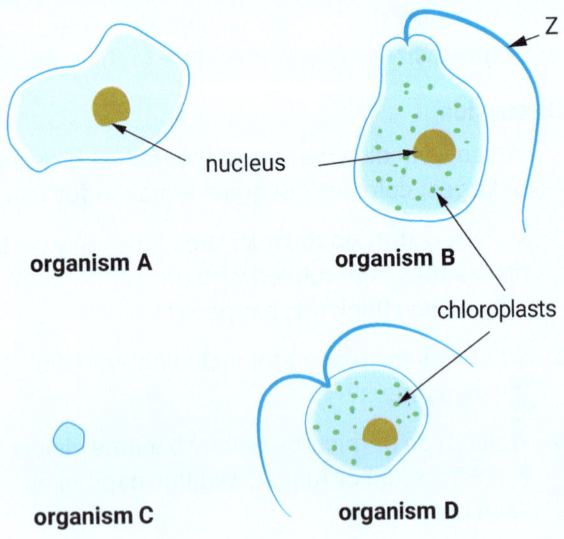

a Which organisms could be classed as protists? Give a reason for your answer.
b Suggest why organisms B and D do not have to rely on other organisms for food.
c Suggest a function for structure Z on organism B.
d Which organisms could be classed as algae? Why?

3 The photo below shows bacteria called *E. coli*, which are found in human waste.
a Into which of the three bacterial groups would you classify *E. coli*?
b Use a ruler to find the size of an average bacterium. Then divide this by 10 000 to find the size of a single bacterium.

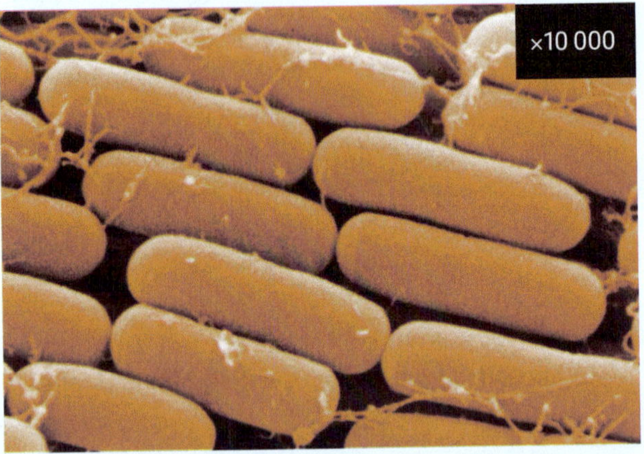

4 Algae have often been called the grass of the ocean. Suggest why.

5 Some algae such as Atlantic Ocean kelp can grow as tall as forest trees. Why is kelp classed as a protist and not a plant? What features do algae have in common with plants?

6 Suggest why the growth of bacteria is much more rapid in tropical rainforests than in areas such as the Simpson Desert.

CHALLENGE

1 The diagrams show two protozoans moving across a microscope slide. Suggest the method that each protozoan uses for movement.

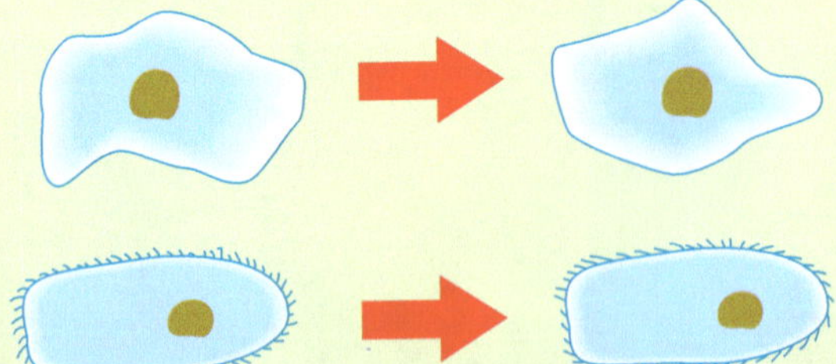

2 Some people who lived more than 400 years ago believed that micro-organisms existed even though they had not seen them directly. Suggest a reason for this.

3 In an investigation similar to the one on page 110, an extra Petri dish was added. The lid of this Petri dish was lifted off for 15 seconds and then replaced and sealed with tape. After 5 days a small number of colonies were observed growing in the Petri dish. Write an inference to explain this observation.

4 Meat that is left out of the fridge 'goes off' quickly. Meat that is put in the fridge is edible for 2 or 3 days. Meat that is kept in a freezer is edible for up to six months.
Explain in terms of microbes the differences between these three situations.

ISBN 978 1 4202 3736 8

5.2 Helpful microbes

Some microbes play an important role in the environment by decomposing dead organisms. But microbes are important in other ways. Certain bacteria are found in your gut and help digest various types of foods. Some bacteria and fungi are used to make foods such as cheese, yoghurt, bread, beer and wine. Fungi are valuable in producing medical drugs such as antibiotics.

Decomposers

All large organisms have microbes on them and inside their bodies. They usually live there quite harmlessly and are kept in check by the body's defences. But when an organism dies, its microbe-fighting ability stops and it is attacked by many different microbes. They decompose the organism by using the carbohydrates, proteins and fats in its cells for food. In the process, these large complex molecules are broken down into smaller ones such as carbon dioxide, water and nitrates. Some of these small molecules are gases, which are given off into the air and smell foul. Gradually the body's cells are destroyed and the organism decays.

Decomposers recycle matter in the environment. Elements such as oxygen, hydrogen, carbon, nitrogen and phosphorus that are found in the body of an organism are released when it dies and decomposes. Molecules containing these elements can be absorbed by plants and algae and re-used for growth. If the plants and algae are eaten by an animal, then the elements are used in the animal's body. When it dies and decomposes, the elements are released once again.

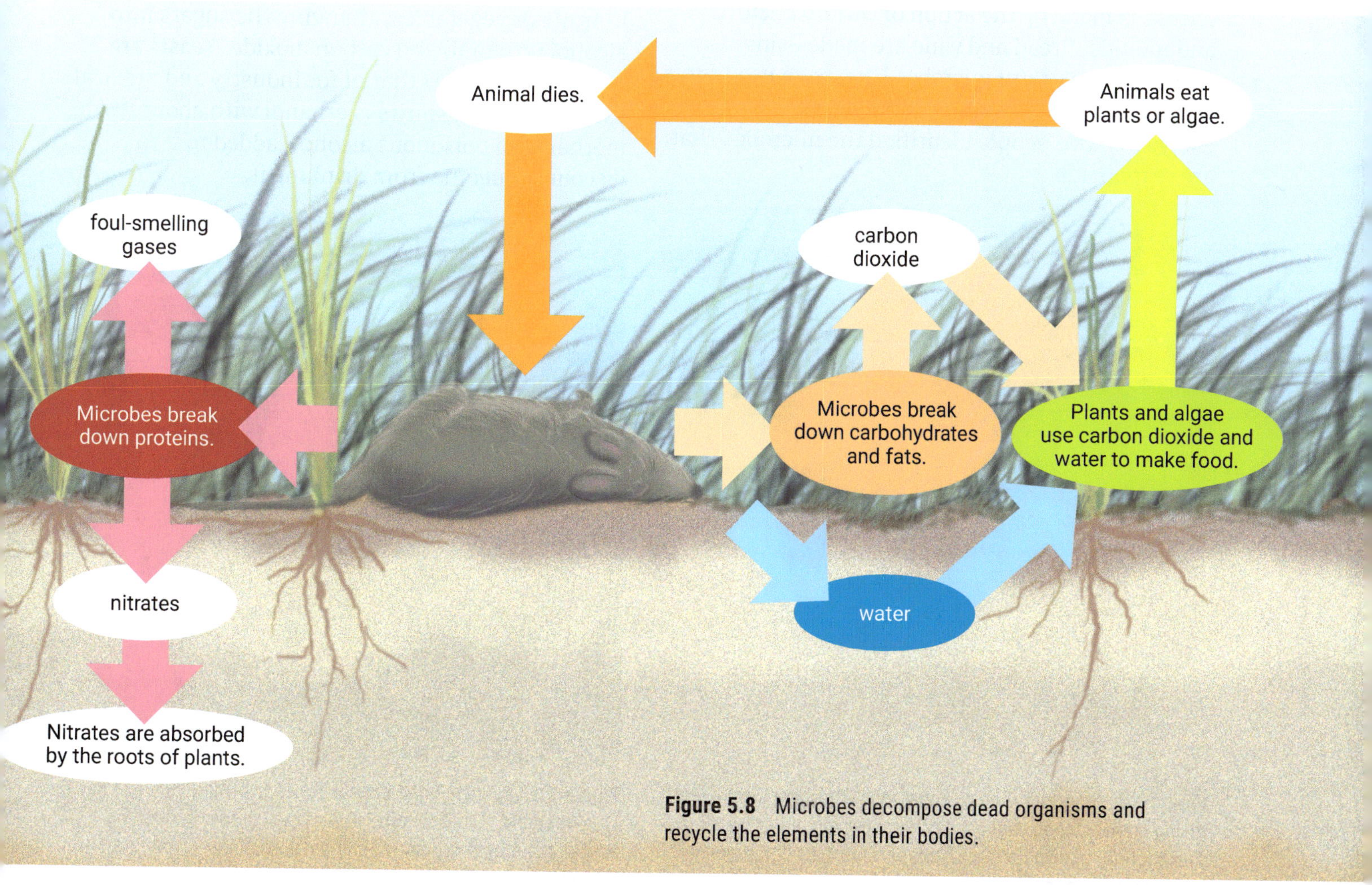

Figure 5.8 Microbes decompose dead organisms and recycle the elements in their bodies.

ISBN 978 1 4202 3736 8

These activities can be done at home or in the lab.

A compost heap

1. Half fill a large cardboard box (e.g. an apple box) with some fresh lawn clippings.
   Record the temperature in the middle of the clippings.
2. Leave the clippings for 1 or 2 days.
   Record the temperature in the middle of the pile. Then open up the pile and describe any changes that have occurred to the grass.
3. Continue observations for about a week.
   Write a summary of your observations.

Decomposing bread

1. Place a small piece of bread in a Petri dish and tape the lid on.
2. Add a similar piece of bread to another Petri dish but this time add a few drops of water to the bread. Tape the lid on.
   Observe the changes to the bread each day for 4 days. Use a hand lens if necessary.
   Suggest why you added water to the bread in the second dish.

Making foods

Various types of bacteria, yeasts and moulds are used in the food-making industry. For example, cheese is made by the action of certain bacteria and moulds. Bread and wine are made using yeasts. These useful microbes have been used for thousands of years, although only in the last 250 years have people identified the microbes that are responsible.

Figure 5.9 Bacteria feed on sugar in milk turning it into acid, giving unsweetened yoghurt its characteristic sharp taste. The bacteria shown here are *Lactobacillus*, one of the many bacteria used in making yoghurt.

Yeasts are used in the making of bread and fermented drinks such as beer, wine, cider and ginger beer. In the process of **fermentation** (FUR-men-TAY-shun), yeasts feed on the sugars in fruits or vegetables, changing the sugars into alcohol (ethanol) and carbon dioxide. Yeasts are also used to make ethanol for industry and medical use. Methylated spirits is ethanol with about 10% methanol (a poisonous alcohol) added to it to discourage people from drinking it.

Figure 5.10 Blue-veined cheese is made by adding a special type of edible mould to milk. During the six-month ripening period, the mould grows in patches or colonies, giving the cheese its characteristic blue streaks.

ISBN 978 1 4202 3736 8

Fermentation occurs in the making of bread as well as in the making of alcoholic drinks. Does this mean there is alcohol in bread? When yeast is added to a mixture of flour, sugar, vegetable oil and water, the yeasts feed on the sugar and produce carbon dioxide and alcohol. The dough rises as the trapped carbon dioxide bubbles expand the dough. When the dough is baked, the heat from the oven evaporates the alcohol, which escapes into the air.

Alcohol from bread

During the baking of a loaf of bread, small amounts of ethanol (alcohol) are given off and released into the air. With millions of loaves of bread being baked each day throughout the world, the amount of ethanol released into the air is considerable. Recently the US Environmental Protection Agency investigated the effect of ethanol on the destruction of ozone in the atmosphere. They suspect that ethanol is an ozone-destroying gas, and suggest that bakeries trap the ethanol so that it is not released into the air.

Microbes in the gut

Humans need small quantities of vitamins in their daily diet to maintain healthy bodies. This is because the human body, unlike plants and microbes, is unable to make all of the vitamins it needs. For example, plants, microbes and many animals are able to make vitamin C, whereas humans, monkeys and guinea pigs have to eat foods containing this vitamin. Fortunately, microbes in your gut help you by making some of the vitamins you need.

At most times of the day your intestine contains digested food. Bacteria that live harmlessly in your intestine feed on some of the sugars and proteins that you have digested. In return, they produce many types of vitamins, which are absorbed by your blood. These useful bacteria are often killed by antibiotics, and it takes many days to repopulate your gut.

Animals that eat grass, such as cows, also rely on microbes for their wellbeing. Cellulose is a carbohydrate found in plants, and this substance cannot be digested by animals. Grass-eating animals have special structures in their gut that store the partly digested grass and allow bacteria to digest the cellulose. The products of this digestion are then absorbed into the bloodstream of the animal (see Figure 5.11).

Stopping the decay of food

Have you ever left some milk out of the fridge on a hot summer's day? It soon goes 'off' and smells sour. This is because microbes have been growing in the milk and giving off waste products that have an unpleasant taste and smell.

Food must be preserved if it is to be left for some time before being eaten. This can be done in the following ways:

- heating
- refrigerating
- freezing
- drying
- adding chemical preservatives
- irradiating.

These methods either kill the microbes or prevent them from causing the food to decay.

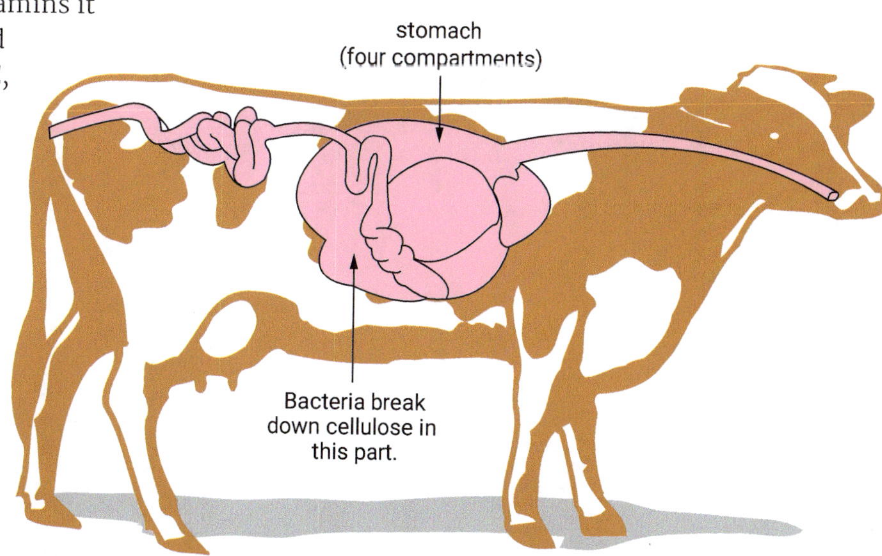

Figure 5.11 The cow, a grass-eating animal, has a stomach with four compartments. In the largest compartment, bacteria and protozoans break down the cellulose in the grass to simple sugars.

ISBN 978 1 4202 3736 8

Heating, refrigerating and freezing

There are two main types of heat treatments. The first is called heat sterilisation and is used in the manufacture of most canned and bottled foods. In this process, the foods are heated under pressure to a temperature of 120 °C. This kills all microbes and their spores.

However, some foods cannot be heated to this temperature because it would destroy the food. Milk, for example, is *pasteurised* to kill harmful bacteria. In this process, the milk is heated to 72 °C for 15 seconds and then cooled to below 10 °C.

Cooling foods in a refrigerator preserves food for short periods. Fermentation slows down as the temperature decreases. Similarly, the speed of the decay reactions in foods slows down as the temperature decreases. Putting food in the refrigerator does not kill microbes; it simply slows down the decay of food.

When food is frozen, the water in the food turns to ice and destroys the cells of most microbes. However, some microbes can grow a hard coating around them to form inactive spores that are resistant to freezing. When the food thaws, these microbes emerge from the spores and can attack the food.

Type of food	Method of preserving
Canned and bottled food	Heat sterilisation: heating under pressure at 120 °C for 15 minutes
Milk, juice	Pasteurisation: heating to 72 °C for 15 seconds, then cooling to below 10 °C
Milk, cheese, butter, eggs, fresh fruit and vegetables, meat	Refrigeration: cooling to 4 °C
Meat, vegetables, processed foods	Freezing: storing between −10 °C and −18 °C

Figure 5.12 In the process of canning and bottling, the food is sealed in the container and heated to kill harmful microbes.

Figure 5.13 This factory produces juices. After the juice is strained, it is pasteurised to kill bacteria as it travels through these heated coils. This means the juice can be stored without refrigeration.

ISBN 978 1 4202 3736 8

Drying

Drying foods to preserve them has been practised for thousands of years by many cultures throughout the world. Meat, fish, vegetables and fruits were traditionally dried in the sun. This technique produced food that could be carried when people travelled over long distances before refrigeration was available. For example, people in northern Africa sun-dried dates, figs and other fruits when travelling in this arid region.

When the food is dried, micro-organisms are unable to grow and reproduce because they need water for these processes.

Adding chemical preservatives

Salt, sugar and vinegar are common household substances used to preserve foods. Microbes cannot grow in foods that have a high salt or sugar content. This is why jams and honey can be stored at room temperature for longer periods than fresh fruit.

Harmful microbes are also killed by acids in vinegar and fruit juices made from lemons and limes. So adding vinegar (pickling) to vegetables such as cucumbers and onions, or to meat, will preserve them for long periods of time. Fermented drinks such as wine and beer also contain chemical preservatives.

Figure 5.14 Different types of fish and seafoods in this Mexican market have been preserved by drying them in the sun.

ACTIVITY

Preserving food with chemicals

How effective are sugar, salt and vinegar in preserving foods? Design an experiment to show how these substances help to preserve food.

Hints:

- Use pieces of bread, slices of apple, or potato.
- Add a few drops of water to the foods to moisten them.
- Use small jars with lids or plastic take-away containers with lids.
- Make sure you include a control.

Work in a group of two or three people and show your draft design to your teacher before doing the test.

Supermarket survey

Which methods of preserving are used for the food you buy in the supermarket? Do a survey of your local supermarket to find out which methods of preserving are used for the food you usually buy each week.

Present your results in a data table.

For each food, suggest an alternative method of preserving it. Include this in your data table.

Also check the 'use by' dates on the foods and work out the period of time people have to buy and consume certain products.

ISBN 978 1 4202 3736 8

Irradiating

Harmful microbes in food can be killed by sterilising them with radiation. Many medical drugs are sterilised in this way. In many countries fruit and vegetables, preserved food and seafood such as oysters are sterilised by radiation before they are exported or sold.

Food irradiation is a process that is promoted by people who believe that it is an effective way to kill micro-organisms in certain foods and to reduce the amount of poisonous chemicals currently used. Other people believe that the radiation that the food is exposed to will destroy nutrients in the food and produce harmful and cancer-forming chemicals in the food.

Food irradiation debate

In November 2000, the Australian and New Zealand Food Authority (ANZFA) received its first application from an Australian company to use radiation to kill micro-organisms in herbs, nuts, oils and tea. In 2003, irradiation of certain tropical fruits was also approved.

ANZFA is a government organisation formed to protect the health and safety of people through the maintenance of a safe and nutritious food supply. Irradiation of foods is currently prohibited in Australia unless it is approved by ANZFA.

Use the websites on this page to find out more about food irradiation.

Some websites have opposing views on irradiation. Use them to list the arguments for and against irradiation of foods.

Your teacher may organise you into groups to prepare notes and discuss this topic, so that you can debate whether or not radiation should be used on foods.

EXPLORE ONLINE

Follow the links to the following websites.

Food irradiation (Better Health)

Food safety

What's wrong with food irradiation

Food irradiation (Food Standards)

You could also search for other websites by typing *food irradiation* in the search box.

1. For each of the words below, write a sentence to show that you understand its meaning.

   decomposer — fermentation

   recycle — preserve

2. What are some of the large, complex molecules broken down by microbes when they decompose a dead organism? What substances are produced? How are these substances used?
3. Using your knowledge of food preservation, explain each of the following statements.
   - a Powdered milk has a 6-month use-by date, whereas fresh milk has a 3-day use-by date.
   - b Pickled meats could be eaten for up to 12 months by sailors on board old sailing ships.
   - c A complete mammoth, 20 000 years old, was found in a Siberian glacier with its skin and hair intact.
   - d An unopened can of beans processed in 1909 was discovered in an old house and was found to be edible.
4. The photo shows a close-up view of a piece of bread. Describe how the action of microbes gives the bread this texture.

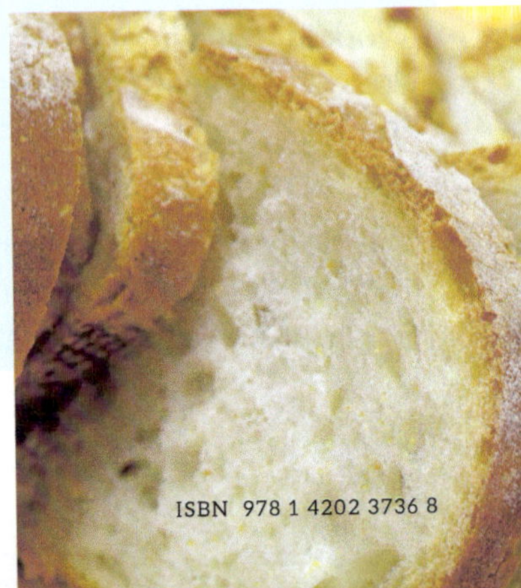

ISBN 978 1 4202 3736 8

5 Zola has a splinter in her finger. Which of the following actions should she take before she uses a sewing needle to remove the splinter? Give reasons for your answer.
- A Wipe the needle with a clean cloth.
- B Wash the needle in hot soapy water.
- C Pass the needle through a flame on the gas stove.

6 Explain the differences between heat sterilisation and pasteurisation. Give examples with your answer.

7 Why are microbes in your gut important for your wellbeing?

8 Write a short story to describe what would happen on Earth if decomposing microbes suddenly vanished.

CHALLENGE

1 Some uncooked chicken is left on the kitchen bench for a number of hours on a warm summer's day.
- Jed says that it will be all right to eat if it is put in the fridge before it is cooked.
- Paige says it should be frozen in the freezer before it is eaten because this kills microbes.
- Jack says it will be all right to eat if it is cooked thoroughly now.
- Kate says that it is not safe to eat and should be thrown away.

Whose suggestion do you agree with? Give reasons why you disagree with the others.

2 Sparkling white wine is made from a type of green grape. However, unlike other white wines, it is allowed to ferment in the bottle. Try to explain each of the following observations about sparkling white wine. Suggest how microbes are involved.
- a The cork is wired onto the mouth of the bottle.

- b The cork pops when it is removed.
- c Bubbly liquid often pours out when the cork is removed.

3 Sophie was experimenting with yeasts. She placed a tube of yeast, sugar and water in a water bath at 35 °C, and a similar tube in a water bath at 20 °C. She used an instrument to measure the amount of carbon dioxide released from each tube. The graphs below show her results.
- a Suggest the aim of Sophie's experiment.
- b In which tube was the average rate of reaction greater? Explain your answer.
- c Suggest why the graph for the tube at 35 °C has a different shape at B than at A.

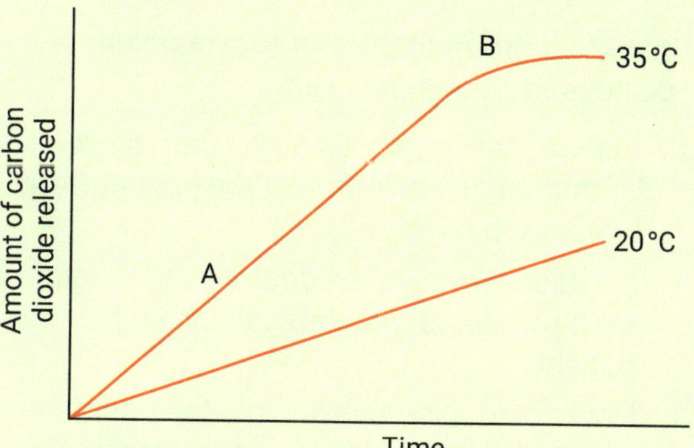

4 Both baby food and milk are treated to remove harmful microbes before they are sold to consumers. Compare and contrast the ways in which the two foods are treated.

5 Give the pros and cons for irradiation of foods. Do you think that foods treated by radiation should be labelled? Why?

ISBN 978 1 4202 3736 8

5.3 Microbes and disease

ACTIVITY

The photo below shows a patient and surgical staff in a hospital operating theatre.

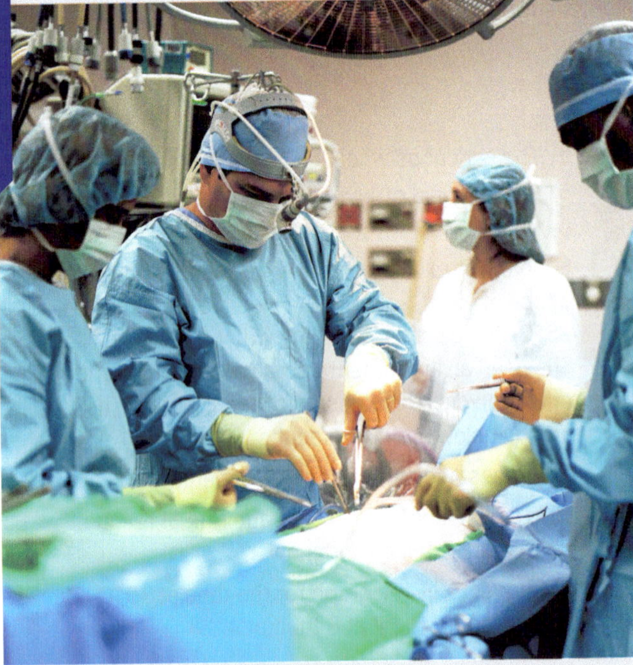

Work in a group of three or four people to discuss the questions below.

1 List as many methods as you can think of that are used to protect the patient from infection by microbes.
2 In which ways are the doctors and surgical staff protected from infection from the patient?
3 How could microbes get into the operating theatre? How would you reduce or stop their entry into the operating theatre?

Causes of disease

Disease is a word used to describe the poor health of your body. There are many different types of diseases, and many different causes.

Diseases such as measles, chickenpox, ringworm, tetanus, malaria and the common cold are caused by microbes. These diseases are called **infectious diseases**. Figure 5.15 shows a child with the infectious disease chickenpox. Infectious diseases are 'caught' by skin contact with an infected person, by breathing air containing the microbes, or by breathing in the water drops in sneezes and coughs from an infected person. Once the disease-causing microbe is inside your body, it multiplies rapidly in the warm, moist conditions there. Often a poison, or **toxin**, given off by the microbe will damage or destroy your body cells and make you sick.

Other diseases (such as haemophilia, asthma, multiple sclerosis, leukaemia and some heart diseases) are not caused by microbes.

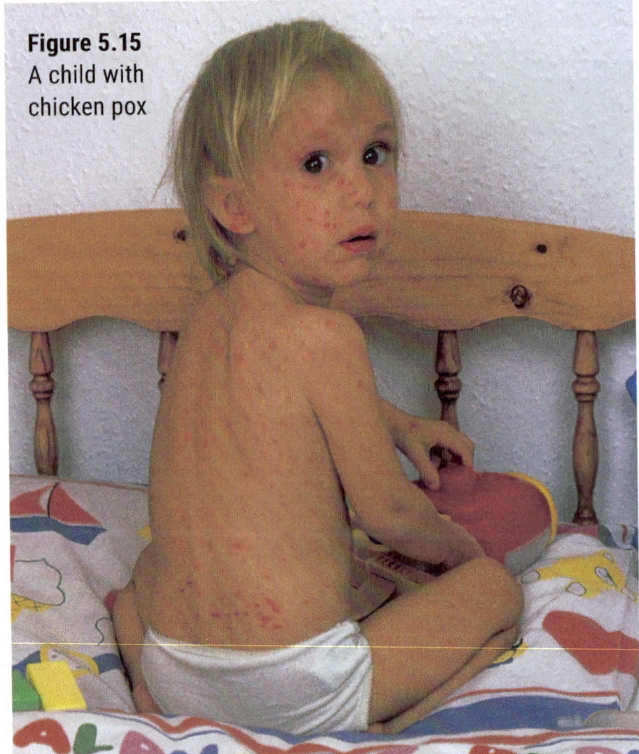

Figure 5.15 A child with chicken pox

Viruses

Most people have heard of viruses, and associate them with common colds and influenza. **Viruses** are extremely small and are not made up of cells. They can form crystals like non-living matter. For these reasons, biologists argue about whether or not they are living organisms, and into which group they should be classified. However, for convenience biologists call viruses microbes.

ISBN 978 1 4202 3736 8

Viruses are completely parasitic—they can only live in or on another organism. They use the materials inside the organism's cells to make new viruses, and often destroy the cells in the process. Viruses infect all living things—even bacteria. They are much smaller than bacteria, so they are easily carried about in the air. They are also found in water and in the soil.

Insects can spread viruses from one organism to another. Most plant diseases are caused by viruses carried on the bodies of insects or in their saliva. For example, when aphids or leafhoppers carrying a certain disease-causing virus bite into a plant to suck the sap, the viruses flow out with the saliva and into the plant's cells and can infect those cells.

Rabies is a deadly viral disease that affects some animals including dogs and humans. When an infected dog bites you, the viruses invade the cells in your body and spread rapidly in the blood. If untreated, the disease is usually fatal. Luckily rabies is not found in Australia.

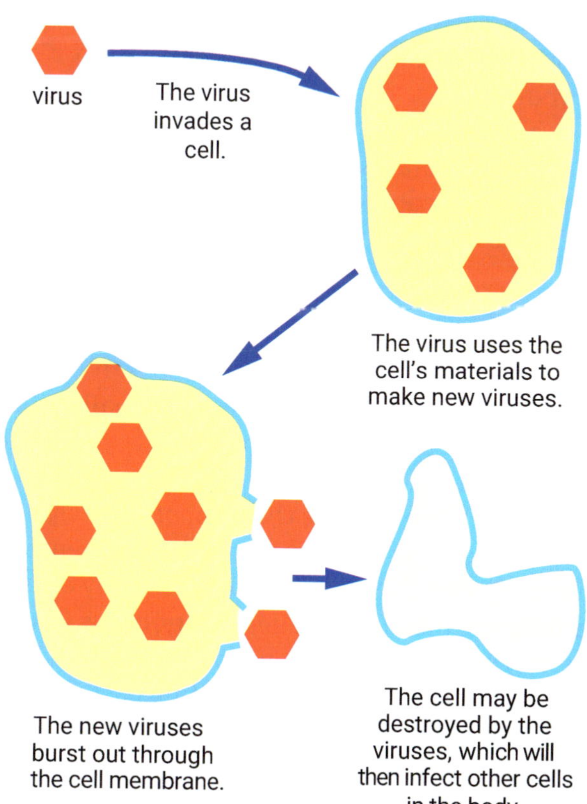

Figure 5.16 A virus reproduces inside a cell by taking control of the cell's DNA and using the cell to make copies of itself.

The table below shows some of the diseases in humans caused by microbes.

Type of microbe	Disease
bacteria	tetanus, pneumonia, cholera, botulism (food poisoning), gonorrhoea, syphilis, gangrene, boils, bubonic plague, tuberculosis, acne, bacterial meningitis
viruses	influenza, common cold, rubella (German measles), measles, chickenpox, warts, cold sores, polio, HIV (AIDS), rabies, smallpox, genital herpes, foot-and-mouth disease
fungi	ringworm, thrush, athlete's foot
protozoa	malaria, sleeping sickness, amoebic dysentery

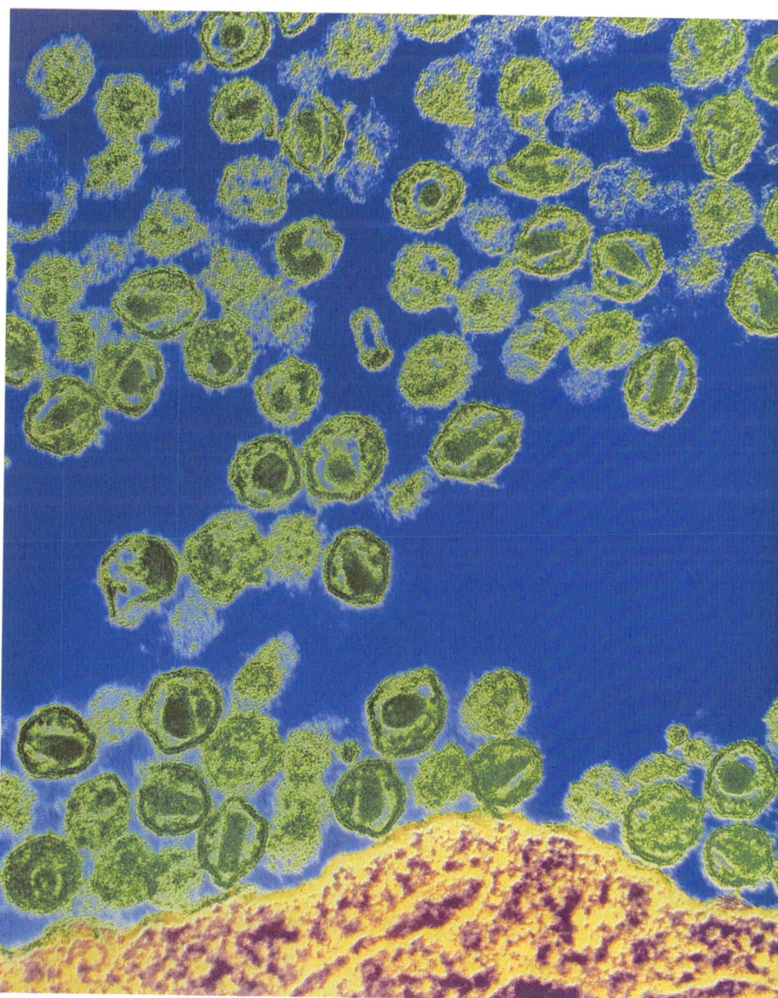

Figure 5.17 An electron micrograph of HIV (the virus that causes AIDS). The core of the virus (dark green) contains the genetic material.

ISBN 978 1 4202 3736 8

SCIENCE AS A HUMAN ENDEAVOUR

Finding the cause of cholera

In 1854, an English doctor, John Snow, was worried by the high death rate of people with the disease cholera (KOL-er-a). This disease causes constant vomiting, cramps and diarrhoea, dehydration and death. Snow thought that the disease could be caused by the drinking water found in the wells throughout London. Others thought that it might be passed from person to person like influenza, or carried by animals such as rats and fleas that carried the plague.

Snow recorded the names and addresses of his patients who had died from cholera on a map of his district in London (•). He also recorded the location of the wells where the people had obtained their drinking water. His map is shown on the right.

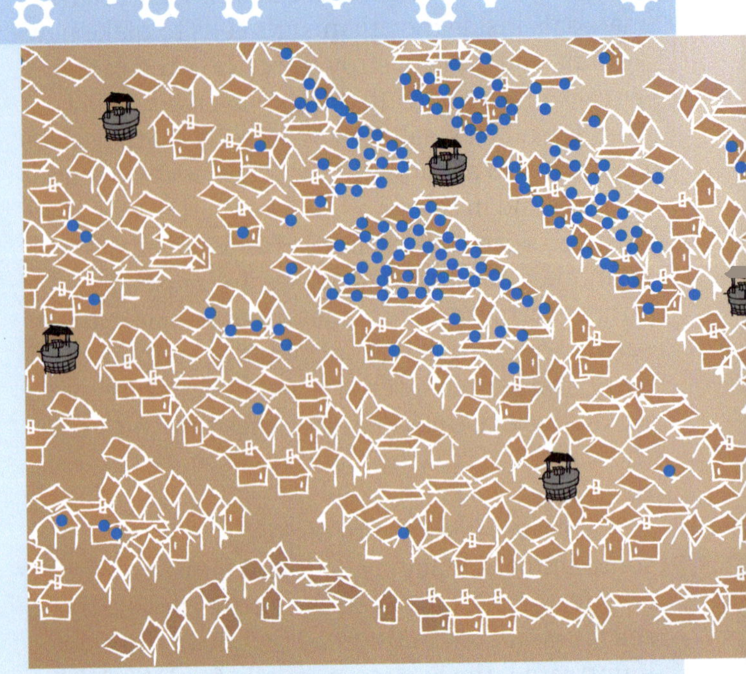

Work in a group of three or four people to discuss the following questions.

Questions

1 What pattern would you expect if Snow was correct?

2 If rats carried the disease, what pattern would you expect?

3 Does the data support Snow's inference that cholera is spread in the drinking water? Is the other inference possible? Give reasons for your answer.

4 What tests could you do to show that the drinking water is responsible for the disease. List the variables that you would need to control in these tests.

Fighting disease

All organisms have ways to fight disease-causing microbes. For humans these ways are fairly well understood.

The first line of defence against invading microbes is our skin. The skin is not a very good habitat for microbes. The sweat that keeps the skin moist contains salt and acids, which most microbes cannot tolerate. Microbes usually enter the body through the nose or mouth or directly into the blood through a wound in the skin. The warm, moist areas in the nose, mouth and lungs are protected from infection by liquids produced by the cells lining the nose and mouth. These substances kill most harmful microbes. If the microbes pass into the gut, they are usually killed by the strong acid produced by special cells in the stomach lining.

If microbes penetrate the skin, other body defences come into action. When bacteria enter a wound, they release toxins that kill body cells. Large white blood cells called **phagocytes** (FAG-o-sites) move out of broken blood vessels and squeeze between the cells, engulfing bacteria and dead body cells. After engulfing and destroying bacteria, many of the phagocytes burst, releasing their contents. This, together with the other liquid around the wound, is called *pus*.

ISBN 978 1 4202 3736 8

The composition of blood

Blood consists of red blood cells, white blood cells and platelets suspended in a pale straw-coloured liquid called *plasma*. Platelets are very small cell fragments that play an important role in blood clotting.

Red blood cells

Red blood cells contain haemoglobin, an iron-containing substance that carries oxygen. They are made in huge numbers in the bone marrow but only live for about 100 days.

White blood cells

White blood cells are part of the body's defence system and are present in smaller numbers than red blood cells. There are different types of white cells. Some, such as phagocytes, move about through the tissues of the body, removing and destroying bacteria. Other white blood cells make antibodies.

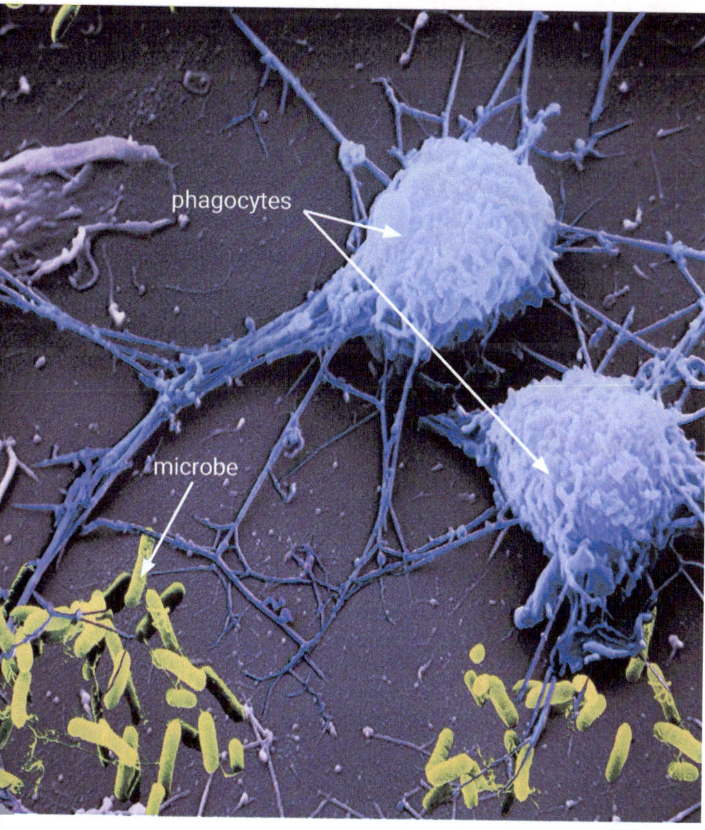

Figure 5.18 Phagocytes are large, mobile white blood cells. They are able to move through blood vessels and between cells. These phagocytes are attacking some *E. coli* bacteria.

Antibodies

If microbes get into the blood, they can be quickly carried to all parts of the body. Other types of white blood cells then swing into action.

These white blood cells make proteins called **antibodies**. There are many types of antibodies, and each is designed to attack the surface of a particular microbe or destroy the toxins it releases. For example, if your body is invaded by the virus that causes measles, antibodies are made to attack it. The antibodies join onto the virus and stop it entering your cells. Phagocytes then engulf and destroy the virus.

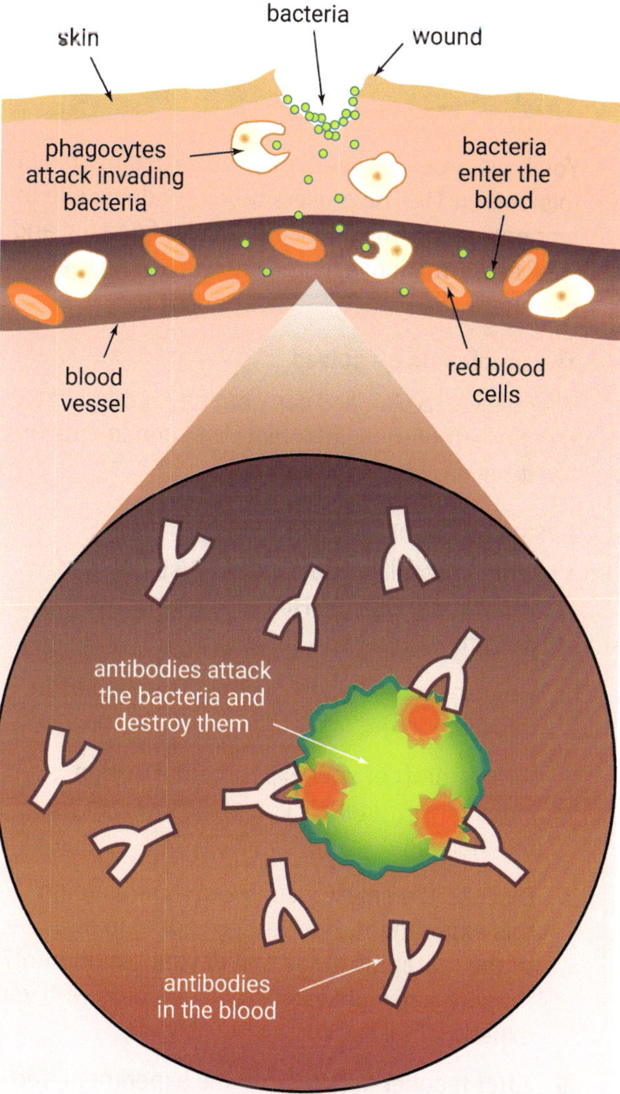

Figure 5.19 How the body reacts to the invasion of microbes

ISBN 978 1 4202 3736 8

EXPERIMENT 5.1 Controlling bacterial growth

You have probably seen many advertisements like this one on TV. The makers of the products like to emphasise the importance of 'scientific tests' and 'new powerful active ingredients' in selling their product.

The problem to be solved

Your task is to plan and carry out an experiment to test the effect of disinfectants and antiseptics on bacterial growth.

Designing the experiment

1. Work in small groups and discuss the tests you will do. Use the Hints to help you with your design.
2. How are you going to control the variables in this experiment?
3. Write a draft of your experimental design, including a materials list, and show it to your teacher.
4. Discuss the safety issues and risks in doing this experiment. How are you going to dispose of the materials at the end of your experiment? Prepare a risk assessment sheet to go with your experimental design.
5. After teacher approval, do the experiment and write a full report of your findings.

Hints

1. Re-read Investigation 5.2 on page 110.
2. Rub a cotton bud over the desk, floor or other surface for 10 seconds to collect bacteria. Then rub the cotton bud over the agar on the dishes as you did in Investigation 5.2.
3. Select two or three antiseptic or disinfectant solutions and read the directions on how to dilute them for use (if necessary).
4. You can use small discs cut out from filter paper soaked in disinfectant solutions. Arrange the soaked filter paper discs on a Petri dish as shown below.
5. Remember to seal the lids on the dishes and store them upside down.

Extending the experiment

You might like to test whether the concentration of a disinfectant has an effect on the growth of bacteria.

Inquiry

The *Mythbusters* TV team investigated the myth that the toilet seat in a house has many more bacteria on it than a kitchen floor.

Design an experiment to prove or disprove the myth.

ISBN 978 1 4202 3736 8

Controlling infections

There are five main ways to control disease-causing micro-organisms.

1 *Immunity and vaccination*

Why is it that some people catch the flu while others don't? Those people who do not contract the disease are said to be immune to it. Immunity to a certain disease is due mainly to the presence of specific antibodies in the blood. The antibodies react rapidly with the invading microbes and reduce their numbers in the blood.

Your body can be 'tricked' into producing antibodies by injecting dead microbes or treated toxins into your blood. The substance is called a *vaccine* and the process is called **vaccination**. The white blood cells that make antibodies recognise the foreign substance and make antibodies to destroy them. These antibodies remain in your blood and protect you in the event of further infection.

Generally, a healthy body is able to defend itself against attack by most disease-causing microbes. You feel unwell for a few days while your body makes the antibodies to combat the infection. However, some diseases can be life-threatening and you have to take drugs.

How vaccines work

Modern vaccines are made from heat-treated bacteria and viruses or from the toxins they produce. The treatment kills the micro-organisms and reduces the risk of infection. However, the body still recognises the foreign organisms or their toxins and makes antibodies to combat them. In this way, your body builds up an immunity against that particular disease. (See Figure 5.20.)

Immunisation programs run by government health departments require preschool children to be immunised against tetanus, whooping cough, diphtheria, poliomyelitis and also measles, mumps and rubella.

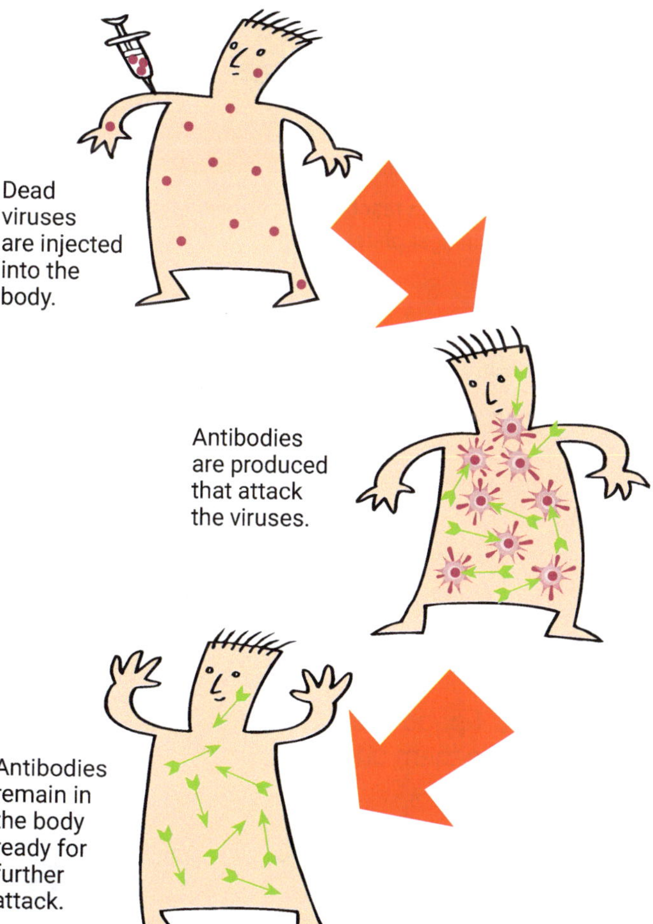

Figure 5.20 How some vaccines work

Discovering immunisation

For centuries smallpox was a dreaded killer disease. In 1796, English country doctor Edward Jenner observed that people who caught cowpox, a fairly harmless disease, rarely caught smallpox. He injected some cowpox pus from a girl into eight-year-old James Phipps. James caught cowpox but recovered normally. He then injected some pus from a patient's smallpox sore into James. This was a very risky thing to do, but James did not catch smallpox, nor did he contract the disease for the rest of his life. Jenner had found a way to increase the body's immunity against disease by vaccination.

ISBN 978 1 4202 3736 8

2 Disinfectants and antiseptics

In Experiment 5.1 you investigated disinfectants and antiseptics. *Disinfectants* are chemicals such as chlorine bleach that kill bacteria on objects. *Antiseptics* are used to kill bacteria on the skin. One of the oldest and most commonly used is alcohol. Others include iodine, soaps and detergents.

3 Hygiene

Many diseases are spread through untreated water or by human waste that is not removed from living places. These diseases include cholera, typhoid fever and hepatitis A.

4 Isolation

Some diseases such as chickenpox and measles are very infectious. A person with chickenpox has to stay at home (isolated from non-infected people) to stop the spread of the virus.

5 Antibiotics

Antibiotics are the most commonly used drugs to fight infections. Antibiotics are substances that actually stop the growth of bacteria inside the body. They are produced naturally by microbes, particularly fungi. Chemists have copied the chemical structure of some of these natural antibiotics, and now they are made artificially. However, there is a concern that some microbes are becoming resistant to antibiotics, and it is important that doctors prescribe them sparingly. See page 127.

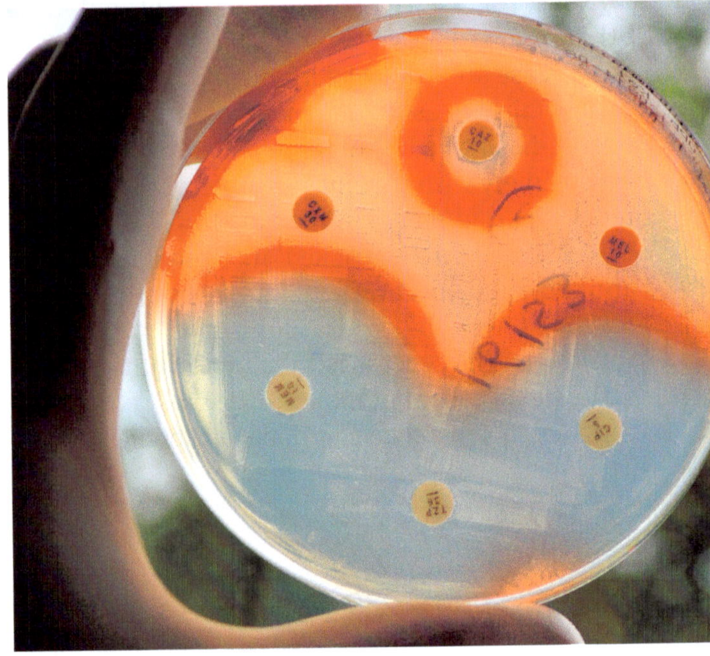

Figure 5.21 Bacteria reacting with different types of antibiotics on an agar plate

ACTIVITY

Glandular fever

Glandular fever or infectious mononucleosis is caused by a virus that can be transmitted from person to person by direct contact with saliva. For this reason it has been called 'kissing disease'.

The virus is very common, and most people have been exposed to it during their life. Infected young children usually show no or few symptoms. Teenagers and young adults are more likely to become sick if they are infected.

Symptoms

The symptoms of glandular fever include a fever, sore throat and swollen lymph glands in the neck, under the arms and in the groin. People usually feel constantly tired and are generally unwell.

How is it diagnosed and treated?

A doctor will perform a blood test to determine whether the virus is in the blood. However, because the disease is caused by a virus, antibiotics will not be effective.

Most healthy people will recover from the infection in about 4 to 6 weeks. However, the virus remains in the body for life.

Task

Produce a 1-page flyer for patients with glandular fever to help them understand more about this disease.

Check out these websites, using the links provided:

Mononucleosis

Glandular fever

ISBN 978 1 4202 3736 8

SCIENCE AS A HUMAN ENDEAVOUR

Antibiotics

Probably the best known substances used against infectious diseases caused by bacteria are antibiotics. They are produced naturally by micro-organisms to stop the growth of other micro-organisms—a bit like chemical warfare. Most antibiotics used in medicine are produced by fungi. For example, the green mould that grows on oranges is a fungus called *Penicillium*. This fungus produces a chemical that kills bacteria and other fungi.

Many antibiotics used in medicine are now synthetic. Chemists use the known molecular structure of natural antibiotics as models to make new ones.

How antibiotics work

Some types of antibiotics, e.g. penicillin, prevent bacteria from making new cell walls. This stops the bacteria from reproducing and therefore stops further infection. The body's antibodies and phagocytes attack the remaining bacteria, and a sick person feels better in a few days.

Other types of antibiotics, e.g. chloromycin, pass into the bacterial cell and interfere with the processes that make proteins. This kills the bacteria.

Antibiotics cannot kill viruses since viruses do not have a cellular structure, nor do they have a cell wall. Also the processes in viruses are different from those in bacteria, and are unaffected by antibiotics.

Discovering penicillin

In 1928, Alexander Fleming returned to England from a short holiday in Scotland to find that *Penicillium* mould had contaminated some Petri dishes containing bacteria. He found that the bacteria had been killed where the mould was growing.

Ten years after this chance discovery, Australian scientist Howard Florey led a large team of scientists who finally purified penicillin. After successful clinical trials, the antibiotics were mass-produced and were used to treat wounded soldiers in the last year of World War II.

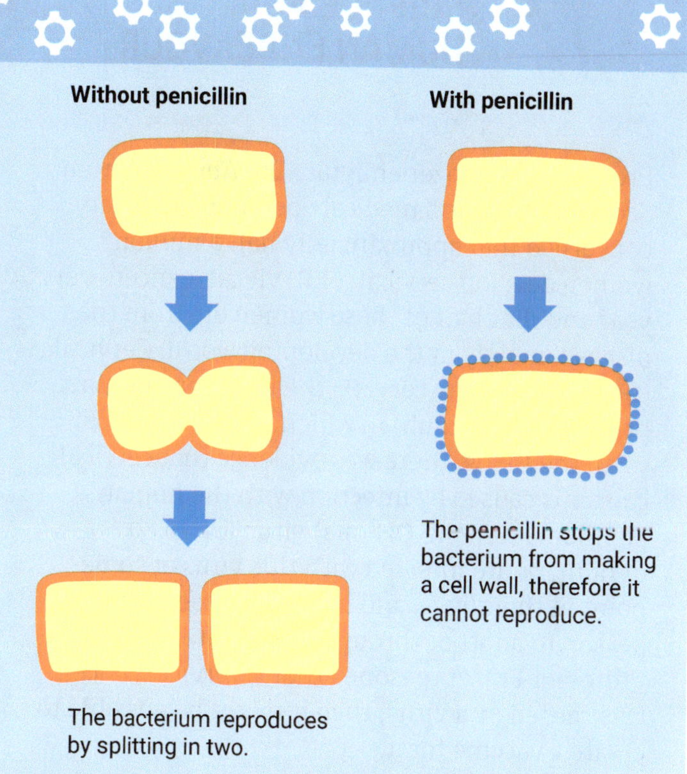

Figure 5.22 Penicillin stops bacteria reproducing. This stops the spread of the bacterial infection.

Howard Florey was knighted in 1944, and he, Fleming and Ernst Chain were awarded the Nobel Prize for Medicine in 1945. Florey was honoured by having a suburb of Canberra named after him.

To find out more about antibiotics, penicillin and the scientists who worked in this field, go to some of the following websites.

EXPLORE ONLINE

Follow the links to the websites below.

Antibiotics
This website contains detailed information about antibiotics as well as questions and answers.

Florey and penicillin
This website describes how penicillin was discovered and developed and how it is used to treat diseases.

Fleming discovers penicillin
This website describes the work of Fleming and Florey in discovering and producing penicillin.

ISBN 978 1 4202 3736 8

SCIENCE AS A HUMAN ENDEAVOUR

Cancer vaccine

Professor Ian Frazer emigrated to Australia from Scotland. He was a medical doctor and was very concerned that approximately half a million women develop cervical (SER-vi-cal) cancer every year and that half of these women die from the disease, mostly in the developing world. Cervical cancer affects the cervix—the lower narrow part of the uterus (womb) in which a baby develops.

In the 1980s, there was evidence that cervical cancer is caused by infection with the human papilloma (PAP-il-OH-ma) virus. Up to 80% of humans are infected with this virus at some point in their lives, and it can be passed from one person to another through sexual intercourse. Professor Frazer reasoned that if cervical cancer was caused by a virus, then it might be possible to create a vaccine for it.

The papilloma virus is a tiny sphere covered in spikes (see the computer art below). The inside of the virus is the part that is infectious, so Frazer didn't want that in the vaccine. He wanted to make a particle that looked like the virus so that the body's immune system would respond as if it had been infected by the real virus.

He worked with another scientist, Dr Jian Zhou, but couldn't get the outer spiky shell of the vaccine to grow in the laboratory. After much trial and error, they were very excited in 1991 to see the bumpy virus particles in electron microscope pictures. Frazer and Zhou recorded their findings and quickly lodged a patent application. The vaccine was tested on laboratory animals, and then on young women throughout the world. It was found to be very successful in preventing infection. The vaccine was then mass-produced by a pharmaceutical company and released as Gardasil™ in August 2006. The photo above shows Professor Frazer giving the first vaccine to Therese Raft in Sydney. Tragically his co-researcher, Dr Zhou, died in 1999 and did not see the end result of his hard work.

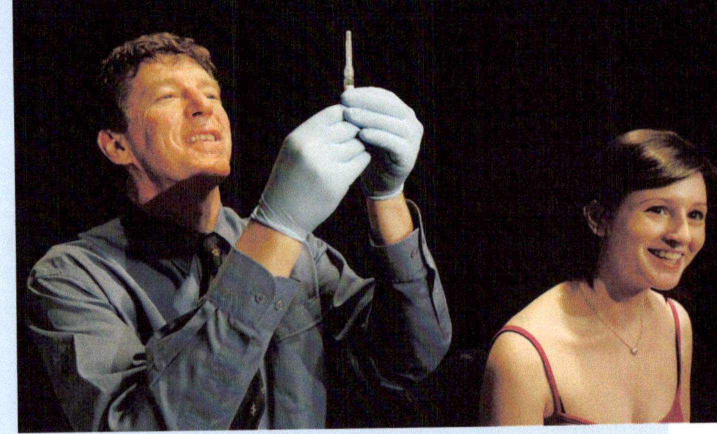

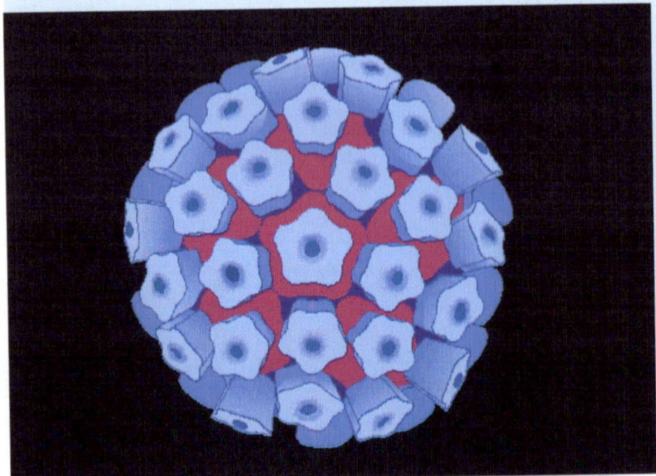

The current vaccine cannot help women who are already infected, but Professor Frazer and his team at the University of Queensland are working to improve the vaccine. They are also working on a vaccine to prevent one form of skin cancer that is caused by the same human papilloma virus.

Questions

1. What development in science led to the development of a vaccine for cervical cancer?
2. Use the information on this page and page 125 to explain how the cervical cancer vaccine works.
3. Why do you think it was important for Frazer and Zhou to lodge a patent application so quickly?
4. Why did it take 15 years from discovery to the marketing of the vaccine?

ISBN 978 1 4202 3736 8

1 Some of the sentences below are *false*. Select the ones that are false and rewrite them to make them correct.
   a Measles, haemophilia and influenza are diseases caused by microbes.
   b Viruses are non-cellular.
   c Viruses can reproduce in the air, water or soil.
   d Antibodies are chemicals made by microbes and they can destroy cells.

2 How do bacteria differ from viruses?

3 Write sentences to show that you know the meaning of the following words:

toxin	engulf	antibody
virus	immune	antibiotic

4 Using a flow chart, describe the events that occur from when bacteria enter a wound to when they are destroyed. The first part of the flow chart is done for you.

Bacteria enter the body through a wound. 

5 Chickenpox and mumps are both viral diseases spread by droplets in the breath, cough or sneeze of an infected person. Your young brother catches chickenpox. Over the next few weeks, your sister catches chickenpox, but you, your older brother and parents do not. Both your parents had chickenpox when they were children.
   a Suggest how your young brother caught chickenpox.
   b Suggest why your young brother and sister caught chickenpox, while you and your older brother did not.
   c Suggest why your parents did not catch chickenpox.
   d Why is it likely that your young brother will not catch chickenpox at some time in the future, but could catch mumps?

1 It is common for influenza to spread rapidly through the population during winter.
   a Suggest why this viral disease spreads so rapidly.
   b Why are older people and certain 'at risk' people advised to have influenza vaccinations? Who would you define as an 'at risk' person?

2 Use the graph below to answer the questions.

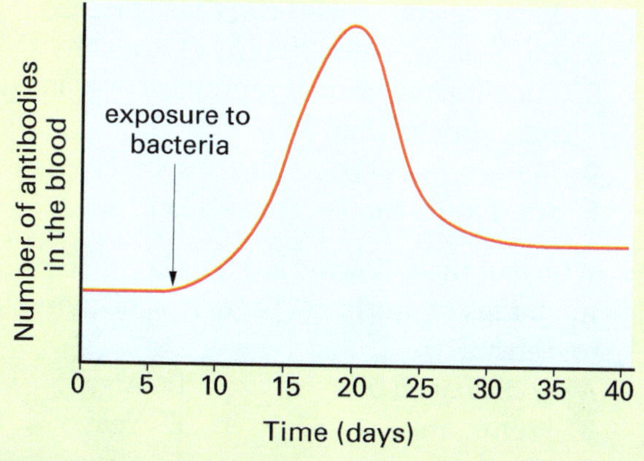

   a Describe the body's reaction when it is exposed to a bacterial infection.
   b How long after exposure to the bacteria is it before the body has built up maximum antibodies in the blood?
   c Suggest why the level of antibodies drops after this time.
   d How does the graph show that the body has an immunity to further attack by this particular bacterium?

3 Some diseases are called *notifiable diseases*, and the doctor treating the infected patient has to contact the government health department with the patient's details. These infectious diseases include HIV (AIDS), hepatitis B and C, rabies, measles and polio. Suggest why these measures are taken in Australia.

4 Use the internet to find out why people who have had hepatitis B and C cannot donate blood. Does this apply to hepatitis A?

ISBN 978 1 4202 3736 8

MAIN IDEAS

Copy and complete these statements to make a summary of this chapter. The missing words are on the right.

1 ______ are microscopic organisms that can be seen only with a microscope. They include ______, single-celled protists and fungi such as ______ and moulds.

2 Most bacteria are harmless and help ______ matter in the environment by ______ dead organisms.

3 Some microbes are used in the ______ industry, for example to make cheese, bread, beer, wine and yoghurt.

4 The ______ of food by microbes can be slowed down or stopped by heating, ______, freezing, drying, adding preservatives or irradiation.

5 ______ such as tetanus, malaria and influenza are caused by microbes.

6 ______ are non-cellular and can grow only inside an organism's cells. They cause diseases such as polio, the common cold, measles and chickenpox.

7 In humans, foreign microbes are destroyed by large white blood cells called ______ and by ______ in the blood.

8 Infectious diseases can be controlled by ______, isolating the infected person, ______ and by the use of drugs such as ______.

antibiotics
microbes
bacteria
refrigeration
decomposing
food-making
yeasts
infectious diseases
antibodies
phagocytes
good hygiene
recycle
vaccinations
decay
viruses

CH•5 REVIEW

1 In which of the following places would you expect to find microbes? (There may be more than one answer.)
A human skin
B garden soil
C volcanic lava
D the air
E the ocean
F human blood
G on a sterile needle
H milk

2 Microbes cannot be seen because they:
A move too rapidly.
B are transparent and you see through them.
C live beneath the soil and in the water.
D are too small to be seen with your eyes.

3 Describe the ways in which harmful bacteria are stopped from infecting your body.

4 Which of these statements about bacteria and viruses are incorrect? (There is more than one answer.)
A Bacteria and viruses can cause disease.
B Bacteria are cellular; viruses are not.
C Bacteria and viruses reproduce only inside other organisms.
D Many types of bacteria are decomposers.
E Antibiotics kill bacteria and viruses.

5 Which of the following are not classified as monerans or protists? (There may be more than one answer.)
A Rod-shaped bacteria
B Protozoans
C Viruses
D Algae
E Yeasts
F Seaweed

ISBN 978 1 4202 3736 8

6 During fermentation, which of the following substances are produced?
   A carbon dioxide and vitamins
   B carbon dioxide and alcohol
   C carbon dioxide and water
   D alcohol and vitamins

7 Look at the items in the photo below. Describe the methods used to preserve each of the foods.

Use the following information to answer questions 8–10.

Debbie was testing the action of yeast on a sugar solution. She covered the test tubes with a balloon to catch any gas given off. The data tables show her tests and her results.

Tests

Tube 1	Tube 2	Tube 3
Yeast + sugar + water Kept at 40 °C	Yeast + sugar + water Kept at 20 °C	Yeast + water Kept at 40 °C

Results

Tube 1	Tube 2	Tube 3
Balloon filled up	Small amount of gas in balloon	No gas in balloon

8 Which of the following is least likely to have affected the rate of reaction?
   A the temperature of the solutions
   B the size of the balloons
   C the amount of sugar in the solution
   D the amount of yeast used

9 Why was there no gas produced in tube 3?

10 What was the aim of Debbie's test? What was the purpose of tube 3?

11 Josep and Helena set up two Petri dishes that contained bacteria that had been mixed with the agar beforehand. Petri dish 1 contained bacteria A, while Petri dish 2 contained type B. They then soaked filter paper discs in two types of antiseptics, X and Y, and placed them on the agar jelly. The diagram below shows their results after four days. There are no bacteria growing in the clear areas around the discs.

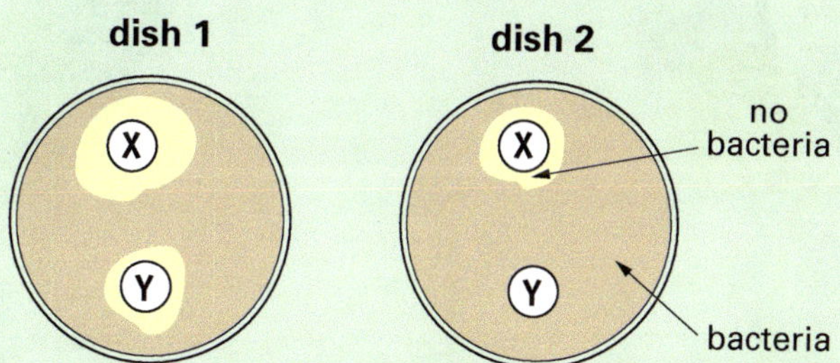

   a Why do you think the bacteria were mixed with the agar beforehand, rather than using the rubbing method you used in Investigation 5.2 (page 110)?
   b Write a conclusion for the experiment.
   c Josep concluded that antiseptic X was more effective than antiseptic Y in killing bacteria. Do you agree with his conclusion? Explain.

12 Your friend Kate tells you that she has chickenpox. You had chickenpox when you were much younger, but your other friend Belinda has never had chickenpox. Why is there a greater chance that Belinda will catch the disease than you?

Check your answers on page 300.

ISBN 978 1 4202 3736 8

IN THIS CHAPTER

Science Understanding

> recognise that the chemical properties of a substance will affect its use

> list ways of recycling and reusing plastics and discuss ways of increasing the amount of recycling

> discuss possible solutions to the problem of plastic shopping bags

Science Inquiry Skills

> investigate the physical and chemical properties of common plastics

> discuss ways of improving an experiment on rusting

> explain the properties of hydrocarbons, detergents, plastics and fibres in terms of their chemical structure

CH•6 Everyday substances

ISBN 978 1 4202 3736 8

GET STARTED: *EXPLORE*

Look carefully at the cartoon below. You will see many different objects made of many different substances.

> Draw up a table like the one below and list as many different objects as you can and the substances you think they are made of. One example has been done for you. You may not be able to complete the right-hand columns yet, but you should be able to by the time you have finished this chapter.

Object	Substance it is made of	Useful properties of this substance	How the substance is made	Can it be recycled?	If recyclable what could it be made into?
kitchen sink	*stainless steel*	*strong, does not rust*	*from iron obtained from iron ore*	*yes*	*steel objects*

ISBN 978 1 4202 3736 8

6.1 Metals

In certain places in the Earth's crust we find metal compounds called ores, which can be mined and the metals extracted. The ways these metals are used depend on their properties, as shown in the photos on this page.

Figure 6.1 Pure copper is easily drawn into wires, and is an excellent conductor of electricity. So it is used for electrical wiring.

Figure 6.2 Because aluminium is strong and light and resists corrosion, it is used in boat building.

Alloys

Pure iron is of little use for making things because it is too soft and stretches easily. It also rusts very easily. When a little carbon (about 0.5%) is mixed with it, it becomes much harder and stronger. It is then called *steel*, and is used in buildings, bridges, ships and car bodies. When nickel and chromium are mixed with iron, it becomes even harder and no longer rusts. This substance is called *stainless steel*, and is used for kitchen sinks, tableware and surgical instruments.

These metal mixtures are called **alloys**. They have properties different from those of the metals they are made from. These properties depend on the proportions of each metal used. There are thousands of different alloys, and some of these are listed in the table on the next page.

Figure 6.3 The shininess of gold makes it an excellent reflector of light and other radiation. This is why gold mirrors are being used in the James Webb space telescope.

ISBN 978 1 4202 3736 8

Alloy	Made from ...	Special properties	Uses
stainless steel	70% iron 20% chromium 10% nickel	does not stain, corrode or rust	kitchen sinks, tableware, surgical instruments
duralium	96% aluminium 4% copper	light and strong	aircraft parts
brass	70% copper 30% zinc	shiny, hard and does not corrode	musical instruments, car radiators, ornaments
aluminium bronze	92% copper 6% aluminium 2% nickel	harder than brass, does not corrode	'gold' coins
cupro–nickel	75% copper 25% nickel	light, hard and does not corrode	'silver' coins
solder	60% tin 40% lead	soft, low melting point	joining wires and pipes
Nitinol	55% nickel 45% titanium	if bent it slowly returns to its original straight shape, does not corrode	dental braces

Use the table to answer these questions.

1. Which metals are used to make duralium?
2. Name a lead alloy.
3. Which of the alloys in the table contain copper?
4. What differences are there between the alloys used to make 'gold' and 'silver' coins?
5. Suggest why stainless steel is more expensive than ordinary steel.
6. Suggest how the Nitinol wire used in dental braces straightens teeth.

Figure 6.4 Nitinol dental wire is used to straighten teeth.

ISBN 978 1 4202 3736 8

Corrosion

When a metal reacts with air, water or other substances in its surroundings, new substances are formed. This process is called **corrosion**. Iron and steel corrode in damp air to form rust (brown iron oxide), which has properties very different from the metals.

When left exposed to the weather for many years, copper becomes coated with a greenish compound. Aluminium becomes coated with aluminium oxide, but this thin coating protects the metal from further corrosion. Gold (impure), silver and brass slowly become tarnished (discoloured), especially in cities where there are acidic gases in the air. This is why metal objects need to be cleaned regularly.

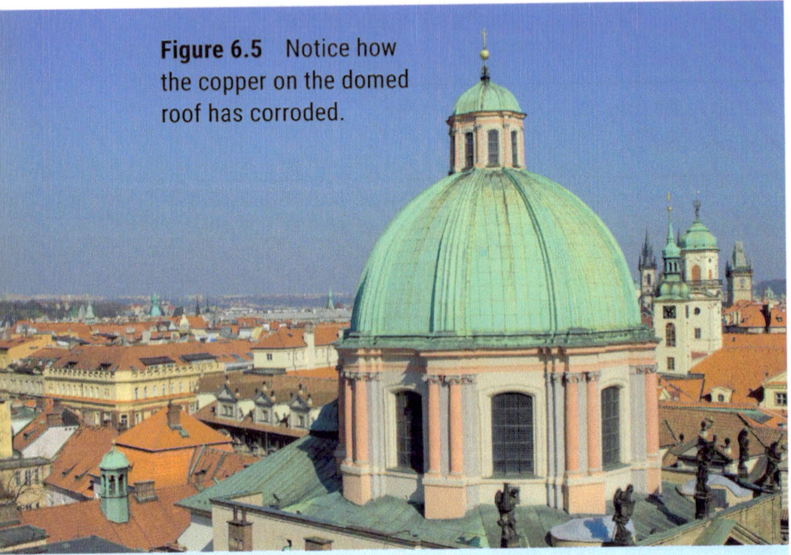

Figure 6.5 Notice how the copper on the domed roof has corroded.

How to stop rust

When iron is made into stainless steel it does not rust. But stainless steel is expensive to use in large quantities, so other methods of protection are used. These usually involve coating the metal with some unreactive substance to keep out air and water.

1 Painting

Steel exposed to the weather has to be painted regularly. Paints that contain zinc are often used.

2 Coating with oil, grease or tar

Objects such as bike chains and the insides of car engines, which cannot be painted, are protected this way.

3 Coating with plastic

Garden chairs, dish racks, tennis court fences and the inside of steel cans are protected in this way.

4 Coating with metal

Cans used to store food are made from steel coated on both sides with a layer of tin. Tin is used because it is fairly unreactive, and non-toxic. If the tin coating becomes damaged, the exposed iron will corrode rapidly, spoiling the contents of the can. This is why it is unwise to use scratched or dented cans.

Bumper bars on older cars are plated with chromium, and cutlery is sometimes coated with nickel.

5 Sacrificial protection

Iron for garden sheds, roofs, guttering and light poles is coated with zinc or a zinc–aluminium alloy. This process is called *galvanising*. The zinc slows down corrosion because it is more reactive than iron. This means it corrodes first and is sacrificed to protect the iron. After many years the iron becomes exposed and starts to rust.

When a bar of zinc is attached to the side of a steel ship, it corrodes instead of the steel. When it is nearly eaten away it is replaced by a fresh bar.

Figure 6.6 Painting the Sydney Harbour Bridge is a never-ending job.

ISBN 978 1 4202 3736 8

EXPERIMENT 6.1 Why do things rust?

Research question

What effect do air, water and salt have on the rusting of iron?

Planning

1 You need to vary the amounts of air, water and salt, but control all other variables. One way to do this is to use four test tubes as shown.

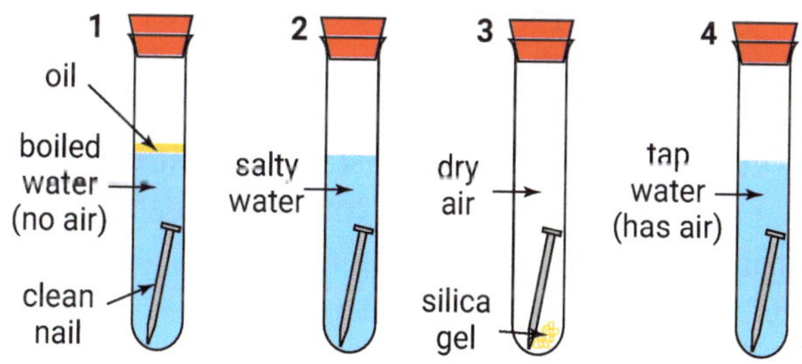

What is the purpose of the oil in test tube 1?

Why use the silica gel in test tube 3?

2 Before you start, predict what will happen.

3 Don't forget to prepare a data table to record your observations over a week or so.

4 You could vary the concentration of the salty water.

5 You could try coating a nail with grease.

6 You could weigh the nails before and after to make your experiment more quantitative.

7 Your teacher could show you how to do the experiment by covering the nails with agar in Petri dishes.

Discussion

1 Try to explain your results. Were your predictions correct?

2 Suggest ways of improving your experiment.

Write your conclusion

Inquiry

Design a controlled experiment to compare the rates of corrosion of aluminium, copper and iron indoors and outdoors. If possible, carry out your experiment and write a report.

CHECK

1 Kris thought their car was rusting faster because they lived near the sea. Do your results in the experiment support this idea?

2 What is an alloy? Why are alloys used instead of pure metals?

3 Suggest why aluminium rather than steel is used to make aircraft bodies.

4 Name three household objects made from stainless steel. Suggest why they are made from this alloy.

5 a What is corrosion? Give some examples.
b Which two substances cause corrosion?

6 Why don't aluminium window frames need to be painted?

CHALLENGE

1 Iron that is tin-plated rusts very slowly. Why?

2 What is the process of galvanising? How does this prevent iron from rusting?

3 Why is it unwise to buy dented cans of food?

4 A sculptor makes two identical statues out of bronze. One is placed in Melbourne and the other in Alice Springs. After 10 years the Melbourne statue is badly corroded, but the other is in good condition. Explain why.

5 Oil rigs often have big lumps of magnesium attached to their legs under water. Suggest a reason for this.

6 An advertisement for garages says *Keep your car inside one of these garages and your car won't rust.* Do you think this is correct? Explain.

7 Why do car exhausts rust so rapidly? What are the advantages and disadvantages of stainless steel exhausts?

ISBN 978 1 4202 3736 8

6.2 Carbon compounds

You walk into a shop selling freshly ground coffee. The aroma of coffee is due to hundreds of different chemical compounds all containing the element carbon. In fact, over 90% of all the compounds on Earth contain carbon. Carbon compounds are made up of carbon and a few other elements—mainly hydrogen, oxygen and nitrogen. All living things are made of carbon compounds; your food is made of them and your clothes are made from them.

The reason there are so many different carbon compounds is that carbon atoms can form four chemical bonds, due to the attractive forces between atoms. Carbon can bond (link) to other atoms such as hydrogen and oxygen. For example, methane is the main ingredient in natural gas, and is found in swamps when plant material decays. A methane molecule consists of a carbon atom bonded to four hydrogen atoms. Its *molecular formula* is CH_4, meaning there are four hydrogen atoms (H) for each carbon atom (C).

If you want to show how the atoms are bonded to each other you can use a **structural formula**. For example, the structural formula for methane is:

```
     H
     |
 H — C — H
     |
     H
```

This is an easy way to draw a three-dimensional molecule in two dimensions. Each line represents a chemical bond.

The second reason there are so many carbon compounds is that carbon can bond to itself. For example, Figure 6.8 shows a molecule of ethanol (a type of alcohol), the most important ingredient in beer, wine and spirits. Notice that each carbon atom is bonded to four other atoms. Each hydrogen atom is bonded to only one atom, and oxygen is bonded to two atoms.

In the activity on the next page you can make models of carbon compounds for yourself.

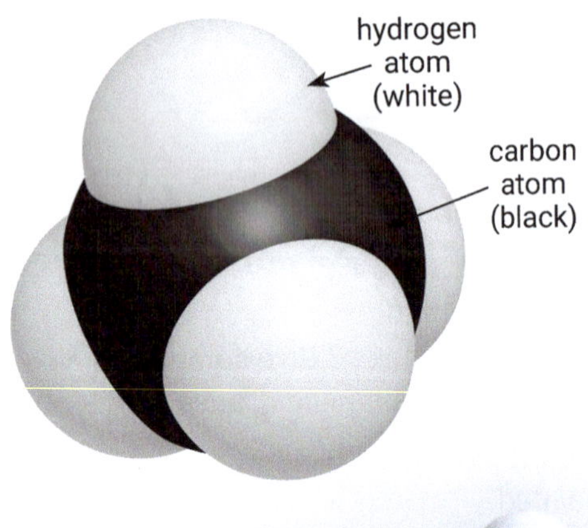

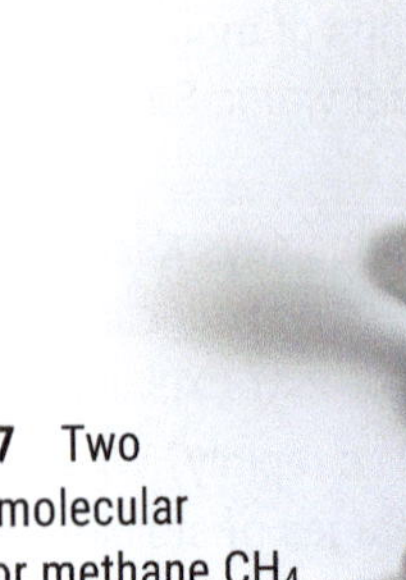

Figure 6.7 Two different molecular models for methane CH_4

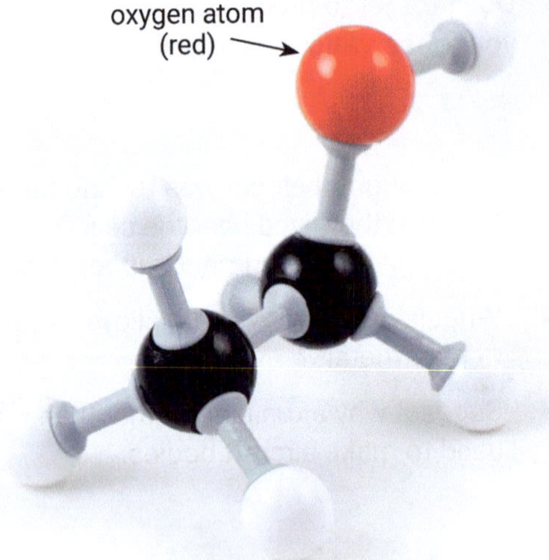


```
     H   H
     |   |
 H — C — C — O — H
     |   |
     H   H
```

Figure 6.8 A molecular model of ethanol (C_2H_5OH), with its structural formula

ISBN 978 1 4202 3736 8

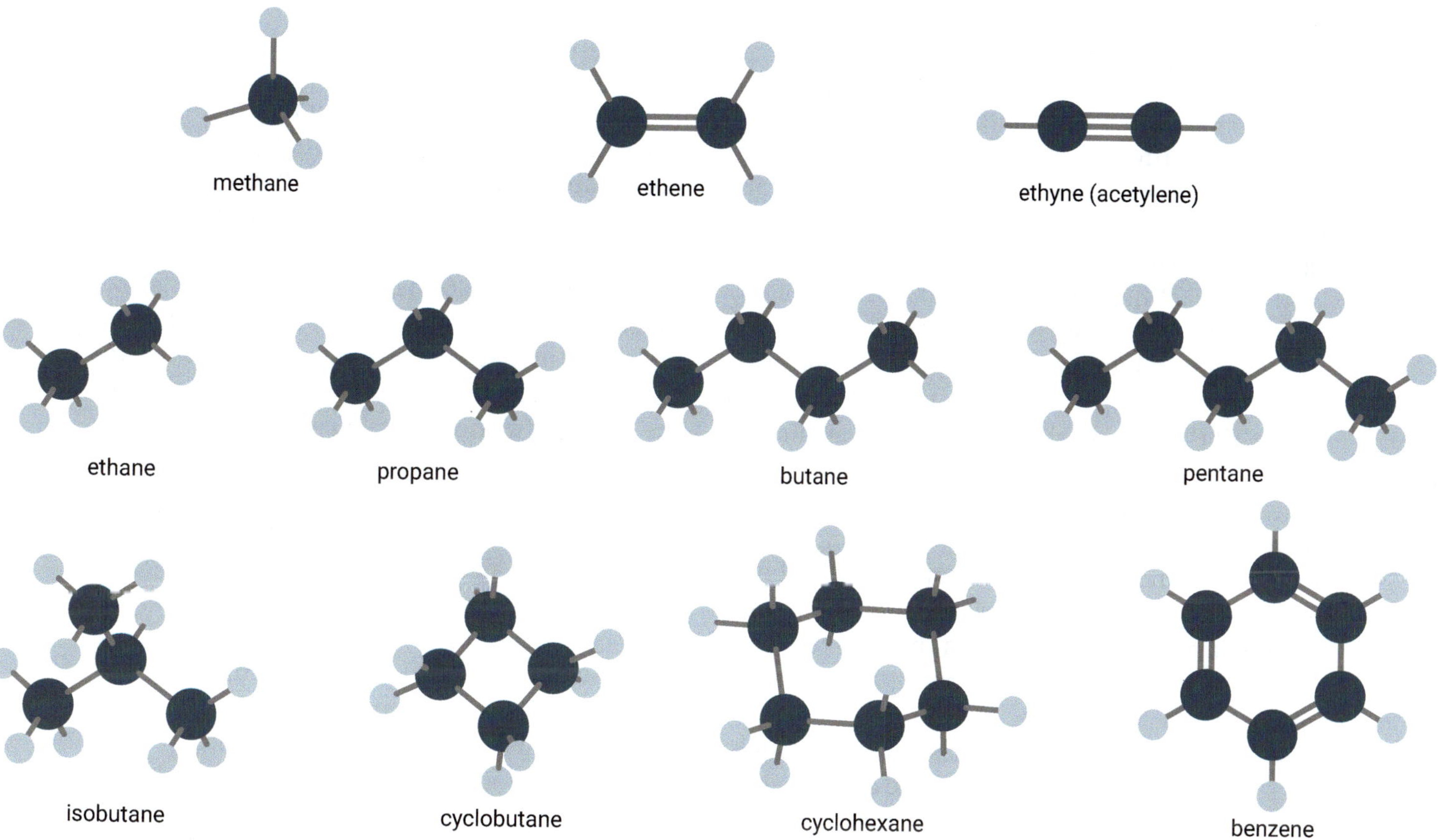

Figure 6.9 Some common carbon-based compounds. Carbon can also form double or triple bonds to itself, or rings.

ACTIVITY

You will need a molecular models kit.

1. Your teacher will show you how to use the kit, or you can read the instructions.
   Which colour balls represent carbon atoms? Hydrogen atoms? Oxygen atoms?
   How are the bonds between atoms represented?
   How many bonds can carbon form?
   How do you know?
   How many bonds can hydrogen and oxygen form?
2. Make a model of a methane molecule CH_4 and sketch it with the different atoms labelled.
3. Join two carbon atoms with one bond then add as many hydrogen atoms as possible. This represents ethane—a gas similar to methane.
   Write down the formula for ethane, and draw its structural formula.
4. Make models to represent propane, C_3H_8, and butane, C_4H_{10}. Compare your models with those made by other students.
   Did you find that there are two different ways of making C_4H_{10}? Draw the two structural formulas.
5. You can use the models kit to have competitions. For example, how many different carbon compounds can you make using 4 carbon atoms, 2 oxygen atoms and 10 hydrogen atoms? You can use all of these atoms or only some of them—but you must obey the bonding rules by filling all the holes in each of the atoms. As you make each compound write down its structural formula.

ISBN 978 1 4202 3736 8

Carbon compounds from oil

Crude oil is a liquid containing a mixture of hundreds of different carbon compounds. Most of these are **hydrocarbons**—compounds of carbon and hydrogen only. These different compounds can be separated by the process of **fractional distillation**.

At an oil refinery the crude oil is heated in a furnace and passed into a large tower called a fractionating column, which is hot at the bottom and cooler at the top (see Figure 6.10). At the bottom almost all the hydrocarbons boil. Their vapours rise in the column, cooling all the way. When they have cooled down enough they condense back to liquids. Hydrocarbons with high boiling points condense in the lower, hotter part of the column. Hydrocarbons with lower boiling points condense further up, in the cooler part of the column. In this way the crude oil is separated into *fractions* with different boiling points.

Some of the fractions are gases, most are liquids, and some are solids. To explain these differences you need to consider the size of the molecules in the fractions. There are attractive forces between molecules, and the more atoms there are in a molecule the stronger these attractive forces are. Therefore large, heavier molecules have very strong attractive forces between them, and are generally very close together.

The fractions from the top of the column consist of small, lighter molecules containing only a few carbon atoms. This is why these fractions are gases, e.g. methane (CH_4), ethane (C_2H_6), propane (C_3H_8) and butane (C_4H_{10}), with low boiling points.

Most of the fractions from crude oil are liquids; for example, octane in petrol is C_8H_{18}. Their molecules are larger, with stronger forces between them, and closer together than the molecules in the gaseous fractions. As you go down the column, the molecules in the liquid fractions become larger, the molecular forces between them are stronger and hence the boiling points increase. The **viscosity** also increases. For example, petrol is quite runny (like water), but lubricating oil is thick (viscous). This is because of the stronger attractive forces between the molecules. The larger molecules also tend to get tangled up and cannot move about as easily as the smaller ones.

The solid fractions from the bottom of the column contain very large molecules, with very strong attractive forces, and packed close together. For example, the paraffin wax used to make candles contains molecules with 20–27 carbon atoms.

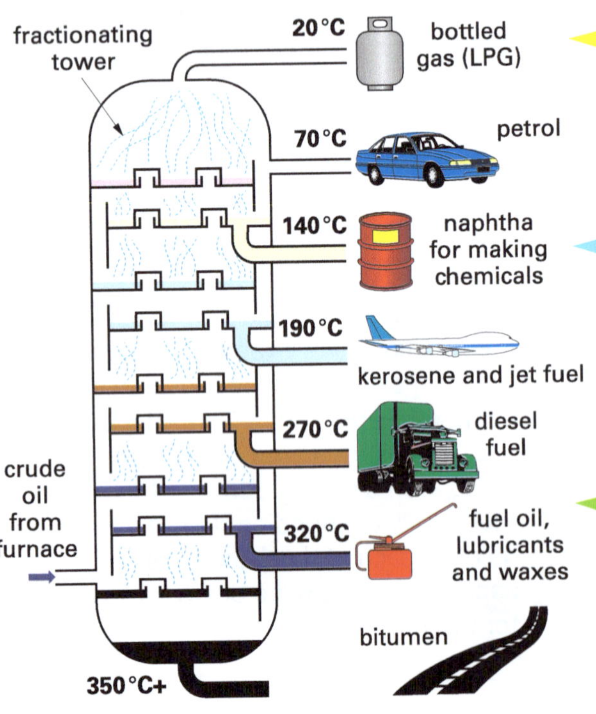

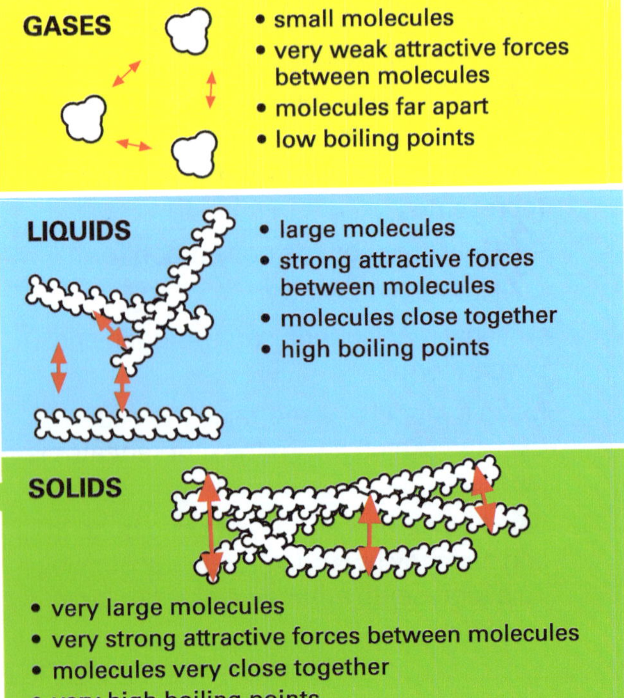

Figure 6.10 Fractional distillation of crude oil

ISBN 978 1 4202 3736 8

Soaps

Water has always been used to clean our bodies and clothing. However, water cannot remove oil and grease because these substances do not dissolve in water. You need a soap or a detergent to remove them.

It is thought that the first soap was made about 2500 years ago, probably by the Egyptians. Soaps are carbon compounds made by boiling vegetable or animal oils with caustic soda (sodium hydroxide). When salt is added to the mixture, hard lumps of the soap form on the surface. Dyes and perfumes can be added to make the soap look and smell better.

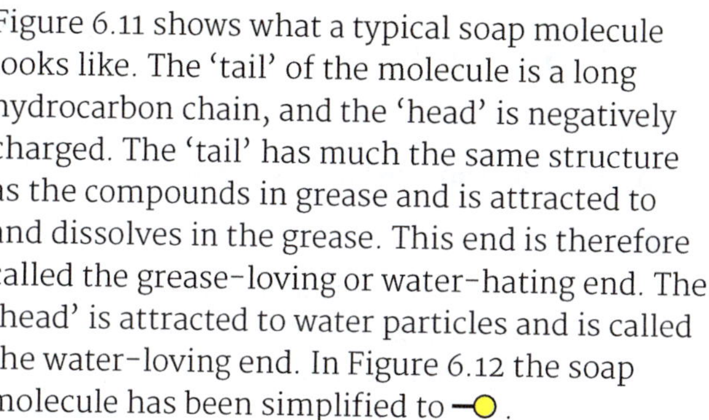

How soap works

Figure 6.11 shows what a typical soap molecule looks like. The 'tail' of the molecule is a long hydrocarbon chain, and the 'head' is negatively charged. The 'tail' has much the same structure as the compounds in grease and is attracted to and dissolves in the grease. This end is therefore called the grease-loving or water-hating end. The 'head' is attracted to water particles and is called the water-loving end. In Figure 6.12 the soap molecule has been simplified to —O.

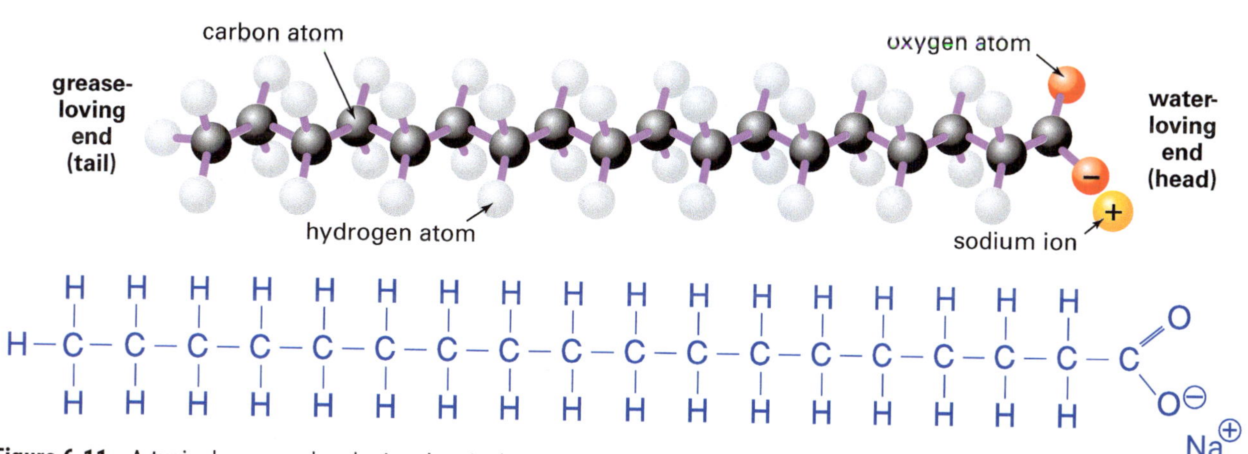

Figure 6.11 A typical soap molecule. Its chemical name is sodium stearate $C_{17}H_{35}COO^{-}Na^{+}$.

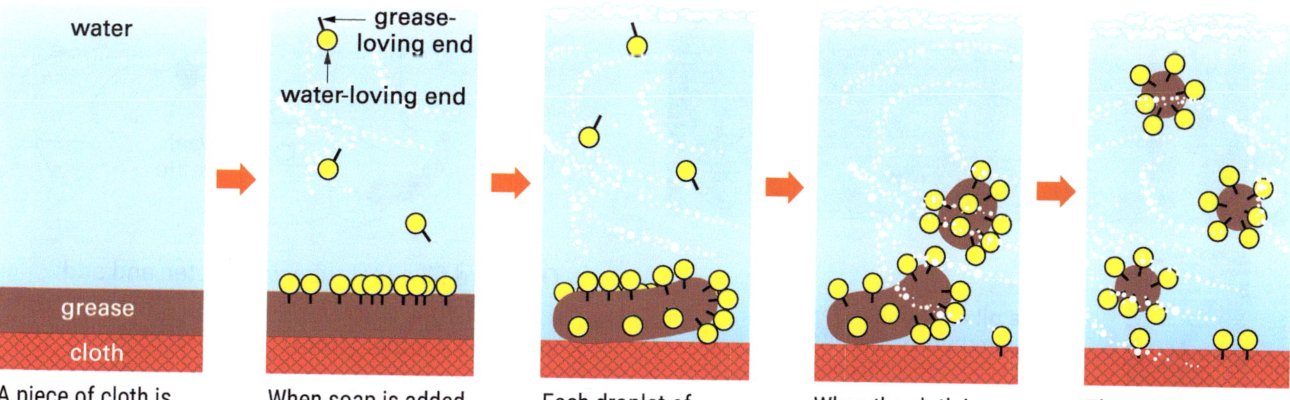

A piece of cloth is covered with grease. Water alone will not remove the grease.

When soap is added, the grease-loving 'tails' stick into the surface of the grease, leaving the charged 'heads' sticking out into the water.

Each droplet of grease becomes surrounded by a layer of soap molecules with their charged heads on the outside.

When the cloth is shaken in a washing machine, the large grease droplets break up into smaller droplets that can be removed from the cloth by further shaking.

The small grease droplets and attached soap molecules repel one another and do not re-attach themselves to the cloth. The excess soap molecules form a protective layer on the cleaned cloth.

Figure 6.12 How soap removes grease from a piece of cloth

ISBN 978 1 4202 3736 8

INVESTIGATION 6.1 Making soap

Aim

To make soap and test how well it works.

Materials

- hotplate (or burner, tripod and gauze mat)
- 2 beakers (approx. 400 mL)
- 2 test tubes with stoppers
- large evaporating basin
- stirring rod
- wash bottle
- spatula
- 90% **ethanol** (or **methylated spirits**)
- kerosene
- olive oil or castor oil
- 6 M **sodium hydroxide** solution (**caustic soda**)
- saturated sodium chloride solution
- piece of filter paper
- measuring cylinder (10 mL)

Risk assessment and planning

- Sodium hydroxide is very corrosive, especially when it is hot. So you will need to take special care that you do not spill any. What should you do if you do spill it? To protect your hands you could wear plastic gloves.
- Why is it safer to use a hotplate than a Bunsen burner in this investigation?
- You should not test the soap on your skin. Why not?

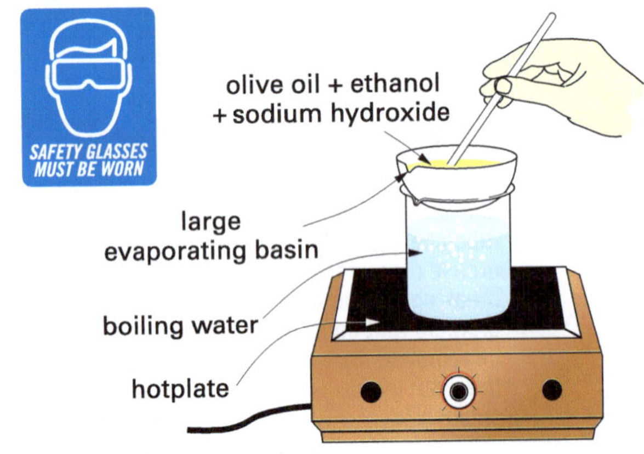

Method

1 Set up a steam bath as shown below left.

2 Add 5 mL of olive oil and 5 mL of ethanol to the evaporating basin. Then carefully add 5 mL of sodium hydroxide solution, while stirring. (The ethanol speeds up the reaction between the oil and the sodium hydroxide by dissolving the oil.)

3 Continue heating for about 30 min, while stirring. The olive oil should 'disappear' and the mixture become pasty.

4 Turn off the steam bath and add 5 mL of sodium chloride solution, while stirring. Soap is insoluble in this salt solution, and should float to the surface as a white curdy solid.

5 Use a spatula to skim off the soap into a beaker. Rinse it several times with a small amount of water, using a wash bottle.

6 Leave the soap to dry on a piece of filter paper.

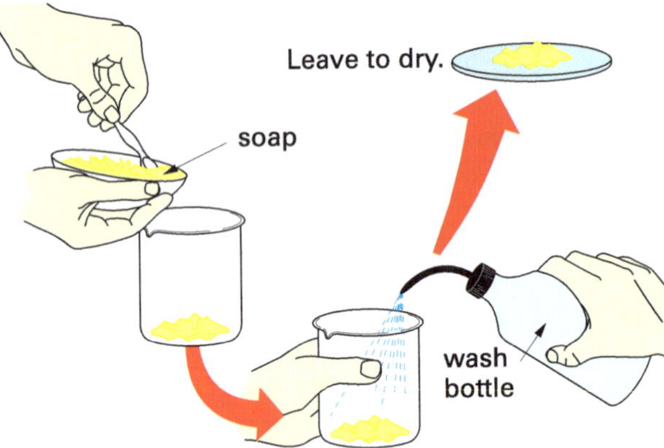

7 One-third fill a test tube with water, and add some of your soap. Then put in a stopper and shake. Is a lather (froth) produced?

8 Add 2 or 3 drops of kerosene to water in a test tube, and shake. (Kerosene acts like grease.)

Describe what happens.

9 Now add some of your soap to the kerosene and water and shake it again.

What happens this time? Explain.

ISBN 978 1 4202 3736 8

Detergents

Detergents are synthetic compounds with a head and tail structure similar to that of soap. They are usually made from chemicals obtained from crude oil, and they are used as dishwashing liquids, washing powders and shampoos. They work better in hard water than soap does. (Hard water is water containing large amounts of dissolved solids that prevent soap from producing a lather.) This is why detergents are more effective cleaners than soaps.

The first detergents were not **biodegradable**. They were not broken down by microbes (micro-organisms) into simpler compounds. As a result they caused foam to appear in rivers and streams. In some polluted areas, even a glass of drinking water would have a head of foam on it. The problem was solved by changing the ingredients used to make the detergents. Detergents are now biodegradable and in most countries the law requires that all detergents be biodegradable.

A second problem is that many detergents contain chemicals called *phosphates*. These are used to 'soften' the water, allowing the detergents to remove more dirt from clothes. These phosphates are washed into drains and then into rivers and lakes where they are good fertilisers for algae and other water plants, which grow quickly and then die. The decay of these plants uses up the oxygen in the water and as a result fish and other aquatic animals may die. For this reason there are limits on how much phosphate manufacturers can put in detergents.

Figure 6.13 Detergents in waterways can be a pollutant.

CHECK

1. Explain why carbon forms so many different compounds.
2. Which elements do hydrocarbons contain?
3. How many bonds can each of these atoms form?
   a carbon
   b oxygen
   c hydrogen
4. In what ways are soaps and detergents similar? In what ways are they different?
5. Describe in your own words how dishwashing detergent removes grease from dishes. Use a diagram if you wish.
6. What is the difference between fractional distillation and ordinary distillation?
7. A crude oil sample was separated into four fractions. These were collected in the temperature ranges shown in the table.

Fraction	Temperature range (°C)
A	5–70
B	70–115
C	115–200
D	200–380

Which fraction would you expect to:
   a be the most like petrol?
   b have molecules with the longest chains?
   c be the most viscous?
   d have the lowest boiling point?

ISBN 978 1 4202 3736 8

CHALLENGE

1 Look at the flow diagram below, which shows what happens in a factory making washing powder.

a What raw materials are used by the factory?

b What is the end product?

c Are there any waste materials?

d Why is a sieve used before packing the washing powder?

e Suggest why the bleach and perfume are not added with the other ingredients at the start.

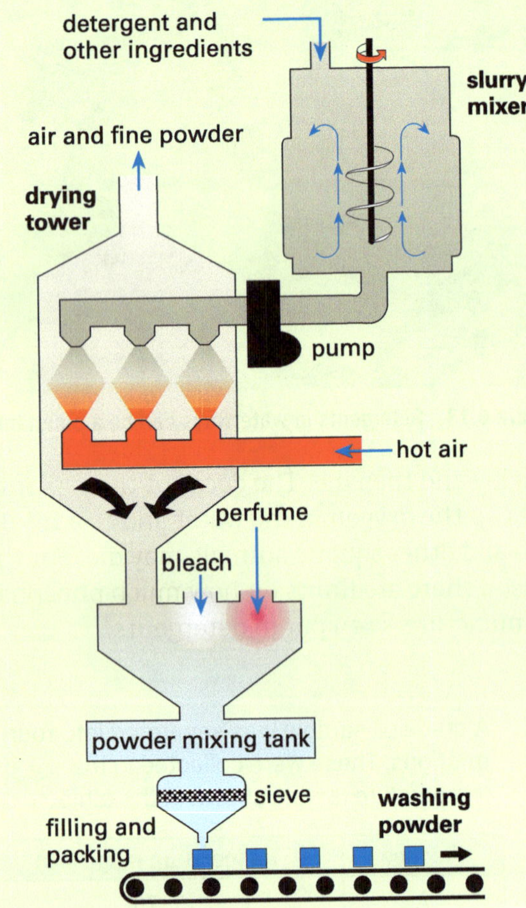

2 Methane gas burns in oxygen to produce carbon dioxide and water. Write a word equation for this reaction.

3 Suppose you had to wash in sea water. Which would it be better to use—soap or detergent? Why?

4 Examine the following formulas. Which formulas are correct according to the bonding rules? For each one that is incorrect, say why. (You could try to construct the molecules using molecular models.)

H−C(H)(H)=C−H

H−C(H)(H)−O−C(H)(H)−H

H−C(H)(H)−H−C(H)(H)−H

H(H)C=C(H)H

= represents a double bond

5 CFC-11 (once used in aerosols) has the molecular formula CCl_3F.

a Use the formula to work out what CFC stands for.

b Draw the structural formula for CCl_3F.

6 Butane and isobutane both have the same formula C_4H_{10}. However, isobutane has a lower boiling point and lower density than butane. Use molecular models to explain these differences in properties.

EXPLORE

1 Suppose you are asked to test whether liquid detergents wash clothes better than powder detergents. Describe how you would do this, making sure you control variables.

2 Do a supermarket survey of laundry detergents. What does the [P] symbol mean? What other symbols are used on the packages?

3 Use a molecular models kit to build a soap molecule (Figure 6.11 on page 141). Use the internet to find out how detergent molecules are different from this.

ISBN 978 1 4202 3736 8

6.3 Plastics and fibres

In 1907 Leo Baekeland, a Belgian working in the United States, was trying to make an artificial substitute for shellac. (Shellac is a material obtained from tiny Indian lac bugs, and was used widely for making varnishes and waxes.) Instead, he made a new substance that kept clogging up his test tubes. Nothing would dissolve this dark brown substance, and on further investigation he found that it could be moulded into shapes. It was a good insulator of heat and electricity.

What Baekeland had done was to mix phenol and formaldehyde, which reacted to form a substance we now call *plastic*. (The word *plastic* means 'easy to mould'.) Baekeland called the new material *bakelite*, from his own name, and it was soon being used for saucepan handles and electrical fittings. Bakelite has now been replaced by newer plastics.

Polymers

Suppose you mix two different chemicals, A and B. Each contains small molecules, and they can react together to form a bigger molecule, A–B. However, it is sometimes possible for the molecules to go on linking up to form giant long-chain molecules called **polymers**. (Poly means 'many' and mer means 'units'.)

All polymers contain big molecules that consist of many small molecules (or *monomers*) linked together. The chemical reaction in which small molecules link to form polymers is called **polymerisation** (pol-IM-er-eyes-AY-shun). Sometimes the monomers are all the same, and sometimes there are two different monomers.

Figure 6.14 Bakelite was used to make radios like these, as well as jewellery, telephones, billiard balls and board game pieces.

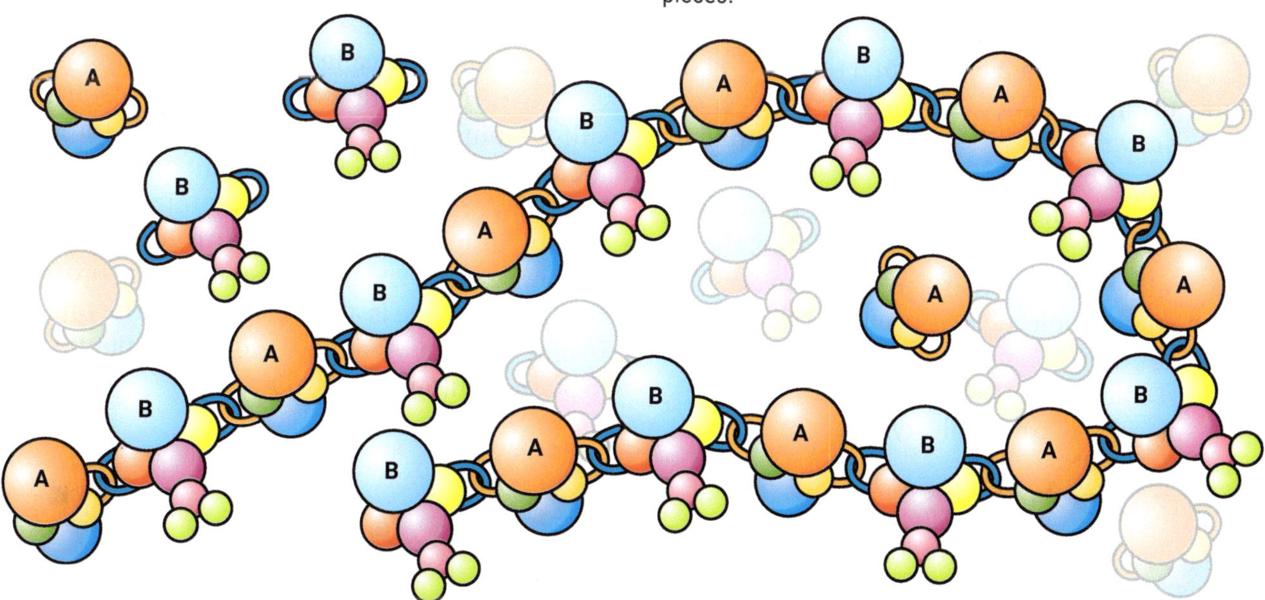

Figure 6.15 Two different small molecules join to form a long-chain molecule called a polymer. The process is a bit like linking paperclips together.

ISBN 978 1 4202 3736 8

Making polythene from ethene

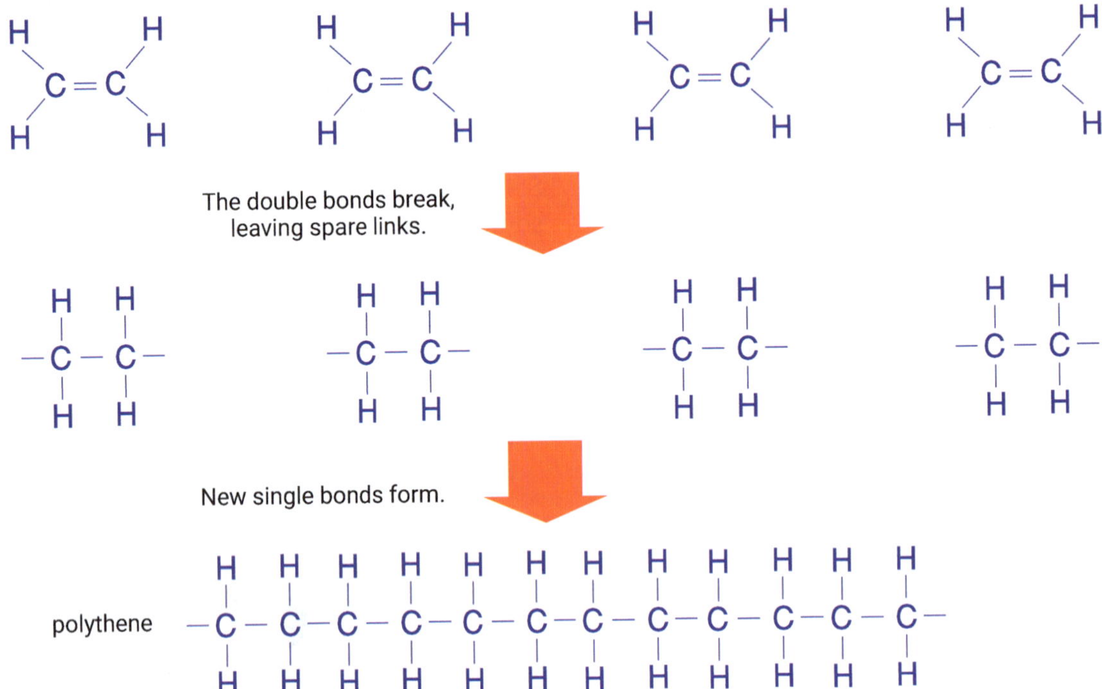

Figure 6.16 Formation of polyethene

Polythene (or polyethene) is formed by the polymerisation of ethene, one of the many chemicals obtained from crude oil. Ethene molecules contain a double bond (represented by a double line), which makes them very reactive. If the double bonds are broken and then linked to neighbouring molecules, a long chain polymer can form as shown above. These chains can contain up to 50 000 c arbon atoms.

Many other monomers besides ethene can be polymerised. For example, vinyl chloride is the same as ethene except that one hydrogen atom in each ethene molecule has been replaced by a chlorine atom. It can be polymerised to form polyvinyl chloride or PVC. Similarly, styrene can be polymerised to form polystyrene.

Natural polymers

Many polymers occur naturally. For example, cellulose and starch in plants are both polymers made from glucose units. If we represent a glucose molecule by the symbol g, then the structure of starch is –gggggggggg–, and the structure of cellulose is –gƃgƃgƃgƃgƃ–. In cellulose every second glucose molecule is upside down.

The cells and tissues in your body are made of proteins, which are polymers made from smaller molecules called *amino acids*. These contain carbon, hydrogen, oxygen, nitrogen and sulfur atoms. Figure 6.17 shows part of a protein molecule made from five different amino acids. Different proteins are made by joining the amino acids in different sequences. With 20 amino acids to choose from, the number of different proteins that can be made is almost limitless. The protein insulin contains 51 amino acid units, but most proteins contain 500 or more.

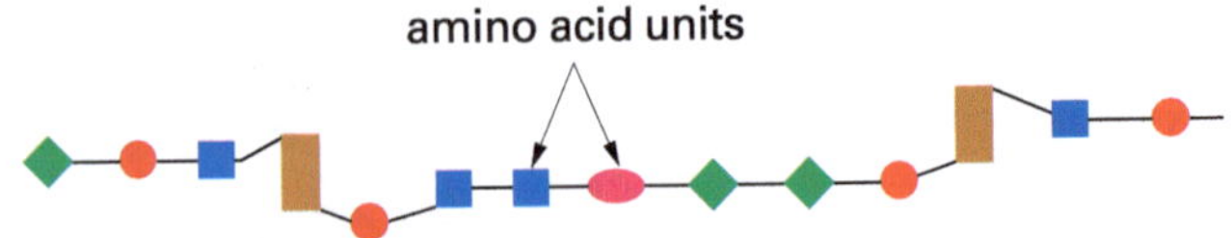

Figure 6.17 This piece of protein chain is formed from five different amino acids.

ISBN 978 1 4202 3736 8

Milk contains a protein called casein (KAY-seen). It can be extracted by adding about 5 mL of vinegar to about 100 mL of warm milk, in much the same way that cheese is formed.

Separate the curds (white solid) from the whey (yellowish liquid) by pouring the mixture through some coarse cloth. Squeeze the cloth with your fingers to remove the whey. Knead the casein (curds) in warm water, and shape it into a ball or some other shape.

Leave the plastic to dry for a day or two.

- Describe the properties of your plastic.
- Suggest what you could use casein plastic for.

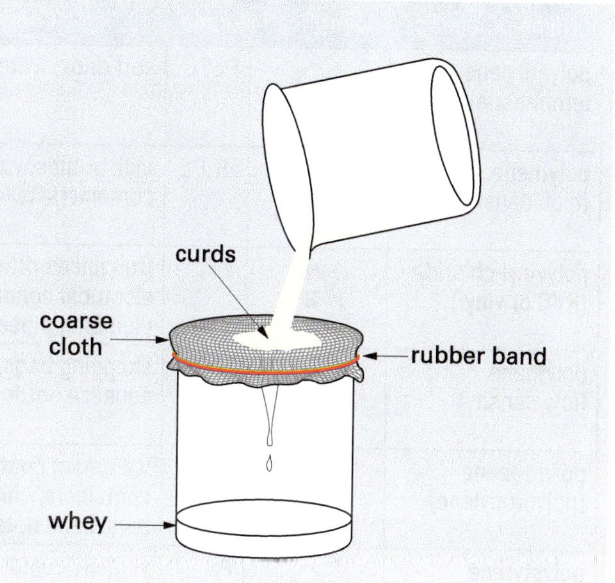

Properties and uses of plastics

There are two different types of plastics. One type contains long thin molecules which form tangled chains. In these plastics there are strong bonds between the atoms along the chain, but much weaker forces between neighbouring chains. Because of this, the chains can slide over each other easily (Figure 6.18). Hence these plastics stretch and bend easily and soften on heating. Such plastics are called **thermoplastics**. Examples are polythene, polystyrene, PVC and nylon.

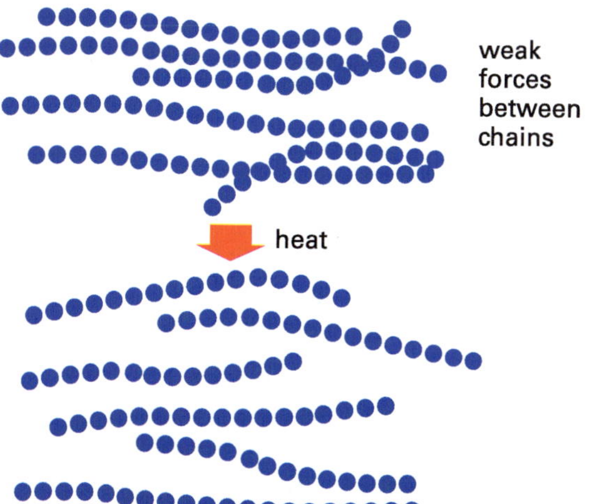

Figure 6.18 The chains of a thermoplastic move apart when the plastic is melted.

Thermoplastics can be shaped in a number of different ways. Each method involves heating the plastic so that it becomes soft, shaping it and then cooling it.

The second type of plastic is made from molecules whose atoms form strong bonds *between* chains as well as along the chains (Figure 6.19). These plastics are called **thermosets**. They are much harder than thermoplastics, and less flexible. They do not melt when heated. Instead they char (blacken) as the cross-links are broken and the polymer starts to decompose. Examples of thermosets are bakelite, fibreglass resin and epoxy glues.

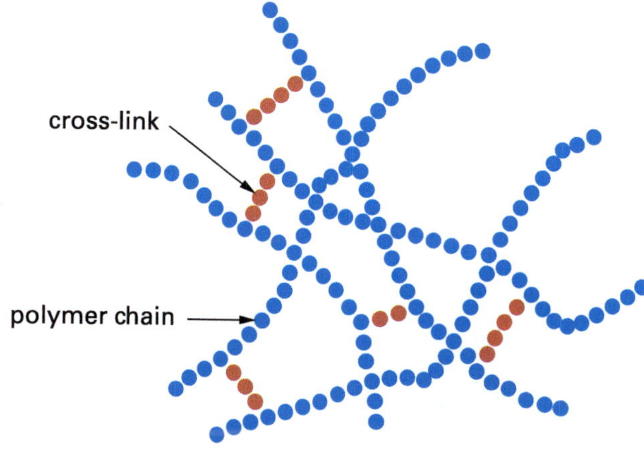

Figure 6.19 The structure of a thermoset

ISBN 978 1 4202 3736 8

Name of plastic	Identification code		Common uses
polyethylene terephthalate	1	PETE	soft drink, water and juice bottles
polythene (high density)	2	HDPE	milk bottles, various household containers, black polypipe
polyvinyl chloride (PVC or vinyl)	3	V	fruit juice bottles, raincoats, electrical conduits, garden hoses, plumbing pipes
polythene (low density)	4	LDPE	shopping bags, food wrap film, squeeze bottles
polypropene (polypropylene)	5	PP	ice-cream containers, take-away containers, chip packets, bumper bars, bank notes
polystyrene	6	PS	yoghurt containers, refrigerator linings, foam packaging

EXPLORE ONLINE

Follow the links to the websites below.

Macrogalleria
A cyberwonderland where you can explore five levels of information on how plastics and fibres are used and made. Includes many photographs.

Plastipedia
Comprehensive information on the history of plastics, plastics processes and applications.

Learning Centre
Information on the uses of plastics and recycled plastics in Australia.

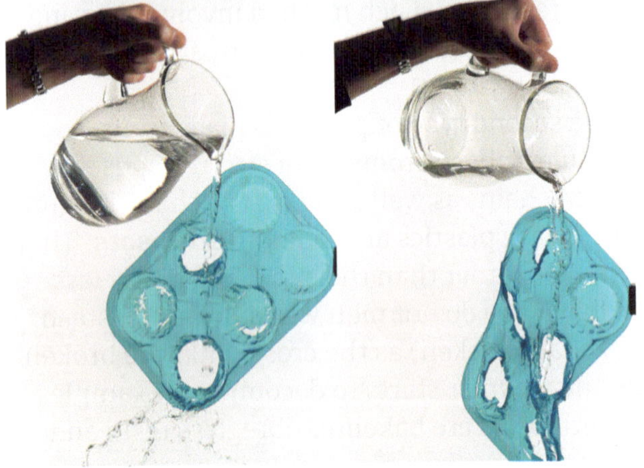

Figure 6.20 This Plantic plastic tray is made from corn starch and degrades rapidly in water.

Recycling and reusing plastics

There are two problems with our increasing use of plastics (more than one million tonnes per year in Australia). First, most plastics are made from oil, a non-renewable resource which will become scarce in the future. Second, what do we do with plastic objects when we are finished with them? There are many different ways of reusing them. For example, you can use old ice-cream containers to store things in. However, most councils now collect plastic items from homes for recycling.

There are three different ways of recycling plastics.

1 Thermoplastics can be melted and remoulded. However, because there are so many different types of plastic, they must first be washed and sorted, otherwise the recycled plastics will not have the properties you want. This costs money, and the recycled plastic may be more expensive than new plastic. However, a number of manufacturing plants in Australia now use recycled plastic. For example, milk bottles are recycled to make products such as garbage bins and detergent bottles, and plastic banknotes are recycled to make compost bins.

ISBN 978 1 4202 3736 8

2 Mixed plastic waste can be crushed and moulded into a range of products such as plastic posts, plant pots and park benches. It can also be mixed with wood to make a composite material. For these uses the lower quality product is not a problem.

3 Because plastics burn, they can be used as fuel. So far this has proved difficult because of the toxic gases produced and the sticky nature of many plastics as they burn. However, technologists have found a way of converting waste plastics into oil. In the system shown in Figure 6.21, some of the gases formed as the plastic decomposes are condensed to give valuable oil products. The rest of the gases are used to heat the plastic waste. Such a system can handle 8000 tonnes of plastic waste in a year. It is an expensive process but a useful way of saving our non-renewable oil reserves.

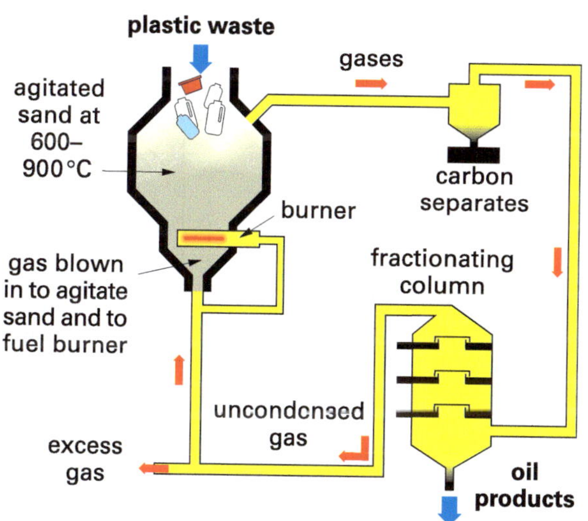

Figure 6.21 Recycling plastic waste into oil

Figure 6.22 Blocks of crushed plastic ready for recycling

Figure 6.23 The crushed plastic (left) and the oil obtained from it

ACTIVITY

At present we recycle only about 20% of the plastics we use. How can we increase the proportion recycled? Discuss this problem in a group using the steps below.

1 Brainstorm all the forces in favour of more plastics recycling. For example, most people are environmentally conscious at present.

2 Brainstorm the forces opposed to recycling. For example, it's too much trouble to separate plastics from other rubbish.

3 Decide which of the forces could be changed. Is it easier to strengthen the 'for' forces or weaken the 'against' forces?

4 On the basis of step 3, write out a plan to solve the problem.

ISBN 978 1 4202 3736 8

Properties of plastics

Planning guide

1 Use the identification codes in the table on page 148 to collect a range of commonly used plastics.

2 Read through the six properties of plastics on this page, then select the properties you want to test. **Don't burn the plastics, since the fumes released pollute the laboratory and cause breathing difficulties for some students.**

3 You will need to work out for yourselves how to do each test. Write down a plan, which should include:
   - how you will do the test, with a diagram if necessary
   - what you will measure, and how you will record and display your data
   - a list of the equipment you will need
   - any safety precautions you will need to take.

   Your teacher must approve your plan before you can start.

4 Write a report of your test, noting the properties you think are useful and those that are not so useful for objects made of the plastics you tested.

Resistance to chemicals

Are plastics affected by acids and bases? You could try to dilute hydrochloric acid and caustic soda. (Caution: corrosive) Do the plastics dissolve in solvents such as turpentine or alcohol? (Caution: flammable)

Resistance to heat

Put the plastic on a heatproof mat. Heat the end of a screwdriver in a burner, then press it against the plastic. Does the plastic melt?

Insulating properties

Set up a simple electric circuit to test whether the plastics are insulators or conductors. Compare them with metals.

Degradability

Do the properties of the plastic change when it is exposed to air, sunlight and water or buried for some time? Is this a good thing or a bad thing?

Density

If a plastic floats in water, it is less dense than water—its density is less than 1 g/cm^3. If it sinks, its density is more than 1 g/cm^3. To calculate the density of a plastic sample, you use the formula:

$$\text{density} = \frac{\text{mass}}{\text{volume}}$$

To measure the volume of a plastic object, use a displacement can as shown. If the plastic floats, you will need to hold it under water with a thin piece of wire.

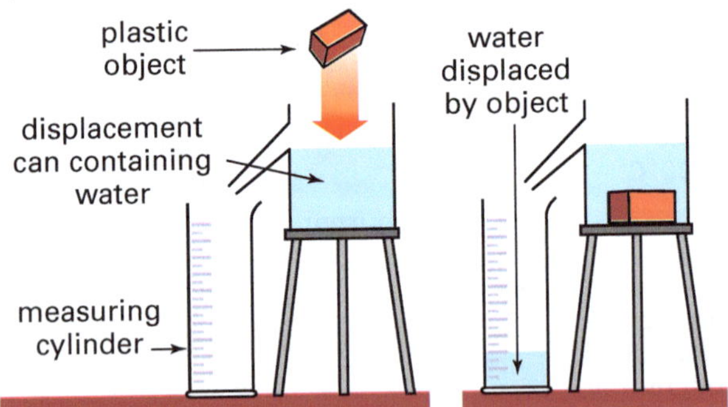

Flexibility and stress

Can the plastic be bent or stretched, and if so does it return to its original shape? To what angle can it be bent before it breaks? Does its flexibility change when the plastic is warmed in hot water?

You can check for strained areas (coloured lines) in plastics as shown.

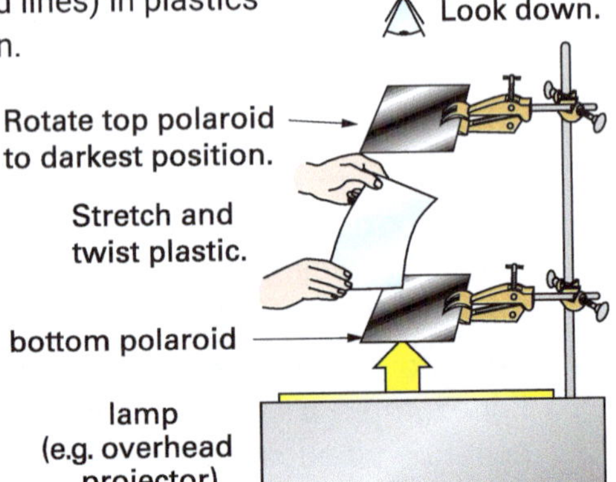

ISBN 978 1 4202 3736 8

SCIENCE AS A HUMAN ENDEAVOUR

The plastic bag problem

Answer the questions below. For some you will need to search on the internet. For others you will need to discuss them in a small group.

1 What are the advantages and disadvantages of the thin plastic shopping bags that you receive at supermarkets?

2 What are these bags made from? What raw material is used to make them?

3 How many shopping bags do Australians use each year?

4 Most plastic bags end up in landfill. How long do they take to break down?

5 Planet Ark estimates that plastic bags littering the ocean kill at least 100 000 birds (including penguins), whales, seals and turtles every year. However, the Plastics and Chemicals Industries Association (PACIA) says 'there is no credible evidence to support the claim that hundreds of seabirds and mammals die each year from ingesting plastic bags'. Why do these two groups disagree so much? What is the truth?

6 How can plastic shopping bags be reused around the home?

7 Ireland had a major litter problem, so in 2002 they introduced a levy of 15 cents per plastic shopping bag. This reduced the use of the bags by 90% in 6 months, but at the same time the use of plastic bin liners increased. Try to explain what happened.

8 Can plastic shopping bags be recycled? How?

9 Many people are now using reusable 'green' shopping bags made of polypropylene or calico. Are these a solution to the problem of our overuse of plastic shopping bags? Explain your answer.

10 Following disastrous floods in 1988 and 1998, which were due in part to drains being blocked by plastic bags, the country of Bangladesh banned plastic bags. South Australia also banned them in 2009. Do you think it would be a good idea to ban plastic shopping bags throughout Australia? Why or why not?

11 What are biodegradable, photodegradable and compostable shopping bags?

12 Scientists first made polythene in 1933 and in 1974 it was used to make plastic shopping bags. Since then the use of these bags has skyrocketed. However, these plastic bags have become a major problem for the environment. From what you have learnt by answering the questions on this page, what do you suggest we do about the problem?

ISBN 978 1 4202 3736 8

Fibres

Clothes are made from fabrics which, in turn, are made from threads spun from fibres.

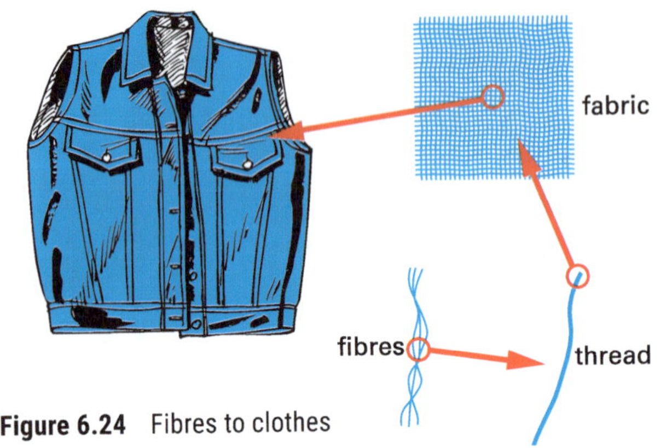

Figure 6.24 Fibres to clothes

All fibres are polymers, and they may be natural or synthetic. Cotton and wool are natural materials obtained from plants and animals. Cotton is made of cellulose polymers, and wool is made of protein molecules. Synthetic fibres such as nylon and polyester are polymers made from chemicals obtained from oil and coal. The key below shows the different types of fibres.

Properties of fibres

The polymer chains in fibres are very long. They are lined up close to each other and attract each other strongly. This makes the fibre very strong and difficult to break. Stretching the fibre helps to line up the molecules, making it even stronger.

Figure 6.25 Synthetic fibres like nylon are made by forcing the melted material through holes in a metal plate called a spinneret, named after the organ through which a spider spins the silk for its web.

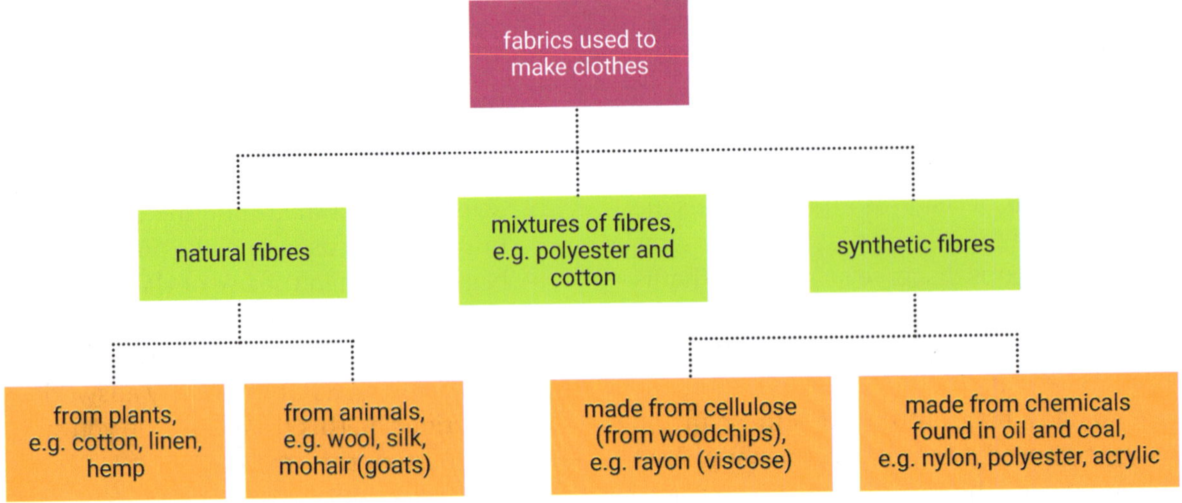

Figure 6.26 A key for classifying fibres

ISBN 978 1 4202 3736 8

Stretchy fibres like nylon or wool have molecules that are normally closely curled like a telephone cord. When you pull on the fibre, it stretches by straightening out the loops, but when you release it, the molecules curl up again.

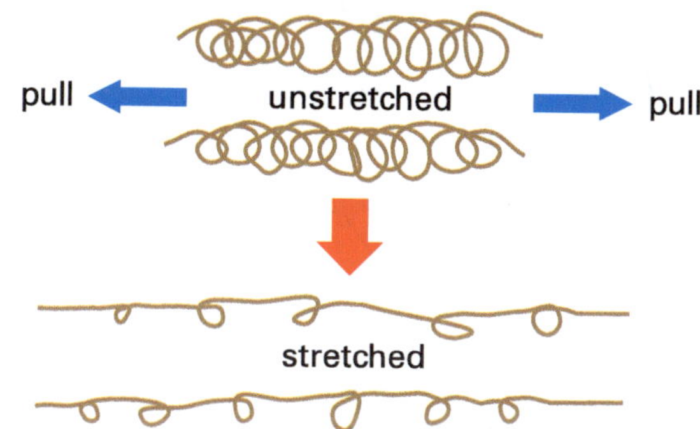

Figure 6.27 How a fibre can be stretched

Many of the properties of synthetic fibres are different from those of natural fibres. Cotton absorbs moisture and allows sweat to evaporate. It is therefore comfortable to wear on hot days. Synthetic polyester does not wear out as easily as cotton, but does not absorb moisture. This makes it sticky and uncomfortable on hot days. By making clothes from a *mixture* of cotton and polyester, you have the properties of both fibres. For example, 65% polyester and 35% cotton is a common mixture. Clothes made from this blend are hard-wearing, comfortable in summer and drip-dry (non-iron).

Perming hair and creasing wool

Each strand of your hair contains many long chains of the protein keratin. These long polymer chains are held together by different types of chemical bonds. The strongest of these bonds is made up of two sulfur atoms.

In curly hair the S–S bonds between the protein polymer chains hold the strands of hair in a curly shape. If you have your hair chemically straightened, solutions are added to your hair that break and re-form the S–S bonds, as in Figure 6.28.

In step 1 the hair is moistened, causing some of the bonds between the protein chains to break.

In step 2 the straightening solution is added. This causes a chemical reaction in which the remaining S–S bonds are broken. Because the cross-links have been broken, the hair is relaxed and can be combed straight. Unfortunately, the straightening solution has the 'rotten egg' smell of sulfur compounds.

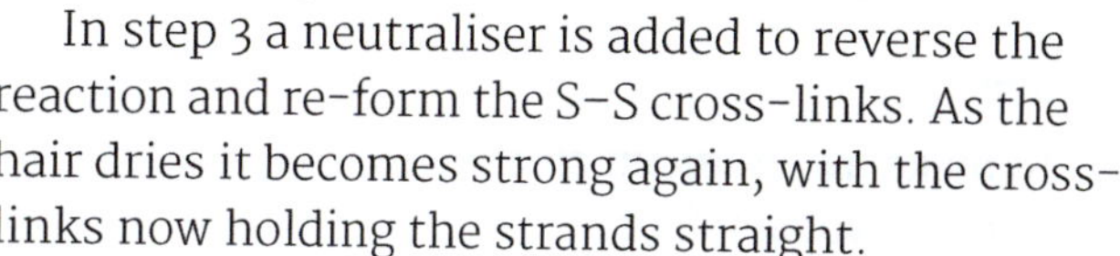

In step 3 a neutraliser is added to reverse the reaction and re-form the S–S cross-links. As the hair dries it becomes strong again, with the cross-links now holding the strands straight.

Wool fibres are also made of protein polymer chains. Australian scientists from CSIRO developed a process to make permanent creases in wool. It works in much the same way as permanent waving of hair, which is the reverse of chemical straightening. Permanent crease skirts or trousers don't lose their crease when worn or washed.

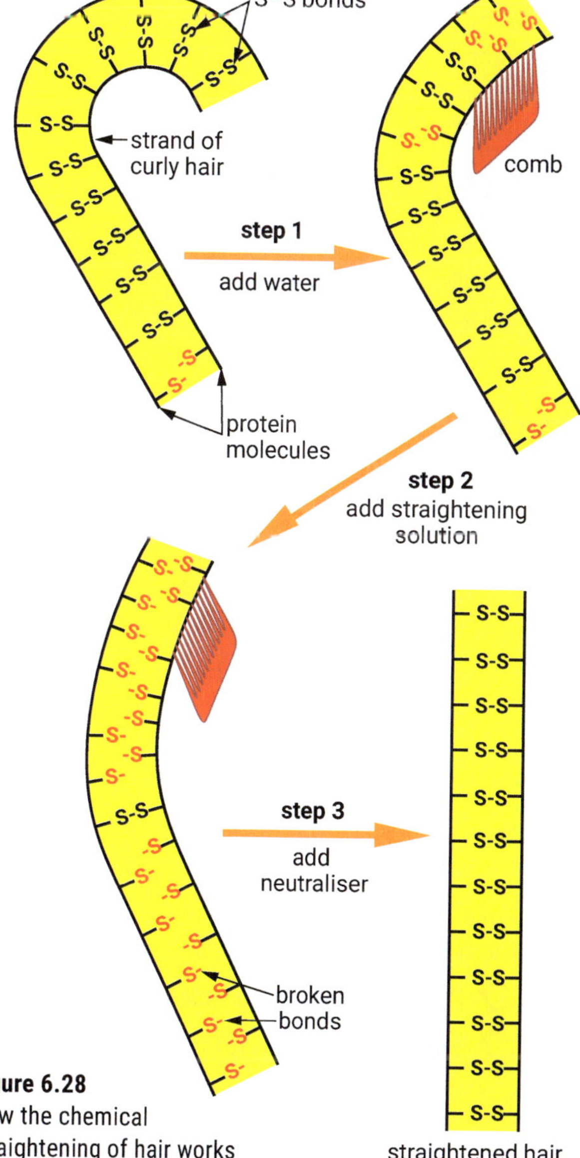

Figure 6.28 How the chemical straightening of hair works

ISBN 978 1 4202 3736 8

INVESTIGATION 6.2 Permanent crease

Aim

To make a permanent crease in a piece of woollen fabric.

Materials

- permanent crease solution (3% **sodium hydrogen sulfite** or **sodium metabisulfite** solution containing a little detergent)
- 2 pieces of woollen fabric
- sponge
- evaporating basin (or similar container)
- container for washing wool samples
- detergent
- hot water
- steam iron

Toxic

Risk assessment and planning

Read the investigation carefully.

What safety precautions will be necessary?

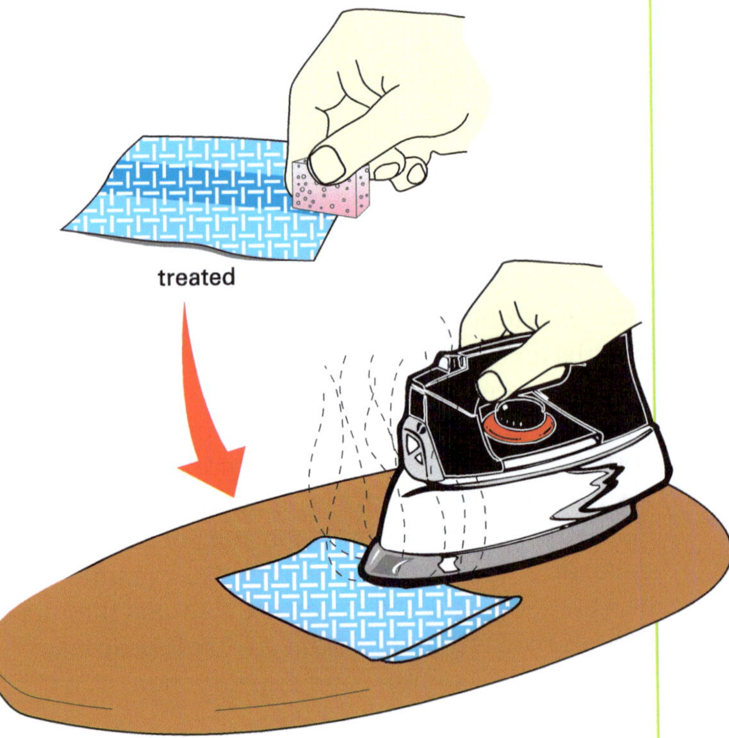

PART A

Method

1 Pour some of the permanent crease solution into the evaporating basin. Sponge a line of solution down the centre of one of the wool samples.

2 Crease the sample along the sponged line and press the crease in with the steam iron for about 30 seconds. Using the steam iron again, press a similar crease in the untreated sample.

3 Wash both samples in hot water (about 70 °C) containing a small amount of detergent.

4 Take the samples out of the water and allow them to dry.

 Compare the two samples.

Write a conclusion for the investigation.

PART B Inquiry

Design an experiment to answer one of these questions.

1 Does the temperature of the washing water in step 3 have any effect?

2 Does this permanent creasing method work for other fabrics, e.g. polyester?

3 What effect do different washing solutions have? Some you could try are liquid and powder detergent, ammonia solution, vinegar, wool wash and fabric softener.

ISBN 978 1 4202 3736 8

SCIENCE AS A HUMAN ENDEAVOUR

Goodyear and rubber

Rubber is a natural polymer consisting of long tangled molecules. Over the years scientists have learnt how to modify its properties and to make synthetic rubbers. But this process has not been straightforward, as this story shows.

For many centuries the native people of Central and South America used a milky liquid (latex) from certain trees to make elastic, bouncing balls. The early European explorers used this latex to waterproof their clothes. This worked well until the sun melted the rubber, making it soft and sticky. And during cold weather the rubberised material became hard and brittle.

Charles Goodyear, an American inventor, became interested in finding a process to make rubber that would have the same properties in hot and cold weather. However, his experimenting soon used up the little money he had and he was put in prison more than once for owing people money. At one time he sold to the government a large order of mailbags that had been coated with his modified rubber to make them waterproof. But the mailbags turned sticky and shapeless from heat even before they left the factory.

Centuries ago, South American people discovered the essential properties and uses of the rubber tree.

After 10 years of unsuccessful experimenting, Goodyear accidentally spilt a mixture of rubber and sulfur on a hot stove. To his surprise the rubber didn't melt, and when he scraped it off and let it cool he found that it was not sticky. To test the rubber he left it outside in the cold overnight, and in the morning it hadn't become hard and brittle.

Goodyear continued experimenting and found the right amounts of rubber and sulfur to use and the best conditions for heating. He applied for a patent for a process he called *vulcanisation* after Vulcan the Roman god of fire.

Goodyear didn't live happily even after his discovery. He had to defend his patent many times against people who wanted to steal his process. He was a poor businessman and never recovered from his huge debts. Not long before he died he said:

Life should not be estimated exclusively by the standard of dollars and cents. I am not disposed to complain that I have planted and others have gathered the fruits. A man has cause for regret only when he sows and no one reaps.

Questions

1 What was wrong with the first rubber made from rubber trees?
2 What did Charles Goodyear mix with rubber to vulcanise it?
3 What is a patent?
4 Suggest how Goodyear could have avoided the disaster with the mailbags.
5 Read carefully what Goodyear said before he died. Rewrite this more clearly in your own words.
6 Using what you have learnt about polymers, explain why rubber is so elastic.

ISBN 978 1 4202 3736 8

CHECK

1 Why is it that the names of so many plastics start with P?

2 In some cars about 30% of the components used are made from plastics. Why do car manufacturers use so much plastic?

3 For each of the items below, list the properties of plastic that make it suitable for making that particular object.
   a hand-held hair dryer
   b packaging foam
   c stackable chairs

4 List the advantages and disadvantages of making drinking cups from plastic.

5 What is the monomer for the natural polymers cellulose and starch?

6 It is possible to recycle thermoplastics but not thermosets. Why is this?

7 a Look at Figure 6.21 on page 149. Explain in your own words how this process works.
   b How could this process be used to generate electricity?

8 Give at least one example of each of the following:
   a a natural fibre from a plant
   b a natural fibre from an animal
   c a synthetic fibre made from oil or coal

9 Which properties would be particularly important in choosing a fibre for each of the following uses?
   a fishing line
   b laboratory coat
   c non-iron shirt or blouse
   d swimwear
   e towel

10 Compare and contrast chemical straightening of hair and perming of hair.

11 Most synthetic fibres are thermoplastic. What does this mean? Why must a cool iron be used on clothes made from synthetic fibres?

12 Some people think that the cost of plastics will rise as the world's oil resources diminish. Why?

CHALLENGE

1 The casein polymer you made in the activity on page 147 is soft at first, but it hardens if you leave it for a day or so. Infer what causes this hardening. Draw a diagram.

2 Why do curls in hair tend to disappear when the hair becomes wet?

3 The table opposite compares the properties of wool and polyester (a synthetic competitor).
   a Make a list of the properties in which wool is clearly superior to polyester.
   b If you were employed by the wool industry to find ways to improve some of the properties of wool, which ones would you choose to work on? Why?
   c Design an experiment which could be used to measure a fabric's resistance to shrinkage.
   d How would you expect the crease resistance of wool to change in humid conditions? Why?
   e Suggest why polyester is so much better than wool at resisting insect attack.

4 In South Australia there is a returnable deposit on plastic drink bottles. Do you think this would significantly reduce the litter created by these items in your state? Explain.

Property	Wool	Polyester
strength	★★	★★★★
ability to accept a dye	★★★★	★
ease of use	★★★	★★
crease resistance	★★	★★★
resists insect attack	★	★★★★
low flammability	★★★★	★
resists shrinking	★	★★★★
does not show dirt	★★★★	★
moisture absorbency	★★★★	★
takes permanent crease	★	★★★
feel	★★★★	★

★ poor ★★ fair ★★★ good ★★★★ best

5 Suppose you have the five different amino acid units shown below. How many different protein fragments could you make by putting these five units together, using each one only once? Draw them.

ISBN 978 1 4202 3736 8

EXPLORE

1 a Various processes are used to shape plastics, e.g. injection moulding, extrusion and blow moulding. Use library resources to find out about these processes. To get you started follow the links to **Plastic processes**.

b The photo below shows how thin plastic film is made. Explain how you think this is done, starting from solid pellets of plastic.

2 Stretch a rubber band around a block of wood and put it in the freezer for about 10 minutes. Predict what you think will happen when you remove the rubber band from the block. Do it and observe what happens. Then try to explain what happened.

3 Check plastic containers on supermarket shelves or at home. Use the recycling symbols in the table on page 148 to identify the type of plastic in each.

a Add more uses to the right-hand column of the table.

b Which are the most common plastics?

c Are plastics recycled in your area? Which types?

4 Use the internet to research one or more of the following:

- What are conducting plastics? When were they first discovered? How are they different from normal plastics? What can they be used for?
- What are bulletproof vests made of? Follow the links to **How body armour works**.
- How do shampoos and conditioners keep your hair soft and shiny?
- What research is being done into spider-web fibres?

5 When you add hardener to a resin, it sets to form a thermosetting plastic. Use a resin kit to embed an insect in resin, like the mosquito in amber in *Jurassic Park*.

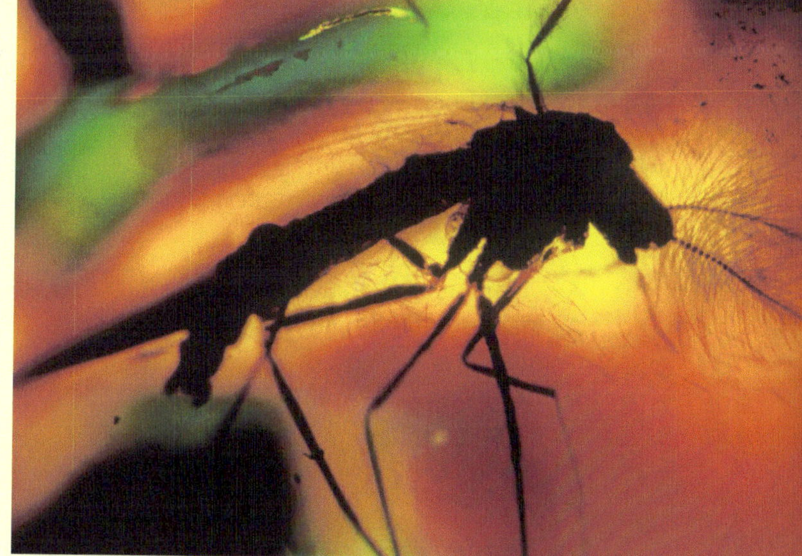

ISBN 978 1 4202 3736 8

MAIN IDEAS

Copy and complete these statements to make a summary of this chapter. The missing words are on the right.

1 The properties of metals can be changed by making ______, which are mixtures of two or more different metals.

2 A metal ______ when it reacts with the air and with water.

3 To prevent iron from ______ you must cover it with some other material, e.g. paint, oil or another metal.

4 Crude oil is a mixture of ______ (compounds containing hydrogen and carbon) that can be separated by fractional ______.

5 Soaps and ______ both contain long molecules with a water-loving end and a grease-loving end.

6 Polymers are very large molecules made by joining many small molecules together in a process called ______.

7 Plastics are synthetic polymers that can be ______ into various shapes.

8 ______ melt when heated, but ______ do not (due to cross-links between the polymer chains).

9 Fibres consist of natural or ______ polymers. Their properties depend on the type of polymer they contain.

distillation
moulded
alloys
synthetic
detergents
polymerisation
hydrocarbons
rusting
thermosets
corrodes
thermoplastics

CH•6 REVIEW

1 Soap works as a cleaning agent because:
A it has long chain-like molecules.
B it is made from grease.
C it dissolves in water.
D its molecules have a water-loving end and a grease-loving end.

2 A polymer is a substance that is:
A made up of many small molecules linked together.
B formed when two or more chemicals react together.
C made soft by heating.
D like bakelite.

Questions 3 and 4 refer to the graph on the right, which shows how the strength of a zinc–copper alloy varies with the percentage of zinc in the alloy.

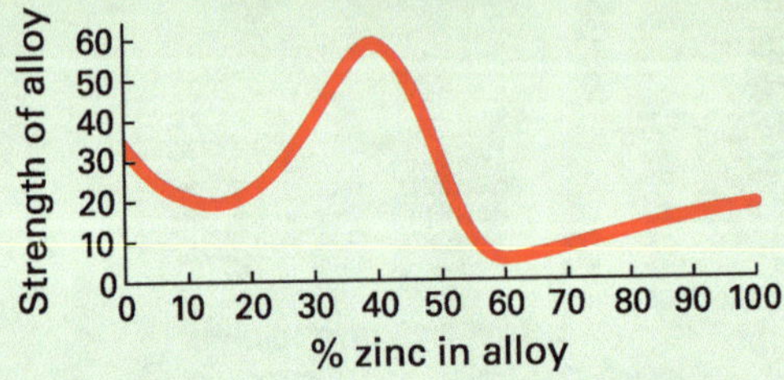

3 At which percentage of zinc is the alloy strongest?
A 20
B 40
C 60
D cannot tell

4 Which is the weakest substance?
A pure copper
B 40% zinc and 60% copper
C 60% zinc and 40% copper
D pure zinc

ISBN 978 1 4202 3736 8

5 Which of the following statements are true and which are false? Rewrite the false ones to make them correct.

- **a** Carbon atoms form four chemical bonds.
- **b** Hydrocarbons contain carbon, hydrogen and oxygen atoms.
- **c** Only thermoplastics can be recycled.
- **d** Hair and wool are both protein polymers.
- **e** Petrol has a higher boiling point than diesel fuel.

6 The diagram shows how plastics are shaped by the process of extrusion.

- **a** Use the diagram to explain how this process works.

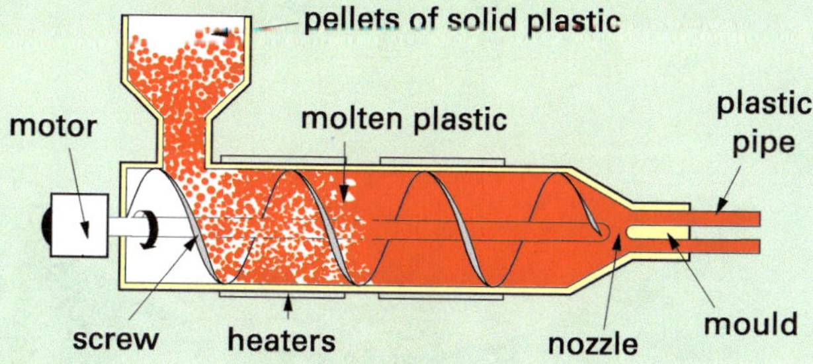

- **b** Which one of the following products could *not* be produced by this method?
  - **A** PVC pipes
  - **B** detergent bottles
  - **C** roof guttering
  - **D** biro casings

7 The boiling points of four liquids are as follows:

	Boiling point (°C)
acetic acid	118
acetone	56
propanol	97
water	100

Suppose you had a mixture of these liquids, and you distilled it.

- **a** Which would be the first substance to distil? Which would be the last?
- **b** Two of the liquids are very difficult to separate by fractional distillation. Which are they, and why are they difficult to separate?

8 How are the following iron or steel objects usually protected from rusting?

- **a** bridge
- **b** bicycle chain
- **c** food cans
- **d** cutlery

9 Polymer A consists of carbon atoms bonded like this:

$$-CH_2-CH_2-CH_2-CH_2-CH_2-CH_2-CH_2-$$

Polymer B consists of carbon atoms bonded like this:

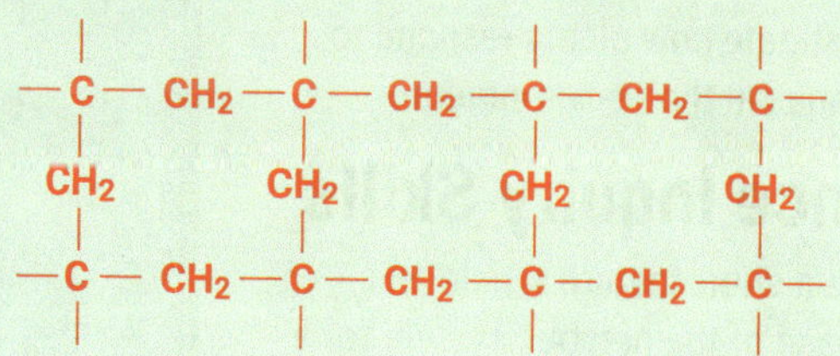

- **a** Which polymer, A or B, is a thermoplastic? Give a reason.
- **b** Which polymer would not melt on heating?
- **c** Which polymer would be stronger? Why?
- **d** Which polymer could be recycled?
- **e** Which polymer is polythene?

10 A major manufacturer of school uniforms is concerned by reports that their windcheater jackets develop holes in the elbows after only a few weeks wear. They insist that threads of the fabric used are able to support very large loads without breaking.

- **a** Write a short explanation of why the breaking strength of the thread is not the problem.
- **b** Describe a test that could be used to decide whether the jackets were any better or worse than those of their competitors.

Check your answers on page 301.

ISBN 978 1 4202 3736 8

IN THIS CHAPTER

Science Understanding

- describe the structure and function of the nervous system
- identify functions for different areas of the brain
- demonstrate reflex actions in the body
- use flow diagrams to show how body systems are controlled and coordinated by the nervous and endocrine systems
- explain how body systems work together to maintain a functioning body
- develop an understanding of negative feedback systems in the body
- investigate how plants respond to changes in their environment

Science Inquiry Skills

- write a short feature article on a disease of the nervous system for a science magazine
- research the cause and effects of spinal cord damage

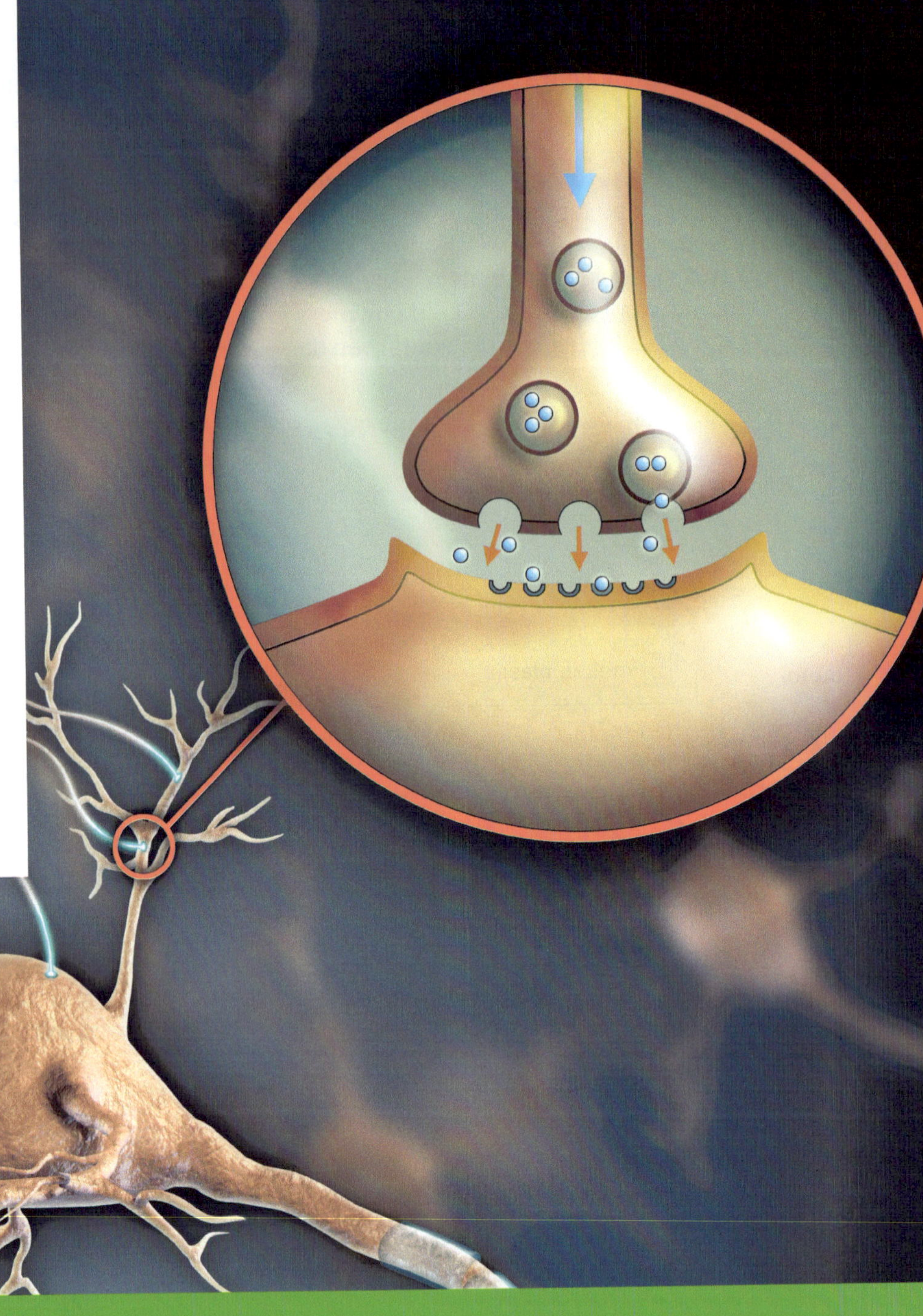

CH•7 Body in balance

ISBN 978 1 4202 3736 8

GET STARTED: *IMAGINE*

> Imagine you are going for a run. As you start to exercise your body is at normal temperature, but after a while you begin to feel hot. This can occur even on a cool day. Despite feeling hot you are able to keep running and keep your body temperature under control. After finishing your run you still feel hot, but your body returns to normal temperature quite quickly.

1 What responses does your body have to keep your temperature from getting too high, and which body systems are involved for each response?

2 Draw a flow chart to show the steps your body goes through to:
  - detect the extra heat in your body
  - respond in various ways to keep your body temperature from going too high
  - respond again as your body temperature returns to normal.

3 What are some of the other things that your body must control to keep you alive?

> Gemma's aunt gave her a pot plant for her room. She placed it near a window. After 2 weeks, the stem had bent towards the window. Gemma rotated the plant 180°. After a week, the plant had straightened up, and after another week it was again bending towards the window.

1 What stimulus made the plant bend?

2 Design an experiment to test your idea.

3 Given that plants have neither bones nor muscles, suggest how they bend towards the light.

ISBN 978 1 4202 3736 8

7.1 The nervous and endocrine systems

You are walking along a path with a friend, eating an apple. Have you ever wondered how you can move, digest food, breathe, think, talk and keep your blood flowing all at the same time without even having to think about it?

All the systems in your body are controlled and coordinated by two other systems—the **nervous system** and the **endocrine system**. The **brain** is the main organ of the nervous system and controls the actions of the nerves and the endocrine system.

The nervous system consists of the brain, the spinal cord and nerves that run to all parts of your body. Messages called nerve impulses travel very quickly along nerves.

The endocrine system consists of a number of endocrine glands throughout your body, which produce chemical messages called **hormones**. These are sent out in the blood, so they take longer to act than nerves, but their effects generally last longer.

The brain is the control centre of your body, and has nerve connections to all parts of the body. At any one time, a huge number of signals are travelling to and from the 10 000 million nerve cells that make up your brain.

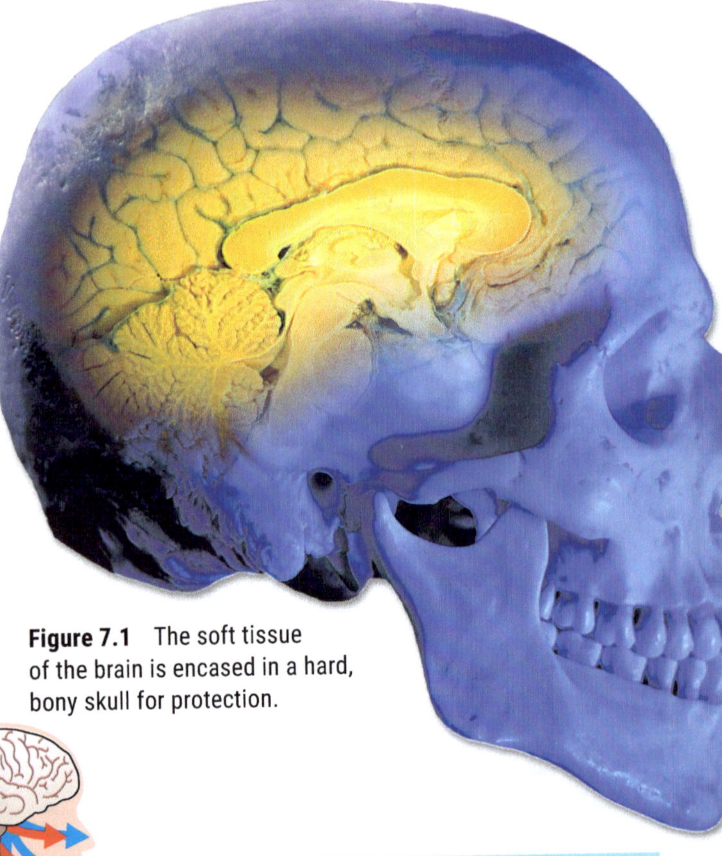

Figure 7.1 The soft tissue of the brain is encased in a hard, bony skull for protection.

Brain—master organ of control

NERVOUS SYSTEM

Controls:
- muscle movement
- senses
  - sight
  - hearing
  - touch
  - pain
  - taste
  - smell
- heartbeat
- breathing
- digestion
- memory
- speech

ENDOCRINE SYSTEM

Controls:
- levels of glucose in blood
- water levels in body
- glucose breakdown in cells
- heat production
- sexual maturity
- sperm and egg production
- growth of cells and tissues

Figure 7.2 Body control

ISBN 978 1 4202 3736 8

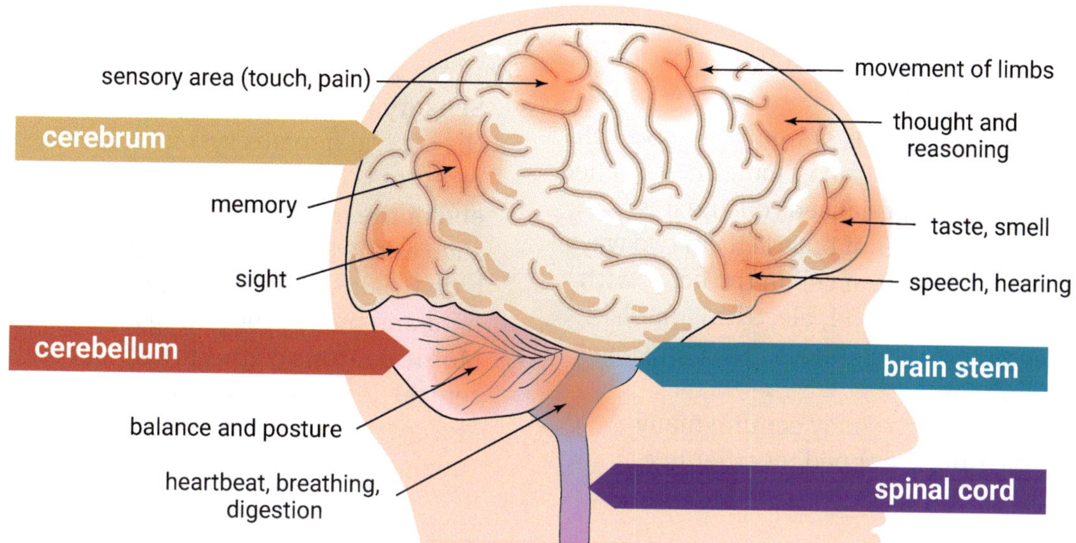

Figure 7.3 The central nervous system (brain and spinal cord)

Parts of the brain

There are three main parts of the brain.

Cerebrum

The **cerebrum** (ser-EE-brum) is the largest part of the brain and controls memory, speech and conscious thought. It receives information from sense **receptors** to give you the sensations of taste, sight, touch, hearing and smell. The cerebrum also controls actions such as walking, running and jumping. All of these actions are called *voluntary actions* because you control them by thinking about them.

Cerebellum

The small part of the brain behind the cerebrum is the **cerebellum** (ser-a-BELL-um). This coordinates muscular activity without you having to think about it (*involuntary actions*). It helps you balance when you ride your bike, surfboard or rollerblades, and it coordinates all your muscles when you walk, run and jump so that you do not fall over.

Brain stem

The **brain stem** at the base of the brain is responsible for other *involuntary actions* such as heartbeat, pulse, digestion and breathing. It also plays a role in controlling other functions, including swallowing, sleep, blood pressure and the startle response.

Dissection of a sheep's brain

You will need a sheep's brain, a scalpel, a cutting board, disposable gloves and newspaper.

Your teacher may show you the features of a brain by using a videoflex microscope camera before you start.

1 Identify the cerebrum, cerebellum and brain stem. Describe their colour and appearance.

2 Notice that the cerebrum consists of two parts called hemispheres. Use the scalpel to separate the two hemispheres, then cut one of the hemispheres in half lengthways to see a cross-section of the cerebrum.

Draw a cross-section of the brain and label the parts.

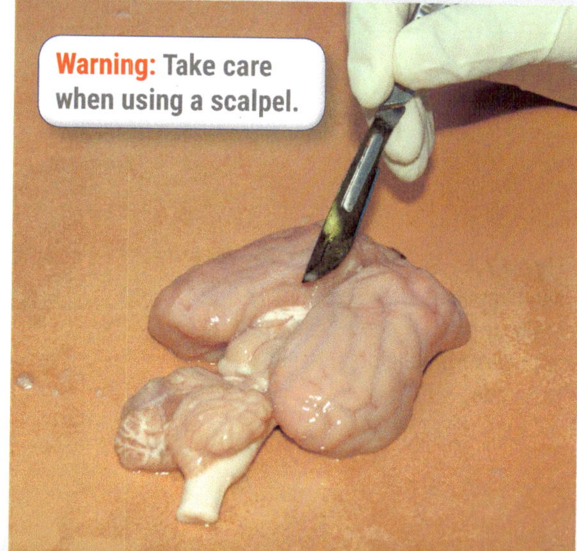

ISBN 978 1 4202 3736 8

Types of nerves

The basic unit in the nervous system is a nerve cell or **neuron** (NEW-ron). Neuron is sometimes spelled *neurone*. A neuron is a specialised cell and it is different from other types of body cells in two ways. First, it is the longest type of cell in the body, with a long branch or fibre. Second, electrical impulses travel along the nerve fibre. These impulses travel in one direction only.

The main nerves in the body contain many individual nerve fibres wrapped together in a sheath.

There are three types of neurons:

- **Sensory neurons** send nerve impulses to the central nervous system from the body's receptors. The receptors are attached to one end of the neuron and detect external stimuli such as the level of carbon dioxide in the blood, pain or touch, fullness of the bladder, light and many other things.
- **Motor neurons** connect the brain and central nervous system to muscles or glands. They carry impulses to the muscle or gland which causes an effect, such as the muscle

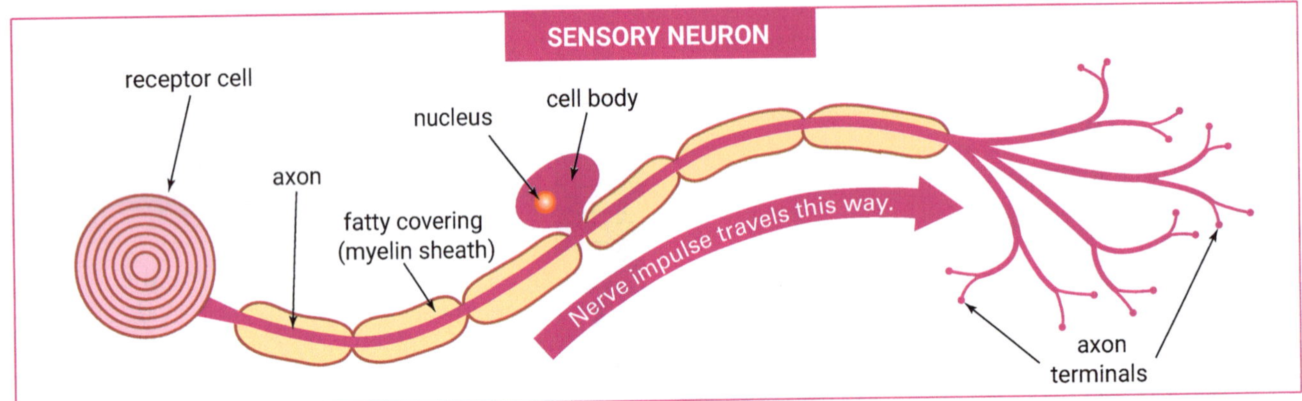

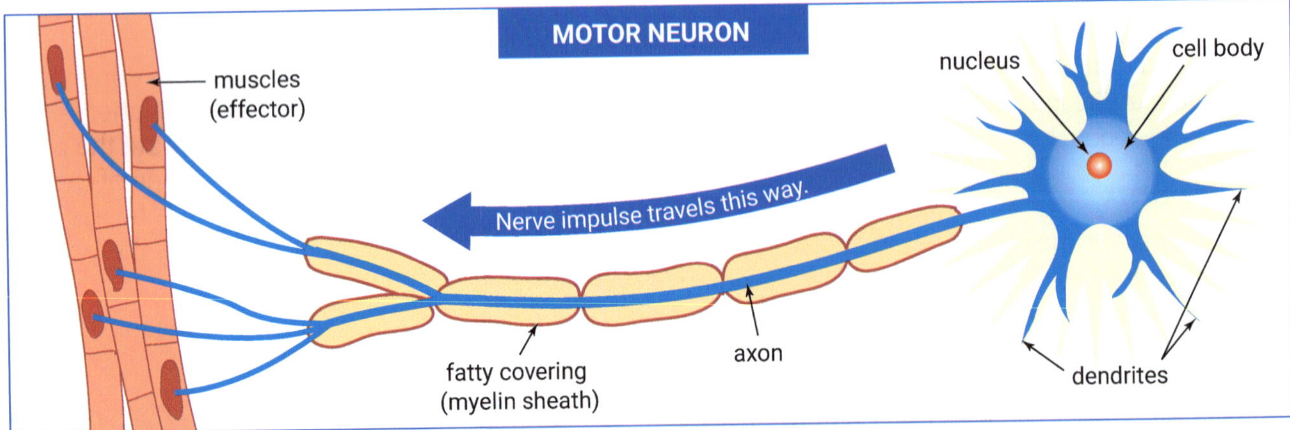

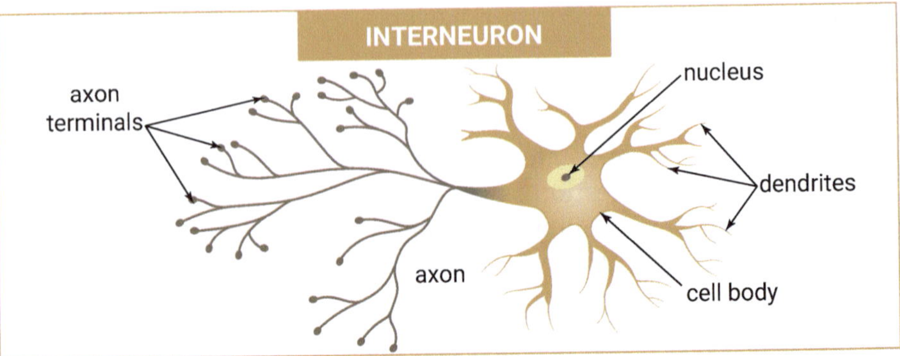

Figure 7.4 Sensory nerve cells send impulses from receptors to the brain, and motor nerve cells send impulses from the brain to muscles or glands.

ISBN 978 1 4202 3736 8

contracting. The muscle or gland is known as an **effector** as it causes or effects (brings about) change.

- **Interneurons** connect motor neurons and sensory neurons. About 90% of neurons are interneurons, and are contained in the brain and spinal cord. Interneurons process, store, retrieve and transmit information.

Nerve cell structure

Neurons consist of three main parts which can be seen in Figure 7.4:

- The *cell body* contains the nucleus and controls the cell functions, such as respiration to produce energy inside the cytoplasm, just like in any cell.
- The *axon* is the long fibre that carries the impulses along the nerve. In motor and sensory neurons it is wrapped in a fatty layer called *myelin* (MY-e-lin) sheath, which insulates the axon, enabling the electrical impulses to travel faster. The Axon ends on many small swellings called axon terminals.
- *Dendrites* are short extensions of the cell body, like branches that go out in all directions. They connect with the axon terminals of other neurons and carry impulses towards the cell body.

The synapse

Nerve impulses travel through neurons as an electrical signal. But the neurons do not touch each other—there is a tiny gap between them called the **synapse** (SIGH-napse). When a nerve impulse reaches the end of one neuron, it cannot jump to the next neuron. In the axon terminal swellings are special chemicals called neurotransmitters. When the impulse reaches the axon terminal it releases the neurotransmitter, which travels across the synapse to receptors on the next neuron, and stimulates a nerve impulse in that neuron. After crossing the synapse, the neurotransmitter is quickly broken down so the neuron can be stimulated again by more neurotransmitters. Scientists have found more than 50 types of neurotransmitters in the human body.

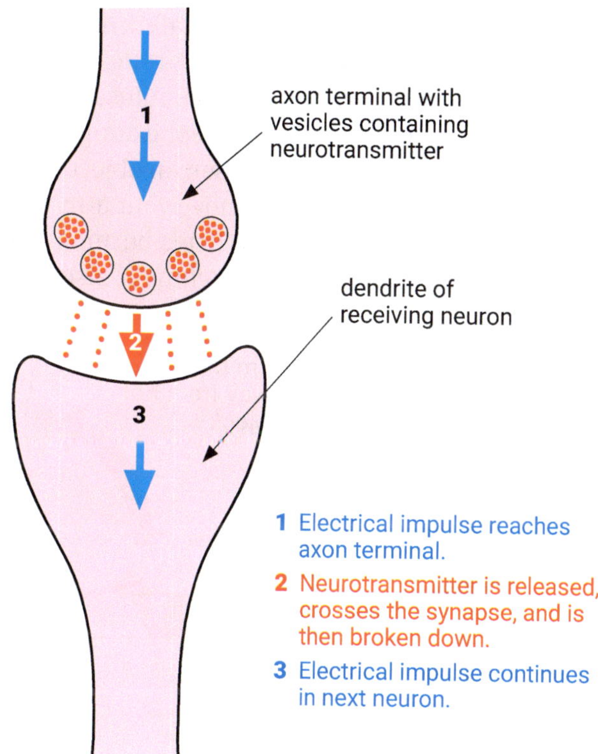

Figure 7.5 How neurotransmitters work

Receptors

The table below lists the sensory receptors in the human body and the stimuli they detect.

Receptor	Stimuli detected
rods and cones in eye	light
cells in cochlea (inner ear)	sound
skin (many receptors)	touch, tissue damage (pain), vibration, pressure, hot and cold
around hairs in skin	touching the hair
taste buds in tongue	chemicals in food
olfactory cells in nose	chemicals in air
cells in inner ear balance organs	movement

ISBN 978 1 4202 3736 8

Reflex action

Not all the information from the receptors is coordinated by the brain. Sometimes a nerve impulse takes a shortcut to the spinal cord and back. This occurs in a **reflex action**. A reflex action is a very fast response to a stimulus, and is generally a response to danger. For example, you blink to protect your eyes when an object approaches your face, and you move your arm away quickly from a frying pan that is burning hot to avoid damaging your skin. Coughing is also a reflex action.

In a reflex action, the nerve impulse travels from a receptor along a sensory neuron to the spinal cord and then back along a motor neuron to a muscle. The whole action takes place very quickly because the brain does not coordinate the action.

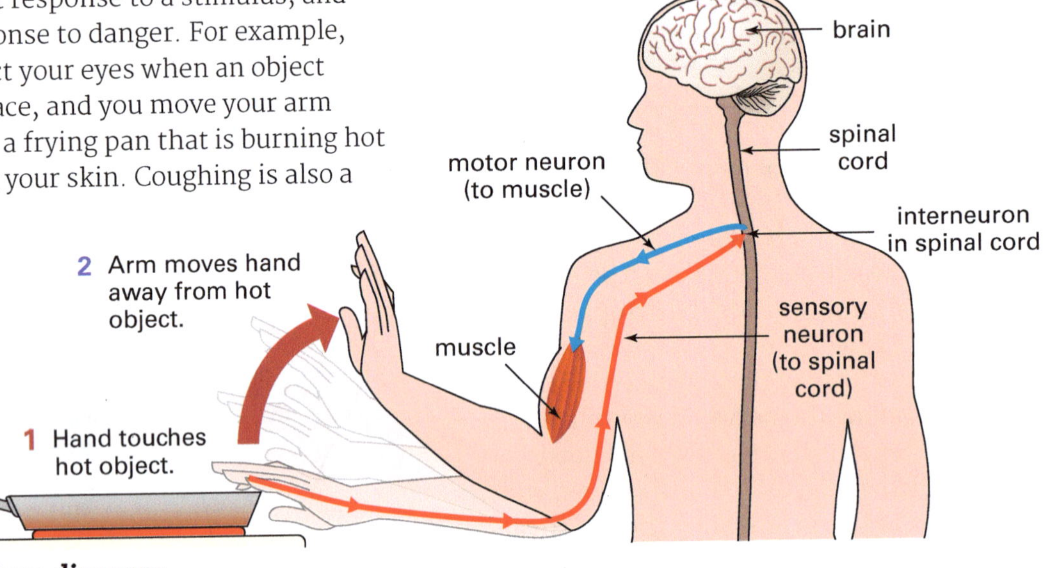

Reflex action flow diagram

Receptors in the skin detect heat or cell damage. → Nerve impulse is sent along sensory neuron to the spinal cord. → Nerve impulse from spinal cord is sent along motor neuron. → Impulse from motor neuron acts on muscle in arm. → Arm moves away from heat.

Figure 7.6 How a reflex action works

Reflex actions

There are a number of reflex actions that you can observe in humans.

1 Have your partner sit on a chair with one leg crossed over the other. With the side of your hand, gently tap their knee just below the knee cap.
   - Describe this reflex action. What type of receptor detects the stimulus?
   - Draw a flow diagram like the one above to show the reflex action.

2 Stand behind a window or a glass door or hold a piece of clear plastic in front of your face. Have your partner throw a crumpled-up piece of paper at the glass.
   - Why do you blink every time the paper ball is thrown, even though you are protected by the glass?
   - Which receptor is used in this reflex action?
   - Draw a flow diagram to show the reflex action.

3 Cover one eye with your hand for at least 30 seconds. After this time, have your partner look at your eye when you take your hand away.
   - What happens to the size of the pupil?
   - What was the stimulus that caused this response?
   - Draw a flow diagram to show the reflex action.

ISBN 978 1 4202 3736 8

Nerves, poisons and drugs

Poisons

There are a number of poisons that react with neurotransmitters causing paralysis and even death. For example, curare (coo-RAR-ray) is a poison extracted from plants in South American forests. Animals that are hunted and hit by arrows dipped in curare become paralysed. Bacterial poisons that cause food poisoning and tetanus also stop neurotransmitters working.

Insecticides

When the neurotransmitter is released and stimulates the other neuron, an enzyme destroys it so that it cannot keep acting and stimulating the other neuron. This process also occurs in insect nerves. The active ingredient in some insecticides reacts with the enzyme and destroys it. This means that nerve impulses fire continuously, causing the insect's muscles to move rapidly and uncontrollably, resulting in death.

Drugs

Nicotine, found in tobacco, is a stimulant because it acts like a neurotransmitter in the brain, giving a pleasurable effect (making the body more active or alert). However, it is addictive and very toxic in large amounts.

Amphetamines ('speed') and cocaine are also stimulants because they increase the release of neurotransmitters. This results in heightened emotions and an increased feeling of alertness and confidence. But later this can lead to anxiety, panic, depression and hostility.

Some other drugs such as alcohol and heroin decrease the release of neurotransmitters, making the person inactive or drowsy. These drugs belong to a group called depressants (the opposite effect of stimulants).

Figure 7.7 There are many types of poison dart frogs in South America which have been used in hunting for centuries. This strawberry poison dart frog, from the Costa Rica tropical rain forest, contains enough poison to kill 10 adults.

Figure 7.8 Long-term over-use of depressants such as alcohol leads to liver and heart damage as well as memory loss and brain damage.

ACTIVITY

1. Compare the action of drugs (stimulants and depressants) on nerve transmission with the action of poisons such as curare.
2. Suppose you want to use the internet to find out more about what is on this page. What search words would you use? How could you guarantee that the information was genuine and accurate?
3. Invite a drug and alcohol consultant to discuss with the class the use and misuse of stimulants and depressants. Prepare some questions to present to the consultant well before the discussion.

ISBN 978 1 4202 3736 8

SCIENCE AS A HUMAN ENDEAVOUR

Nervous system diseases

Two major diseases of the nervous system are multiple sclerosis and motor neuron disease.

Multiple sclerosis (MS)

MS is a disease that attacks the neurons in the brain and spinal cord. It is an unpredictable disease because the symptoms can occur at any time and be mild or severe. MS symptoms range from blurred vision to complete blindness, and from tingling and numbness to paralysis.

MS affects more women than men, and occurs more commonly in people with northern European ancestry. In 2016, there were 2.5 million people worldwide affected with MS, and 23 000 of those were in Australia.

MS is not contagious or directly inherited and the actual cause is at present not known. The disease occurs when the body's own immune system attacks the fatty substance (called myelin) around the neurons, causing a disruption to nerve transmission.

Motor neuron disease (MND)

MND is a group of diseases that affect the motor neurons that control the muscles that enable you to use your arms and legs, breathe, talk and swallow. It does not affect your intellect or your memory.

The cause of MND is not known. In almost all MND sufferers, there is no family history of the disease. The disease is often fatal within two to five years after diagnosis. One well-known exception is Stephen Hawking, the Cambridge University physicist and cosmologist, who has had MND for nearly 50 years.

Symptoms of MND usually occur in people aged 50 to 70. They start with muscle weakness in the arms and legs, which gets progressively worse. Muscles tend to wither and speech becomes slurred. At present there is no cure for MND.

Questions and research

1. Draw up a table listing the causes, symptoms and treatment of MS and MND.
2. Use the table to write a 'compare and contrast' paragraph about MS and MND.
3. Use your table from question 1 to write a short feature article for a science magazine on one or both of the diseases. Include a fictitious interview in your article.

Spinal cord damage

Spinal cord injury is a common occurrence from sporting injuries, diving into shallow water, car accidents, falls from heights, and electrocution. Damage to the spinal cord can result in paralysis or partial loss of control of arms and legs, loss of feeling, and loss of control of body organs such as the bladder or bowels.

Most people with spinal cord injuries need assistance with daily life, and may also need assistive equipment such as wheelchairs or

Figure 7.9 This exoskeleton is being tested by a paralysed woman to help her walk again.

ISBN 978 1 4202 3736 8

walkers. Repairing spinal cord damage is still not possible, although progress has been made in some areas. Stem cells offer some hope for regrowing and repairing spinal cord cells, and robotic technology to produce exoskeletons could help some people 'walk' more independently in future.

Questions and research

Research spinal cord damage to produce a report that includes the following information:

1. What are the main causes of spinal cord injury in Australia?
2. What are the main effects on people who suffer a spinal cord injury? What is paraplegia and quadriplegia?
3. Summarise some of the possible cures and solutions being researched to help repair spinal cord damage. Comment on their potential for improving the quality of life for patients.
4. Include references for your research.

Hormones—chemical controllers

Hormones are produced in **endocrine glands**. The difference between these glands and others in your body, such as sweat glands and glands in the stomach lining, is that the hormones made by endocrine glands pass directly into the blood.

There are many different hormones and each one acts on specific target cells. For example, a hormone released by certain cells in the lining of the first part of the small intestine acts only on cells in the pancreas that make digestive juices. However, insulin, which is produced in specialised cells in the pancreas, has a broader action. It makes the liver cells store glucose and helps muscle cells throughout the body absorb more glucose from the blood.

Hormones are different from nerves in that they can act on the whole body, on body systems or on individual organs. Nerves act only on muscles and glands.

The diagram on the right shows some of the major endocrine glands and the effects the hormones produced have on the body.

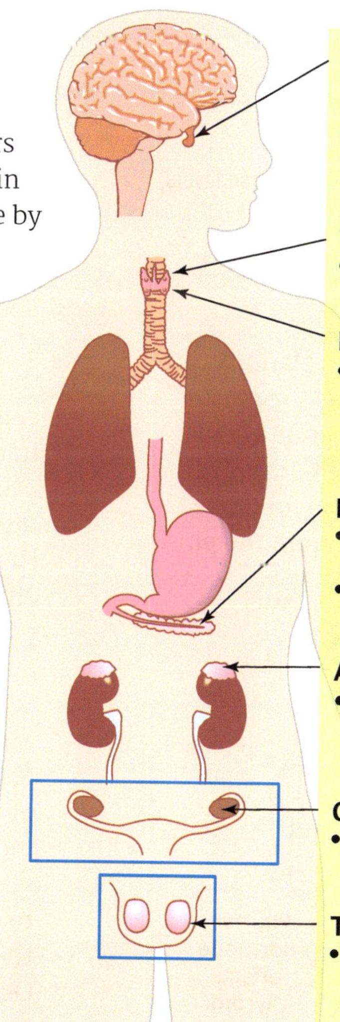

Figure 7.10 The endocrine system produces many hormones that control functions of the body.

ISBN 978 1 4202 3736 8

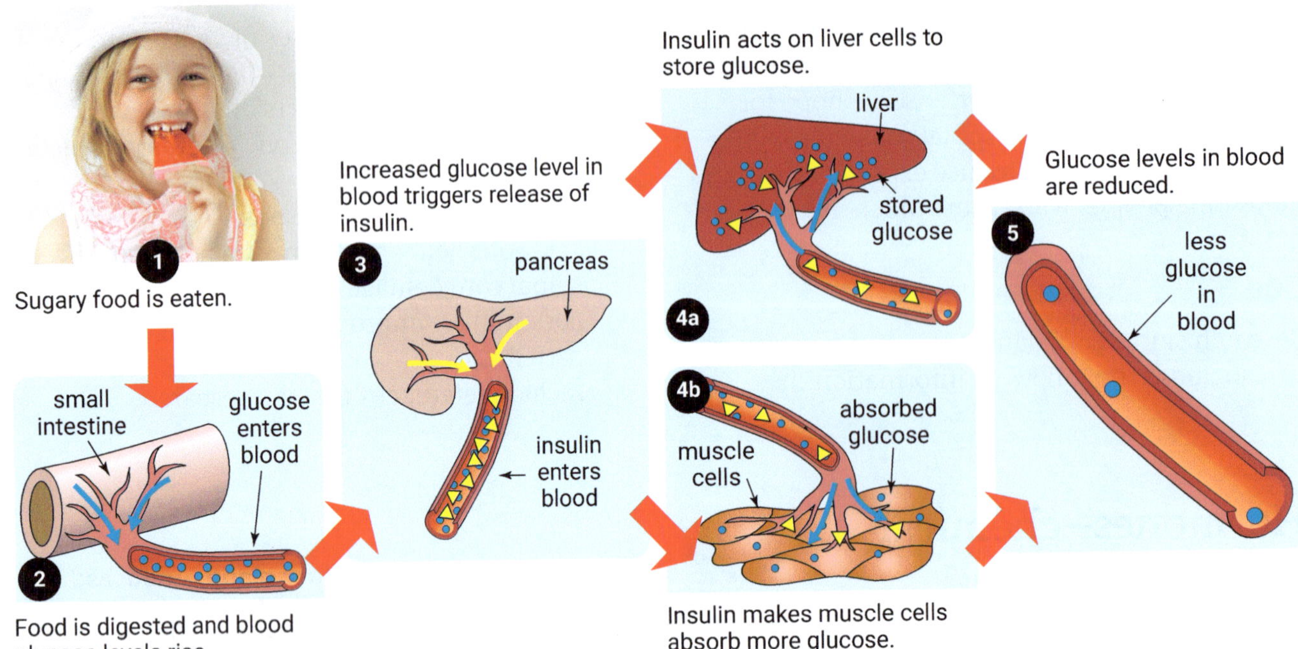

Figure 7.11 How insulin reduces blood glucose levels

The pituitary—the master gland

The **pituitary** (pit-YOU-it-tree) **gland**, located on the underside of the brain, is the master gland that controls other endocrine glands. For example, the release of the hormone from the thyroid gland in your neck is controlled by thyroid-stimulating hormone from the pituitary.

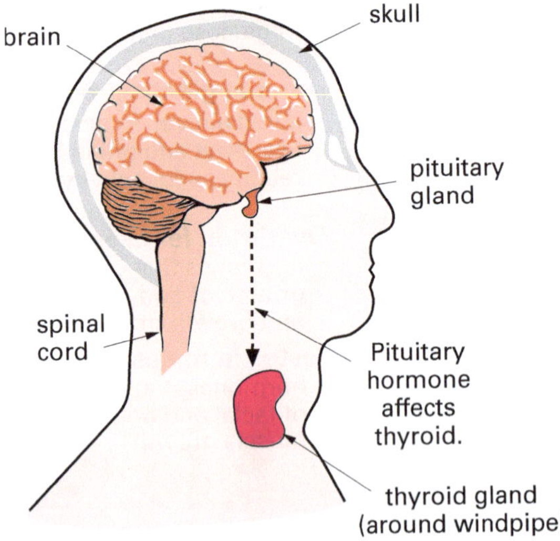

Figure 7.12 The pituitary gland controls other endocrine glands.

Growth and development in humans

The pituitary gland also releases hormones that affect the reproductive organs. Before puberty the main physical growth of a person is under the control of the pituitary, which releases growth hormones.

Between 10 and 15 years after birth, the pituitary begins to release hormones that affect the reproductive organs. This causes major changes to the body and is the beginning of puberty.

In males, one pituitary hormone acts on the cells in the testes, which make sperm. Another acts on the testes to make the hormone testosterone. Testosterone stimulates the growth of facial and body hair and is also responsible for increased muscle growth.

In females, one pituitary hormone leads to egg production in the ovaries and also to the manufacture of the hormone oestrogen. Oestrogen causes an increase in the growth of body hair, the growth of fat cells under the skin and the development of breasts. Another pituitary hormone acts on the ovaries and is responsible for the start of the menstrual cycle.

ISBN 978 1 4202 3736 8

SCIENCE AS A HUMAN ENDEAVOUR

Diabetes

Diabetes is a fairly common disorder. Its main symptoms are glucose in the urine, extreme thirst, hunger and loss of weight. Before the 1930s, fewer than 20% of people lived more than 10 years after developing diabetes.

Diabetes is caused by the pancreas not producing any or enough of the hormone insulin. After a meal is digested, a large amount of glucose is absorbed by the blood and the level of blood glucose rises. Insulin makes liver cells store glucose, and makes muscle cells absorb more glucose from the blood. The net effect is to reduce the amount of glucose in the blood. In people with diabetes, the pancreas does not produce enough insulin, or in some cases none at all, hence the high levels of glucose in their blood.

Types of diabetes

There are two main types of diabetes.

- **Type 1 diabetes** occurs when the body stops making insulin. This type is found in only 10–15% of all diabetes sufferers, and it occurs mainly in young people. It cannot be prevented and is treated by daily insulin injections.
- **Type 2 diabetes** is the most common type and it occurs mainly in people over 40 who are overweight and inactive, have high blood pressure or heart disease, or in women who develop diabetes in pregnancy.

Diabetes is a serious illness for which there is no cure at present. *You cannot cure diabetes but you can control it.*

Parts of the body affected by diabetes

The high levels of glucose in the blood cause serious problems in the body. The blood vessels and nerves are the most affected. The walls of the small blood vessels thicken and block the blood supply. This causes problems in the eyes, kidneys, legs and heart.

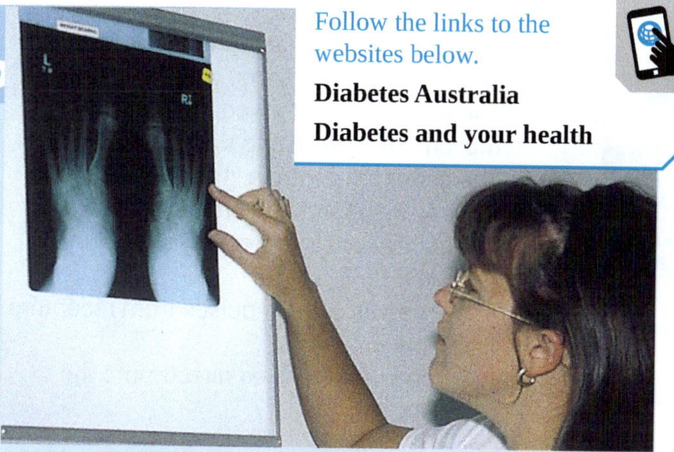

Follow the links to the websites below.

Diabetes Australia

Diabetes and your health

Working with diabetes

Susan Mylne is a podiatrist (a person who examines feet) who works in community health and is very interested in diabetes.

Why is Susan interested in diabetes? Type 2 diabetes sufferers often have nerve damage in their feet. This causes 'pins and needles', a burning sensation and numbness. These people also suffer blood vessel damage, which leads to poor blood circulation, and causes cramps, ulcers and pains in the legs.

Many of Susan's patients who have these symptoms often do not know they have diabetes. She can test for the complications of diabetes in their feet, help them in their treatment, and help educate them about their illness.

Questions and research

For this section you may use the websites in 'Explore online' above or search the internet for further information.

1 What is the importance of insulin in the body?
2 What is the difference between type 1 and type 2 diabetes? Which type would you class as a 'lifestyle illness'? Why?
3 There has been a rapid increase in the number of people with type 2 diabetes in the last five years. Suggest reasons for this.
4 Suppose you are in charge of preparing a brochure about diabetes. What information will you include to inform people about the effects of diabetes and its prevention? Prepare a draft design for the brochure.

ISBN 978 1 4202 3736 8

1 Some of the following statements are false. Rewrite the false ones to make them correct.
   a The endocrine system controls muscle movement, speech and the senses.
   b The cerebrum controls involuntary actions such as heartbeat and breathing.
   c The spinal cord is protected from injury by the skull.
   d Motor nerves carry impulses from receptors to the brain.
   e Hormones are released directly into the bloodstream.

2 Suggest why reflex actions might be useful for an organism's survival.

3 What are the differences between a sensory neuron, a motor neuron and an interneuron?

4 The knee jerk reflex that occurs when the knee is tapped is an example of a simple reflex action. Copy the drawing top right and replace the letters with a description of what occurs at each stage.

5 Describe how hormones control the growth and development of the human body.

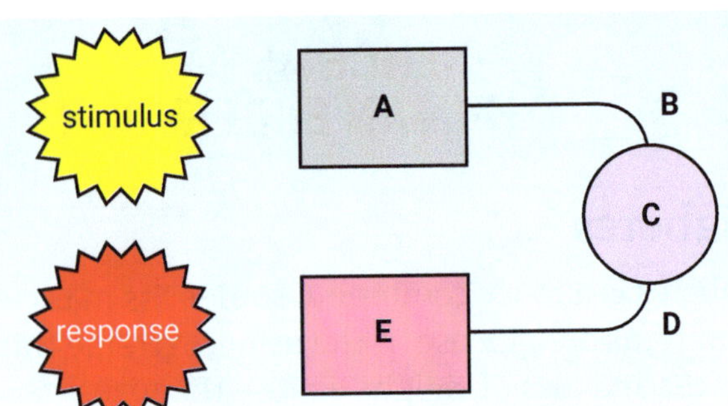

6 How are the actions of nerves different from those of hormones? Give examples.

7 What is the role of the pituitary gland in the functioning of your body?

8 Suppose you hear a sudden, extremely loud noise. What type of responses might occur to this stimulus? Would these responses be caused by nerves or by hormones? Explain.

9 a From the information on page 165, try to explain what the terms *electro-transmission* and *chemical transmission* of nerve signals mean.
   b Use a diagram to describe how a nerve signal can travel over the gap between neurons.

1 A way to test a person's reaction to a stimulus is to drop a ruler between their thumb and fingers. The reaction distance is how far the ruler falls before they catch it.
   a Ruby tested Ben's reaction by dropping a ruler seven times. The table shows his results. Suggest an inference to explain the results.

Trial	1	2	3	4	5	6	7
Reaction distance (cm)	28	21	17	13	11	10	9

   b Ruby then tested Mia's reactions. The ruler fell an average of 16 cm before she caught it with her right hand, and 18 cm with her left hand. Suggest a reason for the different results.
   c Use the formula below to find Mia's reaction time.

$$\text{reaction time (s)} = \sqrt{\frac{\text{reaction distance (cm)}}{500}}$$

2 A person who suffers a fractured neck is often paralysed below the fracture. Why is this?

3 The graph below shows the amount of glucose in a person's blood over a 24-hour period.
   a Explain the reason for the peaks in the graph.
   b At what times during the 24 hours did the amount of insulin being released from the pancreas increase? Suggest what caused this increase. What was the effect of the increase in insulin in the blood?
   c Suggest why the level of blood glucose decreases slightly during sleep.

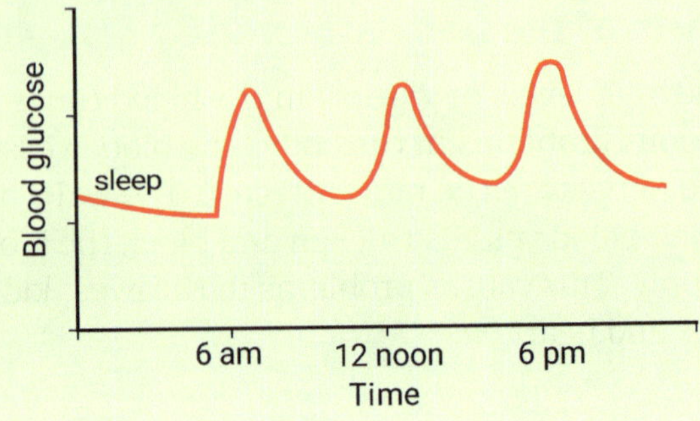

ISBN 978 1 4202 3736 8

7.2 Plant responses

Why do some plants, such as the one in Figure 7.13, close their leaves at night?

The plant closes its leaves in response to light and darkness. In the morning, the sunlight acts as a stimulus for the leaves to open. Plants also respond to other external stimuli such as gravity, temperature, moisture and, in some plants, touch.

Plants do not have a nervous system, muscles or specialised glands, so how do they detect stimuli and how do they react to these stimuli?

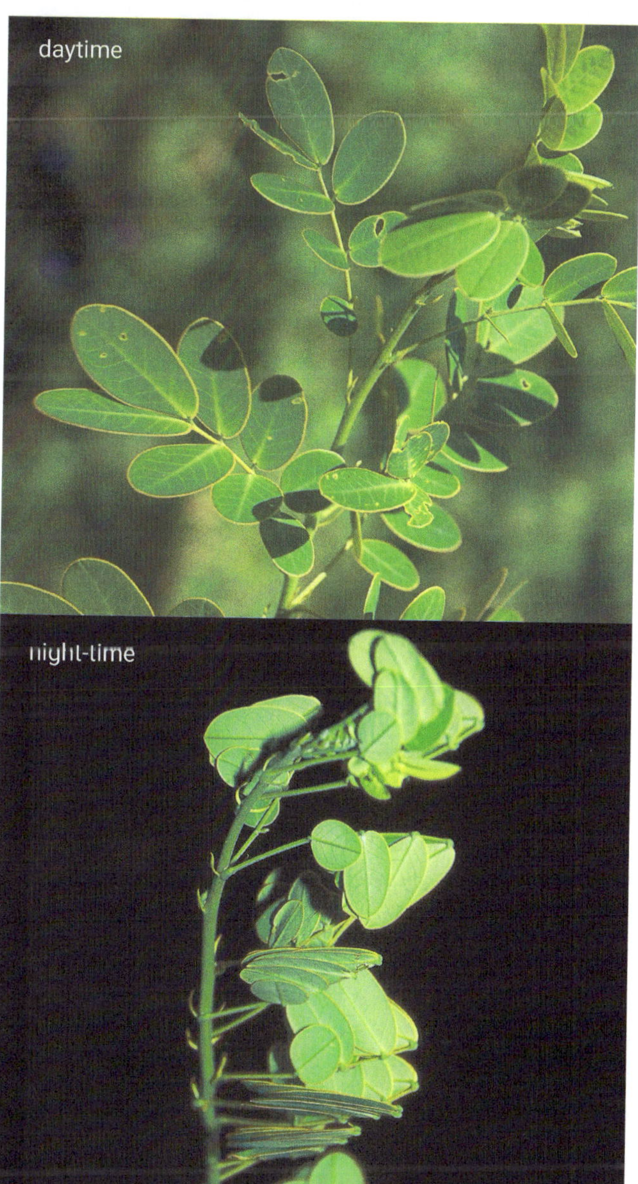

Figure 7.13 A plant response to daylight

Plant hormones

The internal control of activities in a plant are due to hormones. These chemical compounds are made by certain cells and are distributed throughout the plant from cell to cell or by the microscopic tubes that run through the plant. Plant hormones are quite different chemically from those in animals. But, as in animals, very small amounts of the hormones have a large effect on the target cells and the plant as a whole.

Plants have far fewer hormones than animals, and, unlike animals, have no specialised glands such as those in the endocrine system.

Plant hormones are responsible for controlling the growth of stems and roots, the ripening of fruit and the loss of leaves during autumn. Hormones also determine when a plant will flower and when seeds will germinate.

When a seed germinates, the young root of the plant grows downwards. If the seedling is turned upside down, the root will bend and continue to grow downwards. This response to gravity is caused by a number of hormones that are produced in the root cells. You can investigate the response to gravity on the next page.

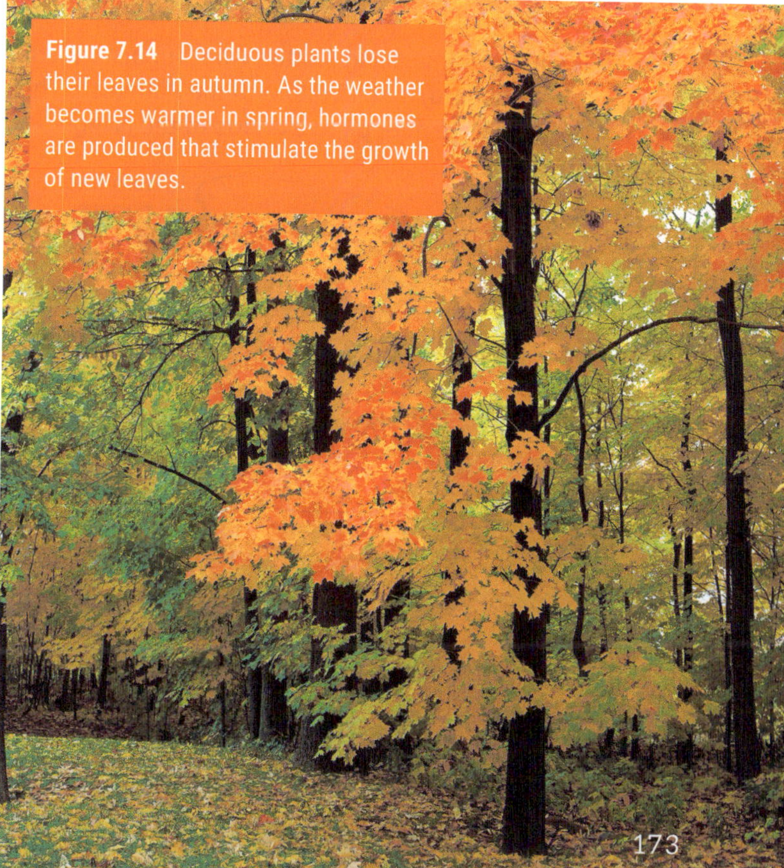

Figure 7.14 Deciduous plants lose their leaves in autumn. As the weather becomes warmer in spring, hormones are produced that stimulate the growth of new leaves.

ISBN 978 1 4202 3736 8

INVESTIGATION 7.1 Responses of plant roots

Aim

To investigate the responses of roots to gravity.

Materials

- 8 small bean seeds, e.g. mung beans (see Risk assessment and planning below)
- glass jar
- Petri dish (11 cm diameter)
- filling, e.g. rubber carpet underlay, cotton wool, newspaper or cardboard
- 2 filter papers (to fit Petri dish)
- Blu-Tack
- adhesive tape

Risk assessment and planning

Prepare the bean seeds in advance. Soak them in a jar of water until they start to germinate. Tip out the water and leave the jar in a cupboard until the roots grow to about 1 cm long. Rinse the sprouts with fresh water each day.

Method

1 Place some filling (cotton wool, rubber underlay, cardboard) in the bottom of the Petri dish. Then place two filter papers on top of the filling.

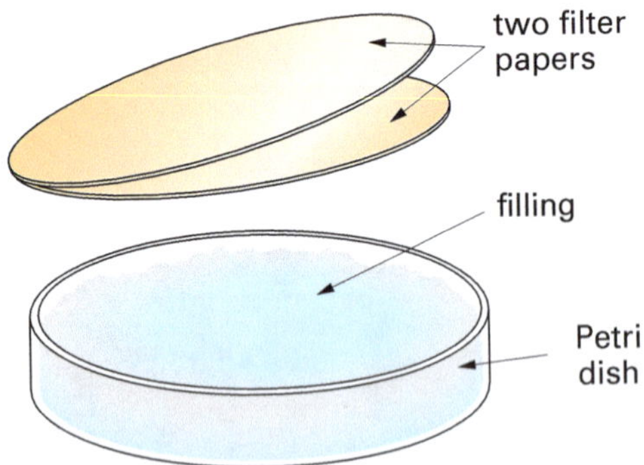

2 Place two bean sprouts on the filter paper with their roots pointing outwards, as shown in the diagram at the top of the next column. Place another two sprouts at the 90°, 180° and 270° positions.

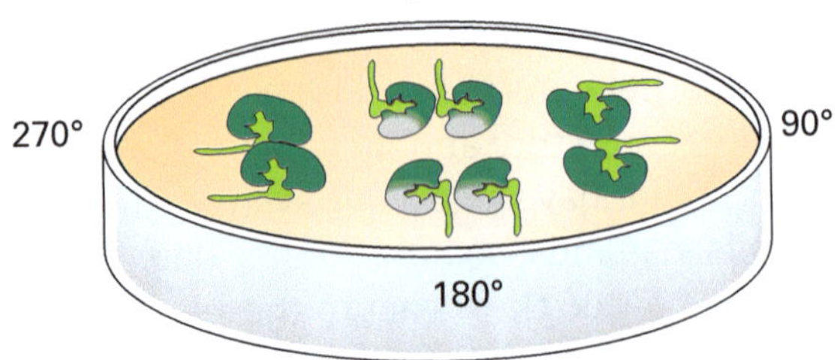

3 Moisten the filter paper with some water. Put the lid on, and tape it securely. Make sure the seeds are jammed in by the thickness of the filling and filter papers. If the seeds move, add more layers of filter paper.

4 Place the Petri dish on a piece of Blu-Tack in a vertical position, as shown. Use a marking pen to mark the position of the roots.

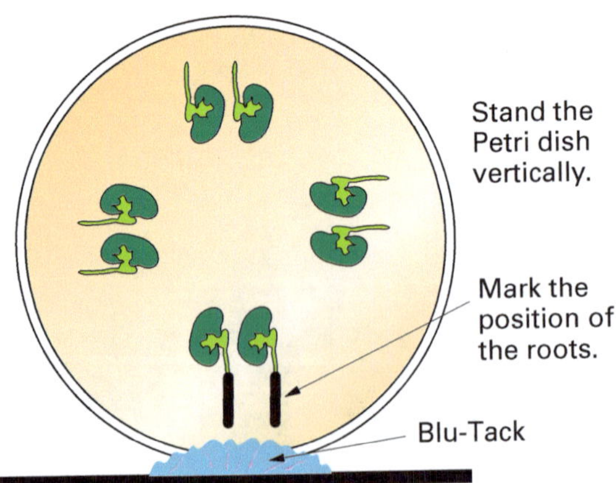

5 Leave the Petri dish in this position for a day. Then rotate the dish 90°.

Record what happens to the direction of growth of the roots.

6 Repeat for another position and again record your observations.

Discussion

1 Write a report of your findings.

2 Suggest what might happen to the roots if the Petri dish was taken into space.

ISBN 978 1 4202 3736 8

Growth hormones

A group of hormones called **auxins** (ORK-sins) are responsible for controlling the growth of stems and roots. They also make plants bend towards the light.

Response to light

One type of auxin is made in the cells in the growing tips of plants. This hormone moves through the cells away from the light.

The hormone passes downwards until it reaches the cells in the growth region just below the tip. It acts on these cells and makes them divide and grow in length; hence, the plant grows taller. The cells in the growing region of the plant are the target cells for auxin. Cells outside this region do not respond to the hormone (see Figure 7.16).

When light comes from one side, more auxin is found on the side away from the light. This causes the growth of cells on the darker side, which bends the plant towards the light (see Figure 7.17).

Figure 7.15
This tomato plant was placed on its side 2 days before the photo was taken.

Weedkiller

The weedkiller *glyphosate* is a compound with a similar structure to the auxin produced in a plant's growing tip. (Roundup® is a brand name of the weedkiller that contains this compound.) When glyphosate is applied in very dilute solutions, it will promote plant growth. However, in stronger solutions, plants grow uncontrollably and eventually die.

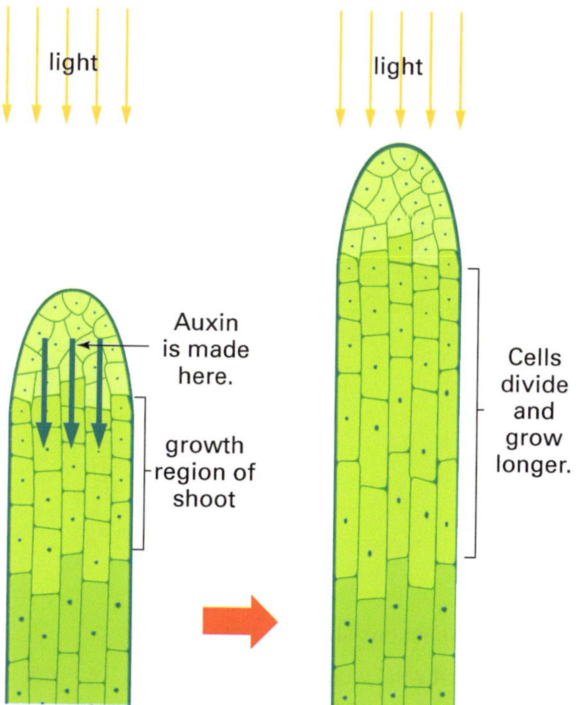

Figure 7.16 Auxin moves to the cells in the growth region where it makes them divide and grow longer.

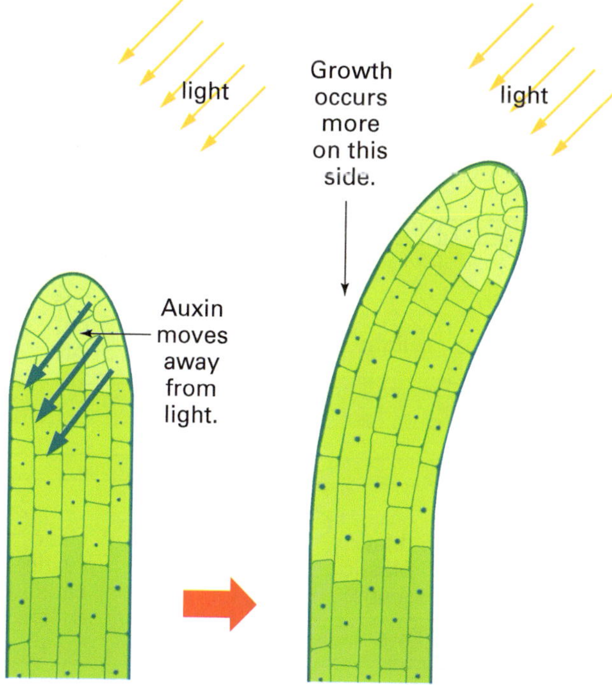

Figure 7.17 When light comes from one side, cells on the darker side divide and grow more than the cells on the other side.

ISBN 978 1 4202 3736 8

1 Describe the changes in plants that hormones are responsible for.

2 Most animals, particularly mammals, show rapid responses to certain stimuli such as hot objects, loud noises and bright light. Plants, however, respond very slowly to stimuli. Suggest reasons for this difference.

3 A pot plant has been growing near a window for a number of weeks and all the leaves face the window.
The plant is turned around 180°. After a few weeks, the leaves of the plant have moved around to face the window again.
   a What external stimulus is the plant responding to?
   b What is the advantage of such a response to the survival of the plant?

4 A bean seed was germinated and left to grow for a number of days. It was then turned sideways as shown.
   a Draw what you would expect to happen to the seedling.
   b What stimulus is the seedling responding to?

1 Biologists believe that another hormone besides auxin is produced in root tips. They think that this hormone stops the action of auxin. Suggest how the two hormones might be responsible for the growth of the bean shoot in Check 4 above.

2 The diagrams below show the results of four tests in an experiment using oat seedlings. Use the diagrams to answer the following questions.
   a Suggest a title for the experiment.
   b Look at the results of C and D. Write an inference to explain the differences.
   c Which were the control seedlings? Explain.

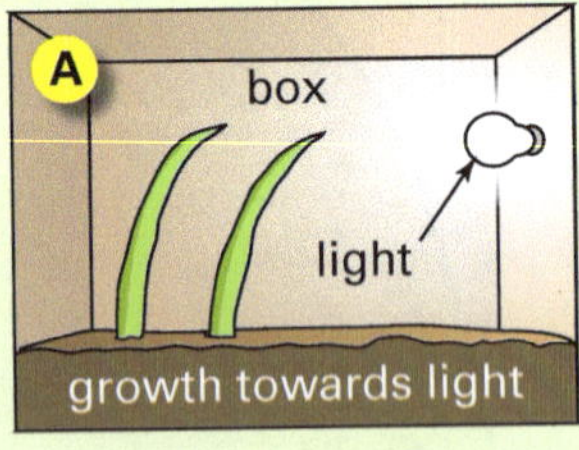

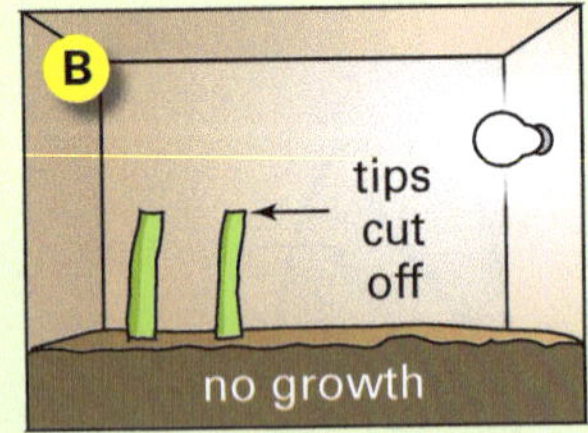

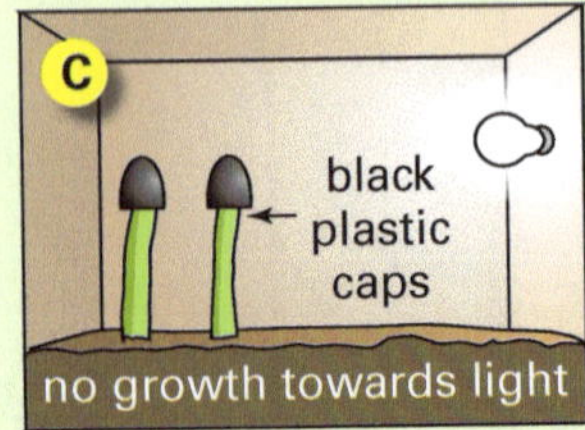

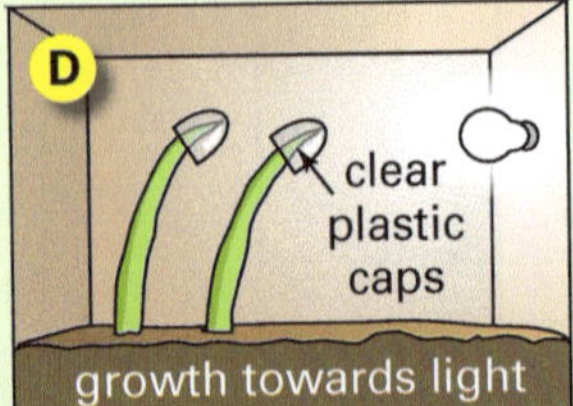

3 Plant fertilisers promote the growth of plants. In which ways are fertilisers different from plant hormones?

4 The root of a 1-day-old bean seedling was marked with equally spaced lines. The root was observed on the second day.

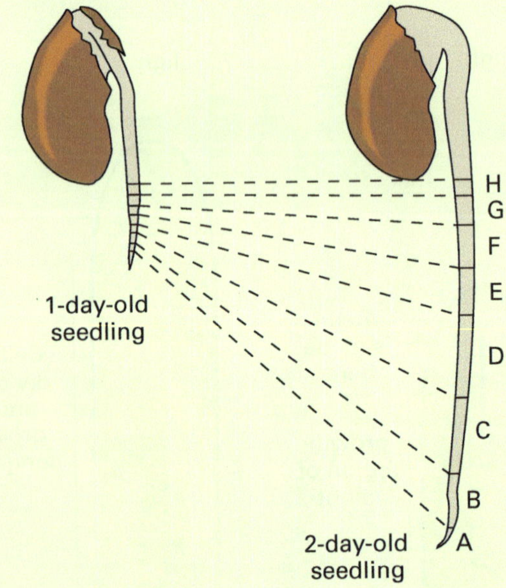

   a Which sections showed most growth?
   b Suggest what would have happened if the lower four sections had been cut off after day one.

ISBN 978 1 4202 3736 8

7.3 Keeping the balance

On page 169 you learnt how the glucose level in the blood is controlled by the hormone insulin. Now let's look at the control of heat and water in the body.

Heat balance

Your body temperature stays at about 37 °C. This is the *set-point temperature* for humans. The chemical reactions in the cells of your body work best at this temperature. All mammals and birds have a constant body temperature, although the set-point temperature varies between groups of these animals. For example, the set-point temperature for a magpie is about 39 °C, while for echidnas it is about 31 °C.

Most of the heat needed to maintain the set-point temperature is generated during the respiration of glucose in your body cells, particularly in the liver, kidney and brain. On the other hand, most of your body's heat loss is by radiation from your skin surface and by evaporation of sweat from your skin. The table at the bottom of the page summarises the ways in which heat is gained and lost by your body.

Control of heat balance

During exercise, your muscles generate heat, which increases the temperature of the blood. A receptor in the brain just above the pituitary gland is sensitive to changes in the temperature of the blood and sends out nerve impulses to the skin and sweat glands.

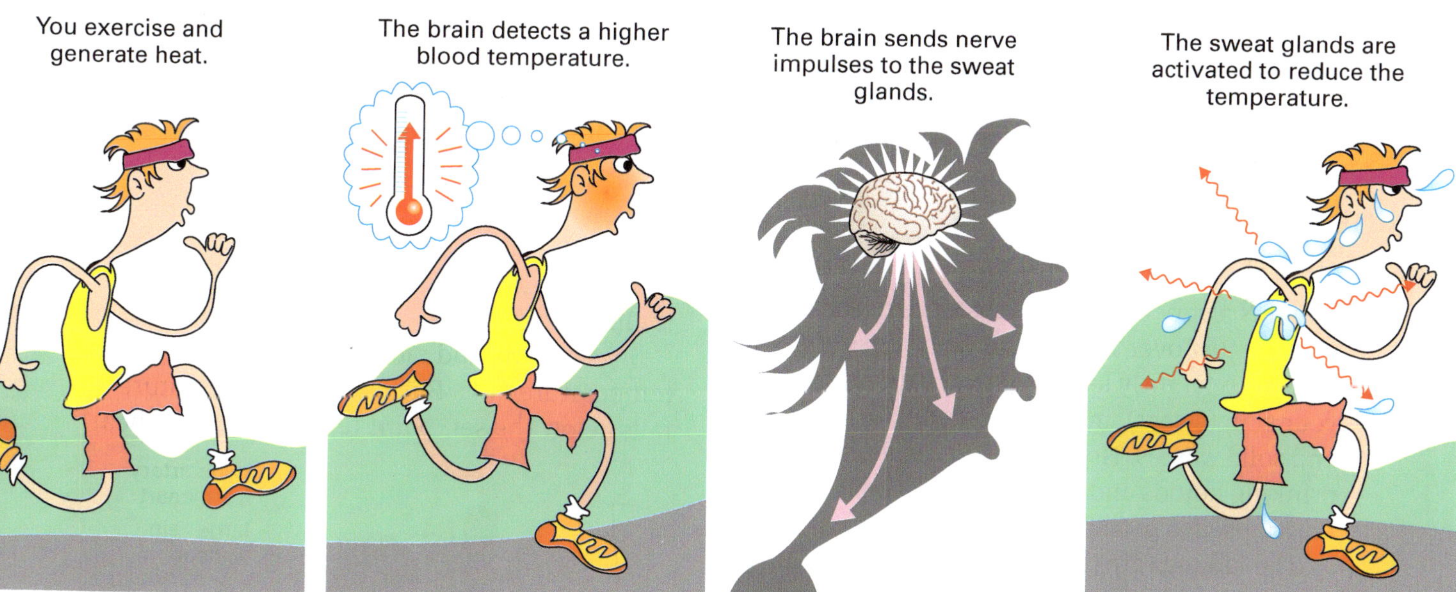

Figure 7.18 The heat receptors in your brain are sensitive to changes in blood temperature and send impulses to the skin and sweat glands.

How heat is gained	How heat is lost
• heat released from respiration in all cells in the body • heat released from respiration in muscle cells during exercise • absorption of heat from the sun and atmosphere	• radiation of heat from blood flowing beneath the skin (the heat loss increases as the outside temperature decreases.) • evaporation of sweat from the skin • heat lost when breathing out • heat lost in urine and faeces

ISBN 978 1 4202 3736 8

ACTIVITY

Evaporation and cooling

When your body gets hot, you sweat to reduce your temperature. How does sweating affect body temperature?

1 Place a small drop of water on your arm. Blow on it to evaporate the drop.
   What do you feel? Blow on your other (dry) arm to compare the sensation.
2 Repeat step 1 using a drop of methylated spirits.
   Suggest a reason for the different sensation with the methylated spirits.
   Your arm feels cooler when the liquids evaporate. Why? Use the particle model and your knowledge of change of state to help you answer this.
   Write an inference to explain why sweating lowers your body temperature.
   Design an experiment to measure the cooling effect of evaporation. (You can use a thermometer or a datalogger with a temperature sensor for this.)

Air conditioners

Like your body's control of temperature, air conditioners also use negative feedback to control temperature.

Use the points below to design a negative feedback flow diagram for an air conditioner, like the flow diagram on the previous page.

1 You set the temperature on the control panel to, say, 23 °C. This is the set-point temperature.
2 There is a temperature sensor in the air conditioner unit.
   Draw the flow diagram showing how the air conditioner works.
   A sensitive thermometer will show that the temperature in the room fluctuates around 23 °C. Why is this? Sketch a graph that shows the temperature in the room over a few hours. Mark on it when you think the air conditioner switches on, and when it switches off.

Water balance

Water is a very important substance in the body because just over 70% of your body mass is water.

You constantly lose water by evaporation from your skin (sweat), in your breath and in liquid and solid wastes. However, water is replaced by drinking liquids and by eating foods. Many cell reactions also produce water.

The kidneys are responsible for water balance in your body. These two bean-shaped organs lie close to your backbone behind the small intestine. They have a rich blood supply from a branch of the large artery that runs from the heart to the lower part of your body.

The kidneys filter wastes from your blood. Your body contains about 5 litres of blood, and about 1 litre of this passes through the kidneys every minute. So in 5 minutes all the blood in your body is filtered. The filtering process occurs as the blood flows through capillaries in the kidney. Water and dissolved wastes pass out of the blood into tiny collecting tubes. A lot of the water is reabsorbed, and a concentrated solution called urine flows into the bladder for storage until it is eliminated.

By sweating and radiating heat, your body temperature falls. But when the temperature falls below the set-point temperature, the heat

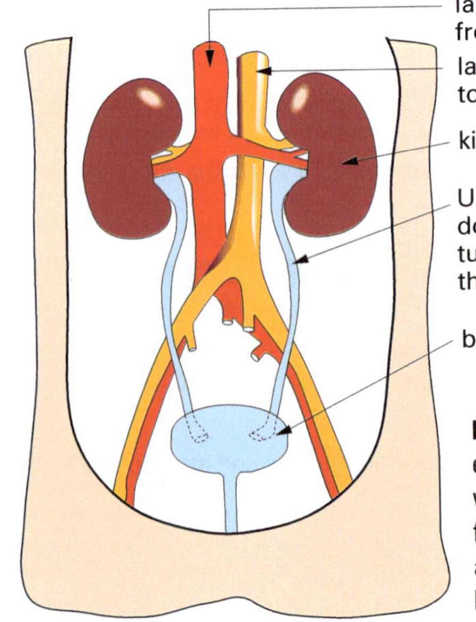

Figure 7.19 In the excretory system, the wastes in the blood are filtered by the kidneys and are then eliminated by the body as urine.

ISBN 978 1 4202 3736 8

receptors in the brain detect a lower blood temperature and send nerve impulses to the skin and sweat glands to reduce heat loss, and the body temperature gradually rises. All of these actions cause the body temperature to fluctuate between 36 °C and 38 °C.

This system of control is called a **negative feedback system** because the response acts as a stimulus to oppose (negative action) the change caused by the original stimulus.

In the example, the original stimulus is the higher body temperature. The body's response is to activate the skin and sweat glands to lower the body temperature. Following this, the lower body temperature acts as a stimulus for the brain to oppose the original action caused by the high body temperature.

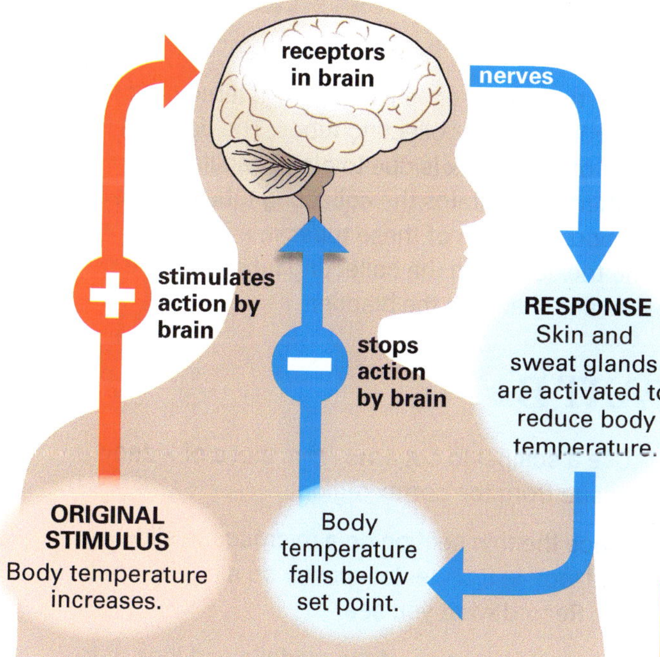

Figure 7.20 How a negative feedback system controls body temperature

Inputs and outputs

The table top right shows the water inputs and outputs that might occur in a person on a mild day. The outputs and inputs are generally balanced.

The amount of water that is filtered out of the blood changes with the heat and humidity of the day, the amount of

Water outputs (mL)		Water inputs(mL)	
urine	1500	drinking	1500
sweat	600	food	800
breath	400	from cell reactions	300
faeces	100		
total	**2600**	**total**	**2600**

sweat you produce and the amount of water you drink.

How much water is lost from the kidneys is controlled by nerves and hormones. Receptors in the brain are sensitive to changes in the amount of water in the blood. For example, on a hot day a lot of water is lost in sweat. This means that the water content of body fluids, including blood, drops. The brain detects a lower water content in the blood and sends a nerve impulse to the pituitary gland. A hormone called ADH is released that acts on the tiny tubes in the kidney. These tiny tubes filter out less water and the volume of water lost decreases.

Water balance and negative feedback

Water balance in the body is contolled by a negative feedback system (see Figure 7.21). When the kidneys reduce the amount of water in the urine (because of the water lost by sweating), the water content of the blood gradually increases. This increase is detected by the receptors in the brain which in turn stimulate the pituitary gland to release less hormone, and so the reverse of the original action occurs.

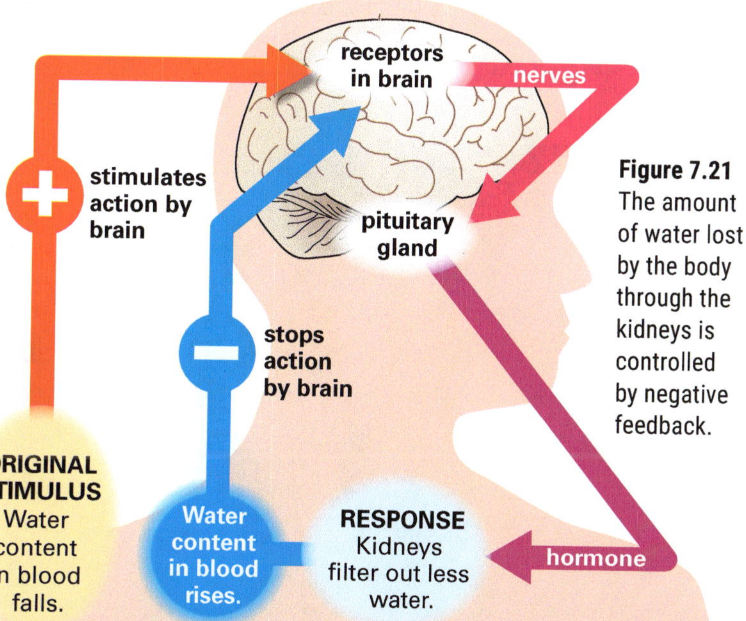

Figure 7.21 The amount of water lost by the body through the kidneys is controlled by negative feedback.

ISBN 978 1 4202 3736 8

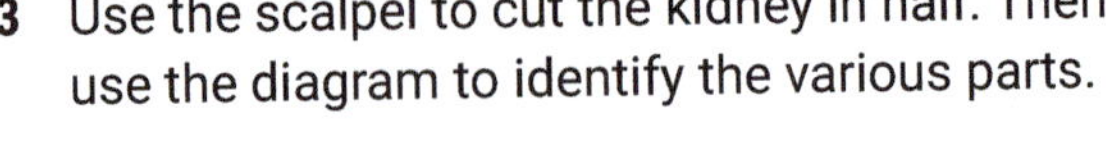

INVESTIGATION 7.2 Kidney dissection

Aim

To examine a sheep's kidney.

Materials

- sheep's kidney
- scalpel, scissors and tweezers (forceps)
- dissecting board
- disposable gloves and lab coat
- newspaper
- microscope and slides
- disinfectant and towel

Risk assessment and planning

- Read through the investigation so that you know exactly what to do.
- What safety precautions will you take when handling the kidney? How will you dispose of the kidney when you have finished with it?

PART A

1 Lay a few pages of a newspaper on the bench and put the dissecting board on them.

2 If the kidney has fat around it, peel it off. Then look for the blood vessels attached to the concave side of the kidney.

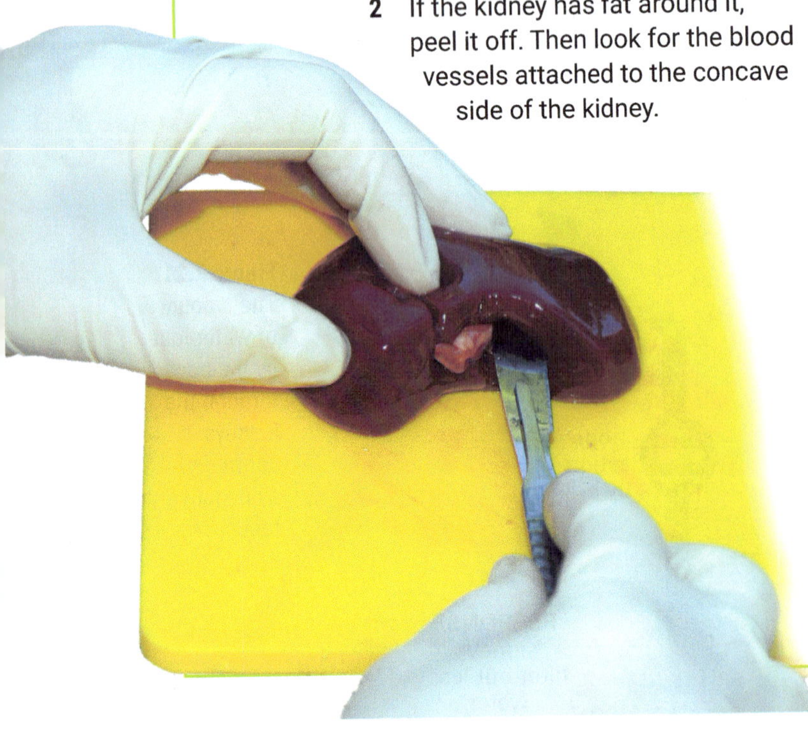

3 Use the scalpel to cut the kidney in half. Then use the diagram to identify the various parts.

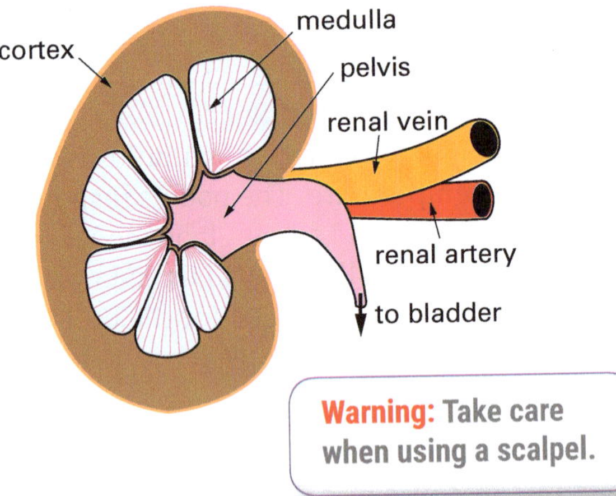

Warning: Take care when using a scalpel.

The cortex is where the wastes are filtered out of the blood into tiny collecting tubes. The dark red colour of the cortex is due to blood capillaries. The medulla contains the collecting tubes. There are about a million of these tubes in a kidney. Urine trickles down the collecting tubes into the pelvis and then into the bladder.

PART B

1 Use a scalpel to cut a very thin piece of kidney tissue from the cortex region.

2 Place the thin section on a microscope slide and look at it under low power on a microscope. Record what you see.

3 Take a thin section of the medulla and look at it under a microscope.

4 When you have finished, dispose of the kidney and scraps, disinfect the dissecting board and wash your hands thoroughly.

Discussion

1 Suggest why there is usually a thick layer of fat around the kidneys.

2 Suggest the advantages in having two kidneys and not just one.

ISBN 978 1 4202 3736 8

SCIENCE AS A HUMAN ENDEAVOUR

Professor Zee Upton

Zee Upton grew up in a small country town in South Australia. She went to university but failed her first year. Zee got a job with CSIRO and finished her degree part-time. She then went on to obtain a PhD in biochemistry from the University of Adelaide. As part of her PhD, she studied growth factors in the blood of chickens and made an unexpected observation. Her supervisors told her to ignore her observation because it wasn't important, but she felt it was.

Six years later, Dr Upton received funding to continue her research. However, she didn't make much progress, and in 2000 moved to the Queensland University of Technology. Dr Upton worked with a PhD student, Jennifer Kricker, and they soon worked out that the unexpected observation was due to a special combination of proteins that speeds up the rate at which certain skin cells grow. They called this patented combination of proteins VitroGro®.

Dr Upton and others formed a company called Tissue Therapies (www.tissuetherapies.com), which was listed on the Australian Stock Exchange in 2004. This company has an exclusive international licence to commercialise VitroGro. Dr Upton is now a professor and has a team of 52 researchers at the Institute of Health and Biomedical Innovation. Her team has used human skin removed during surgery to study how effective VitroGro is in helping wounds repair themselves. In 2008, trials of liquid VitroGro began in Fremantle Hospital for patients with chronic ulcers. These ulcers are common in people with diabetes and in bedridden patients, and can lead to amputations if they don't heal. Hopefully, VitroGro will reduce healing times, with obvious benefits for patients, and a reduction in health costs.

Professor Upton's team would like to develop a bioactive bandage—like an adhesive bandage, but containing VitroGro for faster healing. They are also hoping to improve the spray-on skin

developed by Dr Fiona Wood. This would be particularly useful for burns victims like those from the Bali bombings or the Victorian bushfires. Professor Upton's team is also developing a human skin equivalent that can be used for testing detergents and cosmetics, rather than using animals such as pigs.

Questions

1 What do you think would have happened if Zee Upton had taken the advice of her supervisors and ignored her unexpected observation?

2 Do you think Dr Upton did the right thing in forming a company rather than continuing full time with her research? Explain your answer.

3 Professor Upton says it is essential to have people on her team with different qualifications—physics, chemistry, medicine, mathematics and engineering, as well as biology. Suggest a reason for this.

4 Would you like to join Professor Upton's team? Explain your answer.

ISBN 978 1 4202 3736 8

1 How is heat lost from the human body? In which ways can this loss be reduced?

2 a What is a set-point temperature?
b Which groups of organisms have a constant body temperature?
c What is the advantage to organisms that have a constant body temperature?

3 Some of the following sentences are false. Rewrite the false ones to make them correct.
a Most water is lost from the body as sweat.
b Body heat is lost only by evaporation of water from the skin.
c Heat energy is released during cell respiration.
d Urine is made in the bladder and stored in the kidneys.
e The pituitary gland detects changes in body temperature and sends hormones to the skin and sweat glands.

4 How is water lost by your body? How is it replaced?

5 Use the table of water inputs and outputs on page 179 to infer the changes that would occur if:
a the measurements had been taken on a very cold day
b the measurements had been taken over a period which included exercise.

6 The boxes and arrows below represent the negative feedback system involved in the control of body temperature in Figure 7.20 on page 179.
Use the labels on the diagram in Figure 7.20 to replace the letters in the diagram below.

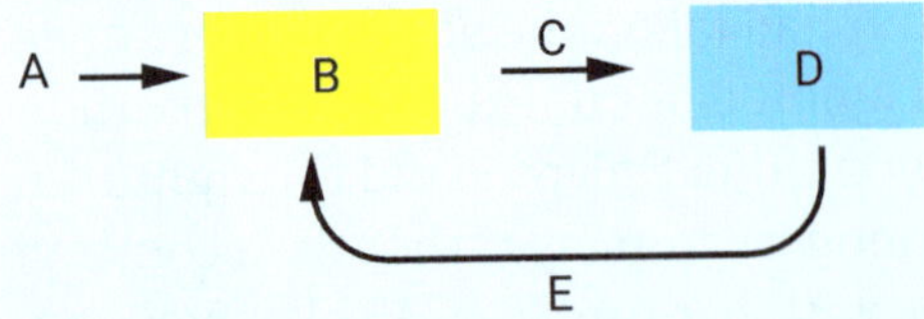

7 The diagram shows the parts of a heating unit that control the temperature in a house.

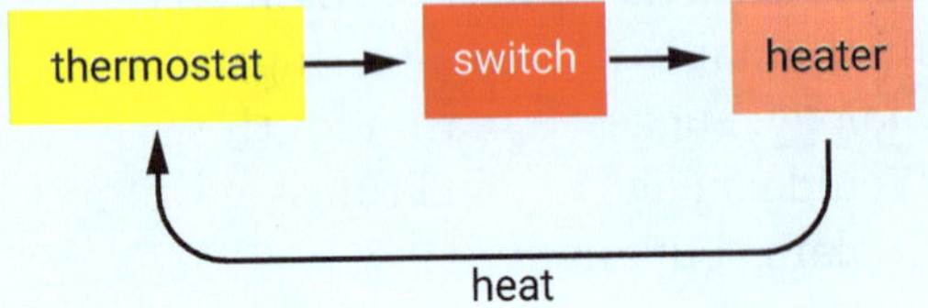

a What is the function of the thermostat?
b What is the function of the switch?
c Why is this an example of a negative feedback system?

8 The graph below refers to the information in question 7. It shows the temperature inside a house over a 2-hour period.
a What is the set-point temperature?
b At what time did the heater turn on? When did it turn off?
c By how many degrees did the house temperature vary?
d Why do you get small rises and falls of temperature in systems that operate by negative feedback?

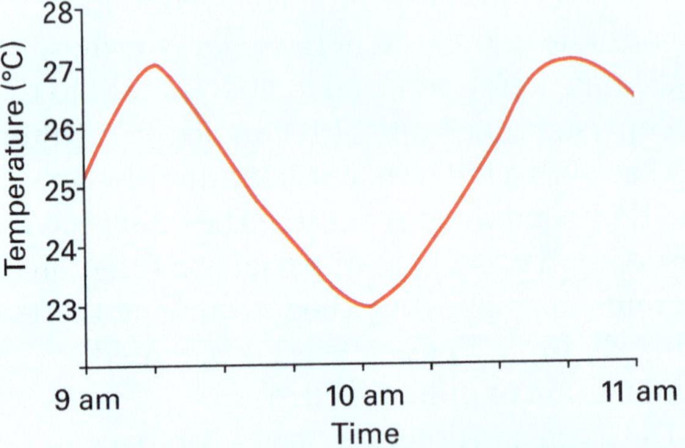

9 Your body contains about 5 litres of blood.
a If the kidneys filter 1 litre of blood per minute, how much blood is filtered in one day?
b How many times is the total volume of blood filtered in a day?
c Suppose your body produced 1500 mL of urine in a day. Express this volume of urine as a percentage of the total volume of blood filtered in a day.

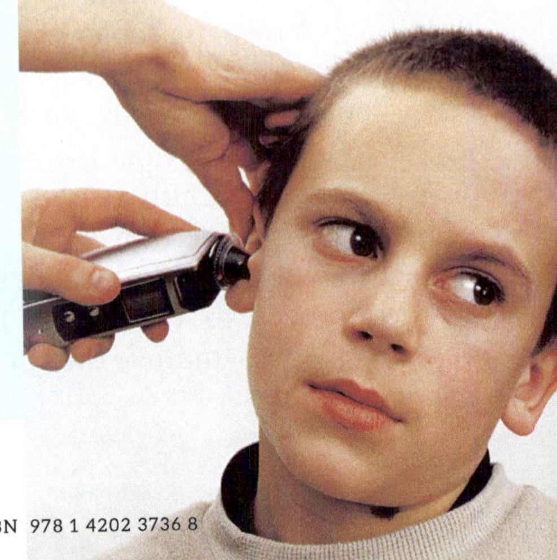

Figure 7.22
A body temperature greater than 6 °C above set point can lead to death. This is why the temperature of patients suffering from fever is closely monitored.

ISBN 978 1 4202 3736 8

CHALLENGE

1 Sweat contains about 99% water and 1% salt, mainly sodium chloride. It is made in the sweat glands in the dermis of the skin and is released over the surface of the skin through the sweat pores. There are about 2 million sweat glands in human skin.

a Use your knowledge of change of state of matter to explain in terms of the particle model how sweat reduces the temperature of the skin.

b Some other mammals such as dogs have very few sweat glands. During exercise or hot weather, dogs pant. Suggest how they might lose heat by this action.

c Why do you think that athletes, after a very vigorous workout, drink liquids containing salts and minerals?

2 To lose heat, the capillaries in your skin dilate (become much larger in diameter) and carry more blood. To retain heat, the capillaries constrict (become smaller in diameter).

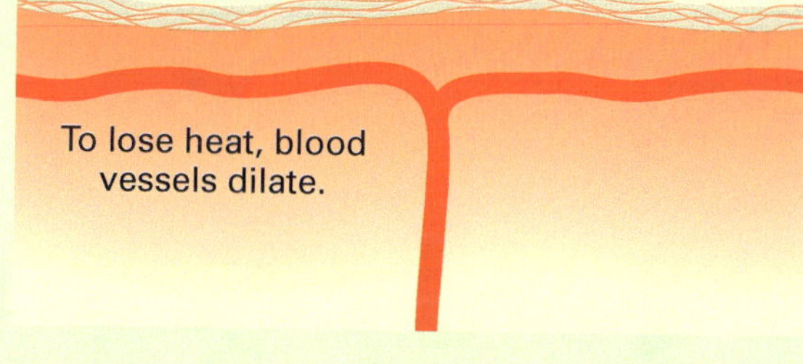

a Suggest how the actions of the blood capillaries can increase and decrease heat loss from the body.

b Why does your skin look much redder during or just after exercise?

c Suggest how a wet suit helps divers reduce heat loss when they are under water.

3 A car is being driven at 60 km/h, the speed limit around town. The car goes up a hill and starts to slow down. The driver presses harder on the accelerator pedal. As the car goes over the top of the hill, it speeds up.

Draw a flow diagram to explain how keeping the car at 60 km/h in this story is controlled by negative feedback.

4 Your breathing rate at rest is about 18 breaths per minute, but when you breathe into and out of a paper bag, your breathing rate increases.

To find out whether it is the lack of oxygen or the rise in carbon dioxide that acts as the stimulus to increase breathing rate, an experiment was carried out. The tables below show the results.

% of O_2 in air breathed in	10	15	20	25	30
Breathing rate (breaths/min)	18	19	18	18	19

% of CO_2 in air breathed in	1	3	6	9	12
Breathing rate (breaths/min)	18	19	25	35	50

a Which gas seems to affect the rate of breathing?

b A receptor at the base of the brain is sensitive to levels of carbon dioxide in the blood. Suggest how negative feedback might control your breathing rate. Draw a flow diagram to help your explanation.

ISBN 978 1 4202 3736 8

MAIN IDEAS

Copy and complete these statements to make a summary of this chapter. The missing words are on the right.

1 All of the body's functions are controlled and coordinated by the _____ system and _____ system.

2 The _____ is the main organ of the nervous system and consists of three main parts: the _____ controls voluntary actions, while the cerebellum and brain stem control _____ actions.

3 Sensory neurons relay _____ from _____ to the brain, while _____ neurons relay impulses from the brain to muscles or glands.

4 A _____ is an automatic response that occurs when an impulse travels to the spinal cord then straight back to a muscle.

5 Hormones are _____ that are made in endocrine glands and are released directly into the blood.

6 _____ are responsible for controlling growth, ripening of fruit and the timing of flowering.

7 Many of the body's processes such as water and _____ balance are controlled by _____.

8 _____ are only found in the central nervous system and do not have a _____ sheath.

9 The gap between neurons is called the _____. Chemicals called _____ are released from axon terminals and cross this gap.

interneurons
reflex action
impulses
brain
heat
neurotransmitters
involuntary
cerebrum
synapse
motor
negative feedback
chemical controllers
nervous
endocrine
myelin
plant hormones
receptors

CH•7 REVIEW

1 In a darkened room, Joel felt a sticky cobweb suddenly cover his face, and he immediately pulled away.

Which type of receptor was used for this action?

A vision
B taste
C touch
D sound

2 The action of Joel pulling away from the cobweb was probably:

A caused by muscles activated by hormones.
B a reflex action.
C the stimulus to the response.
D caused by a negative feedback system.

3 Parts of the body are under voluntary control and others are under involuntary control.

a Which body actions are involuntary?
b Which parts of the nervous system coordinate involuntary actions?

4 The flow diagram below shows a typical reflex action after a bright light has been shone in your eye. Match the letters in the diagram with the words in the following list.

receptors in eye
flash of light
spinal cord
sensory neuron
pupil
motor neuron
pupil decreases

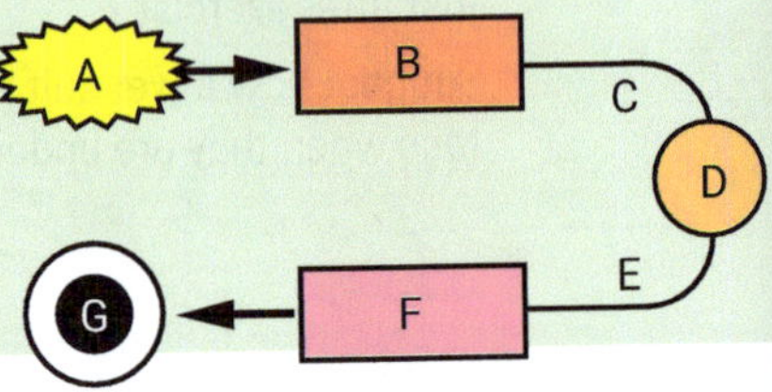

ISBN 978 1 4202 3736 8

5 Nerves and hormones are both used by the body to relay messages from one point to another. How does the action of nerves differ from that of hormones?

6 The body temperatures of four animals were recorded over a 12-hour period. The results are shown in the graph below.

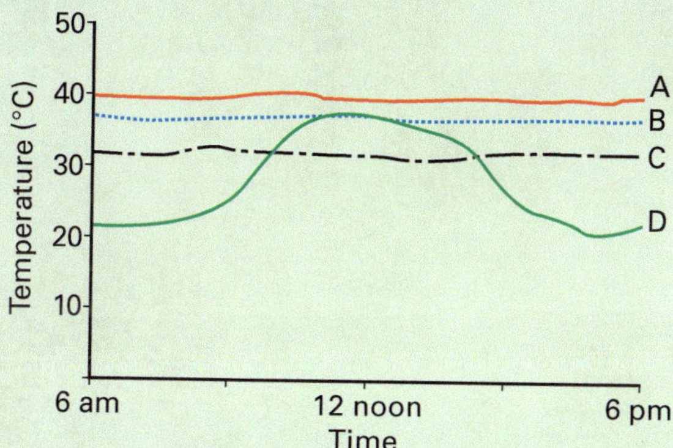

a Which of the animals are likely to be mammals or birds. Explain.

b What is the set-point temperature for animal A?

c Which animal is probably a human?

d Suggest an inference to explain the shape of the graph for animal D.

7 The following three experiments on plant hormones were done with oat seedlings.

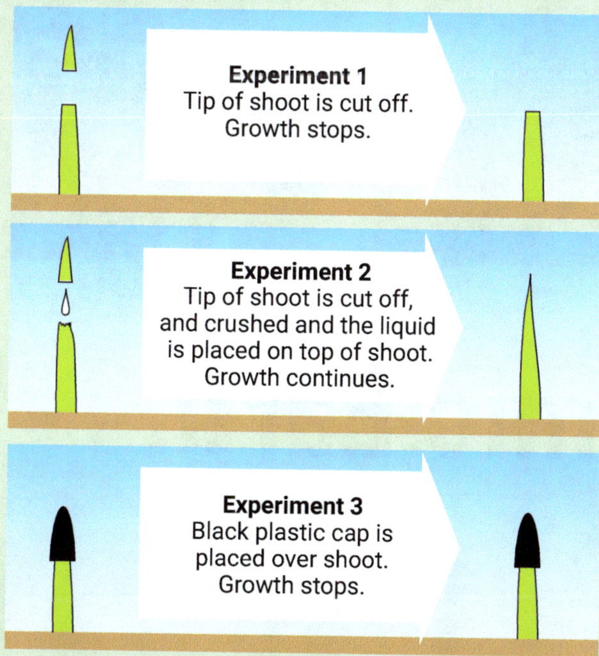

a Write an inference to explain the results of each of the three experiments.

b Design an experiment to show that the plant hormone will act only on cells just below the tip of the seedling and not on cells further down.

8 Read the following information about a hormone that controls the sodium levels in your body.

> The adrenal glands are situated on top of each kidney and produce a hormone (we will call this hormone X) that regulates the amount of sodium in the blood. It does this by acting on the kidney to reduce the amount of sodium that is filtered out into the urine.
>
> The adrenal glands are under the control of the brain and the pituitary gland. Receptors in the brain detect a low blood sodium level. Nerve impulses from the brain are sent to the pituitary gland, which releases a hormone (called hormone Y) that stimulates the adrenal glands to release hormone X.

a What happens to the blood when hormone X is released from the adrenal glands?

b Which part of your body is sensitive to levels of sodium in the blood?

c Negative feedback is used to control the amount of sodium in the blood. Use the information above to replace each of the letters in the flow diagram below.

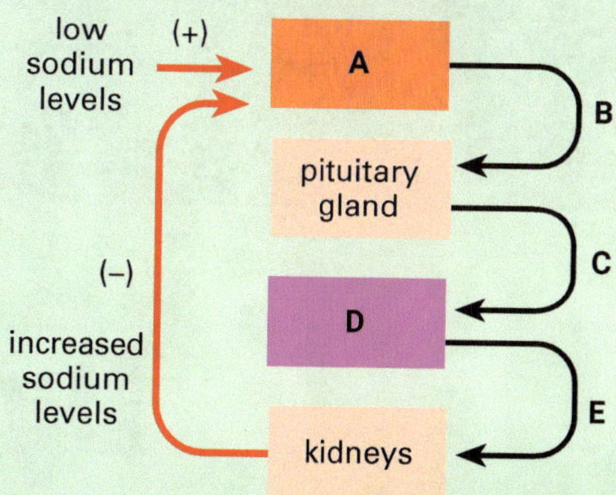

9 Draw a table that shows the features and differences between the three types of neurons.

Check your answers on page 302.

ISBN 978 1 4202 3736 8

IN THIS CHAPTER

Science Understanding

- model earthquake waves using a slinky, and build a model seismograph
- explain the movement of tectonic plates in terms of convection currents in the Earth's mantle
- recognise the Earth's major tectonic plates on a world map
- relate the occurrence of earthquakes and volcanic activity to plate boundaries
- relate the extreme age and stability of the Australian continent to tectonic plate history

Science Inquiry Skills

- locate the epicentre of an earthquake using seismograms from three different places

CH•8 Dynamic Earth

ISBN 978 1 4202 3736 8

GET STARTED: *QUESTION*

Look at the two photos on this page and discuss the questions below.

- What observations can you make from these photos?
- Infer what might have caused these natural disasters.
- What do your observations and inferences suggest about the structure of the Earth's interior?

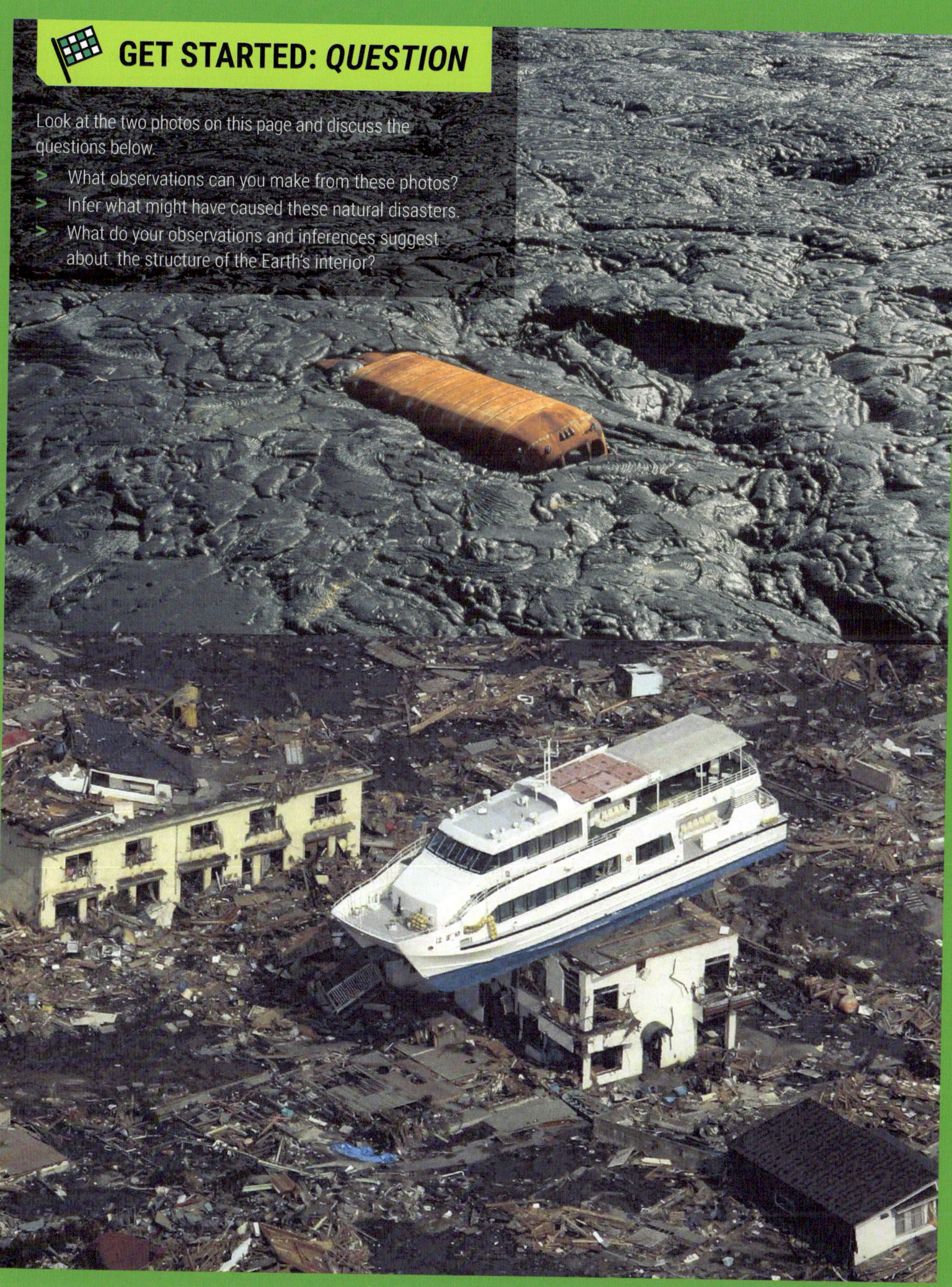

ISBN 978 1 4202 3736 8

8.1 Inside the Earth

Structure of the Earth

Volcanoes are evidence that the Earth is constantly changing. They form when molten rock from inside the Earth forces its way to the surface through a crack in the ground. This molten rock indicates the nature of the material in the Earth's interior. The Earth is made up of three main layers—the crust, the mantle and the core, as shown below.

The layered structure of the Earth can be compared to that of a soft-boiled egg. The outermost layer is called the **crust**. It is rigid and very thin compared with the other two layers and thicker under the continents than under the oceans. Like the shell of an egg, the Earth's crust is brittle and can break.

Below the crust is the **mantle**, which can be thought of as the white of the egg. It is a layer of semi-solid rock almost 3000 km thick. It is hotter and denser than the crust because temperature and pressure inside the Earth increase with depth.

At the centre of the Earth is the **core**. Unlike the yolk of an egg, however, the core is made of two distinct parts—a liquid outer core and a solid inner core.

The crust and the upper part of the mantle form a rigid layer called the *lithosphere* (LITH-os-fear). Scientists infer that below this is the *asthenosphere* (ass-THEN-os-fear). This layer is composed of hot, semi-solid material that can soften and flow due to the high temperatures and pressures over millions of years. The heat causes convection currents that cause the asthenosphere to flow. The rigid lithosphere is thought to 'float' on top of the asthenosphere.

When the molten rock (magma) is forced upwards from the asthenosphere, it may cause cracks in weaker regions of the crust. When this happens, volcanoes form.

Folds in rocks

Some of the largest mountain ranges in the world contain sedimentary rocks. For example, the rocks in Mt Everest were originally laid down under water, and fossil marine shells have been found

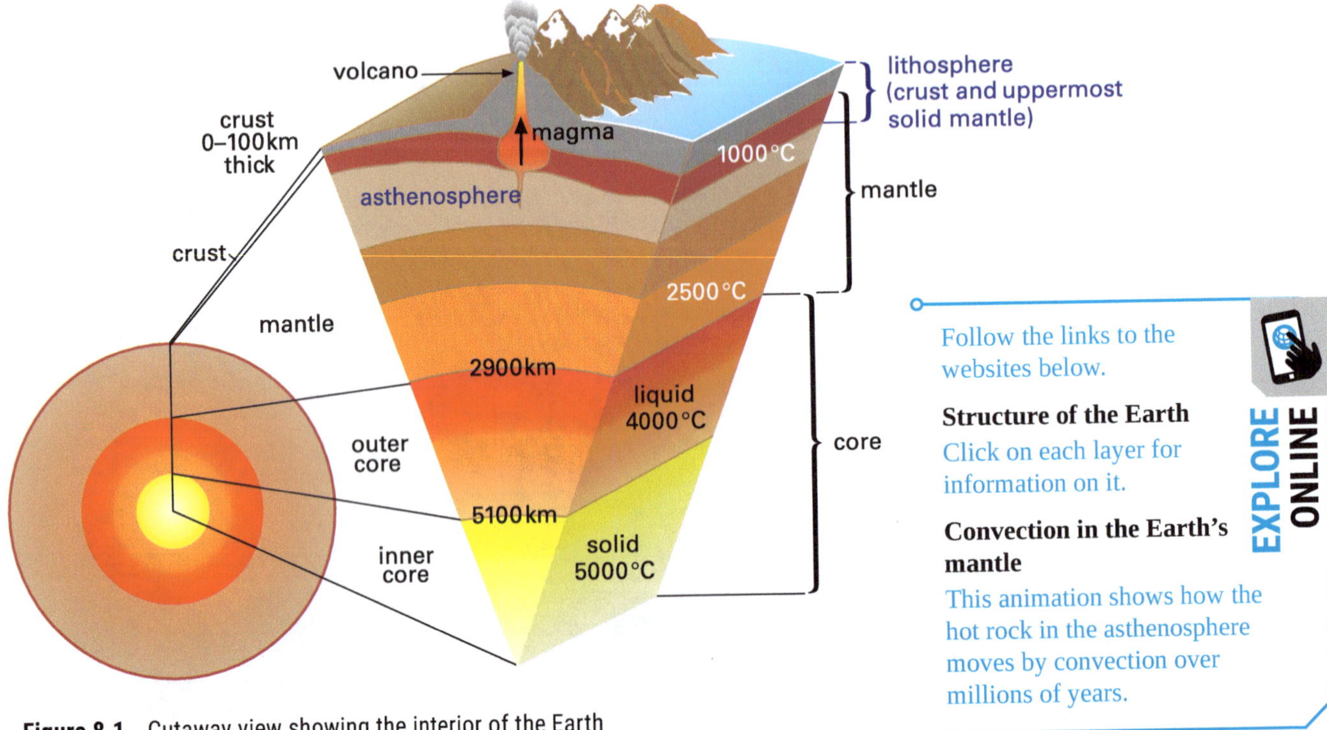

Figure 8.1 Cutaway view showing the interior of the Earth

EXPLORE ONLINE

Follow the links to the websites below.

Structure of the Earth
Click on each layer for information on it.

Convection in the Earth's mantle
This animation shows how the hot rock in the asthenosphere moves by convection over millions of years.

ISBN 978 1 4202 3736 8

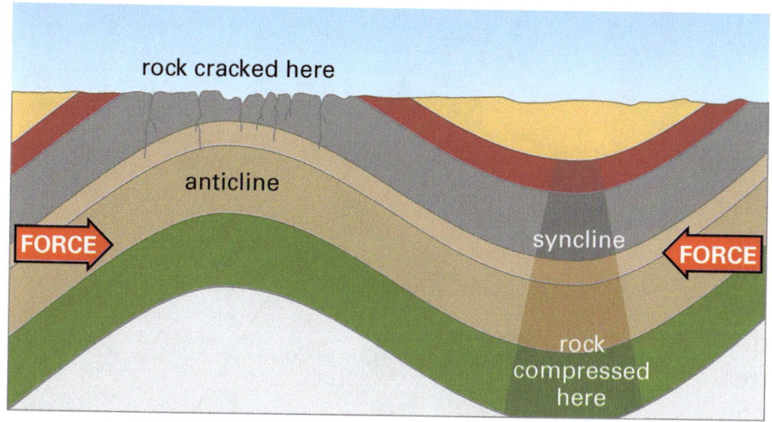

Figure 8.2 Folds are formed when rocks are squeezed by sideways forces.

Figure 8.3 Can you identify the synclines and anticlines in this rock formation?

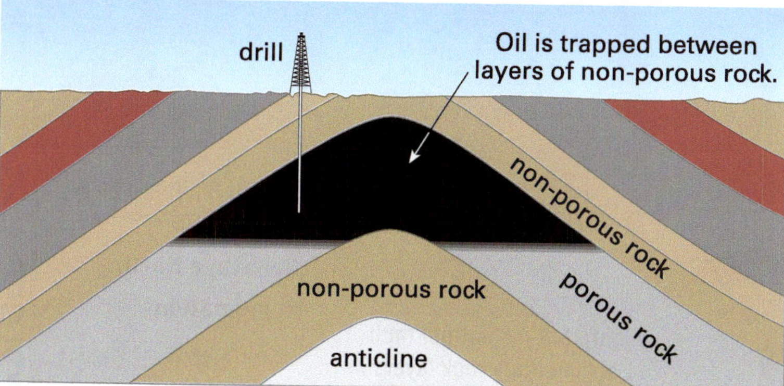

Figure 8.4 Oil is sometimes trapped in anticlines.

in some of the rocks on the mountain. Also, many of the rock layers in these mountains are bent and buckled. How could sedimentary rocks be found on the highest mountains on Earth? And how could they be bent and buckled?

Huge forces caused by movements in the mantle cause changes in the Earth's crust. These forces result in the formation of volcanoes, and the uplifting and buckling of the solid crustal rocks. Scientists have experimented with models of rocks under enormous heat and pressure, and have shown that these solid rocks can soften and bend slowly without breaking. When this happens, the rocks form **folds**, as shown in Figure 8.2. The folds that bend downwards to form a U-shape are called **synclines**. Those that bend upwards like a dome are called **anticlines**.

Minerals are often found near folds. During folding, the rocks in a syncline are subject to enormous compressive forces. As a result, the rock may be metamorphosed (changed) to form new minerals, often containing large crystals. The gold at Bendigo in Victoria was formed this way. The compressive forces may also cause the top of an anticline to crack. Later on, various mineral solutions may flow through these rocks and crystallisation may occur.

Anticlines can also form a trap for oil, which tends to move upwards through porous rocks such as sandstone. However, it cannot move through non-porous rocks such as shale. As a result, oil is sometimes trapped in an anticline, as shown in Figure 8.4. This oil can be tapped by drilling into the anticline. If the drill is too far to the left or right of the anticline, no oil will be struck. Hence, it is important to know exactly where anticlines are when searching for oil.

ISBN 978 1 4202 3736 8

Faults in rocks

If the pressures in the Earth's crust are very intense, the rocks may break, especially if they are near the surface where they are more brittle than the deeper rocks. When this happens, a crack appears and the rocks slide along the crack. A crack in the Earth's crust along which rocks move is called a **fault**. This movement can be vertical (Figures 8.5 and 8.6), horizontal or both, and can be hundreds of metres or a few millimetres. Sometimes a huge amount of rock moves along a fault relative to the rock on the other side, forming mountains and valleys.

Geologists looking for minerals often look along fault lines. The reason for this is that mineral solutions can flow along a fault and crystallise to form minerals.

Oil can also be trapped against faults, as shown in Figure 8.7. It tends to move towards the surface through porous rocks and it sometimes becomes trapped under non-porous rocks.

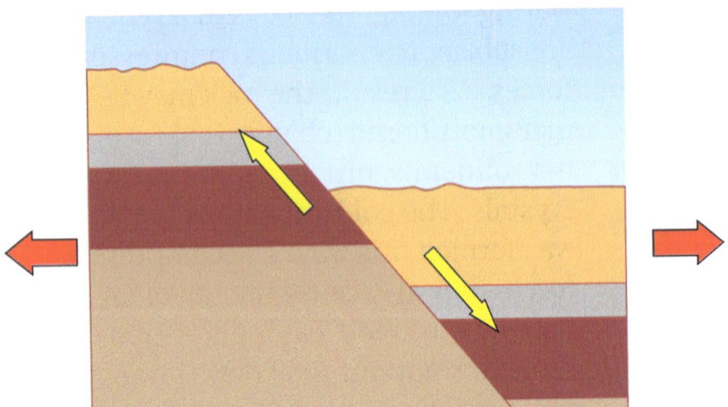

Figure 8.5 A normal fault caused by stretching forces

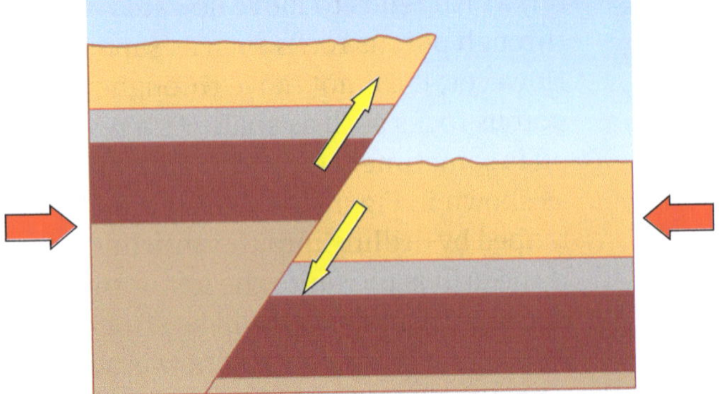

Figure 8.6 A reverse or thrust fault caused by pushing forces

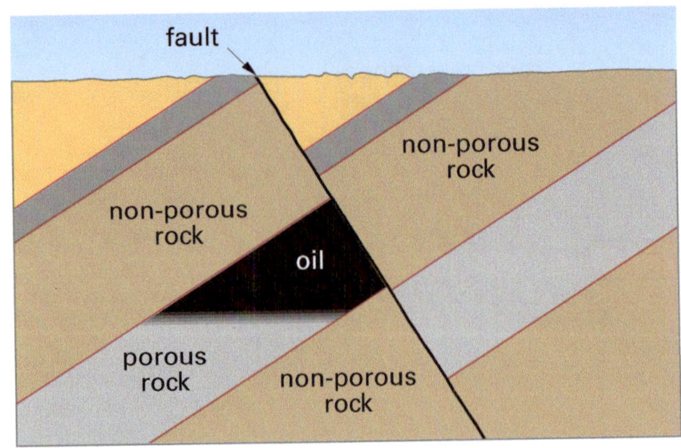

Figure 8.7 Oil can be trapped against a fault.

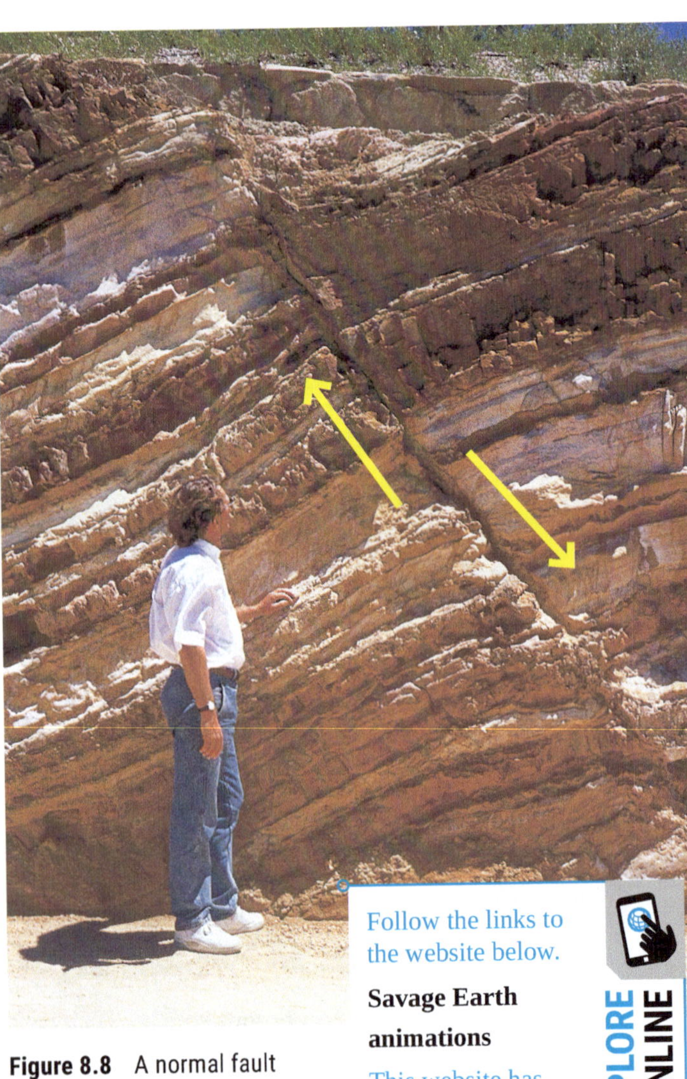

Figure 8.8 A normal fault in sedimentary rock layers in Canberra. Notice that the rock layers on the right have slipped downwards, as in Figure 8.5.

EXPLORE ONLINE

Follow the links to the website below.

Savage Earth animations

This website has many excellent animations, not just of faults.

ISBN 978 1 4202 3736 8

Modelling Earth changes

Suppose layers of sediments were laid down over millions of years in a shallow sea. During this time, these layers changed into sedimentary rocks. The cross-section below shows a block of these rocks.

1. Your task in this activity is to draw models of what might have happened to the sedimentary layers in the block during the course of a number of changes in the Earth over 300 million years.
2. Copy (or photocopy) the sedimentary layers in the block below. Colour the layers. Label this step 1.
3. Read step 2 opposite. Then underneath the step 1 block, draw and colour a cross-section showing what you think might have happened to the block as a result of these events. Label this step 2. Use arrows to indicate the forces that caused the changes.
4. Now read steps 3–6. Draw more cross-sections, each time making changes to the previous cross-section. You may need to do some cutting and pasting.

Teacher note: You may want to give each student a photocopied sheet with six cross-sections on it. They can then cut and paste to make the six steps.

Step 1 (400–300 million years ago): Layers of sediments have been changed to sedimentary rock.

Step 2 (300–250 million years ago): Enormous sideways forces squeezed the rocks, creating two complete synclines in the block.

Step 3 (250–200 million years ago): The pushing forces continued. A reverse fault appeared through the left-hand syncline, pushing up the left-hand side.

Step 4 (200–150 million years ago): The earth movements stopped for a period of time and erosion occurred, eroding the mountains (anticlines).

Step 5 (150–100 million years ago): The valleys filled up with water and sediment until the top surface of the block was fairly level.

Step 6 (100 million years ago to present): A river formed on the right-hand side of the block. Over millions of years, the running water eroded the sedimentary rocks, forming a steep-sided canyon that reaches the bottom of the block.

What would the block in step 6 look like viewed from above? Use your cross-section to sketch a plan view of the rock layers. Mark which rocks are the youngest and which are the oldest.

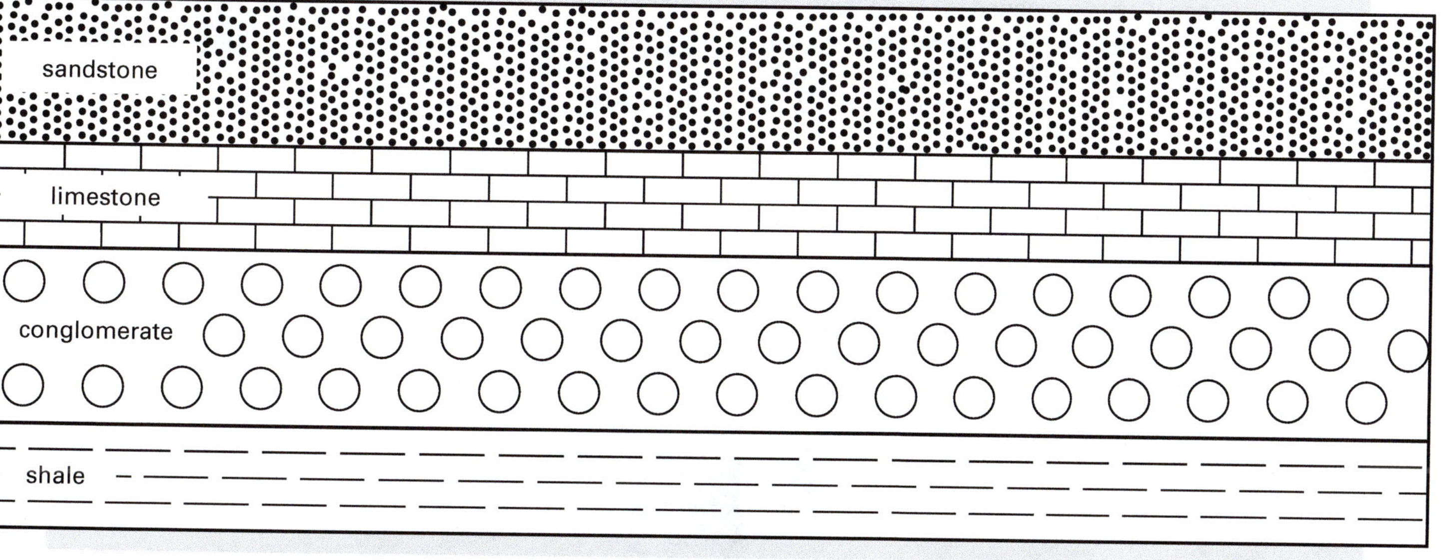

ISBN 978 1 4202 3736 8

1 Write in your own words the meanings of the following words. Then use the page numbers to check the meaning in the text. Change your answers if necessary.

a mantle page 188
b anticline page 189
c fold page 189
d fault page 190

2 a How are sedimentary rocks formed?
b How is it that they are often found above sea level?

3 In which ways is the inferred structure of the Earth like a soft-boiled egg? In which ways is it different?

4 Figure 8.9 shows part of the MacDonnell Ranges just south of Alice Springs. Suggest why the layers are almost vertical rather than horizontal.

5 Look at Figure 8.10. Sketch the patterns in the layers and mark the position of an anticline and a syncline.

6 Oil was found at A but not at B. How could you explain this?

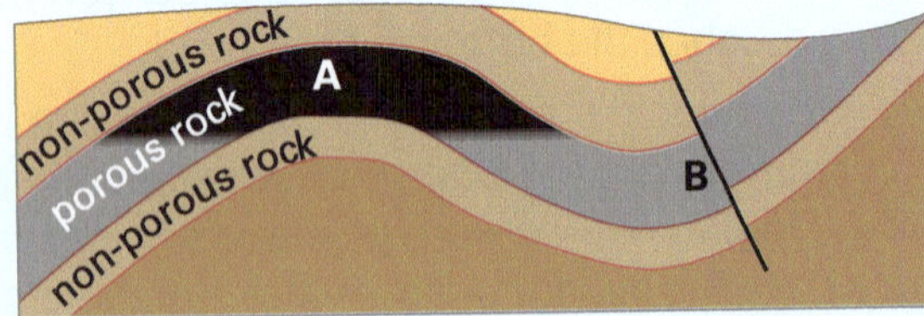

7 Figure 8.11 below shows a fault.
a Suggest what happened to cause this.
b Suppose the land eroded away over a long period of time until it was once again flat. How could you tell where the fault line was?

Figure 8.9 MacDonnell Ranges

Figure 8.10 Rock formation after folding

Figure 8.11 A fault line

ISBN 978 1 4202 3736 8

1 You and a friend are standing at a lookout overlooking a valley with a mountain in the middle of it. Your friend says that the mountain is actually an old volcano that has been eroded. You disagree and say that you think that the land around the mountain was formed under the sea and earth movements pushed the land up to form the mountain. How could you find out who was correct?

2 Infer how the sequence of rock layers in the diagram below was formed. List the rocks from youngest to oldest.

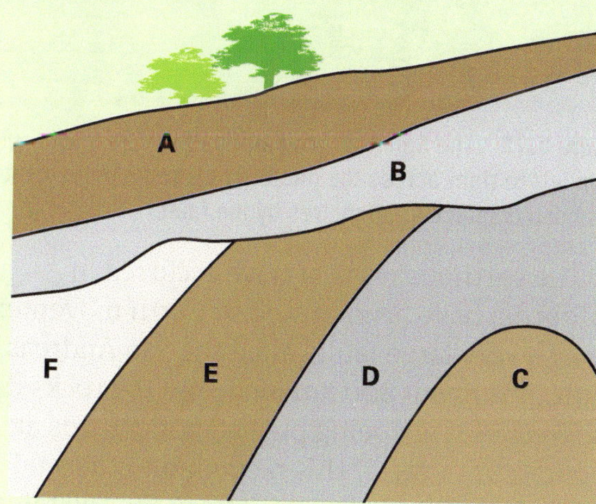

3 The photo below shows one of the Glasshouse Mountains in Queensland.

a What part of a volcano might this structure have been?

b Why did the outer parts of the volcano erode more easily than this structure?

4 The table shows the years from 79 CE to 1944 in which Mt Vesuvius in Italy has erupted. The dates marked in bold show the biggest eruptions. Can you tell when the volcano will erupt again? Write a brief report.

Years when Mt Vesuvius erupted			
79	1631	**1794**	1891
203	1660	1804	1895
472	1682	1805	1900
512	1689	**1822**	1903
685	**1694**	**1838**	1904
993	1707	**1850**	**1906**
1036	**1737**	1858	1913
1139	1760	1861	1926
1306	**1767**	**1871–72**	1929
1500	1779	1875	1944

5 The graph below shows the estimated temperature, density and melting point of the materials in the layers inside the Earth.

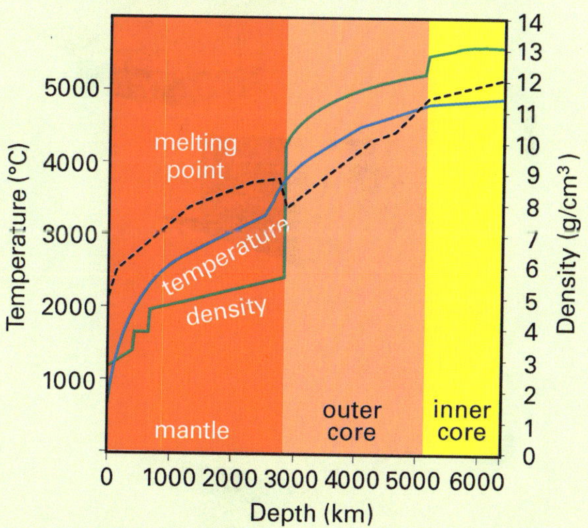

a What is the radius of the Earth?

b How thick is the mantle?

c Which layer has the highest average density?

d Make a generalisation about the temperature changes in the mantle.

e Which layer in the Earth's interior is liquid? Explain how you arrived at this answer.

ISBN 978 1 4202 3736 8

8.2 Earthquakes

Forces caused by movements in the Earth's mantle can change the shape of the rocks in the crust. Generally these movements are very slow. Sometimes, however, there is a sudden movement that can result in large cracks in the ground, and can make buildings collapse and break water and gas pipes.

Earthquakes occur when intense pressures in the Earth's crust cause the rocks to break and move along a fault. The stored-up energy in the rocks is released as shock waves, which spread out in all directions, like the waves produced in a pond when a pebble is thrown in.

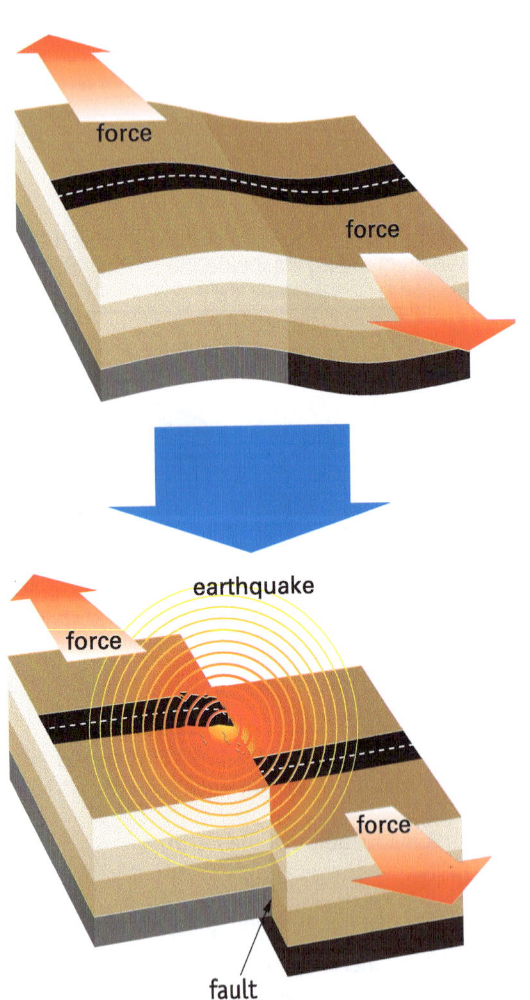

Figure 8.12 Crustal pressure causes the rocks to slide along each other at a fault, causing an earthquake.

Figure 8.13 The San Andreas Fault appears as a scar running from left to right across the photo. The stream in the centre has been displaced 130 metres by the fault.

The earthquakes that have occurred in California have been caused by Earth movements along a very large fault called the San Andreas Fault. It is about 800 km long, and the rocks to the west of it are being pushed northwards at about 5 cm a year. If this movement is slowed or stopped, the pressure builds up along the fault. When the rocks move again, the energy is released and an earthquake occurs.

Figure 8.14 The Christchurch earthquake in New Zealand in February 2011 killed 181 people.

ISBN 978 1 4202 3736 8

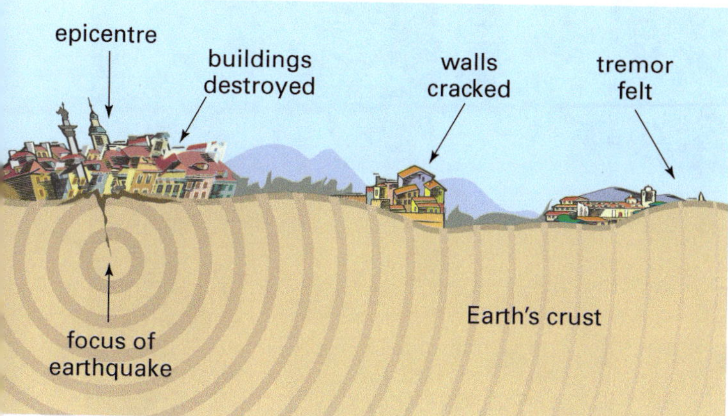

Figure 8.15 Cut-away view of the Earth's crust showing shock waves spreading out in all directions from the focus of an earthquake.

How are earthquakes recorded?

The vibrations or waves caused by an earthquake spread out in all directions. The place inside the Earth where the movement of the rocks occurs is called the *focus* of the earthquake. The **epicentre** is the point on the Earth's surface directly above the focus.

As the shock waves spread out from the focus, they lose energy and the vibrations become smaller. On the surface, the damage is greatest at the epicentre since this is the closest point to the focus, and the damage decreases as you go further away from the epicentre.

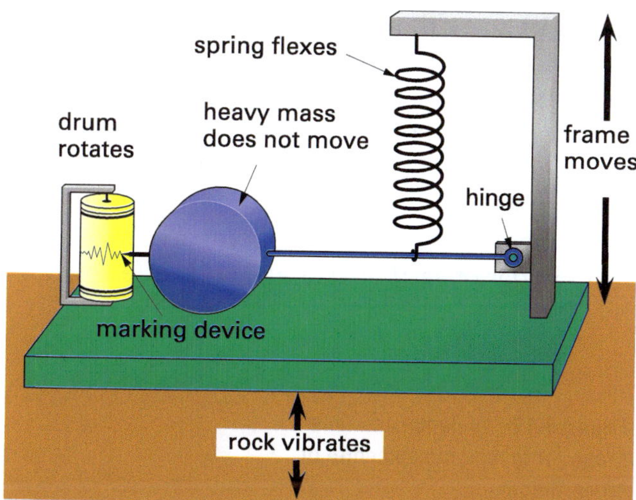

Figure 8.16 A simplified diagram of a seismograph

Earthquakes are recorded on instruments called **seismographs** (SIZE-mo-graphs). The frame of the seismograph is set on solid rock. In an earthquake, the solid mass tends to remain motionless while the rest of the seismograph shakes, and a pen or laser beam makes marks on a rotating drum. Any movements of the solid rock show up as a series of troughs and peaks on the recording. The seismograph in Figure 8.16 measures vertical movements, while other types measure horizontal movement.

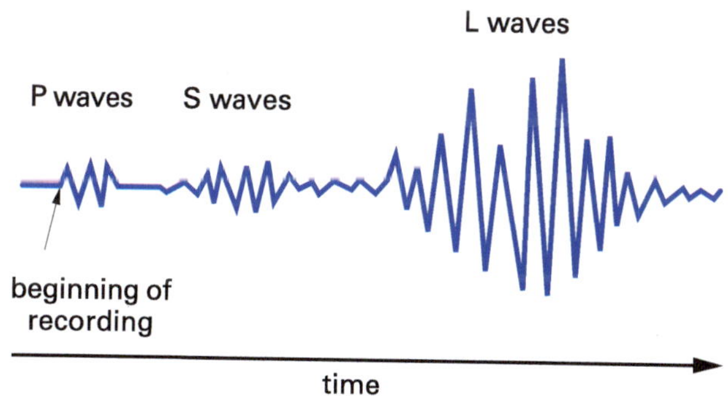

Figure 8.17 A seismogram

A record of the waves caused by an earthquake is called a *seismogram*. Figure 8.17 above shows a simplified seismogram. Notice that there are three types of waves. The first to arrive at the recording station are the primary waves or P waves. These waves are also called *compression waves* and are formed when matter in the rocks is pushed together by the Earth's movement—in much the same way as sound waves form when air particles are compressed by a vibrating object. P waves can travel through solids and liquids.

The P waves are followed by secondary waves or S waves. These are also called *shear waves* and are due to the sideways motion of matter. S waves can travel only through solids.

L waves are the last waves to arrive at an earthquake recording station, but they have the most impact. The energy in L waves causes the surface of the Earth to vibrate like a shaken bowl of jelly. These waves produce the largest movement on a seismogram (see Figure 8.17). L waves, which are also called *surface waves*, cause the most damage to objects on the surface. Buildings, trees and other structures can be completely flattened.

ISBN 978 1 4202 3736 8

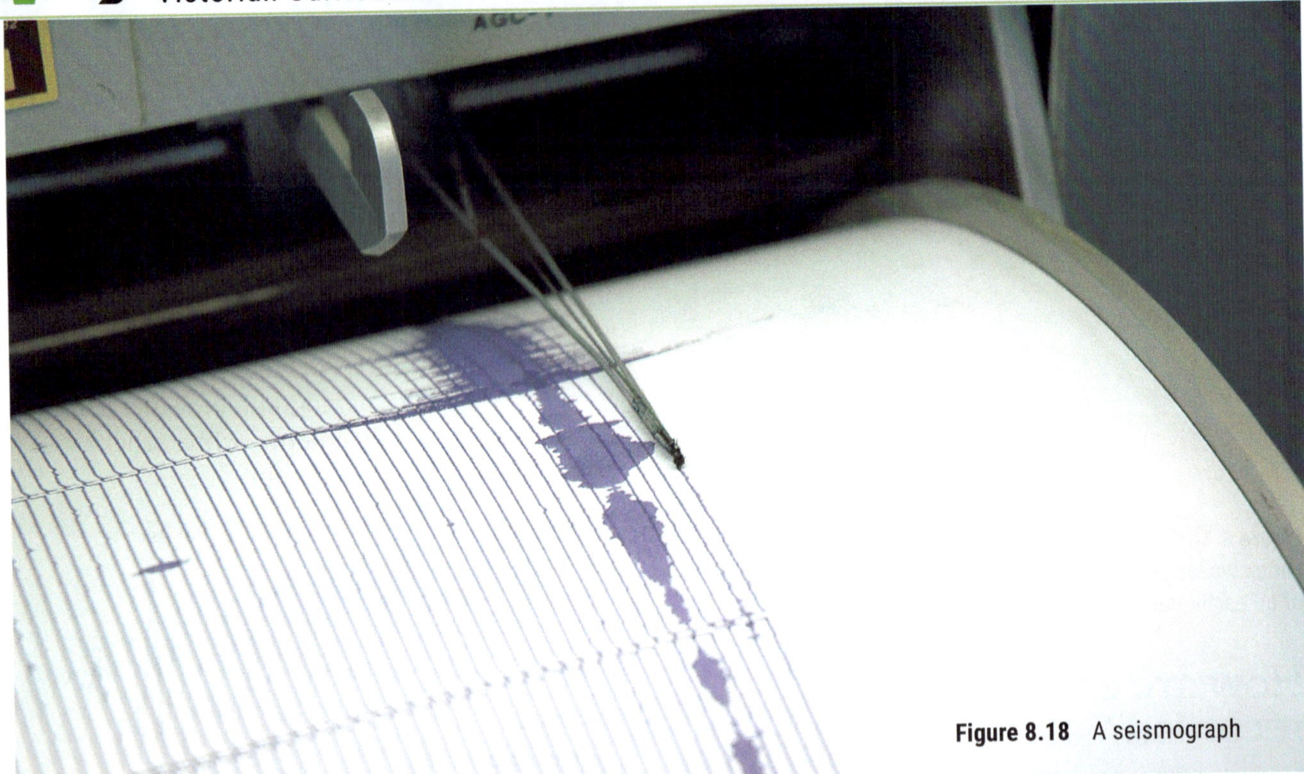

Figure 8.18 A seismograph

The differences in the times taken by earthquake waves to travel certain distances are used by scientists to estimate the exact location of the epicentre.

The graph in Figure 8.19 is a distance–time graph for P and S waves. Suppose the time between the arrival of P waves and S waves is 196 seconds. From the graph, the distance to the epicentre of the earthquake is 2000 kilometres.

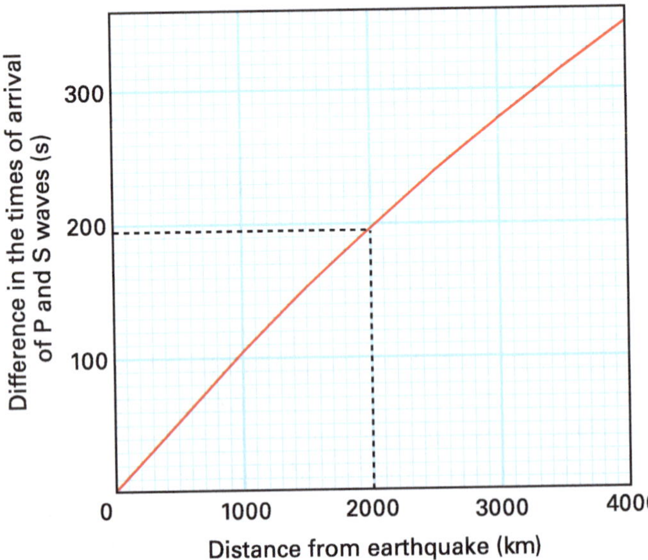

Figure 8.19 Distance and time graph for finding the epicentre of an earthquake

The Richter scale

Scientists can use seismograms to work out the strength or *magnitude* of an earthquake. The magnitude of an earthquake is how much energy it has, and is measured on a scale from 0 to 10. This scale is called the **Richter** (RICK-ter) **scale** after its inventor Charles Richter. A shock of magnitude 2 is the smallest normally felt by humans, and magnitudes of less than this can only be detected by seismographs.

On the Richter scale, the intensity or energy of an earthquake increases tenfold for a single increase in magnitude. For example, an earthquake of magnitude 6 causes ten times more ground motion (potential damage) than one of magnitude 5, and 100 times more than one of magnitude 4.

The table on the next page compares the magnitudes of earthquakes and their effects at the epicentre.

ISBN 978 1 4202 3736 8

Magnitude on Richter scale	Effects at the epicentre
2.5 to 2.9	Detected only by instruments
3.0 to 3.9	Suspended objects may swing, vibrations like those caused by passing trucks
4.0 to 4.9	Wakes sleeping people; dishes and windows rattle
5.0 to 5.9	Felt by all; furniture moves, walls crack and chimneys topple
6.0 to 6.9	Most houses damaged
7.0 to 7.7	Ground cracks, foundations damaged, pipes burst, landslides
7.8 to 8.6	Disastrous; few structures remain, large cracks in ground
greater than 8.6	Total destruction; waves seen on the ground, magnitude 9.5 most severe ever recorded (Iran 1972)

Figure 8.20 In September 1999, the strongest earthquake to hit Taiwan for more than 100 years had its focus 1.1 km below the middle of the island. The photo shows rescue workers searching for survivors in the ruins of this hotel, which collapsed when the earthquake struck at 2 am. It measured 7.6 on the Richter scale.

Earthquakes in Australia

There have been various minor earthquakes in Australia and they occur regularly. Australia has many very minor geological faults that can cause earthquakes, but compared to many other countries Australia is very stable. The largest earthquake recorded in Australia was a magnitude 6.6 near Tennant Creek in 1988.

The most damaging earthquake recorded in Australia was in Newcastle, New South Wales, with a magnitude of 5.6. In December 1989 this earthquake killed 13 people, and more than 100 were injured. It was felt in a 200 000 sq km area of New South Wales and the Australian Capital Territory, from Albury and Cooma to Coffs Harbour and Inverell and as far west as Narromine. It was also felt by people in high-rise buildings in the Gold Coast and Melbourne.

The Cadell Fault scarp is in southern NSW and has had at least five earthquakes in the past of between 7.0 and 7.3 in magnitude. Research this fault and answer the following questions:

1. How large is the fault?
2. Is another earthquake likely to occur along this fault?
3. What could be the impact on Victoria in such an event?
4. Find out about other areas in Australia where earthquakes may occur. Summarise your findings.

To find out more about earthquakes in Australia, check out the **Geosciences Australia Earthquake** page. Also check out the **Cadell Fault**.

EXPLORE ONLINE

INVESTIGATION 8.1 Earthquake waves

Aim

To model earthquake waves and to construct a simple seismograph.

PART A Earthquake waves

Materials

spiral spring (slinky)

Method

1 To model primary or compression waves, stretch the spring on the floor between you and your partner. Bunch up 10 coils, then let them go.

Record what happens as the pulse moves along the spring. In which direction do the coils move? In which direction does the wave move?

2 To model secondary or shear waves, stretch out the spring on the floor as before. This time flick the spring sideways as shown.

How do the coils move this time? In which direction does the wave move?

Discussion

1 How do the two types of waves differ?

2 Which type of wave produces an up and down movement of the Earth? What type of movement would the other produce?

PART B Making a seismograph

Method

1 Use the diagram of the seismograph below as a guide to make a model seismograph (as a class activity).

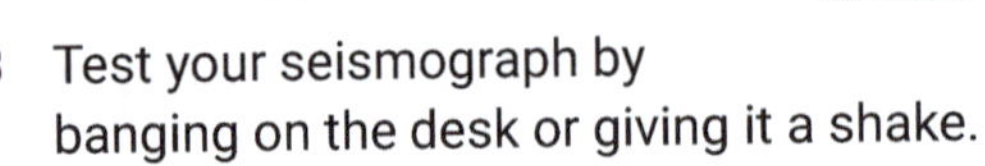

2 To attach the pen to the metal can, first glue the stopper to the base of the can. Then push the blunt end of the pen firmly into the hole in the stopper.

3 Test your seismograph by banging on the desk or giving it a shake.

4 Tear off your seismogram and test your seismograph again.

Discussion

1 Is your seismograph better at recording horizontal or vertical movement?

2 What is the relationship between the intensity of the movement (the earthquake) and the size of the waves?

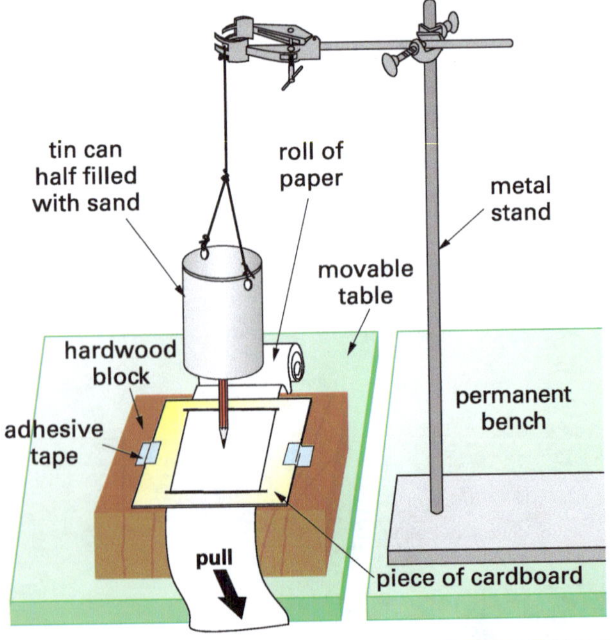

ISBN 978 1 4202 3736 8

ACTIVITY

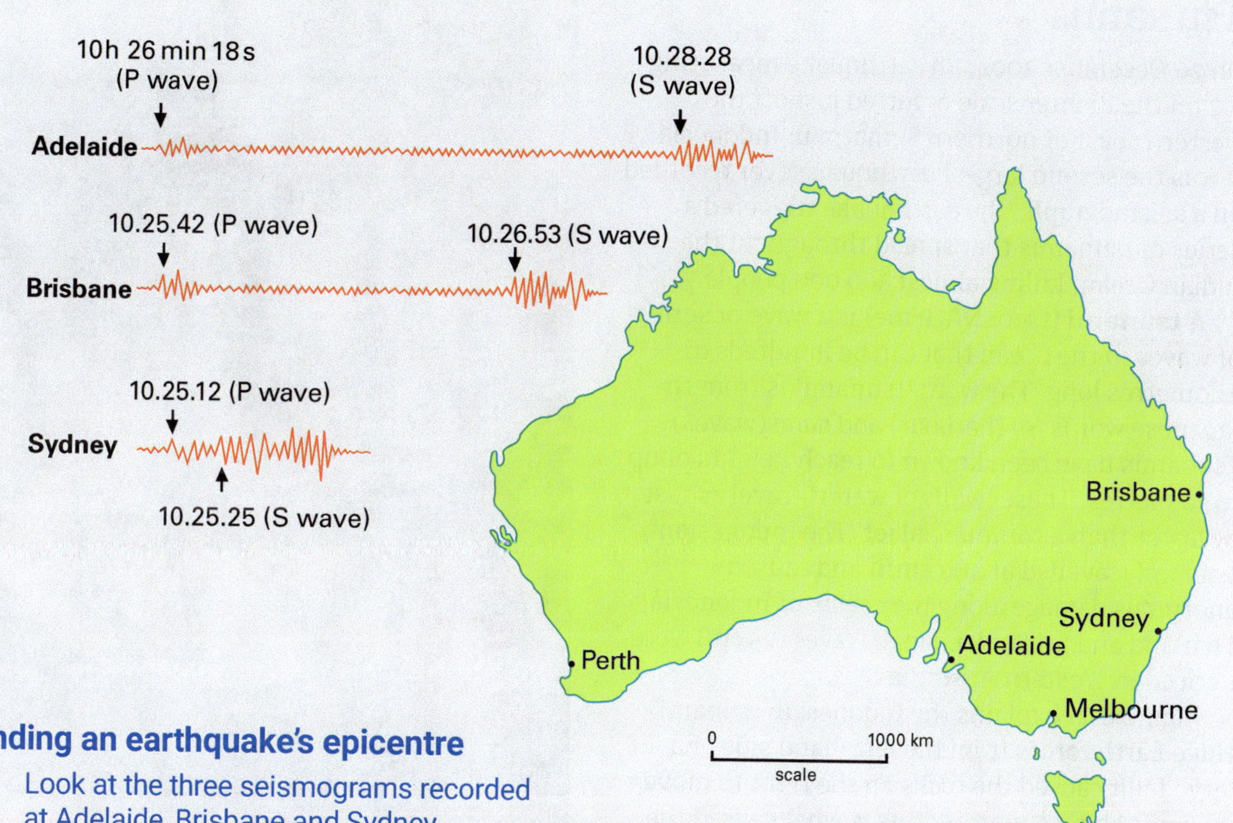

Finding an earthquake's epicentre

1. Look at the three seismograms recorded at Adelaide, Brisbane and Sydney.
   - Which city felt the earthquake first?
   - In which city were the waves strongest?
   - Which city can you infer was nearest to the epicentre of the earthquake?
2. Calculate the difference between the times of arrival of the P and S waves in each city, in seconds.
   - Record the three times.
3. Use the graph on page 196 to work out how far the earthquake waves travelled to each of the three seismographs.
   - Record the three distances.
4. Photocopy and enlarge the map of Australia. Then use the scale to convert the three distances to the distances on the map.
5. Select a city, say Adelaide first. Open out a compass to the scale distance, place the point on Adelaide and draw a circle with this distance as the radius.
   Do the same for the other two cities.
6. The three circles should intersect at one point. In practice they rarely do, but instead enclose a small area as shown, whose central point can be taken as the epicentre of the earthquake.
   - Where was the epicentre?
   - Why are three seismograms needed for this calculation?

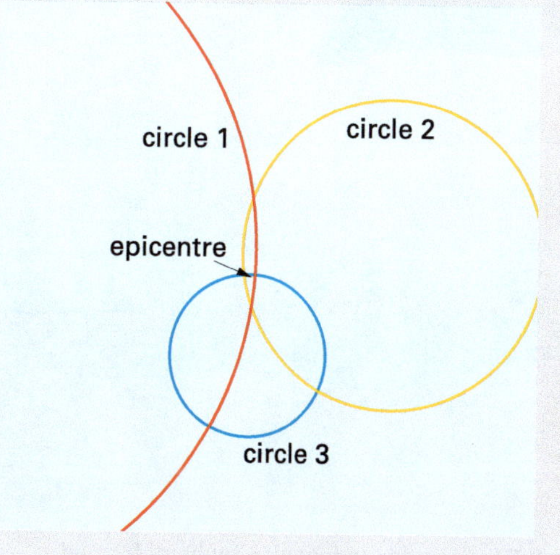

ISBN 978 1 4202 3736 8

Tsunamis

On 26 December 2004, an earthquake measuring 9.2 on the Richter scale occurred just off the western coast of northern Sumatra in Indonesia. It was the second largest earthquake ever recorded on a seismograph. The earthquake triggered a series of tsunamis that spread throughout the Indian Ocean, killing almost 300 000 people.

A **tsunami** (tsoo-NAH-me) is a wave or series of waves in the ocean that can be hundreds of kilometres long. The word 'tsunami' is from the Japanese words *tsu* (harbour) and *nami* (waves). Tsunamis have been known to reach heights of up to 10 metres. These 'walls of water' travel as fast or faster than a commercial jet. The Indonesian tsunami travelled at 480 km/h and caused enormous damage along the coasts of Indonesia, Thailand and Sri Lanka. Small waves (25 cm) were noticed in Western Australia.

Figure 8.21 explains the Indonesian tsunami. Huge Earth forces from the left-hand side of a large fault caused the rocks on the right to move upwards about 5 metres. This is what caused the wave.

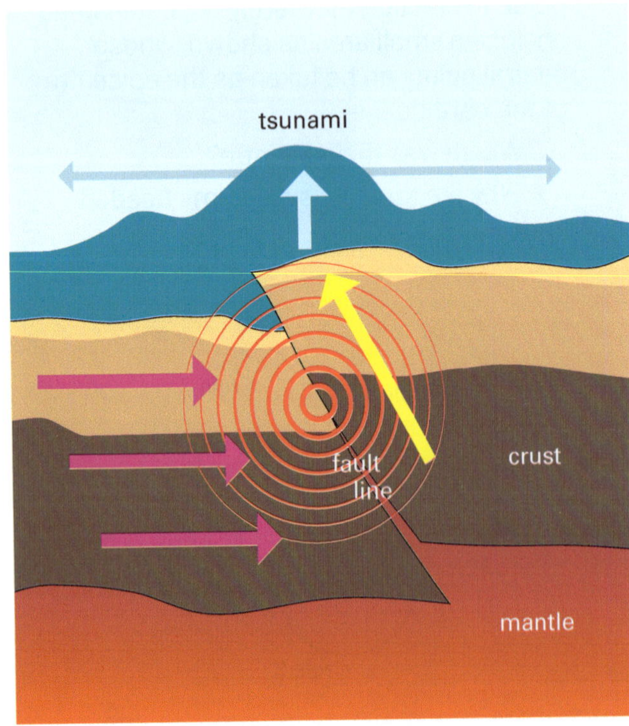

Figure 8.21 How the Indonesian tsunami formed

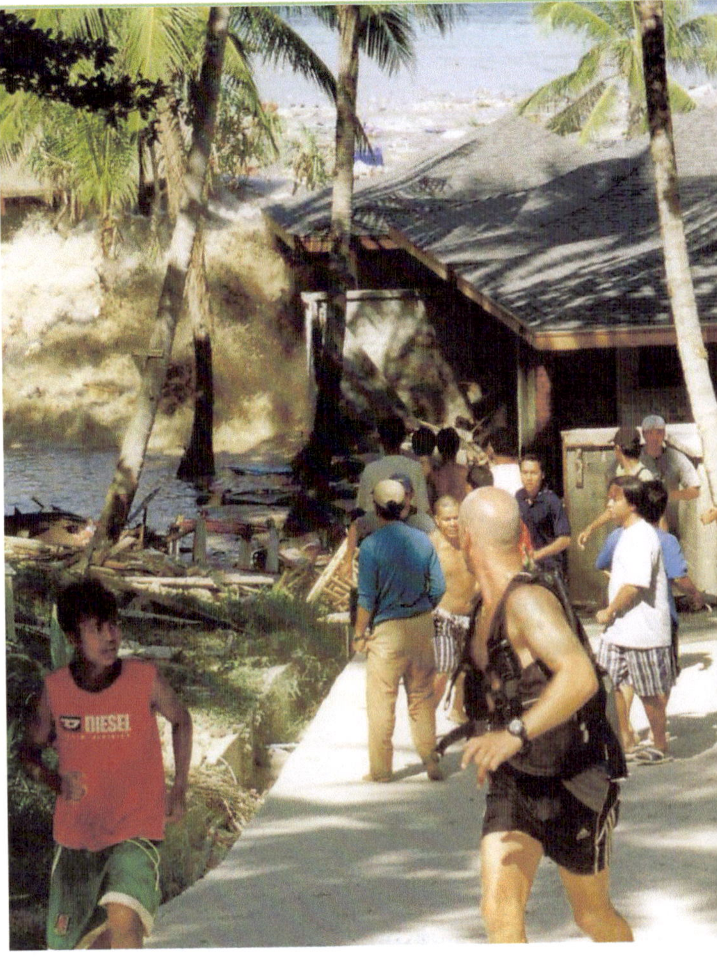

Figure 8.22 People running from the tsunami that devastated Indonesia, Thailand and Sri Lanka in 2004

Figure 8.23 A magnitude 9.0 undersea 'megathrust' earthquake off the coast of Japan in 2011 caused a devastating tsunami. Some of the devastation can been seen in the photo on page 187.

ISBN 978 1 4202 3736 8

CHECK

1. How can a fault cause an earthquake?
2. What are P, S and L waves? Which type causes the most damage?
3. What is the Richter scale? What is the difference, in terms of energy, between a magnitude 4 and a magnitude 5 earthquake?
4. What is the smallest magnitude earthquake that can be felt by humans? What is the magnitude of an earthquake that would cause trees to sway?
5. Explain in your own words how a seismograph works.
6. Use a diagram to explain the terms *focus* and *epicentre* of an earthquake.
7. The diagram below shows a seismogram recorded at a particular location. Use the graph on page 196 to find out how far away the recording station was from the epicentre of the earthquake.

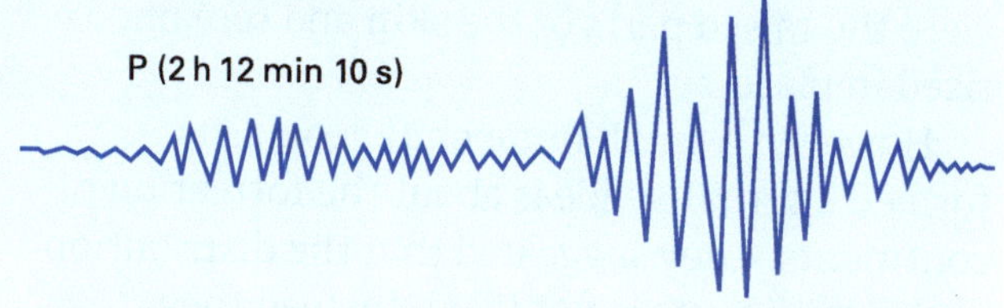

CHALLENGE

1. Three seismograph stations in Brisbane, Adelaide and Perth recorded an earthquake. The times at which the P and S waves were detected are shown below.

Brisbane:	P wave 7 h 20 min 10 s
	S wave 7 h 24 min 10 s
Adelaide:	P wave 7 h 21 min 46 s
	S wave 7 h 26 min 34 s
Perth:	P wave 7 h 23 min 22 s
	S wave 7 h 28 min 59 s

   a. Which city first detected the earthquake?
   b. Which seismograph station detected the earthquake first? What does this tell you about the location of the earthquake?
   c. Using the map on page 199, find the epicentre of the earthquake.

2. P, S and L waves travel at different speeds depending on the composition of the rocks in the crust. P waves travel at about 10 km/s, S waves at 5 km/s and L waves at 3 km/s. Suppose there was an earthquake 700 km north-west of Melbourne.
   a. How long will it take P, S and L waves to reach Melbourne?
   b. Calculate the difference in arrival times of the P and S waves.
   c. Use the graph on page 196 to find the difference between the arrival times of P and S waves after travelling 700 km. Does this agree with your answer from b? If it doesn't, suggest reasons for the difference.

3. California has had a number of very strong earthquakes, compared with Australia. Suggest why.
4. What patterns can you see in Figure 8.28 on page 204? It shows where major earthquakes have occurred around the world.
5. In February 2011, a magnitude 6.3 earthquake occurred in Christchurch, New Zealand.

   The photo below shows damage to a railway line caused by the earthquake. Suggest what sort of earth movement occurred here.

ISBN 978 1 4202 3736 8

8.3 Earth plates

During the late 1800s and early 1900s, most geologists believed that the Earth was still in the process of cooling down because the crust was shrinking. They thought it was buckling like the skin of an old, dried-out apple. The continents were the raised parts of the skin and remained fixed in place.

However, a small number of geologists formed alternative ideas about the formation of continents. They suggested that the distribution of the continents is not fixed. Instead they suggested that some continents, such as Africa and South America, had been joined at some stage in the Earth's history (see Figure 8.24).

In 1915, the German scientist Alfred Wegener suggested that Africa, South America, Australia, Antarctica and India had been joined about 200 million years ago. This supercontinent, called Gondwana, gradually broke apart and the continents separated. Wegener's hypothesis was called **continental drift**.

There was considerable evidence to support Wegener's ideas. Similar rocks are found in Australia, Antarctica and southern Africa. Fossils of an ancient plant called *Glossopteris* are widespread in southern continents (see Figure 8.25) and absent from Europe, Asia and North America. Fossils of *Mesosaurus*, a crocodile-like reptile, are found in southern Africa and in South America.

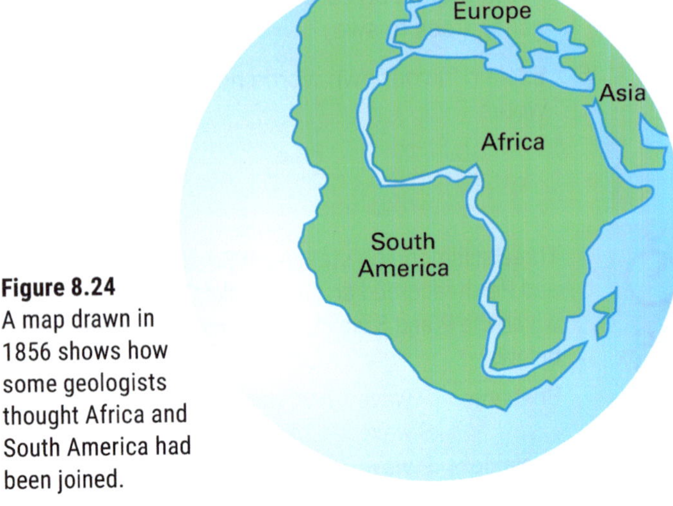

Figure 8.24 A map drawn in 1856 shows how some geologists thought Africa and South America had been joined.

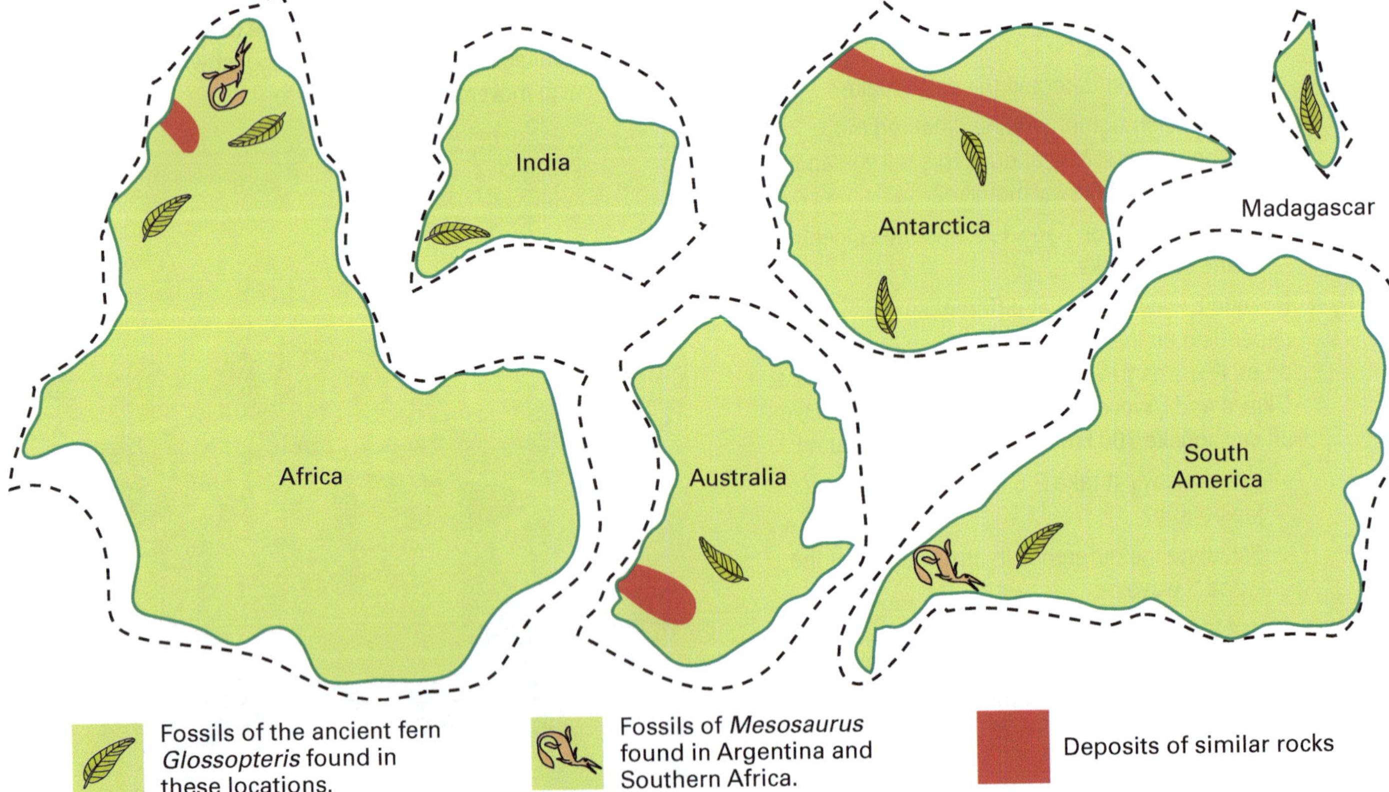

Figure 8.25 A collage of the southern continents showing the locations of similar fossils and rocks (for Activity on next page)

ISBN 978 1 4202 3736 8

1 Photocopy (and enlarge if possible) the map of the southern continents on the previous page.
2 Cut carefully around the dotted lines (the edges of the continental shelves).
3 Try to fit the six pieces together, leaving as few gaps as possible, to make the supercontinent of Gondwana. Try to match up the fossils and similar rocks.
4 When you are happy with the positions of the continents, glue them into your notebook.

Which continents did Australia touch?

You can try this activity by following the links to **Wegener: continental drift**.

Plate tectonics

Wegener's continental drift idea was not well accepted, because few scientists were convinced that there was enough evidence to support it. However, in the 1950s the American scientists Maurice Ewing and Marie Tharp used new technology called *echo-sounding* to map the depth of parts of the Atlantic Ocean. (Echo-sounding uses sound waves to find the depth of the ocean floor.)

In the middle of the ocean, a mountain range higher than any on the continents was mapped. This was called the Mid-Atlantic Ridge and it was later found to run from the Arctic, right down the middle of the Atlantic Ocean.

In the middle of the ridge, Tharp discovered a deep trench. When rock samples from this trench were taken and analysed, they were found to be very much younger than samples taken further away from the ridge. Another surprising piece of evidence was that the temperature of the ocean bed along the Mid-Atlantic Ridge was eight times higher than anywhere else. From these two observations, Ewing inferred that new rocks are being made at the Mid-Atlantic Ridge as hot material rises from the Earth's mantle and pushes the older rocks away from the ridge line.

This new idea created a dilemma. Scientists knew that the Earth is not expanding, but in fact is getting smaller. How could this happen if new crustal rocks were being made along the ridge? In 1960, Professor Harry Hess, using echo-sounding data he had collected from the Pacific Ocean in the 1950s, suggested that as new crust is being made along the ocean ridges, older crust is slowly being pushed outwards towards the continents and then down into the mantle, as shown in Figure 8.26. The idea that giant sections or plates of crustal rock move over the Earth's surface became known as the theory of **plate tectonics**, a modification of the old continental drift hypothesis.

Figure 8.26 Scientists infer that the Earth's crust is made of slowly moving plates that spread out from the mid-ocean ridges. Because the oceanic plates are heavier than the continental plates, they slide under them and melt into the asthenosphere.

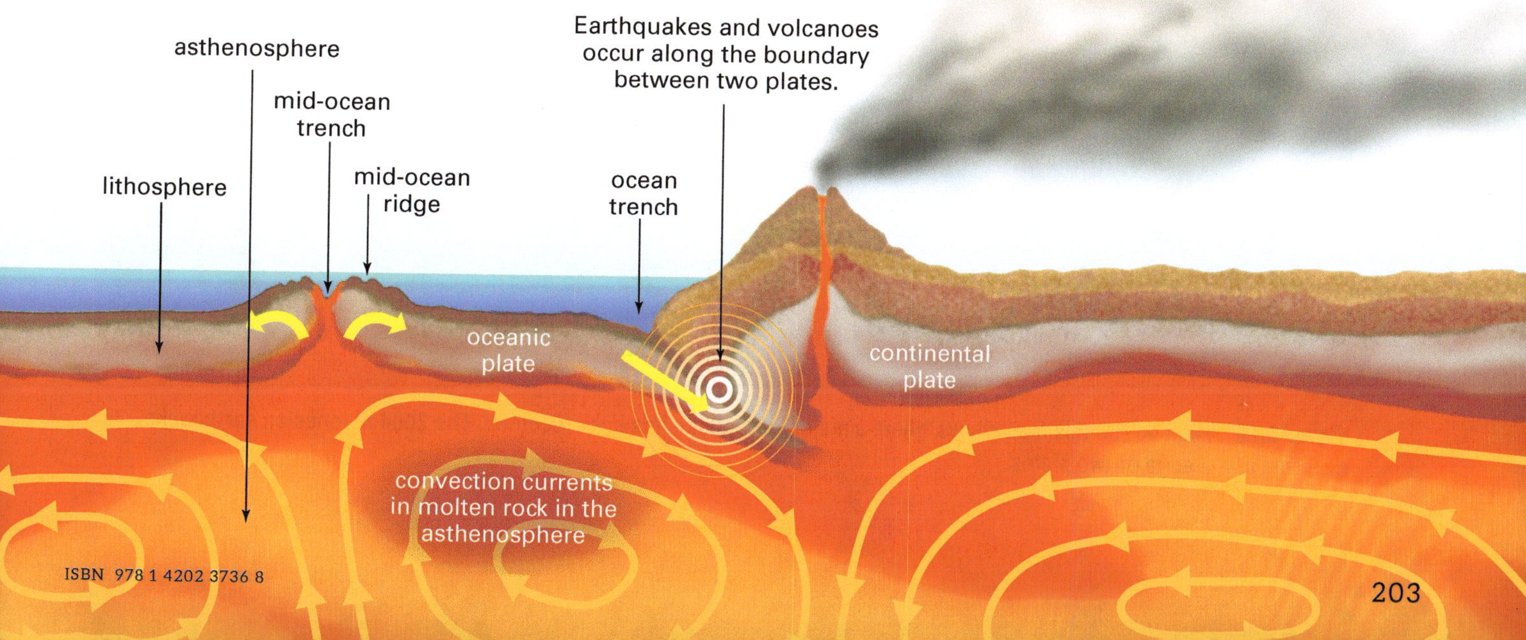

ISBN 978 1 4202 3736 8

The theory of plate tectonics suggests that the Earth's crust is made up of separate plates that float on top of the mantle. Convection currents in the asthenosphere push the molten material to the surface along the mid-ocean ridges (Figure 8.26). The newly formed crustal rock becomes part of the oceanic plates, which are forced away from the ridges in the directions shown in Figure 8.27.

Figure 8.28 shows the world's major earthquake zones. Notice that they correspond to the plate boundaries in Figure 8.27. When an oceanic plate collides with a continental plate, earthquakes and volcanoes are produced along this line. Australia's location is in the centre of the Indo-Australian plate. Being away from the plate edges explains why Australia is very stable geologically, having few and smaller earthquakes and no active volcanoes. It also explains why Australia has remained relatively unchanged for a very long time. Compare this to our neighbour New Zealand, which sits on the edge of two plates and where one plate is sliding under the other, resulting in regular earthquakes and many active volcanoes.

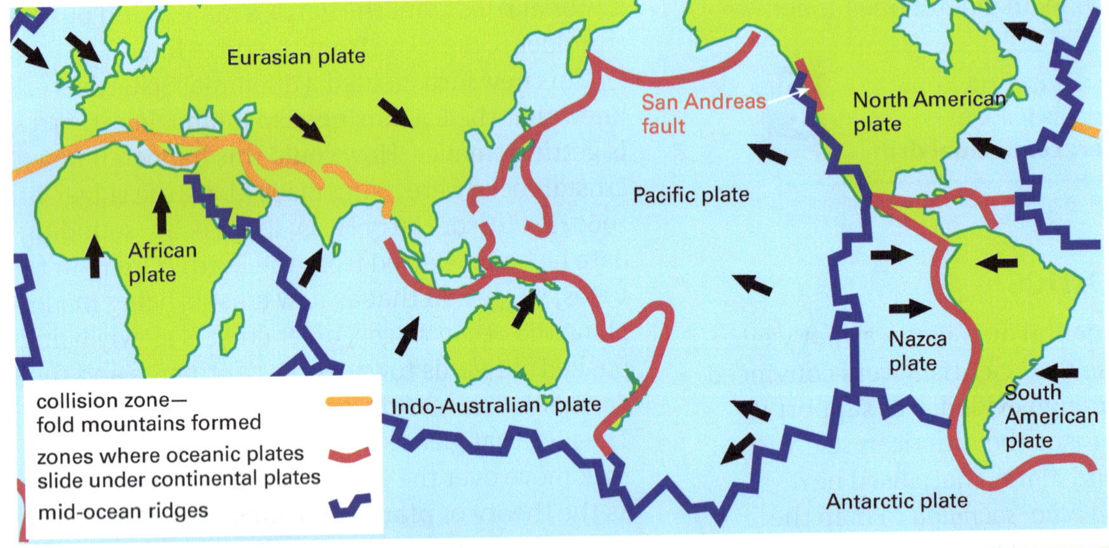

Figure 8.27 The Earth's crust is made up of large plates, which move an average of 2 cm per year.

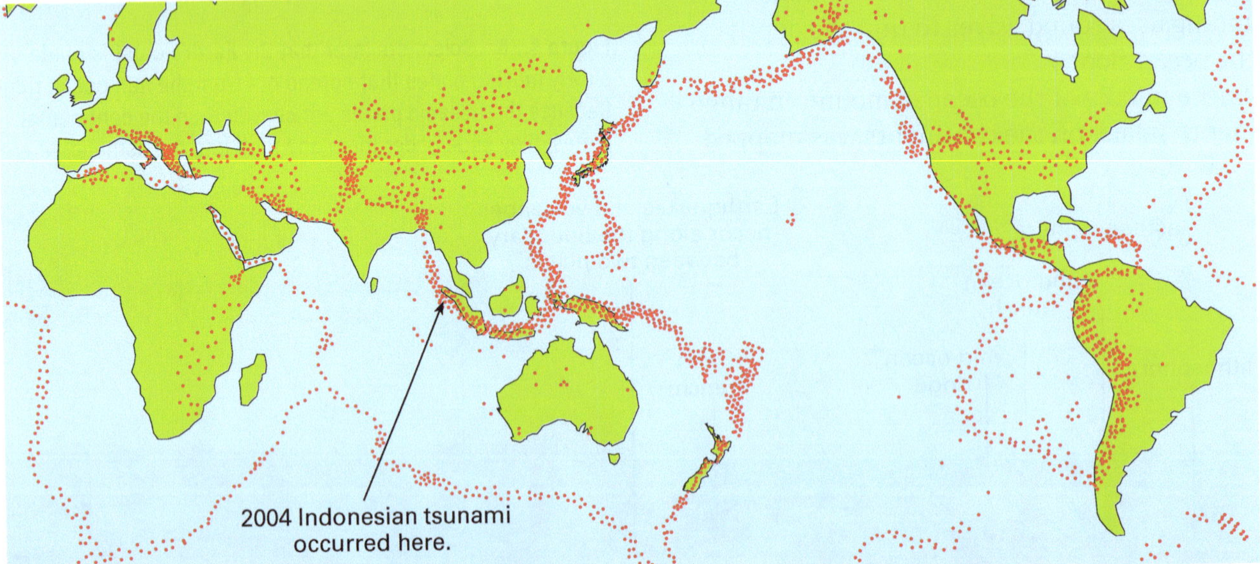

Figure 8.28 Each dot on this map shows where a major earthquake has been recorded. The 2004 Indonesian earthquake occurred at the edge of two plates.

ISBN 978 1 4202 3736 8

CHECK

1. At the start of the 20th century, most scientists accepted a theory about the formation of the continents.
   a. Describe this theory.
   b. How is it different from the theory proposed by Alfred Wegener?
2. Describe the evidence used to support the theory of continental drift.
3. Why is Gondwana called a supercontinent?
4. Suggest why certain fossils have been found in Australia, India and Antarctica but not in Europe or North America.
5. What discoveries during the mid-20th century added new evidence to the idea that the Earth's crust is not fixed in place?
6. Suppose a continental plate moves 5 cm a year. Predict how far it will move in 200 years, and in 20 million years.
7. Ewing and Tharp echo-sounded vast areas of the Atlantic Ocean during the 1950s. They concluded from their results that the seafloor was spreading. Why was this a dilemma for scientists at the time?
8. Copy the drawing below and add the following labels:
   - continental plate
   - oceanic plate
   - region of colliding plates
   - mid-oceanic ridge
   - convection currents in the mantle
   - regions of earthquake activity
   - regions of volcanic activity

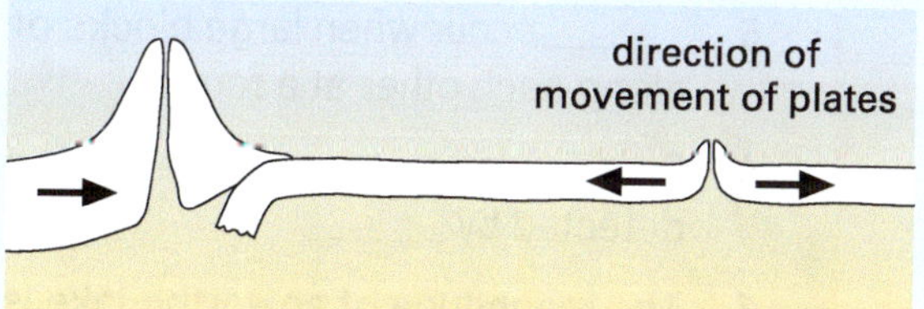

9. Is Australia a stable continent geologically? Explain your answer.

CHALLENGE

1. Some of the highest mountains in the Andes in South America are made mostly of sedimentary rocks containing marine fossils.
   Use your knowledge of plate tectonics to explain how this mountain range might have formed.
2. Use Figure 8.26 on page 203 and the maps on page 204 to explain why a deep-sea trench more than 10 000 m deep is found on the eastern side of Japan.
3. Which way is Australia moving on the surface of the Earth? If it is moving at an average of 4 cm a year, what is its likely position 100 million years from now?
4. There are two types of plates—oceanic plates and continental plates.
   a. Briefly describe what happens when an oceanic plate collides with a continental plate.
   b. Use library resources to find out what happens when two continental plates collide.
5. Use an atlas to locate the following volcanoes: Mt Fuji, Mt St Helens, Mt Vesuvius, Kilauea, Mt Pinatubo, Krakatoa.
   Then use a pencil to mark them on a copy of the map of the Earth's plates on page 204. Why do you think the volcanoes are found in these places?
6. The photo shows a hydrothermal vent or 'black smoker' on the ocean floor. Use the internet to find out what black smokers are, where they are found, and why they are important.

ISBN 978 1 4202 3736 8

MAIN IDEAS

Copy and complete these statements to make a summary of this chapter. The missing words are on the right.

1 Scientists infer that the Earth is made up of three main layers: the ______, the ______ and the core.

2 The mantle consists of very hot, semi-solid material called ______. When this breaks through the Earth's surface, ______ form.

3 ______ and fossils fuels are often found near folds and faults in rocks.

4 Movements in the mantle cause changes in rocks in the crust: rocks that bend slowly form ______ and those that break suddenly form ______.

5 ______ occur when large blocks of rock move suddenly and slide along each other at a fault.

6 Earthquakes produce three types of ______: P, S and L. These are detected by ______.

7 The magnitude of an earthquake is usually measured on the ______ (from 0 to 10).

8 According to the theory of ______, the Earth's crust is made up of a number of large plates that move slowly relative to one another. Earthquakes and volcanoes occur where these plates ______.

9 Australia is a relatively ______ continent compared to many others, as it is located in the ______ of a tectonic plate.

collide
waves
earthquakes
magma
crust
centre
faults
mantle
seismographs
minerals
plate tectonics
volcanoes
folds
Richter scale
stable

CH•8 REVIEW

1 The diagram below shows a cross-section through an area of the Earth's crust.

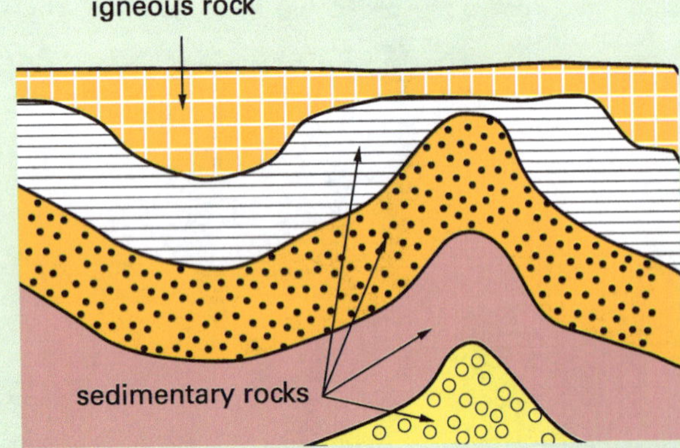

a The shape of the sedimentary rocks is most likely the result of:
   A a fault.
   B horizontal squeezing forces.
   C the way in which the sediments were originally laid down.
   D the pushing up of magma from below.

b Copy the diagram in your notebook. Mark the position of an anticline and a syncline on the diagram.

c Account for the layer of igneous rock on top of the sedimentary rocks. Suggest why this layer is relatively flat on top.

d Infer which of the rock layers is the oldest, and which is the youngest? Give reasons for your inferences.

ISBN 978 1 4202 3736 8

2 Use the maps of the world's earthquake zones and the Earth's plates on page 204 to explain why there is a greater chance of an earthquake occurring in Wellington, New Zealand, than in Melbourne.

3 The seismogram below is a record of an earthquake.

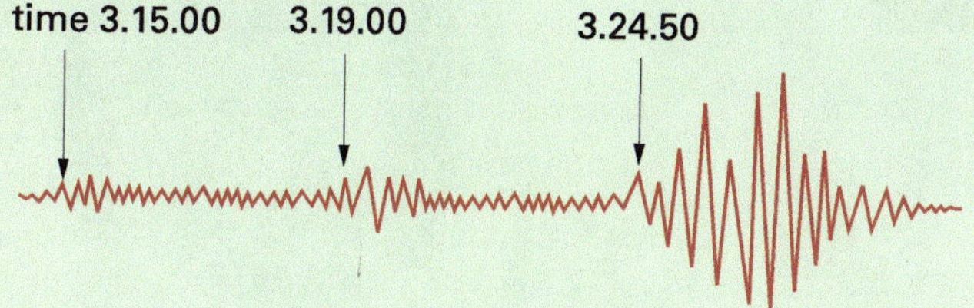

a Which is the P wave and which is the L wave?
b Which waves are due to the compression of matter in the crustal rocks?
c Which type of wave has the most energy?
d Which type of wave causes the most damage to structures on the Earth's surface?
e What is the difference in the arrival times of the P and S waves?
f Use the graph on page 196 to calculate how far away the earthquake was.
g Can you use your answer in **f** to find the epicentre of the earthquake? Explain your answer.

4 Forces inside the Earth act on rocks in the crust as shown in the diagram below. As a result of these forces, the rocks move. Draw labelled diagrams to show what the rock layers could look like after this movement. There are three possibilities.

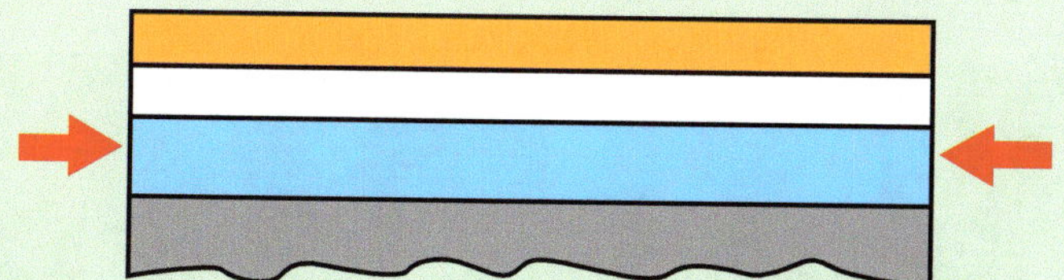

5 Scientists have calculated that the Mediterranean Sea between Africa and Europe is becoming narrower over time. Use the map on page 204 to suggest why.

6 An earthquake of magnitude 6 on the Richter scale is recorded at location A and another of magnitude 3 is recorded at location B.
a Use the table on page 197 to describe the damage that could be caused by each earthquake.
b How many times more energy is released by the earthquake at A than the one at B?

7 The data below summarises the three types of earthquake waves:
- **Primary (P) waves**: speed 5–13 km/s
- **Secondary (S) waves**: speed 4–8 km/s, travel through solids only
- **Long (L) waves**: speed 3–4 km/s, travel only on the Earth's surface, most destructive

a Which type of waves are sometimes called surface waves? Why?
b What is the shortest time an S wave will take to travel 1000 km?
c The diagram below is a cross-section of the Earth. An earthquake occurred at point A. A seismograph at location B detected P, S and L waves. Other seismographs at location C and D detected only P and L waves. Suggest a reason for this.

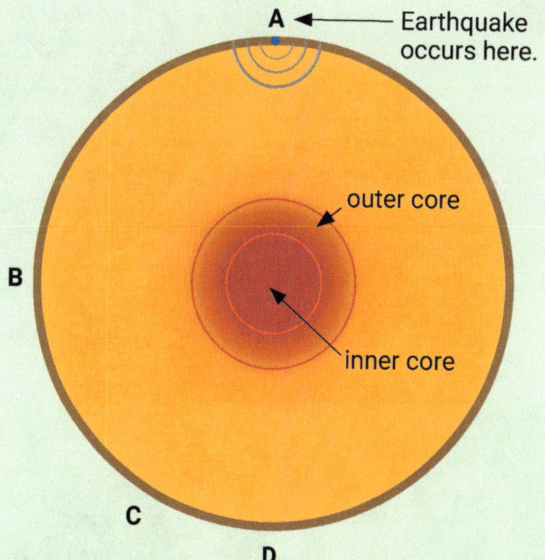

8 Explain why a country like Indonesia or New Zealand has many earthquakes and volcanoes, but Australia has fewer earthquakes and no active volcanoes. How does this relate to Australia's geological history?

Check your answers on page 303.

ISBN 978 1 4202 3736 8

IN THIS CHAPTER

Science Understanding

> describe chemical reactions using word equations
> investigate reactions of acids with metals, bases and carbonates
> give examples of how chemical reactions are important in both living and non-living systems and involve energy transfer
> compare respiration and photosynthesis and their roles in biological processes
> describe how the products of combustion reactions affect the environment
> consider the impact of the Mount Isa mine on the local community

Science Inquiry Skills

> work safely in the laboratory

CH•9 Everyday reactions

ISBN 978 1 4202 3736 8

GET STARTED: *EXPLORE*

When you put sherbet in your mouth, you experience one of the chemical reactions of acids. You can do this at home or at school, provided all your equipment is perfectly clean.

Add the following to a paper cup:

- 3 teaspoons icing sugar
- ½ teaspoon citric acid
- ¼ teaspoon baking soda

Mix the ingredients thoroughly, then taste the mixture.

> Describe the taste of your homemade sherbet.

> What happens when you add a few drops of water to the sherbet?

> Suggest a reason for the distinctive taste of sherbet.

> Why do you think the reaction between citric acid and baking soda occurs only when you add water?

> What chemical substance is baking soda made of?

ISBN 978 1 4202 3736 8

9.1 Acids and bases

Figure 9.1 Common acids

What are acids?

Acids are very common substances and are used widely in everyday life. Some occur naturally, and some are synthetic. Some of them are potentially dangerous because they are **corrosive**—they can 'eat away' metal, wood and clothing, and burn your skin. For example, battery acid contains sulfuric acid (H_2SO_4), which will burn your skin; and hydrochloric acid (HCl) is used to clean mortar from bricks.

The photo above shows some of the many natural acids found in food and drink. They all have a sour taste. The sherbet you made in Getting started contains citric acid, which is in all citrus fruits and tomatoes. Yoghurt contains lactic acid, vinegar contains acetic acid, and grapes contain tartaric acid. The bubbles in soft drinks are due to carbon dioxide, which dissolves in water to form carbonic acid. The hydrochloric acid in your stomach is essential for digestion, and the DNA that makes you different from everybody else is deoxyribonucleic *acid*.

Acids that are corrosive attack your body tissues. This is why lemon juice stings if you get it in a cut on your finger; and bees and ants sting because they inject formic acid into you. You can eat fruit that contains acids because the acid is very dilute. A *dilute* acid is one that contains a large amount of water and a small amount of acid. The opposite of dilute is *concentrated*, and concentrated acids such as the sulfuric acid in a car battery must be handled with great care.

What are bases?

Bases neutralise (cancel out the effect of) acids. For example, some toothpastes contain a weak base to neutralise the acids formed by plaque bacteria on your teeth. Bases are also used to dissolve grease and dirt. Oven cleaners and drain cleaners usually contain caustic soda (sodium hydroxide NaOH), which dissolves grease. Other household cleaners contain ammonia, which can be used to remove grease from floors or to clean windows. Bases that are soluble in water are called **alkalis**. The reason they feel soapy is because they turn the oils on your skin into soap.

ISBN 978 1 4202 3736 8

Indicators

Some solutions are *acidic* and some are *basic*, while others are *neutral* (not acidic or basic). For example, tap water is usually neutral.

A quick way to tell whether a solution is acidic or basic is to use an acid–base **indicator**. Such substances *indicate* when an acid or base is present by changing their colour.

Some indicators occur naturally in dyes in plants. For example, litmus comes from lichens, which grow on the bark of trees and on rocks. In an acidic solution, litmus turns red; in a basic solution, it turns blue. There are also a number of synthetic indicators. One of these is called bromothymol (bro-mo-THY-mol) blue. If bromothymol blue is added to an acidic solution, it changes colour from blue to yellow. If an alkali such as sodium hydroxide is added, it turns blue again.

Figure 9.2 Household cleaners contain alkalis such as ammonia and caustic soda.

If you add bromothymol blue to a basic solution, it stays blue. But if you add acid, it turns yellow. To be sure that a substance is an acid (or a base), you must observe a *change* in the colour of an indicator. Suppose you test a solution with bromothymol blue and it stays blue. You cannot say from this that the solution is basic. It could be water. You would need to use another indicator, such as red litmus, and see if it changes colour in the solution.

Most indicators have only two colours. *Universal indicator* is a mixture of several different indicators. Because of this, universal indicator can be many different colours depending on how acidic or basic the solution is. (See the colour chart below.)

Indicator	Colour when solution is acidic	Colour when solution is neutral	Colour when solution is basic
bromothymol blue			
litmus			
methyl orange			
phenolphthalein			
phenol red			
universal indicator			

pH 0 1 2 3 4 5 6 7 8 9 10 11 12 13 14

Figure 9.3 Indicators show how acidic or basic a solution is. (There is more on this in section 9.2.)

ISBN 978 1 4202 3736 8

INVESTIGATION 9.1 Red cabbage indicator

Aim

To extract the coloured substance from red cabbage, and use it to test acids and bases.

Materials

- 2 or 3 large leaves from a fresh red cabbage
- sharp knife and chopping board
- two 250 mL beakers
- hotplate (or burner, tripod and gauze mat)
- stirring rod
- 6 test tubes and test tube rack
- dilute **hydrochloric acid** (0.5 M)
- dilute **sodium hydroxide** solution (0.5 M)
- various household substances, e.g.:

window cleaner	baking soda
shampoo and conditioner	cream of tartar
antacid tablet	lemon juice
vitamin C	milk
vinegar	

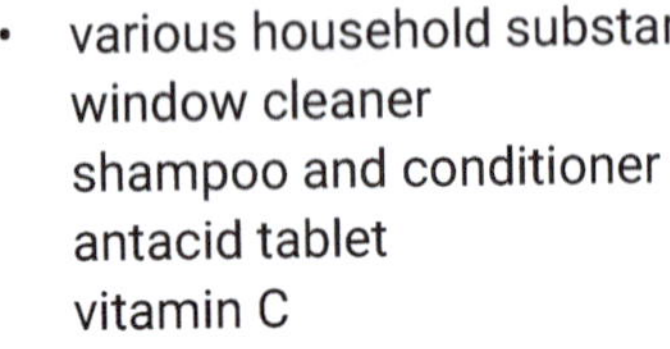

SAFETY GLASSES MUST BE WORN

Risk assessment and planning

- What special precautions are needed when handling acids and bases?
- What should you do if you spill a corrosive liquid?
- Design a data table to record your results.

PART A

Method

1 Cut two or three large red cabbage leaves into small pieces. Put them in a beaker.

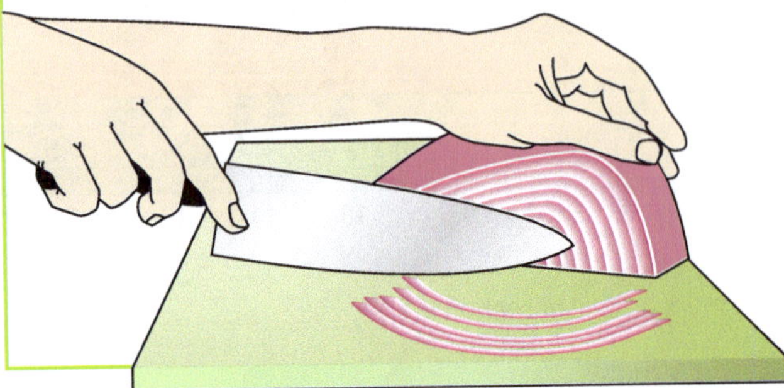

2 Add water to just cover the cabbage pieces.

3 Boil the cabbage mixture for 5–10 minutes. The water should turn a dark colour, and the leaves should almost lose their colour.

4 Let the mixture cool. Then carefully decant the coloured solution into another beaker. Alternatively, you could strain the mixture through a sieve.

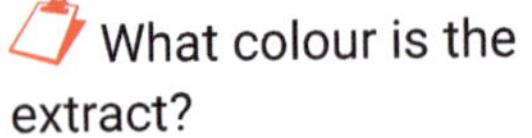

What colour is the extract?

5 Add a small amount of dilute hydrochloric acid to a test tube and label it. Add some sodium hydroxide solution to another test tube. Now add a few drops of red cabbage extract to each tube.

What colours are the solutions?

6 In the same way, use the red cabbage extract to test the various household substances you have collected.

For each substance, record any colour change.

Discussion

1 What colour is your red cabbage extract in a neutral solution (water)? What colour is it in acidic solutions? In basic solutions?

2 Which household substances are the most acidic? Which are the most basic? How do you know?

3 Suggest why the red cabbage extract is called an acid–base indicator.

PART B Inquiry

You may wish to extract and test the colour from flower petals, e.g. tibouchina, hydrangea, hibiscus, yellow pansies. You could also try fresh beetroot or tea leaves.

ISBN 978 1 4202 3736 8

Acids and teeth

When you eat, food often remains between your teeth. Bacteria in your mouth then feed on this food. Chemical reactions occur and the bacteria produce weak acids as waste products. These acids react slowly with your teeth, causing tooth decay. The mixture of bits of food, bacteria, acids and saliva that sticks to your teeth is called *plaque* (PLARK).

The top of a tooth is covered with enamel, which is the hardest substance in the body. The inside of the tooth is made of a softer substance called dentine. If the bacteria and acids cause this hard enamel to decay, then the tooth can be damaged very rapidly.

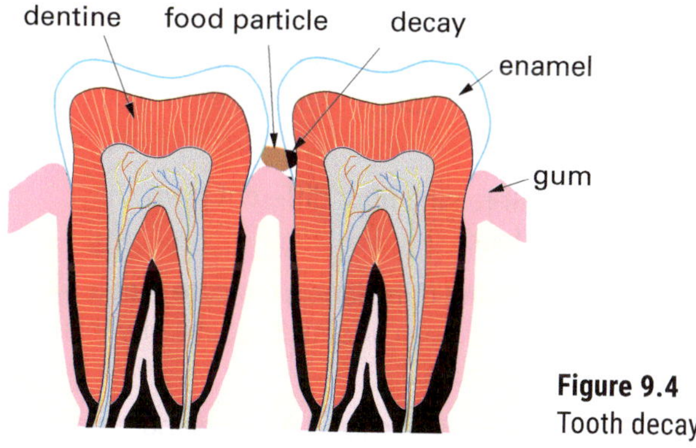

Figure 9.4
Tooth decay

The best way of getting rid of plaque and food particles from your teeth is by brushing them. Toothpastes contain abrasives such as finely powdered chalk that help scrape food particles from your teeth. They also usually contain a small amount of soap or mild detergent. This makes a foam which helps brushing.

Some toothpastes are slightly basic to neutralise the acids produced by decaying food. They may also contain fluoride compounds. These react with tooth enamel in young people's teeth to form a substance that is more resistant to acid attack and less likely to decay.

Stomach acid

Gastric juice is produced in your stomach to help you digest, or break down, the food you eat. This gastric juice consists of dilute hydrochloric acid, the enzymes pepsin and rennin, and water.

The hydrochloric acid helps to kill most microbes. It also helps the enzymes to work, since they will function only in the presence of an acid. These enzymes break down the proteins in food into amino acids, which are needed for growth.

Your stomach is protected from the acid in the gastric juice by a sticky fluid called *mucus*. This protects the stomach wall and acts as a lubricant so that food passes through smoothly. If bacteria called *Helicobacter pylori* get into the lining of the stomach, they can weaken the mucous layer, allowing the acid to attack the lining. This can lead to a painful stomach ulcer (sore).

Acids in food and drink

Baking powder and self-raising flour contain baking soda (sodium hydrogen carbonate $NaHCO_3$) and an acidic substance such as cream of tartar. While the baking powder or flour is dry, no reaction occurs. But when it becomes moist, the baking soda reacts with the acid to form carbon dioxide gas. The bubbles of gas are trapped inside the cake, and when the cake is placed in an oven, the carbon dioxide gas expands, making the cake 'rise'.

Acids are also used to preserve food. For example, when vinegar (dilute acetic acid) is used in the making of pickles or sauces, the acid prevents the growth of bacteria, some of which may be harmful.

Figure 9.5
The holes in this cake are produced by bubbles of carbon dioxide gas.

ISBN 978 1 4202 3736 8

1 Copy and complete the following sentences.
   a Litmus turns _____ in basic solutions, and _____ in acidic solutions.
   b A soluble base is called an _____.
   c Solutions may be acidic, basic or _____.
   d To be sure that a solution is acidic, you must observe a _____ in the colour of an indicator.
   e Bromothymol blue turns yellow in _____ solutions.
   f Vinegar is _____, and ammonia solution is _____.

2 Give the names and uses of two alkalis found in the home.

3 Why is it that the acids in food and drink do not harm your stomach?

4 Use the information in the table below to answer these questions.
   a What acid is present in:
      - sour milk?
      - oil of wintergreen?
      - ants?
   b Where would you find:
      - citric acid?
      - tartaric acid?
   c Apart from its value as vitamin C, why else is ascorbic acid added to fruit juices?
   d If oxalic acid is a poison, why is it possible to eat rhubarb and spinach?
   e Acetic acid can be formed from one of the other acids in the table. Which one?

5 Why are some toothpastes basic?

6 Use the colour chart of acid–base indicators on page 211 to answer these questions.
   a What is the colour of phenol red in a basic solution?
   b What colour is litmus in:
      - vinegar?
      - ammonia solution?
   c What colour would you expect if you added methyl orange to tap water?
   d Jerry added some phenolphthalein to a solution and it remained colourless. When he added bromothymol blue to the solution, it turned yellow. Is the solution acidic, basic or neutral? How do you know?
   e Universal indicator is added to some dilute hydrochloric acid and excess sodium hydroxide (more than enough to neutralise the acid) is then added. What would be the order of the colour changes shown by the indicator?

7 How is self-raising flour different from ordinary flour?

8 Some food and drink labels indicate that a food acid has been added. Suggest why this food acid is added.

9 Sodium hydroxide is used for cleaning greasy ovens. Gavin spilt some on his hand and Fiona suggested that he rinse his hand with vinegar. Is Fiona's suggestion sensible? Explain.

Acid	Found in . . .	Uses
acetic	fermented grapes	vinegar, making PVA glue
ascorbic	fresh fruit and vegetables	vitamin C, food preservative
citric	citrus fruits, e.g. oranges and lemons	fruit drinks, medicines
formic	ants and stinging nettles	preservative and antibacterial agent in livestock feed
lactic	sour milk, tired muscles	yoghurt and cheese, cosmetics
oxalic	rhubarb leaves and spinach leaves	wood bleaching agent, rust remover
salicylic	oil of wintergreen	aspirin, acne creams, heat rubs
tartaric	grape juice	flavouring agent, cream of tartar

ISBN 978 1 4202 3736 8

CHALLENGE

1 When you vomit, you get a sour taste in your throat and mouth. Suggest a reason for this.

2 Water containing bromothymol blue turns yellow when carbon dioxide is bubbled into it. Suggest why this happens.

3 Huw read somewhere that a stinging nettle contains an acid that is injected into your skin if you touch it. How could he show that a stinging nettle contains an acid?

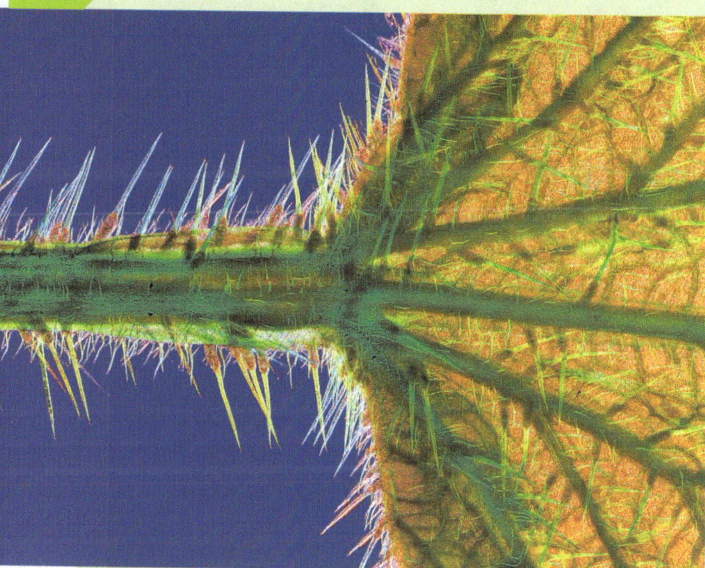

Figure 9.6 A close-up view of the barbs on the leaf stem of a stinging nettle (leaf on the right)

4 When Nemika put a piece of red litmus paper in an unknown liquid, nothing happened. When she added bromothymol blue, it stayed blue. What inference can Nemika make from these observations?

5 Why does boiling a dilute acid make it more concentrated?

6 The dyes in some fabrics change colour when washed with certain detergents. Suggest a reason for this.

7 Andre did an experiment to investigate how well mould grows under different conditions. He placed cubes of cooked pumpkin in three jars. He half filled the first jar with water, the second with vinegar and the third with a solution of baking soda. He then covered the jars and left them in a warm, dark cupboard for 2 days. His results are illustrated below.

What conclusion can Andre draw from his experiment?

EXPLORE

1 Mix equal amounts of cream of tartar and baking soda in a container and then add a little water. Explain your observations.

2 **Disappearing ink**

You will need phenolphthalein solution (0.1 g in 10 mL ethanol, then add 90 mL water). Add 3 M sodium hydroxide, a drop at a time, until the solution turns dark red. Put the solution in a spray bottle.

Accidentally on purpose, spill the red 'ink' on a white cloth or shirt. Blow on the cloth or wave it in the air.

You could try dabbing the ink spot on the cloth with a cotton ball dampened with ammonia or vinegar.

The red colour is due to the sodium hydroxide, which is basic. Carbon dioxide in the air forms carbonic acid, which neutralises the sodium hydroxide. As a result, the phenolphthalein turns colourless.

3 Hayley asks her mother why she squeezes lemon juice onto freshly made fruit salad. Her mother says that the acid in the lemon juice stops the apple and banana going brown.

Design and perform tests to find out whether it is the acid in the lemon juice that stops the browning, or some other substance.

ISBN 978 1 4202 3736 8

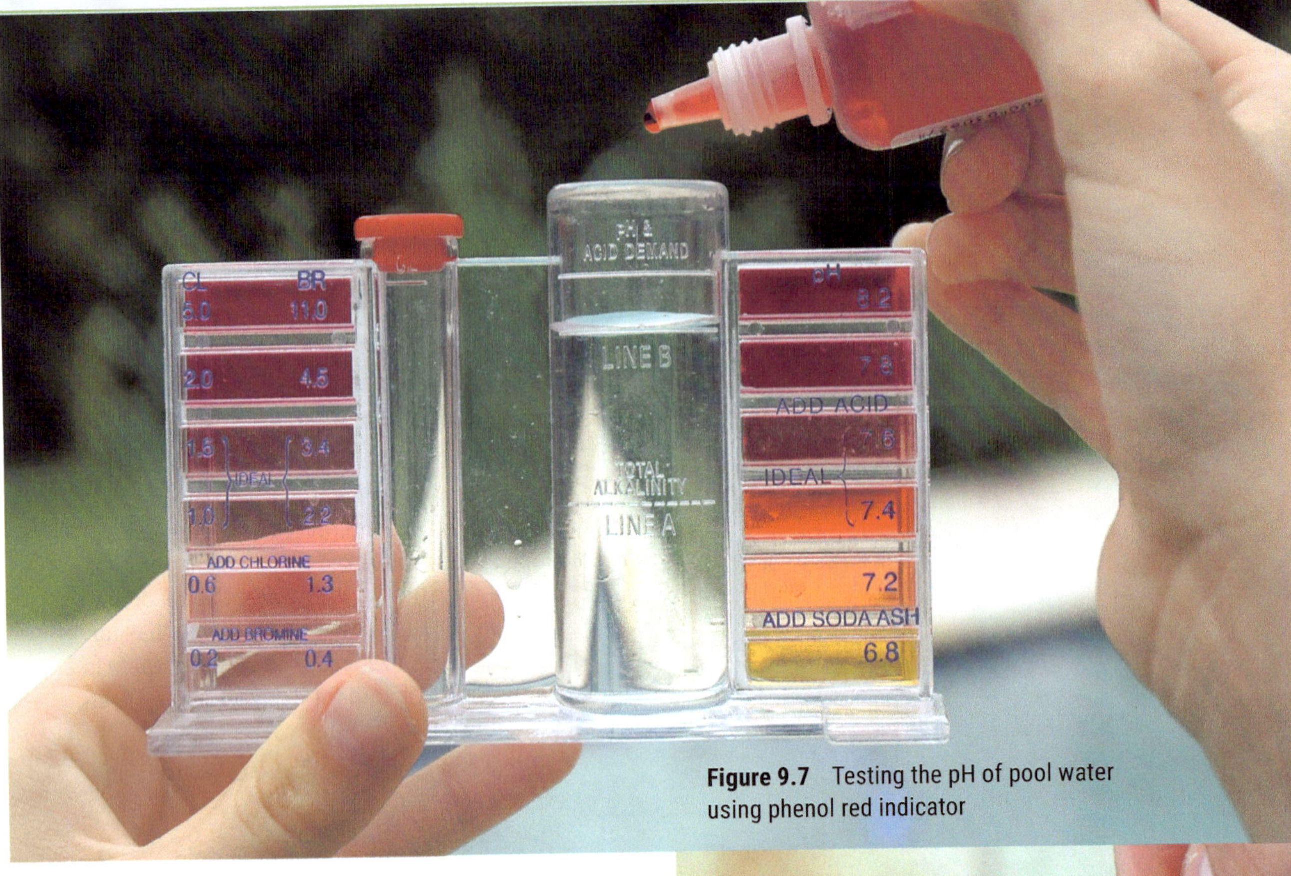

Figure 9.7 Testing the pH of pool water using phenol red indicator

9.2 The pH scale

Jodie is checking the pH of the family swimming pool. She takes a sample from the pool and adds it to the test kit as shown above. She then adds a few drops of phenol red indicator and checks the colours on the scale. The pH is a little bit higher than the ideal 7.4–7.6, so she needs to add some acid to lower the pH slightly. Just what is pH?

pH is simply a scale from 0 to 14 that tells you how acidic or basic a solution is. Acidic solutions have a pH less than 7, and basic solutions have a pH greater than 7. A pH of 7 tells you that the solution is neutral.

To measure the pH of a solution, you can use an indicator. For example, the phenol red that Jodie used to test the pool water changes from red to yellow at a pH of 7.4–7.6. If you use universal indicator solution, you note the colour and read the pH on the special colour chart. If you use pH paper, you simply dip the paper into the solution and note the colour, as shown in Figure 9.8. For example, if it is orange, then the pH is about 4, and if it is green, it is about 8. You can measure pH more accurately using an electronic pH meter.

Figure 9.8 Using pH paper to measure pH

ISBN 978 1 4202 3736 8

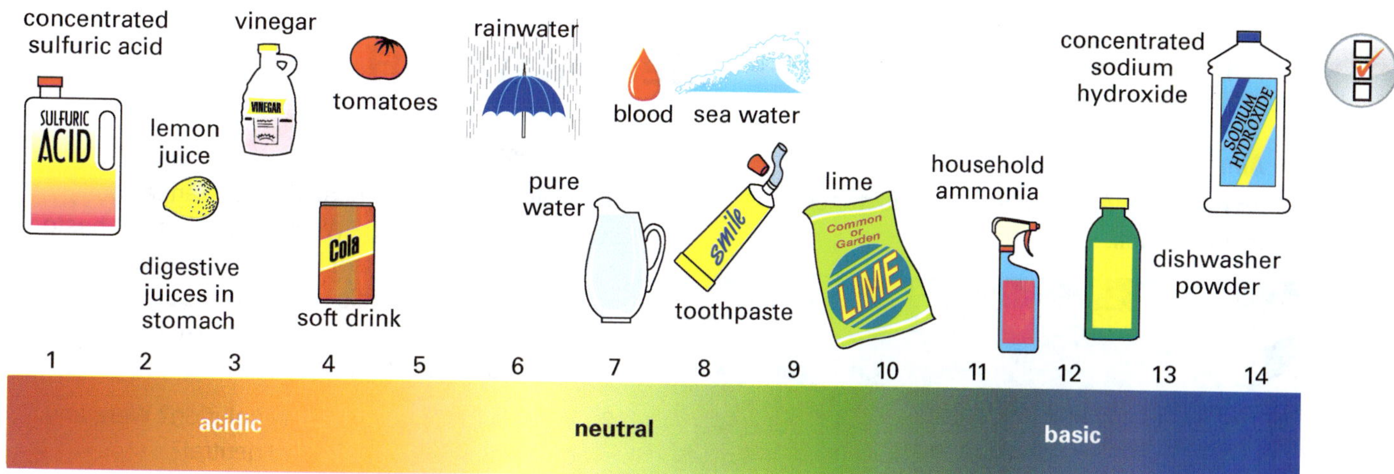

Figure 9.9 The pH scale, with the pH of some common substances

pH in your life

The pH of the liquids in your body varies from one organ to another. In your stomach, acidic conditions (pH 1.5) are needed for the digestion of proteins. In your small intestine, alkaline conditions (pH 8) are needed for the further digestion of food. Your blood has a pH of 7.4, and the pH of urine can vary from 6.5 to 8.

Each strand of your hair consists of a central core surrounded by a scaly covering called the cuticle. At a pH of 4–6, the scales of the cuticle lie flat. They reflect light evenly, making the hair look shiny. If the pH is higher than 6, the cuticle becomes 'ruffled'. Light is reflected in all directions and the hair looks dull. For this reason, shampoos and conditioners contain substances to keep the pH in the range 4–6.

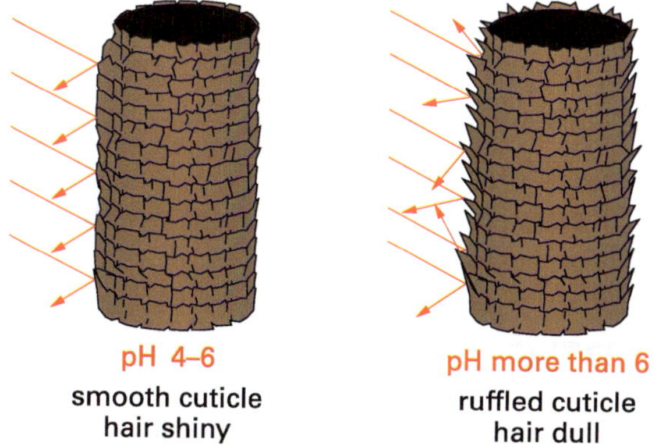

Figure 9.10 The pH of your hair affects how it looks.

The pH of soil is important for the growth of plants, and soil test kits can be bought to test the acidity of the soil. Some plants grow better in acidic soils and some prefer alkaline soils. If the soil is too acidic, you can add basic solutions such as powdered limestone or dolomite. This is often necessary in agricultural areas where nitrogen fertilisers have been used. If the soil is too alkaline, you can add compost, manure or a soluble fertiliser such as ammonium sulfate, which is acidic.

Figure 9.11 In acid soil hydrangea flowers are white or blue. In basic soil the flowers are pink.

ISBN 978 1 4202 3736 8

INVESTIGATION 9.2 Measuring pH

Aim

To measure the pH of various substances, including soil.

Risk assessment and planning

Read both parts of the experiment.

- What safety precautions will be needed?
- Design a data table for Part A.

PART A Household substances

Materials

- various household substances in dropper bottles, e.g.:
  window cleaner
  shampoo and conditioner
  antacid tablet
  vitamin C
  baking soda
  vinegar
  lemon juice
  milk
- universal indicator solution or pH paper
- laboratory acids and bases
- spotting tile

Method

Use the diagram and notes below to measure the pH of various household substances.

SAFETY GLASSES MUST BE WORN

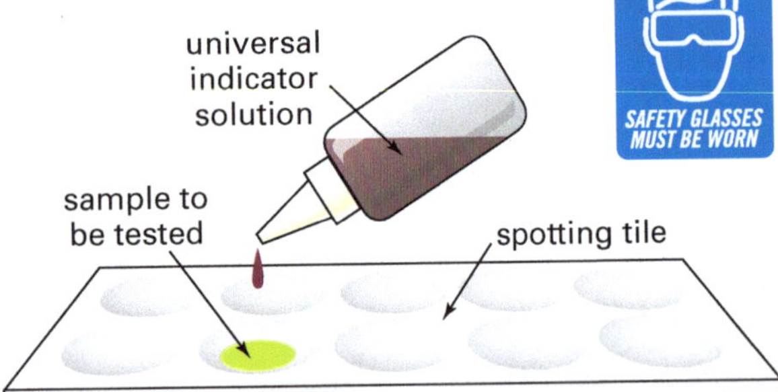

- The samples to be tested must be in solution or wet.
- Put a few drops of the sample in a cavity on a spotting tile and add a drop of indicator.
- If you are using pH paper, add a drop of the sample to a 1 cm piece of paper.

Discussion

1 Which was the most acidic substance you tested (lowest pH)? Which was the most basic (highest pH)? Make a list of the substances from the most acidic to the most basic. How do your results compare with those from Investigation 9.1 on page 212?

2 What is the pH of a neutral solution? Were any of the solutions you tested neutral?

3 Predict what will happen to the pH of an acid when you dilute it with water. Will it be more or less acidic? (You could test this.)

PART B Soils

Materials

- universal indicator solution
- barium sulfate powder for soil testing
- Petri dishes
- soil samples
- iceblock stick or spatula

Method

1 Place half a teaspoon of soil in a Petri dish.

2 Add enough universal indicator solution to make a thick paste. Stir with the iceblock stick.

3 Sprinkle this paste with a thin layer of white barium sulfate powder.

4 After 2 or 3 minutes, match the colour of the powder with the colours on the indicator colour card.

   Record the pH of the soil.

5 Repeat for other soils.

   Which soils were acidic, which were basic and which were neutral?

PART C Inquiry

Design a similar experiment to see if the soil acidity can be changed by adding powdered limestone or ammonium sulfate.

ISBN 978 1 4202 3736 8

EXTRA FOR EXPERTS

Explaining acids and bases

Acids have special properties. For example, they are corrosive. In the activity below, you can test whether they conduct an electric current.

ACTIVITY

1 Half fill a small test tube with dilute hydrochloric acid.
2 Use a conductivity kit to test whether the acid will conduct an electric current; that is, light up the bulb.
3 Wash the electrodes with distilled water, then repeat the test with distilled water instead of dilute hydrochloric acid. What happens?
4 Repeat the test with sodium chloride (salt) solution.

Try to *explain* your observations.

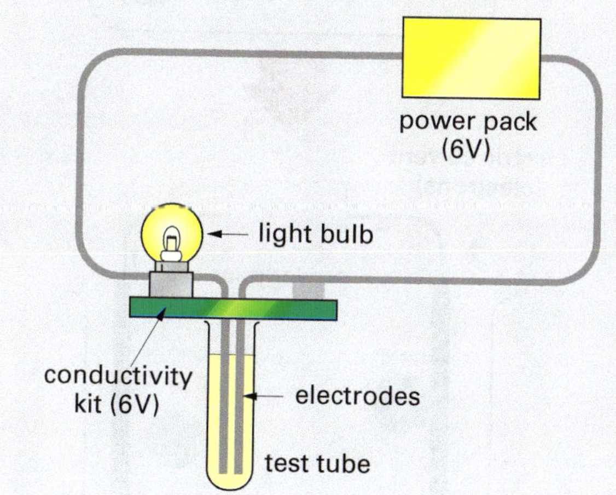

Ions in acids and bases

In the 17th century, one scientist suggested that the particles that make up acids had sharp spikes. He said that these spikes were the reason for the sharp biting feeling of acids on your skin. Scientists have since found that it is because they contain *hydrogen ions*.

As you probably know, an atom is a sort of ball-shaped cloud with a tiny nucleus at its centre. The nucleus is positively charged. The rest of the atom is mostly empty space containing rapidly moving electrons, which are negatively charged. Some of these electrons are close to the nucleus and others are further away.

The number of positively charged protons in the nucleus is the same as the number of negatively charged electrons surrounding the nucleus. So the atom is neutral. However, some atoms can lose electrons (usually the outermost ones), while others can gain electrons. In either case, the atom is no longer neutral. An atom that has lost or gained electrons is called an **ion** (EYE-on).

A hydrogen atom has a single proton in its nucleus, with a single electron orbiting it. This electron is easily lost to form a hydrogen ion with a single positive charge, as shown below.

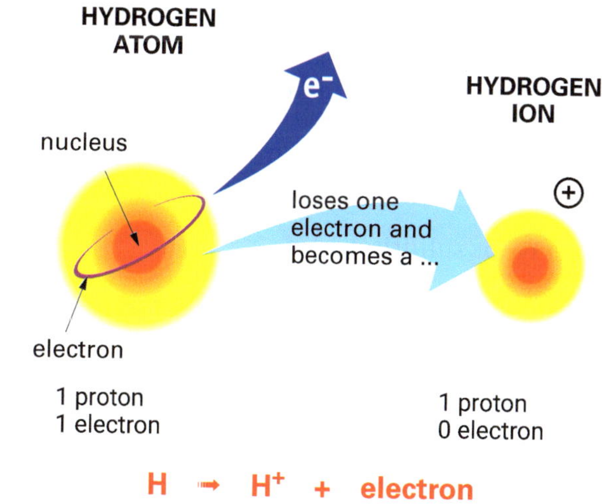

Figure 9.12 How a hydrogen ion is formed

Atoms of metals tend to lose electrons. For example, a copper atom can lose two electrons to form an ion with two positive charges (Cu^{2+}). In contrast to metals, atoms of non-metals tend to form ions by gaining one or more electrons. For example, chlorine atoms form negative chloride ions Cl^-, as shown in Figure 9.13 on the next page.

ISBN 978 1 4202 3736 8

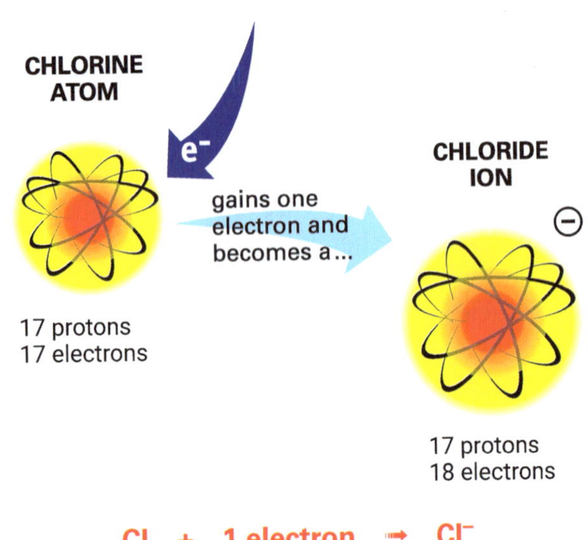

$$Cl + 1\text{ electron} \rightarrow Cl^-$$

Figure 9.13 How a chloride ion is formed

The reason distilled water does not conduct electricity is because the water molecules are neutral. However, hydrochloric acid contains ions that can carry the electric current through the solution.

The formula for hydrochloric acid is HCl. In water, it forms H^+ ions and Cl^- ions:

$$HCl \rightarrow H^+ + Cl^-$$

Similarly for sulfuric acid:

$$H_2SO_4 \rightarrow 2H^+ + SO_4^{2-}$$

So, *if a substance forms H^+ ions when dissolved in water, then it is an acid.*

The pH of a solution is a measure of the concentration of H^+ ions. The p stands for 'power', so pH means the 'power of hydrogen'. Note that pH is always written with a small p and a large H.

Basic solutions contain hydroxide OH^- ions. For example, sodium hydroxide (NaOH) forms Na^+ ions and OH^- ions when dissolved in water.

$$NaOH \rightarrow Na^+ + OH^-$$

Why salt solution conducts electricity

Positive and negative ions attract each other, so many compounds consist of positive and negative ions held together by ionic bonds. For example, sodium chloride consists of Na^+ ions and Cl^- ions. When you dissolve sodium chloride in water, the sodium ions and the chloride ions break apart and spread throughout the water, as shown in Figure 9.14. This is why a salt solution also conducts electricity.

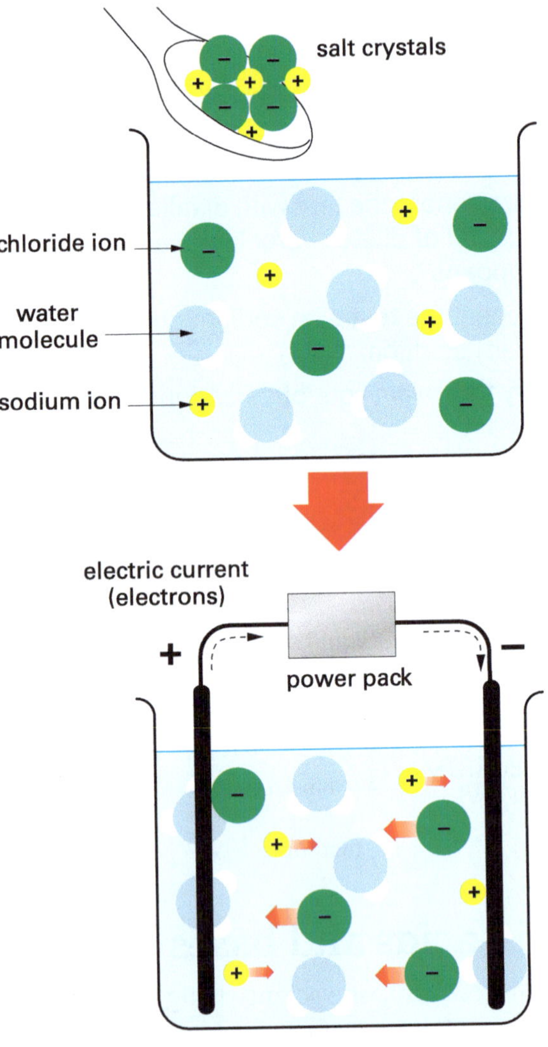

Figure 9.14 Sodium and chloride ions break apart in the solution. The ions carry the electric current through the solution, and electrons carry it through the wires to and from the power pack.

ISBN 978 1 4202 3736 8

1 Match the following in your notebook:

pH 4	neutral
pH 7	moderately acidic
pH 1	moderately basic
pH 8	very acidic
pH 10	slightly basic

2 The words in the following sentences have been jumbled up. Rewrite the sentences correctly.
  a A pH solution has a neutral of 7.
  b An acidic solution is an example of vinegar.
  c More than 7 solutions have a basic pH.
  d H^+ ions tells you of the pH concentration.

3 a Blood has a pH of 7.4. Is this acidic, basic or neutral?
  b When you exercise, your muscles produce lactic acid. What effect might this have on the pH of your blood?

4 The pH of water in a swimming pool is 7.9. The ideal pH level is 7.4–7.6. What should you add to the pool to lower its pH—water, alkali or acid?

5 Farron is reading a booklet on the maintenance of swimming pools. He reads that if the pH falls below 7, the pool may become corroded. Suggest a reason for this.

6 An alkaline solution has a pH of 12. If it is diluted by adding water, will the pH increase, decrease or stay the same?

7 The table below shows the most favourable pH ranges for the growth of some common plants.

Flowers		Crops	
azalea	4.5–5.5	barley	6.0–8.0
calendula	6.0–7.5	clover	5.5–7.0
daffodil	6.0–6.5	wheat	5.5–7.0
hibiscus	6.0–7.0	cotton	5.5–6.5
sweet pea	7.0–8.0	rice	5.0–6.5

The pH of a number of soils was measured:

Soil A pH = 4.0
Soil B pH = 5.0
Soil C pH = 6.0
Soil D pH = 7.5
Soil E pH = 8.5

  a Which of the soils is the most acidic?
  b In which soil would sweet peas probably grow best?
  c Which of the crops would probably grow best in soil E? Explain your choice.
  d What could be added to soil C to make it more suitable for azaleas?
  e Of soils A, B and C, which would need most lime added to it to give a pH of 7?

Figure 9.15 Farmers add lime to the soil to decrease the acidity.

8 Copy and complete these sentences.
  a A hydrogen atom loses one electron to form a _____ hydrogen ion.
  b A chlorine atom _____ one electron to form a negative chloride ion.
  c In water, hydrochloric acid forms positive _____ ions and negative _____ ions.
  d Sodium hydroxide forms positive sodium ions and negative _____ ions when dissolved in water.

9 Suggest why tap water conducts electricity whereas distilled water does not.

10 In terms of ions, how are hydrochloric acid and sulfuric acid similar? How are they different?

ISBN 978 1 4202 3736 8

CHALLENGE

1 Pepsin is an enzyme in the human stomach that speeds up the digestion of proteins. What does the graph below tell you about the activity of pepsin?

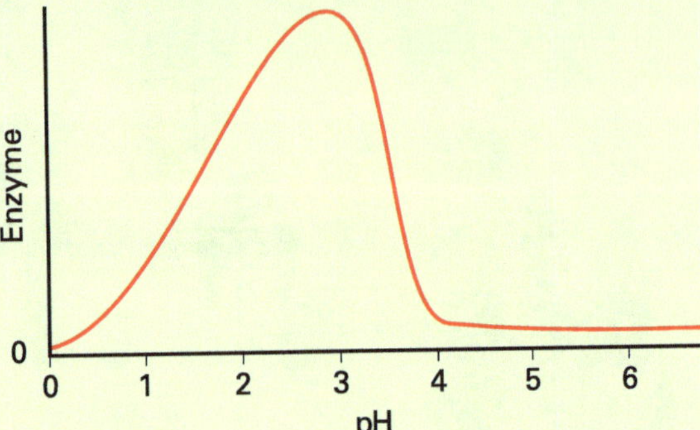

2 The pH of a creek in an industrial area was monitored by a group of Year 9 students over a period of 4 weeks. They collected these data.

Day	1	4	7	10	13	16	19	22	25	28
pH	7.4	7.3	7.4	7.4	7.2	5.1	3.9	3.9	6.5	7.2

a Plot their data on a graph.

b Infer what might have caused the dip in the graph.

3 Garden soil usually becomes more acidic as time passes. Suggest reasons for this.

4 Electricity is passed through dilute hydrochloric acid containing hydrogen ions and chloride ions.

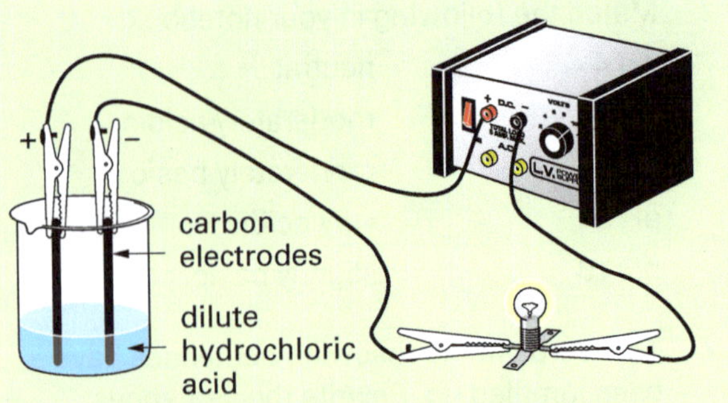

a Draw a diagram showing what you predict will happen to the hydrogen and chloride ions in the solution. Use Figure 9.14 on page 220 as a guide.

b What will happen if the connections to the power pack are reversed?

5 The symbol for a silver ion is Ag^+. How many electrons does a silver atom lose to become a silver ion?

6 What would need to happen for a chloride ion Cl^- to become a chlorine atom?

7 Hydrogen ions are never found on their own. Why is this?

8 Explain why ionic compounds, which consist of electrically charged ions packed together, are electrically neutral. Use sodium chloride as an example.

EXPLORE

1 Test the effect of pH on the cooking of carrots. Put about 100 mL of water into each of three beakers. Add vinegar to the second beaker until the pH is about 4 (use indicator paper). To the third beaker, add baking soda until the pH is about 9. Add thin slices of carrot to each beaker and boil the carrot for about 5 minutes.
Record your results.

2 Design and carry out an experiment to test the effect of pH on the growth of seeds. You could grow the seeds on filter paper or perlite in Petri dishes. Be careful to control all variables except the one you are purposely varying.

3 Design an experiment to see if the pH of different brands of hair shampoo is different. Is there any relationship between their pH and the type of hair for which they are recommended?

ISBN 978 1 4202 3736 8

9.3 Reactions of acids and bases

INVESTIGATION 9.3 Reactions with acids

Aim

To investigate the reactions of acids with metals and with carbonates.

Risk assessment and planning

Read Part A and Part B (on the next page).

- What safety precautions will be needed?
- Design data tables for both parts.

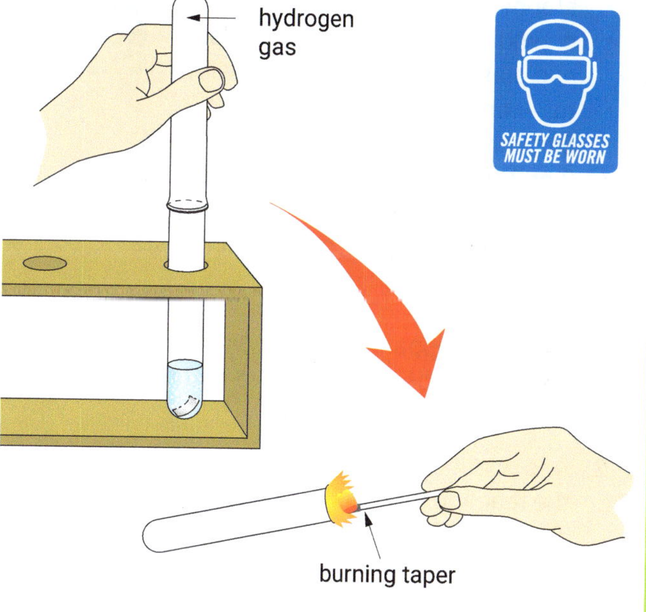

PART A Reaction with metals

Materials

- pyrex test tubes and test tube rack
- dilute **hydrochloric acid** (1 M)
- universal indicator solution or paper
- piece of magnesium ribbon (about 2 cm)
- taper and matches
- samples of other metals, e.g. aluminium, copper, iron, tin, zinc

Corrosive

Method

1 Add about 2 mL of dilute hydrochloric acid to a test tube and add a few drops of universal indicator.

Record the pH.

2 Add a piece of magnesium ribbon to the acid. To trap the gas released, hold a second dry test tube upside down over the mouth of the tube as shown.

3 When the magnesium has reacted, light a taper, tilt the test tube upwards as shown and put the burning taper near its mouth. A 'pop' indicates that the gas is hydrogen.

4 Feel the test tube containing the acid and magnesium.

Write an inference to explain your observation.

5 Note the pH in the tube.

How has the pH changed as a result of the reaction?

6 Test the reactions of other metals with dilute hydrochloric acid.

Record your observations in your data table.

Which metals are the most reactive? Which are the least reactive?

Discussion

1 Is hydrogen gas lighter or heavier than air? How do you know?

2 Write a word equation for the reaction between magnesium and hydrochloric acid. (The colourless solution left in the test tube contains the compound magnesium chloride.)

3 Suggest why the pH increases during the reaction.

ISBN 978 1 4202 3736 8

PART B Reaction with carbonates

Materials

- dilute **hydrochloric acid** (1 M)
- test tubes and test tube rack
- stopper for test tube
- one-holed stopper fitted with glass and plastic tubing, as shown below
- taper and matches
- limewater (calcium hydroxide solution)
- 2 or 3 marble chips (calcium carbonate)
- other carbonates, e.g. sodium carbonate, sodium hydrogen carbonate, copper carbonate

Corrosive

Method

1 Set up the apparatus below. Make sure the collecting tube is dry. Put two or three marble chips into the reaction tube.

2 Add about 5 mL of dilute hydrochloric acid to the reaction tube, then quickly fit the stopper and tubing.

3 After about 2 minutes remove the collecting tube and put a stopper in it. Replace it with another tube one-third full of limewater. Allow the gas to bubble into the limewater while you do step 4.

4 Light the taper, remove the stopper from the first collecting tube, and put the taper into the tube as shown. The taper going out indicates that the gas is carbon dioxide.

5 Go back and observe the limewater from step 3. If it has turned milky, this also indicates the presence of carbon dioxide.

6 Test the reaction of other carbonates with hydrochloric acid.

Record your observations in your data table.

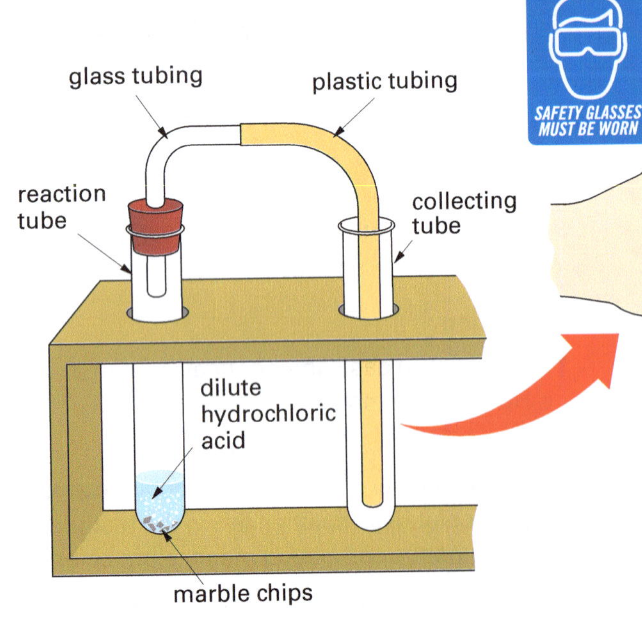

SAFETY GLASSES MUST BE WORN

burning taper

Discussion

1 Is carbon dioxide lighter or heavier than air? How do you know?

2 Suggest why carbon dioxide is used in fire extinguishers.

3 Complete this word equation for the reaction that occurred when you added hydrochloric acid to calcium carbonate:

__________ + __________

→ calcium chloride + water + ________

4 Write an inference to explain why the limewater goes milky when you bubble carbon dioxide into it.

PART C Inquiry

Design and carry out an experiment to find out whether the concentration of the hydrochloric acid affects how quickly it reacts with metals and carbonates.

You must check with your teacher before you start.

ISBN 978 1 4202 3736 8

Salts

In Investigation 9.3 you found that dilute hydrochloric acid reacts rapidly with magnesium to produce two new substances. Hydrogen gas is released and magnesium chloride stays in solution.

magnesium + hydrochloric acid
→ magnesium chloride + hydrogen

Similarly, dilute hydrochloric acid reacts violently with the metal sodium to form hydrogen and sodium chloride.

sodium + hydrochloric acid
→ sodium chloride + hydrogen

The magnesium chloride and sodium chloride belong to a group of compounds called **salts**.

If you had continued the investigation, you would have found that dilute hydrochloric acid and all other dilute acids react with most metals. When the reaction is slow, its rate can be increased by using a more concentrated acid or by heating. You can write a general equation to describe all these reactions.

metal + dilute acid → salt + hydrogen

There are hundreds of different salts. Some are shown in the table below. Note that they are named after the acids they are made from. The most common salt is sodium chloride or table salt. Other examples include Epsom salts (magnesium sulfate) used in bath salts, ammonium nitrate used as fertiliser, and baking soda (sodium hydrogen carbonate) used in cooking.

In Investigation 9.3 you also found that acids react with carbonates to produce carbon dioxide gas.

carbonate + acid
→ salt + water + carbon dioxide

Most 'health salts' consist of sodium hydrogen carbonate plus a solid acid, such as citric acid, and flavouring. When the mixture is stirred with water, carbon dioxide is given off. This produces the bubbles and 'sparkle' of the drink. Carbon dioxide is also responsible for the 'fizz' when you put sherbet in your mouth.

Figure 9.16 A 'fizzy drink' is the result of the reaction between an acid and a carbonate.

Name of acid	Name of salts	Examples
hydrochloric acid HCl	chlorides	sodium chloride $NaCl$ calcium chloride $CaCl_2$
nitric acid HNO_3	nitrates	potassium nitrate KNO_3 ammonium nitrate NH_4NO_3
sulfuric acid H_2SO_4	sulfates	copper sulfate $CuSO_4$ magnesium sulfate $MgSO_4$
carbonic acid H_2CO_3	carbonates	calcium carbonate $CaCO_3$ sodium hydrogen carbonate $NaHCO_3$

ISBN 978 1 4202 3736 8

Neutralisation

Figure 9.17 is a close-up photo of the jaws of an ant. Ants use them to hold you while they inject you with formic acid using a spike on their abdomen. This can be quite painful. To treat the sting, you can add a weak base such as baking soda solution, which neutralises the acid. Bee stings can be neutralised in the same way. In contrast, wasp stings contain a basic substance, which can be neutralised using vinegar, a weak acid. However, if you are not sure what has bitten you, it is best to treat the sting with ice, which numbs the pain.

Neutralisation is a reaction in which an acid and a base cancel each other out to form a salt and water. To neutralise an acid, you add a base, and to neutralise a base, you add an acid.

acid + base → salt + water

Figure 9.17 Many insects inject formic acid which can be neutralised with a weak base.

There are many applications of neutralisation in everyday life. For example, the odour of fish and other seafood is due to bases called amines. Adding lemon juice or vinegar, which are acidic, neutralises the amines, giving a more pleasant smell.

Your stomach contains dilute hydrochloric acid to break down the food you eat. If the contents of your stomach become too acidic, you get *indigestion*. This can happen when you have eaten too much or too quickly. To neutralise the excess stomach acid, people take *antacid*. These tablets or powders contain a weak base, such as baking soda (sodium hydrogen carbonate), magnesium hydroxide or aluminium hydroxide, which neutralises the hydrochloric acid. The baking soda also produces carbon dioxide gas. This makes you burp, releasing the gas trapped in your stomach.

Figure 9.18 The smell of fish can be neutralised with lemon juice. This is an acid–base reaction.

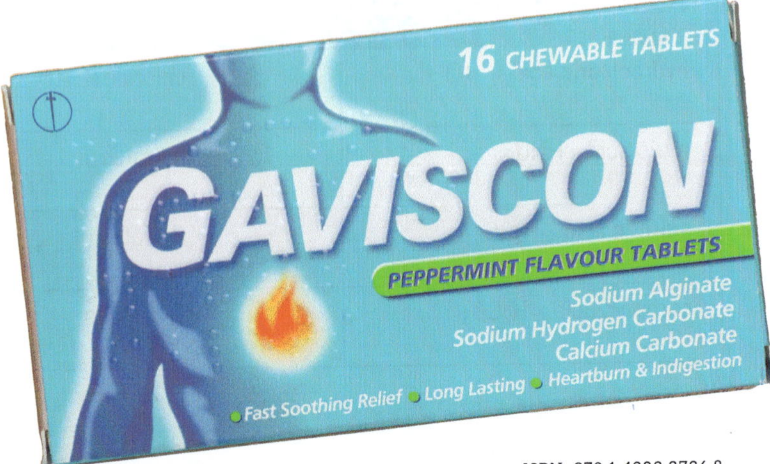

Figure 9.19 Antacids contain bases such as calcium carbonate to neutralise stomach acid.

ISBN 978 1 4202 3736 8

INVESTIGATION 9.4 Antacid

Aim

To measure how much antacid is needed to neutralise some hydrochloric acid.

Materials

- dilute **hydrochloric acid** (0.1 M)
- small flask, e.g. 250 mL
- 50 mL measuring cylinder
- methyl orange indicator
- spatula
- antacid powder or crushed tablet
- plastic Petri dish
- balance

Risk assessment and planning

- Read the investigation, then describe to your partner what you will be doing and why.
- What do you think is the purpose of the sheet of white paper in step 3?
- Draw up a data table like this:

mass of Petri dish + antacid	=	_____ (m_1)
mass of dish + unused antacid	=	_____ (m_2)
mass of antacid used	=	$m_1 - m_2$
	=	_____

PART A

Method

1 Put a spatula of antacid powder in a Petri dish.

Use the balance to measure the mass of the Petri dish plus antacid. Record this in your data table.

2 Use a measuring cylinder to measure out 50 mL of dilute hydrochloric acid. This is similar to the hydrochloric acid in your stomach.
Pour the acid into the flask and add 3 or 4 drops of methyl orange indicator.

3 Place a sheet of white paper under the flask and use the spatula to add antacid *bit by bit* to the acid. Swirl the flask gently to stir the mixture.

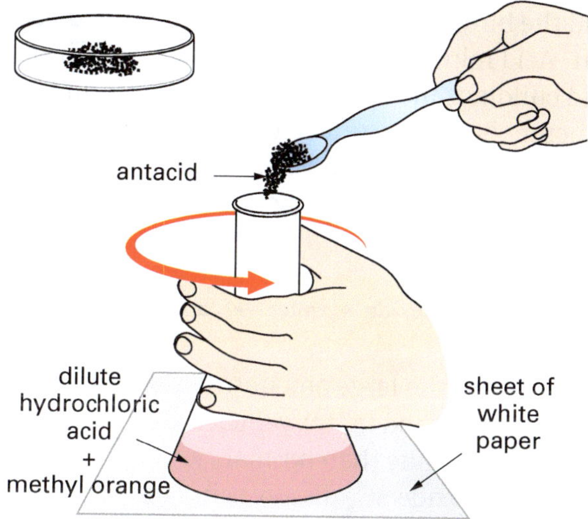

Stop adding antacid when the colour changes from red to orange.

What evidence was there of a chemical reaction as the antacid was added?

4 Measure the mass of the Petri dish and the unused antacid in it.

By subtraction, find the mass of antacid used to neutralise 50 mL of dilute hydrochloric acid.

Discussion

1 How much antacid was needed to neutralise 50 mL of dilute hydrochloric acid?

2 Compare your results with those of other groups. How accurate do you think your measurement was? Explain your answer.

3 If your stomach contained 1 litre of dilute hydrochloric acid, how much antacid powder would you need to neutralise it?

PART B Inquiry

Design a test to compare the effectiveness of several different antacid powders or tablets.

ISBN 978 1 4202 3736 8

Acid rain

In northern Europe and North America, millions of trees have died due to **acid rain**. Some lakes contain so much acid that all the fish have died and the birds that relied on the fish for food have left. Acid rain also speeds up the rusting of iron, and buildings made of marble, limestone and concrete have been affected.

Normal rain is usually slightly acidic because carbon dioxide in the air dissolves in raindrops to form carbonic acid.

carbon dioxide + water → carbonic acid

However, the large amounts of waste gases from industry and motor vehicles are making rain much more acidic than normal.

Sulfur dioxide SO_2 dissolves in rainwater to form sulfurous acid. It also reacts with oxygen in the air to form sulfur trioxide SO_3, which dissolves in rainwater to form sulfuric acid. Nitrogen dioxide NO_2 also reacts with rainwater to form acids.

Scientists are not sure about the most important cause of acid rain. Power stations certainly produce sulfur dioxide, although some of them are now beginning to remove the sulfur dioxide from the waste gases they produce, so that it does not go into the air. At present, however, it looks as though the nitrogen dioxide from car exhausts is as much to blame as sulfur dioxide.

Acid rain is not as big a problem in Australia as it is in Europe and North America. One reason for this is that the coal we burn in power stations does not contain as much sulfur as the coal used overseas. However, acid rain with a pH as low as 3.6 has been recorded in Sydney.

Figure 9.20 This forest in Poland has been damaged by acid rain.

ACTIVITY

Because sulfur dioxide is poisonous, especially for asthmatics, this activity can be done only as a teacher demonstration.

1. Prepare a gas jar containing about 5 mL of water and a few drops of universal indicator.
2. Place a small amount of sulfur in a deflagrating spoon, and use a Bunsen burner to light it.
3. Quickly place the burning sulfur in the gas jar.
4. When the sulfur has finished burning, shake the jar to dissolve the sulfur dioxide gas.
   Note any change in the colour of the indicator. Why has this happened?
5. Add a coloured flower petal, a piece of fruit peel or a piece of coloured paper to the water in the jar.
   What do you observe?
6. Also test the acidity of the water in the jar by adding it to magnesium or marble chips.

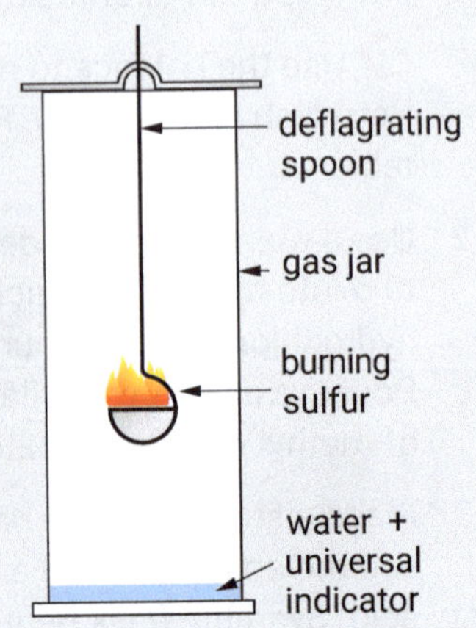

ISBN 978 1 4202 3736 8

1 How can you test for the following gases?
  a hydrogen
  b carbon dioxide

2 Rosie has a white powder that she thinks is calcium carbonate. Mary thinks it is calcium chloride. How could they decide which it is?

3 How does milk of magnesia (magnesium hydroxide) cure an upset stomach?

4 It is unwise to take too many antacid tablets. Why do you think this is so?

5 Why is rain slightly acidic even without air pollution?

6 Marble statues are made of calcium carbonate. Would they be affected by acid rain? How?

7 Which substances are always formed when:
  a an acid reacts with a metal?
  b an acid reacts with a carbonate?
  c an acid reacts with a base?

8 When Anika was stung by an ant, she rubbed the bite with vinegar, but this only made it worse. What should she have done?

9 Some copper jewellery has become tarnished with greenish copper carbonate. Which of the following would you use to clean it without dissolving away the metal itself—baking soda, lemon juice, nitric acid or water? Explain your answer.

10 X and Y are white powders. X is insoluble in water, but Y is soluble and its solution has a pH of 3. When X is added to a solution of Y, bubbles form and a gas is produced.
  a One of the white powders is an acid. Is it X or Y? How do you know?
  b The other white powder is calcium carbonate. What is the gas produced in the reaction?

11 Which two chemicals would you mix to produce:
  a hydrogen?
  b carbon dioxide?

12 The map on this page shows a power station and the average pH of the rain that falls on the countryside around it.
  a Where does the most acidic rain fall?
  b Suggest why the water is more acidic at B than at A.
  c From which direction does the wind normally blow? How do you know?

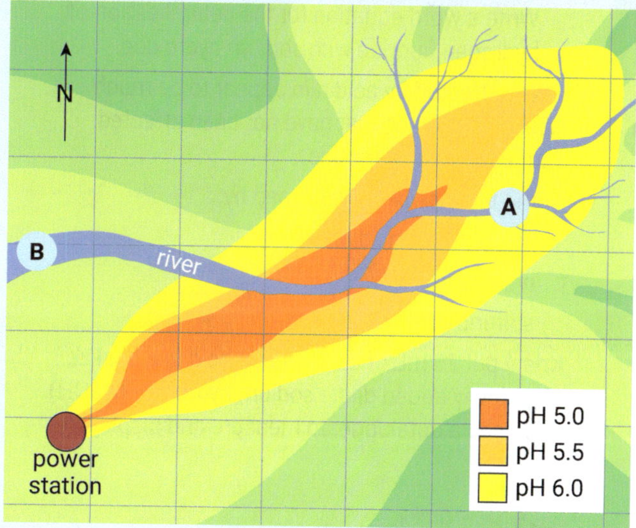

13 Soo-Hong investigated the reaction between magnesium ribbon and acetic acid. In each test, he used 15 mL of dilute acetic acid. Here are his results.

Temperature of acid (°C)	Length of magnesium (cm)	Reaction time (s)
10	2	60
10	4	79
10	6	102
20	2	31
40	2	15

  a What was the aim of the experiment?
  b Predict how long it would take for a 3 cm piece of magnesium ribbon to dissolve at 10 °C.
  c Write a hypothesis linking reaction time to temperature.
  d Use a graph to predict the temperature at which a 2 cm piece of magnesium would react in 45 seconds.

ISBN 978 1 4202 3736 8

1 Hydrogen fluoride (HF) is an acid that reacts with sodium hydroxide to produce a salt that is used to help prevent tooth decay.

   a What is the name of the salt that forms in this reaction?

   b Write a word equation for the neutralisation of hydrogen fluoride with sodium hydroxide.

2 Lakes affected by acid rain appear to be much clearer than lakes that have not been affected. Suggest a reason for this.

3 Name the type of salts formed by:

   a hydrochloric acid

   b nitric acid

   c sulfuric acid.

4 Kristy put 25 mL of dilute acetic acid in a beaker. She slowly added dilute sodium hydroxide (NaOH) and used a datalogger to measure and display the pH.

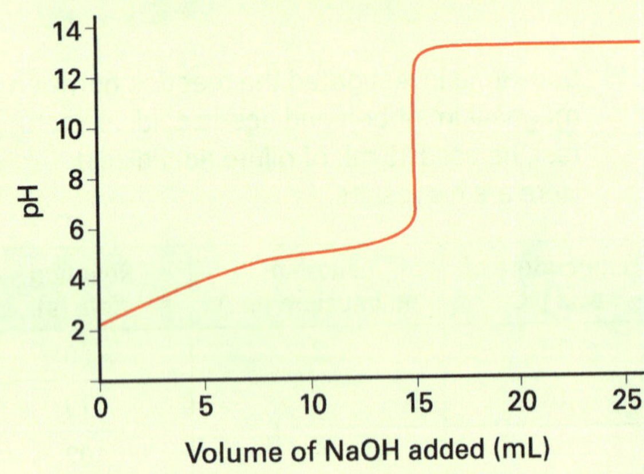

   a What was the pH after 20 mL of NaOH had been added?

   b What volume of NaOH was needed to produce a pH of 5?

   c What volume of NaOH was needed to neutralise the acetic acid?

   d Suppose Kristy had used a more concentrated acetic acid. What effect would this have on the shape of the graph?

   e Suppose Kristy had used more concentrated NaOH. Would the volume she used be more or less than the volume in c?

   f Explain the shape of the graph, relating it to the neutralisation reaction.

5 Write word equations for the reactions that you would expect to occur between:

   a calcium metal and hydrochloric acid

   b zinc carbonate and nitric acid

   c calcium hydroxide and carbonic acid.

6 The diagram below shows how sulfur dioxide can be removed from the waste gases produced in a coal-burning power station. Use the information in the diagram and the equations to explain how the process works.

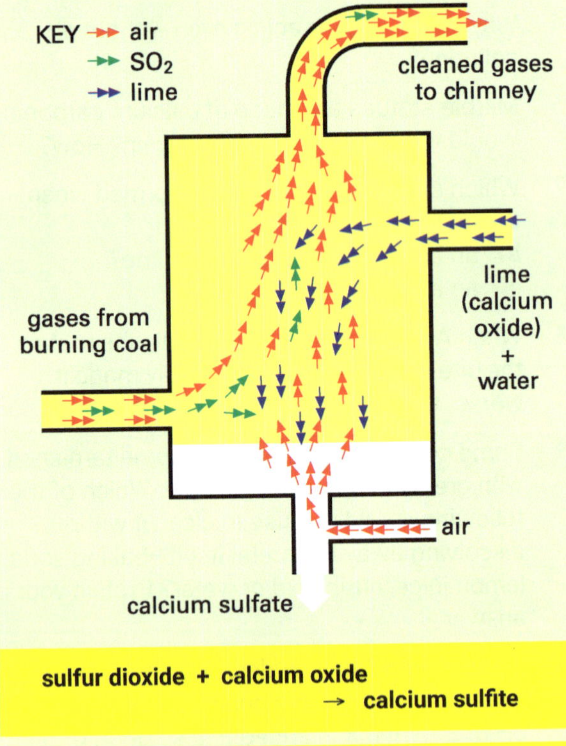

sulfur dioxide + calcium oxide → calcium sulfite

calcium sulfite + oxygen → calcium sulfate

1 Check the labels on antacid medications. What are the active ingredients?

2 In a group, discuss who should pay for the damage caused by acid rain.

3 Use the internet or the library to find out about the damage caused by acid rain in Europe and North America.

ISBN 978 1 4202 3736 8

9.4 Energy in reactions

In this chapter, you have learnt about various everyday reactions involving acids. However, there are many other types of chemical reactions. For example, what do you remember about the chemical reactions that take place in and around all living things? Do the activity below to test your knowledge.

Which type of chemical reaction:

1. do plants use to make food in their leaves?
2. increases levels of carbon in our atmosphere from the burning of fossil fuels?
3. is taking place in your body to give you the energy you need for living?
4. is the basis of all food chains and food webs?
5. occurs when yeast causes bread to rise?
6. occurs when you light up the barbecue?
7. releases carbon dioxide into the atmosphere?

Turn to page 233 to see if your answers are correct.

Energy in and energy out

Reactions sometimes require energy input for them to occur. At other times they release energy.

In Investigation 9.3 you will have felt the test tube heat-up when acid reacted with metals or carbonates. When you did the pop test for hydrogen you would have seen energy released as heat and light. A reaction that releases energy is known as an **exothermic reaction** (*exo* means 'out').

A reaction such as photosynthesis is an **endothermic reaction**, as it requires an input of energy from the sun (*endo* means 'in'). A 'cold pack' used for treating injuries also uses an endothermic reaction. The cold pack feels cold because it is absorbing energy from your body. The reaction in a cold pack is usually ammonium nitrate (NH_4NO_3) dissolving in water, which requires energy to occur.

Photosynthesis

Plants use the energy in sunlight to produce food in the form of a simple sugar called glucose. This can be written as the following word equation:

carbon dioxide + water + sunlight
↓
glucose + oxygen

Plant cells contain structures called *chloroplasts*. These contain chlorophyll, which gives plants their green colour. Chlorophyll absorbs energy from the sun to start the process of photosynthesis. During this process, small molecules—water (H_2O) and carbon dioxide (CO_2)—combine to form large molecules—glucose ($C_6H_{12}O_6$).

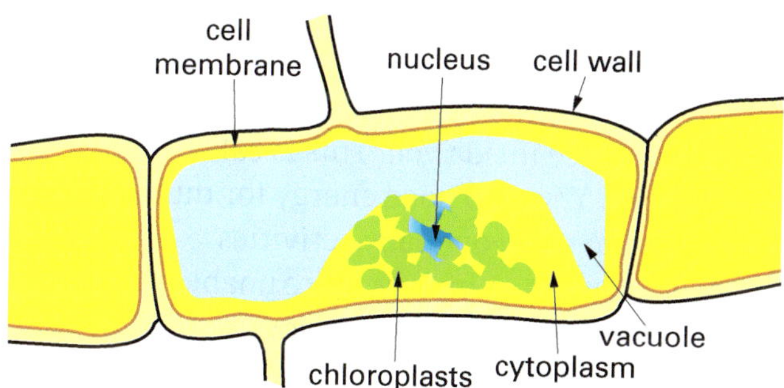

Figure 9.21 Photosynthesis takes place inside chloroplasts.

Life on Earth would not be possible without photosynthesis. It is vital for producing oxygen in the atmosphere. Half of the world's oxygen is produced by land plants such as trees, shrubs and grasses. The other half is produced by phytoplankton (tiny microscopic organisms) that drift around in our oceans. Scientists infer that it wasn't until organisms in the ocean started to photosynthesise and release oxygen 2.7 billion years ago that other life forms could exist.

Organisms that carry out photosynthesis (producers) are the basis of all food chains and food webs. When they are eaten by consumers, the energy that they captured from the sun and used to make large food molecules is passed along the food chain.

ISBN 978 1 4202 3736 8

Cellular respiration

As well as photosynthesis, the plant in the forest is also carrying out respiration. Some of the glucose and oxygen made in photosynthesis is used to make energy for the plant. This energy is needed for other processes such as growth and reproduction of cells and tissues. This is done by breaking down the large glucose molecules to small water and carbon dioxide molecules.

glucose + oxygen
↓
water + carbon dioxide + energy

This process is called **cellular respiration** because it takes place inside cells. Amazingly, the cells in the forest plant carry out thousands of chemical reactions every second. Without the energy made from cellular respiration these would not be possible.

Your body also carries out cellular respiration. You need energy for all the chemical reactions that take place inside you. This is called your *metabolism*. You also need energy for muscles to move and to carry out daily activities.

As you are an animal, you are unable to carry out photosynthesis and make your own food, so you have to eat and digest food to be able to use it. Some of the energy released from cellular respiration is used in digestion, and some is used by your body for heat. This heat keeps your body at a constant temperature. Growth and reproduction, removal of wastes, transportation of substances around your body and repair of cells are all processes that require energy.

Combustion

Combustion is the uncontrolled burning of a fuel with oxygen to produce heat and light. This is also an exothermic reaction. However, it is very different from cellular respiration, which occurs slowly and in a controlled way.

The combustion of wood or petrol requires a large amount of energy to get it started. For example, combustion of petrol in a car engine requires a spark from a spark plug to ignite the petrol and oxygen mixture in the pistons. The large molecules in the fuel are broken down into small molecules, so a large amount of energy is released, mainly in the form of heat.

In many combustion reactions, a flame or explosion occurs, producing heat and light.

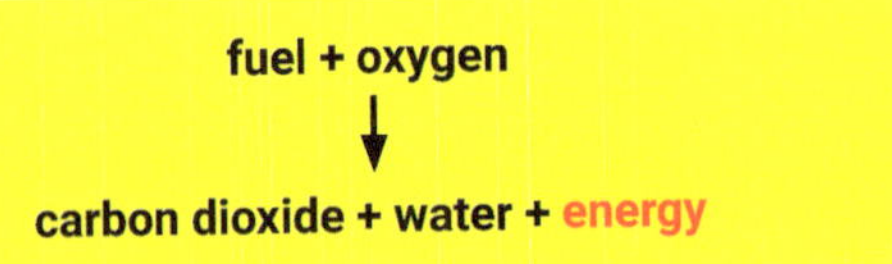

In complete combustion, carbon dioxide and water are produced. However, more often than not, *incomplete* combustion occurs when fuels are burnt in air. This produces other substances too, such as soot and carbon monoxide. Nitrogen oxides are also produced because air contains nitrogen as well as oxygen. These substances all contribute to air pollution.

Figure 9.22 Combustion (burning) is an everyday reaction.

ISBN 978 1 4202 3736 8

Reactions in the atmosphere

Living things rely on the non-living world to survive. For example, air, the weather, the availability of water and minerals in the soil can all affect life on Earth.

Living things rely on chemical reactions that take place in non-living systems, for example the reaction of ozone in the atmosphere with ultraviolet (UV) radiation from the sun. The ozone layer protects all living things on Earth from the harmful effects of ultraviolet radiation. It also helps prevent extreme temperature changes from day to night.

Oxygen molecules in our atmosphere are constantly being broken down by UV radiation to produce single oxygen atoms ($O_2 \rightarrow O + O$). These atoms are then free to combine with other oxygen molecules to produce ozone: $O_2 + O \rightarrow O_3$ (ozone).

Ozone absorbs the UV radiation and stops it reaching the lower levels of the Earth's atmosphere. The ozone layer therefore shields all living things on Earth from the effects of UV radiation. Too much UV radiation can cause changes in the DNA or genetic code of living things. These changes are called **mutations** and may lead to cancer. UV radiation can also cause eye and skin damage and affect the immune system of living things. So the chemical reactions that occur to make ozone are vital to life on Earth.

Scientists have also discovered that nitric oxide (NO) and nitrogen dioxide (NO_2) in the atmosphere also help with ozone formation. These nitrogen oxides are formed by natural processes such as lightning, combustion in forest fires, and chemical processes that take place in the soil.

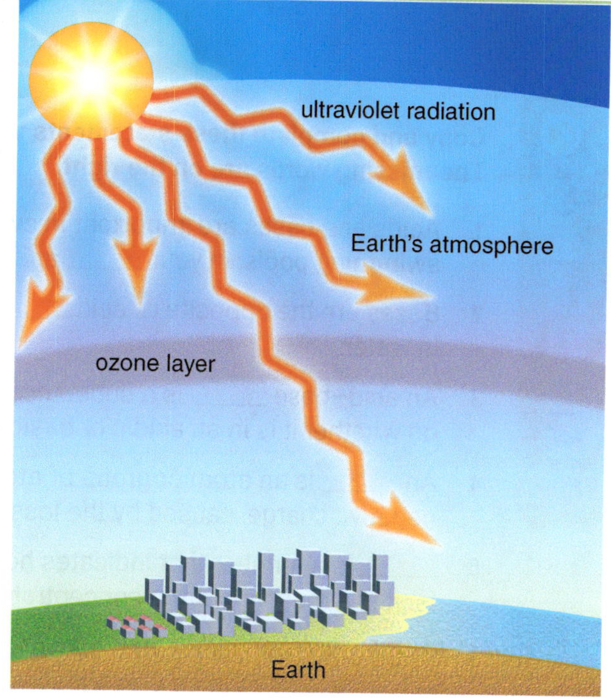

Figure 9.23 How the ozone layer protects us from harmful UV radiation from the sun.

Living systems could not exist without such important chemical reactions.

Living systems are also being affected by chemical reactions caused by humans. For example, as you learnt on page 228, gases produced by power stations and cars can cause acid rain. CFCs (chlorofluorocarbons), once used as refrigerants and propellants, have destroyed some of the ozone layer. UV radiation broke up the CFCs, releasing chlorine atoms. These chlorine atoms then reacted with the ozone molecules and broke them apart. This caused a 'hole' in the ozone layer above Antarctica, allowing dangerous UV radiation to reach the Earth. The use of CFCs is now banned in most parts of the world, but the ozone layer has not yet repaired itself.

CHECK

1. How is a living plant like a factory?
2. What is cellular respiration?
3. Explain the difference between photosynthesis and cellular respiration.
4. What is the difference between an exothermic reaction and an endothermic reaction? Provide examples of each.
5. Draw diagrams to show the difference between an oxygen atom, an oxygen molecule and an ozone molecule.
6. Why is UV radiation harmful to humans?
7. How does the ozone layer stop UV radiation from reaching the Earth?

Answers for Activity
1 photosynthesis **2** combustion **3** respiration **4** photosynthesis **5** fermentation **6** combustion **7** respiration and combustion

ISBN 978 1 4202 3736 8

SW•9 Victorian Curriculum

MAIN IDEAS

Copy and complete these statements to make a summary of this chapter. The missing words are on the right.

1 Acids and bases are important in everyday life, for example in swimming pools, in your ______ and in gardening.

2 Bases are the opposite of acids. ______ are bases that are soluble in water.

3 An acid–base ______ is a substance that changes colour depending on whether it is in an acidic or basic solution.

4 An ______ is an atom or group of atoms that has a positive or negative charge, caused by the loss or gain of ______.

5 ______ is a number that indicates how acidic or basic a solution is. It is a measure of the concentration of ______ ions in solution.

6 An acid is a substance that releases hydrogen ions (H^+) in solution. A base is a substance that releases ______ ions (OH^-).

7 Dilute acids react in a predictable pattern.
- They react with most __________ to produce hydrogen gas.
- They react with carbonates to produce ______ gas.

8 ______ is the process in which an acid reacts with a base to produce a salt (neutral) and water.

9 Endothermic reactions need ______ to make them go, for example photosynthesis. ______ reactions produce energy, for example respiration.

hydrogen
electrons
alkalis
stomach
metals
exothermic
ion
hydroxide
indicator
neutralisation
pH
carbon dioxide
energy

CH•9 REVIEW

1 Which of these household substances contain bases? (There may be more than one.)
- **A** vinegar
- **B** ammonia
- **C** lemon juice
- **D** oven cleaner

2 To work the homemade fire extinguisher on the right, you let the string go so the tea bag falls into the liquid, producing carbon dioxide gas. X and Y are most likely to be:
- **A** baking soda and vinegar.
- **B** baking soda and water.
- **C** baking soda and window cleaner.
- **D** salt and vinegar.

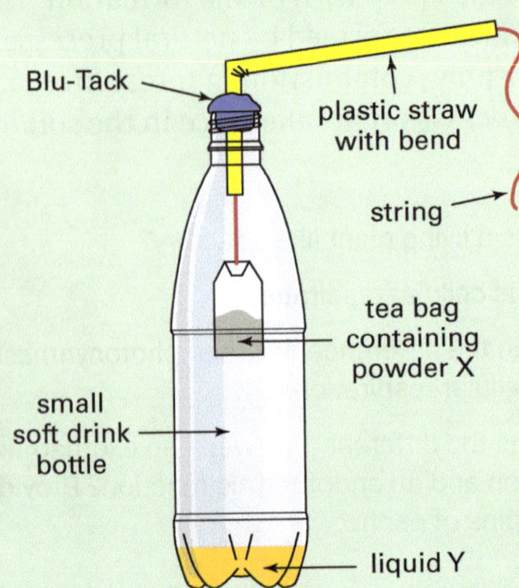

ISBN 978 1 4202 3736 8

3 Which one of the following is *not* an acid (does not form hydrogen ions in water)?

A HF

B H_3PO_4

C HCl

D NaCl

4 Which of the following are *endothermic* reactions, and which are *exothermic*?

a combustion

b photosynthesis

c cellular respiration

d breakdown of oxygen molecules into oxygen atoms

5 In the laboratory, you are given three solutions marked X, Y and Z. You are also given red and blue litmus paper. Copy and complete the following table with the results that would show that:

- X is neutral
- Y is acidic
- Z is basic.

	Unknown solutions		
	X	Y	Z
blue litmus			
red litmus			

6 Shanthi has been given a number of solutions, and she has found the pH of each using indicator paper. Here are her results.

Substance	P	Q	R	S	T	U	V
pH	3	4	9	6	8	5	7

a Which solution is the most acidic?

b Which solution is the most basic?

c Which solution is neutral?

d To make solution Q more acidic, you need to add:

A water.

B sodium hydroxide.

C solution P.

D solution T.

7 Aluminium saucepans lose their shine when they develop a dull coating of aluminium oxide, which is a base. Explain why boiling lemon juice in an aluminium saucepan will leave it shiny.

8 Normal rainwater has a pH of about 6, distilled water has a pH of 7, and acid rain can have a pH as low as 2. Write a short paragraph to explain these differences.

9 Write a paragraph explaining why knowing the pH of soil is important.

10 Using the set-up shown below, Gerard tested the electrical conductivity of three different acids, all of the same concentration. Here are his results.

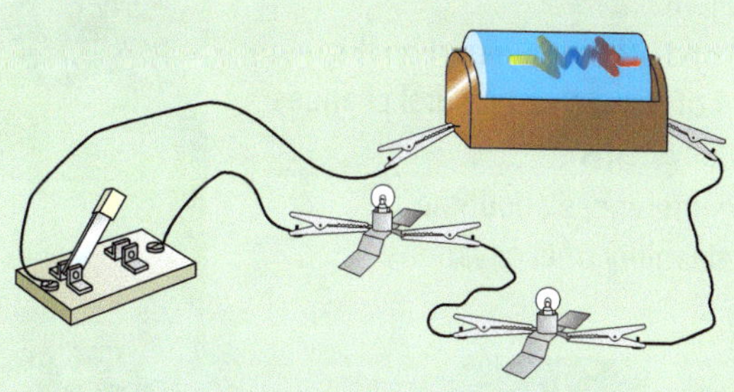

Acid	Bulb
A	glowed brightly
B	did not glow
C	dull glow

How can you explain the differences between the acids?

LAB REVIEW

Collect unlabelled bottles of dilute hydrochloric acid, dilute caustic soda solution, sodium chloride solution and water.

Your task is to work out which is which, using the correct safety procedures.

Collect any chemicals and equipment you will need to do your tests.

When you have finished, briefly describe the tests you did and check your answers with your teacher.

Check your answers on page 304.

ISBN 978 1 4202 3736 8

IN THIS CHAPTER

Science Understanding

- explain how biotic and abiotic factors in the environment can affect the survival of organisms
- describe the flow of matter and energy through ecosystems
- examine the effects on ecosystems of events such as droughts, floods, bushfires, destruction of habitat and introduced species
- present views from different interest groups on the effect of human activity on the Murray–Darling river system

Science Inquiry Skills

- explore interactions between organisms
- write and present a report on the effect of human and natural changes on ecosystems
- make inferences about the sustainability of ecosystems

CH•10 Energy in ecosystems

ISBN 978 1 4202 3736 8

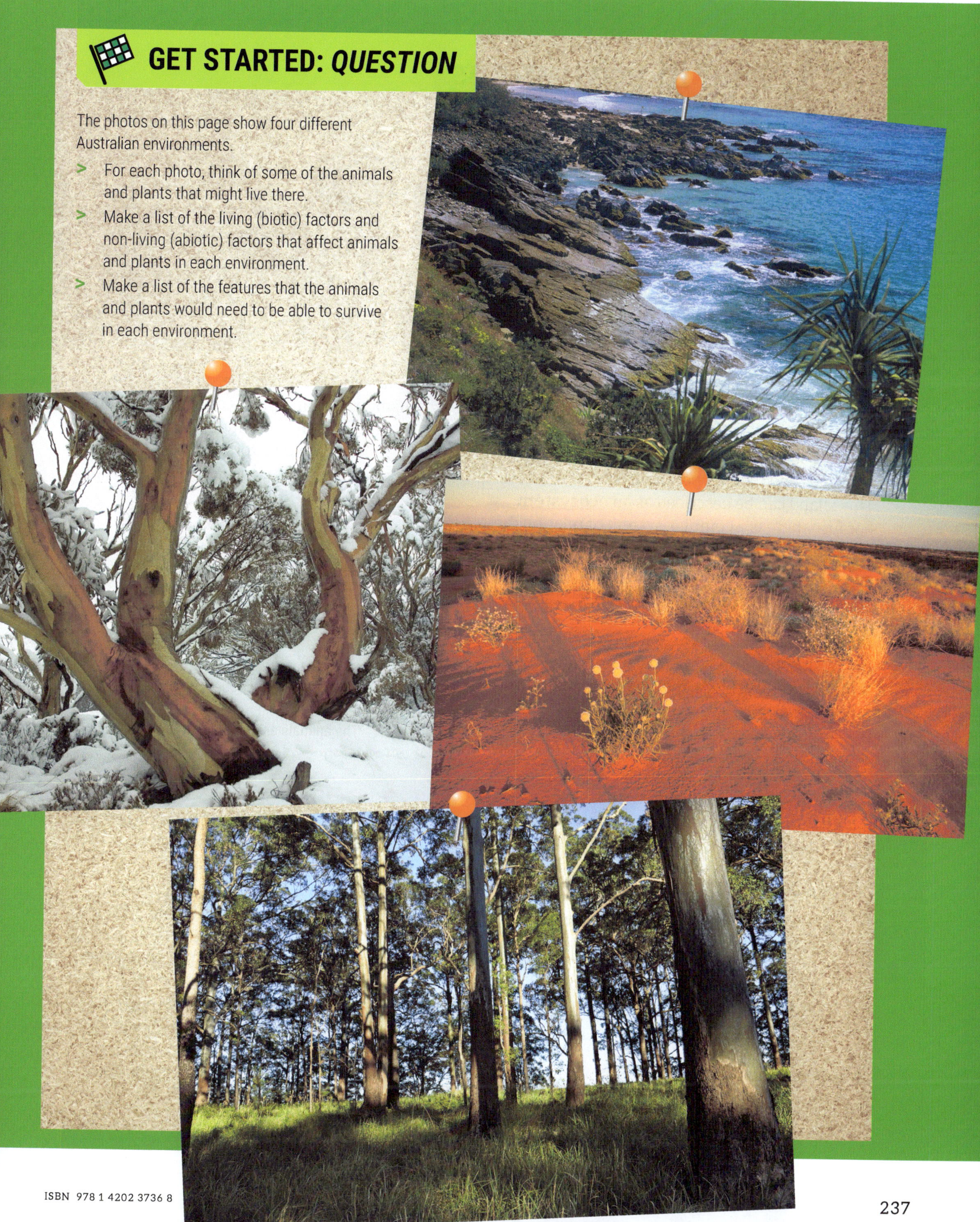

GET STARTED: *QUESTION*

The photos on this page show four different Australian environments.

> For each photo, think of some of the animals and plants that might live there.
> Make a list of the living (biotic) factors and non-living (abiotic) factors that affect animals and plants in each environment.
> Make a list of the features that the animals and plants would need to be able to survive in each environment.

ISBN 978 1 4202 3736 8

10.1 Living in ecosystems

The island in the photo above is a few kilometres from the mainland. Over the years it has been colonised by many types of plants, as well as animals such as lizards, native mice, small wallabies and many types of insects and birds. This island can be described as an **ecosystem** because there is a complex system of relationships between the organisms and with the non-living part of the environment.

The survival of an organism depends not only on its ability to get food and be protected from predators, competitors and disease-causing organisms, but also on the supply of water and air, a suitable temperature and weather conditions, and good soil.

The factors that affect the survival of an organism in its living place can be grouped into two categories—**biotic** or living factors and **abiotic** or non-living factors.

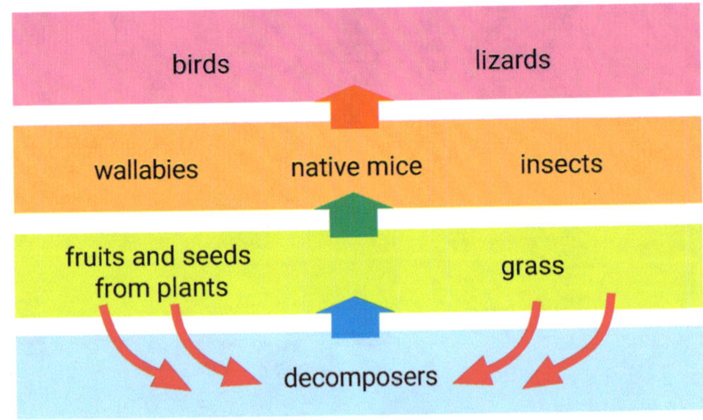

Figure 10.1 The animals and plants that live on the island

ACTIVITY

1. How much do you know about food webs? Use Figure 10.1 to answer the following questions.
   - Which of the organisms are producers? Write a definition of producer.
   - Which organisms are competitors of native mice? Which animals are predators of native mice? Explain your answer and include definitions of competitor and predator.
   - Draw a food web for the organisms in Figure 10.1.
2. Form a group of three or four people and discuss these questions about the island.
   - The island probably formed as the sea level changed many thousands of years ago. Suggest how the island might have been colonised by organisms.
   - Why can the island be described as an isolated ecosystem? How is this different from other ecosystems such as those that might exist in each of the four photos on the previous page?
   - Most of the food chains you could draw for the organisms on the island would contain a producer and no more than one or two consumers. Suggest why the food chains are short.

ISBN 978 1 4202 3736 8

The biotic factors in an ecosystem are all the living things that interact with an organism—the availability of food, the presence of predators and competitors, and the organism's ability to ward off disease-causing organisms.

The abiotic factors include temperature, light, humidity, the availability of air and water, and soil fertility. These factors are extremely important for the survival of any organism. For example, microscopic algae (plankton) are found only in the surface waters of the ocean where there is sufficient light for photosynthesis. And many reptiles and amphibians will hide away in logs or holes in the ground when the air temperature falls in winter.

Figure 10.2 Microscopic algae are found near the surface of the ocean where there is enough light for photosynthesis.

ACTIVITY

Work in a small group for this activity.

Look at the photo of the river. Write a brief report on the survival of an organism in this ecosystem, using the following three points as your structure.

1. Choose an organism that lives in or around the river. Make a list of all the biotic factors that affects its survival. Give examples if possible.
2. Construct a food web that contains the organism.
3. Describe the abiotic factors that may affect the survival of your organism. Give examples.

Swap reports with another group. Read their report and assess its good points and poor points. Make some brief notes.

Give the report back to the other group and read their notes about your report.

ISBN 978 1 4202 3736 8

Adaptations

The survival of an organism also depends on the characteristics of the organism itself. For example, the organisms in the photos in Figures 10.3–10.5 live in quite different habitats, and each organism has characteristics that enable it to survive in its own particular habitat. These characteristics are called **adaptations** (ADD-ap-TAY-shuns).

For example, the jabiru in Figure 10.3 lives in wetland areas of northern Australia. It has very long legs that enable it to walk through the swampy areas where it finds food. Its beak is long and pointed so it can collect snails, worms and fish from the water and mud. It also has large, strong wings to help it escape from enemies.

Figure 10.3 Jabiru **Figure 10.4** Dolphin **Figure 10.5** Kookaburra

ACTIVITY

1 Look at the animals in Figures 10.3–10.5. For each animal, list all the physical and biological factors that may affect its survival in its habitat. Suggest how the animal's adaptations help its survival.

2 Your teacher will supply you with three or four photos of animals. Alternatively, select animals from those shown here. Work in a group for this part of the activity.

- Use observations and your knowledge of the animals to make inferences about how well their characteristics help them survive.
- For each animal, record your observations about its size, shape, colour and other characteristics that you think are important in its survival.
- Decide where each animal lives and describe its habitat. Then infer how the characteristics help it survive in its habitat.

ISBN 978 1 4202 3736 8

Types of adaptations

The katydid (KAY-tee-did) in Figure 10.6 is similar to a grasshopper. It eats the leaves and shoots of plants. Birds and carnivorous insects such as preying mantises feed on katydids.

A katydid has a number of adaptations that ensure its survival. Its body is sideways flattened and is leaf-green in colour. This helps to camouflage it on plants. It also quivers, making it appear like a leaf moving in the wind. It has very keen eyesight and long, strong legs that help it escape quickly when threatened by predators. The katydid lays a very large number of eggs in the soil.

For convenience, we can classify adaptations into three groups—structural, functional and behavioural.

Structural adaptations refer to the shape and size of the organism and how the various parts of its body are put together; for example, the katydid's flattened body, its colour and the shape and size of its legs.

Figure 10.6 A leaf or an insect?

Functional adaptations refer to the working of an organism's body; for example, the katydid's egg-laying ability and the way it can digest plant leaves and shoots.

Behavioural adaptations are to do with how the organism behaves; for example, the quivering of the katydid mimics the movement of leaves and makes it hard to see in the bushes.

INVESTIGATION 10.1 Colour adaptations

Aim

To use a model to explain the effect of colour on the survival of organisms in different habitats.

Materials

- 60 coloured plastic discs, toothpicks or beads (20 green, 20 red and 20 yellow)

Risk assessment and planning

- Work in groups of three. One member will be the scatterer, the other two will be the predators.
- Carefully read through the Method and prepare data tables for steps 3 and 4.
- You will need to do this experiment on at least two different surfaces or 'habitats'; for example, grass, dirt, sand, concrete, carpet or leaf litter.

Method

1 Measure out a 3 m × 3 m area on your selected surface. Mark the corners of the square with pieces of paper, sticks or rocks. You could mark the area with string if you have some.

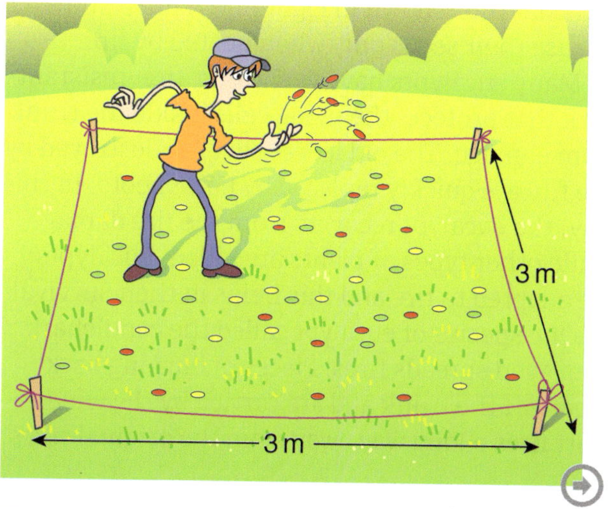

ISBN 978 1 4202 3736 8

2 Ask the 'predators' not to look, then scatter the discs randomly over the marked area.

3 Give the 'predators' 15 seconds to find as many discs as they can.

Count the numbers of each colour of disc found and record the data.

4 Collect all the discs and repeat steps 1 to 3 on the other surfaces.

Record the results in your data table.

Discussion

1 For each colour, calculate the survival rate as a percentage of the original 20.

$$\% \text{ survival rate} = \frac{\text{number remaining}}{20} \times 100$$

2 Draw a bar graph of the percentage survival rates for the three different colours.

3 Compare the survival rates for the different surfaces. Suggest why they are different.

4 Compare your survival rates with those of other groups. Your teacher may organise a class discussion.

5 Suppose the three different coloured discs were part of a large disc population in a particular 'habitat'. Assume the same 'predators' were present. Predict what might happen to the disc population in the area over a period of time. Give reasons for your prediction.

6 Do you think your model was a good one? Suggest ways in which you could improve it.

7 Using the results of your model, write a generalisation about the effect of camouflage (colour) on the survival of organisms in a particular habitat.

Adapted to fire

During a hot, dry summer, the chance of bushfires anywhere throughout Australia is quite high. Bushfires destroy houses and other property and burn out hectares of bush. The fires also kill animals that cannot escape from the flames.

However, fire is part of the Australian environment and many native plants are fire-tolerant. Some even need fire for their survival. For example, the seeds of some wattles need the heat from fires to germinate, and the thick woody banksia fruit (shown in Figure 10.7) open and release their seeds only when heated by fire.

Many eucalypts have very thick, fire-resistant bark that protects the living cells inside the trunk from damage. The old leaves that are destroyed by the fire are quickly replaced by new shoots. In this way, the eucalypt recovers from the fire damage while other types of plants are killed. Eucalypts are adapted to fire and this helps in their survival.

One species of eucalypt called the candlebark gum even spreads fires. Pieces of burning bark break off the trunk and are carried by the wind to start fires a long way away from the original trees.

Figure 10.7 The thick, woody fruit of the banksia opens and releases its seeds when heated by fire.

ISBN 978 1 4202 3736 8

Why populations change

In Investigation 10.1 you probably found that of the three colours of discs, the green ones were the most difficult to find on grass, while the yellow or red ones were easily seen and picked up by the 'predators'. As a result, the green discs had a higher survival rate on grass.

In any population of organisms, there are *variations* among the individuals. For example, in a population of field mice, you might see dark-coloured mice and light-coloured mice, short mice and long mice, mice with larger ears and mice with shorter ears.

Figure 10.8 In a population of field mice, you often see variations in colour, size and shape.

In the disc model, there were colour variations in the disc population. When equal numbers were placed on grass, more of the green discs 'survived' than either of the other colours. In this case, biological factors (the 'predators') caused a change in the make-up of the population. The green discs had the most favourable characteristics for a grass habitat and are said to be *selected*.

In a natural ecosystem, this selection of favourable characteristics is called **natural selection**. The organisms in a population that have favourable characteristics survive in a particular habitat, breed and pass on their characteristics to their offspring.

What happens if the conditions change?

Suppose there is a drought and the grass in our model dies, leaving bits of dead grass and sand-coloured soil. The green discs will now be more easily seen by the 'predators' than the yellow ones. Under these conditions, the yellow discs have a higher survival rate than the green ones. The yellow discs are better adapted to this habitat, and after some time the make-up of this population will be different from the disc population on the green grass.

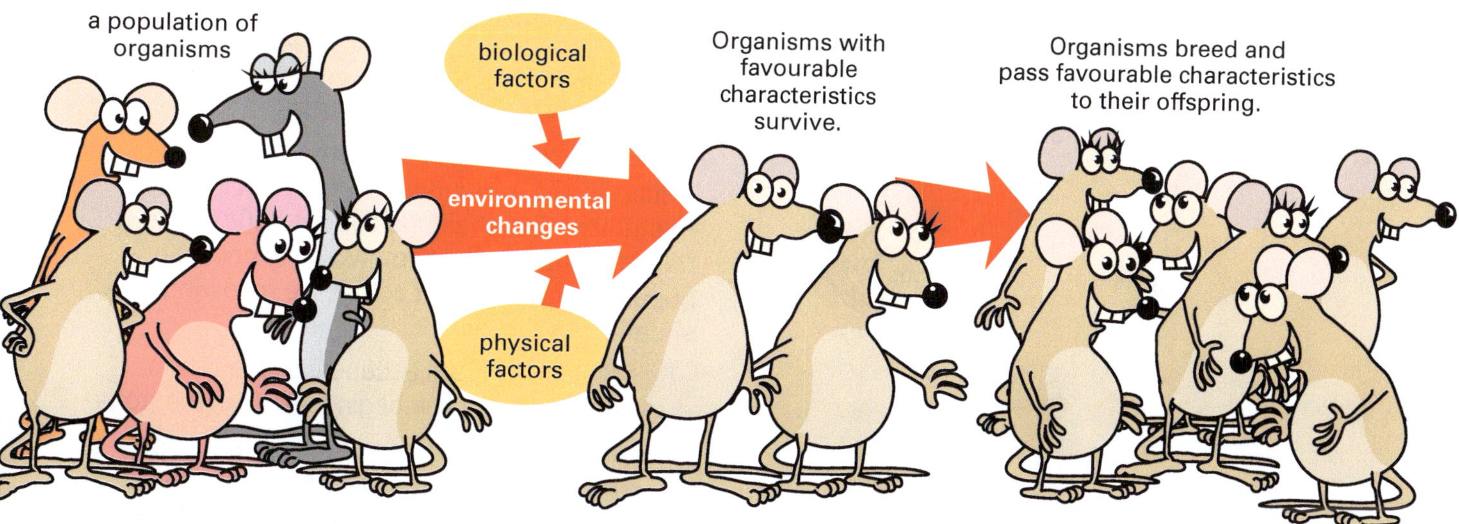

Figure 10.9 How natural selection works

ISBN 978 1 4202 3736 8

Situation 1: Breeding like rabbits

Describe the changes you would expect in a population of rabbits in each of the following situations:

1 As the weather warms and spring arrives, new grass sprouts and there is plenty of food available.

2 The government, in an attempt to control the rabbit population, introduces a virus that makes the female rabbits sterile and unable to have babies.

3 A small number of the rabbits are not affected by the virus, so survive and continue to breed.

4 After the next spring, which again is warm and wet with plenty of food, there is a summer drought.

Situation 2: Mosquitos attack

Study the life cycle of the mosquito. The eggs are laid in water, where they grow into larvae and pupae before they turn into mosquitos. Describe the changes you would expect in a population of mosquitos in each of the following situations.

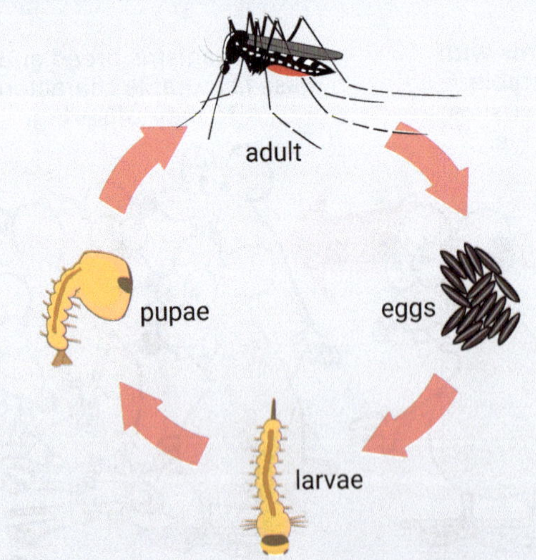

1 Wet weather arrives, and in your backyard old tin cans and a wheelbarrow fill with some water. Your swimming pool has not been cleaned either and has stagnant water.

2 You decide to get some trout fish and put them into your swimming pool to grow them as food.

3 In the heat of summer the water dries up in all the puddles and tin cans. But once autumn arrives the water returns.

Situation 3: Snow gum survival

Snow gums grow high up in the alpine areas of Victoria at altitudes of 1300–1800 metres. They are adapted to growing in this region at the temperatures that exist there. Describe the changes you would expect in a population of snow gums in each of the following situations.

1 Global warming increases the average temperature and this means that at altitudes of up to 1500 metres the temperature is now outside the range at which snow gums survive. What will happen to the distribution of the snow gums, and at what altitudes will they now grow? What will happen to the size of the population?

2 Some of the snow gums survive at the higher temperatures. What do you think will happen next? Use the concepts of natural selection and adaptation in your answer.

Situation 4: Corroboree frog

The corroboree frog lives in small seasonal wetlands and surrounding vegetation in the Australian Alps at alitudes above 750 metres. Describe the changes you would expect in a population of frogs in each of the following situations.

1 Wild or feral cats escape into the alpine national parks and begin to prey on the frogs.

2 The wetlands are protected by fences to ensure the cats cannot get in, and a pest control program is put in place to reduce cat numbers.

ISBN 978 1 4202 3736 8

3 The frogs are susceptible to a disease caused by the amphibian chytrid fungus. The fungus affects a population of frogs in one area of the wetlands.

4 Some frogs survive the fungus and become immune, and are able to start breeding again. How can this be explained in terms of the survival of the fittest?

CHECK

1 Classify the following statements according to whether they refer to structural, functional or behavioural characteristics.
   a Frill-necked lizards raise the large spiny layer of skin behind their head when they are threatened.
   b Sharks have a very streamlined shape.
   c Sea turtles lay up to 100 eggs in the breeding season.
   d When sea turtle eggs hatch, the young turtles dig through the sand and head directly for the water.
   e Many plants that live on the rainforest floor have very large leaves.
   f Fungi release enzymes that are able to break down the dead organism they are growing on.
   g The large front legs of a preying mantis have spines on them.

2 Certain plants have prickles or thorns on them.
   a What is the advantage to the plant of having these structures?
   b Name three plants that have these structures.

3 Look at the three types of birds' feet in the diagram below.
   a Describe the habitat in which each bird might live.
   b How does the structure of its feet help the survival of each bird in its habitat?

4 a What are biotic and abiotic environmental factors? Give examples of each.
   b Make a list of the biotic and abiotic factors that might affect the survival of a dingo in its natural habitat. Are these factors the same for a pet dog in a city suburb, or a working dog on a sheep farm? Explain your answer.

5 The biotic factors in the environment do not affect humans and domestic animals such as dogs and cats as much as they affect other animals in natural ecosystems. Suggest why.

6 The sugar glider is a small possum-like animal that lives in eucalypt forests. At night it feeds on the nectar in the flowers in the forest canopy. It has a thin layer of skin that stretches from its front legs to its back legs.

   a Suggest a reason for the skin between the sugar glider's legs.
   b Suppose the animal did not have the skin between its legs. What problems would the animal then have to face?
   c Suggest why the animal feeds at night.

7 Explain the process of natural selection in your own words. Infer what might happen in the long term to a population of a particular type of animal whose individuals looked, functioned and behaved identically.

ISBN 978 1 4202 3736 8

CHALLENGE

1 A certain type of moth called the peppered moth has two main colour variations—a light form and a dark form.

Figure 10.10 The light and dark forms of the peppered moth on a lichen-covered tree

During the day, the light form rests on light-coloured trees and rocks, while the dark form rests in cavities in trees, on dark-coloured bark of trees and in caves.

a What do you think would be the main predators of the peppered moth?

b Suggest why the moths rest during the day. Which type of adaptation is this?

c In an experiment, students caught and counted the moths in a particular place. Over 3 nights, they caught five light-coloured moths and 46 dark-coloured ones. Use this information to infer what the place was like where the students caught the moths, and why they caught so many dark-coloured ones.

2 The European rabbit is native to Spain. It was introduced into Australia in 1859 and on several occasions after this to 'enrich the country'. By 1890, rabbits were in plague proportions in south-eastern Australia.

a Suggest why rabbits spread so quickly in Australia.

b Make inferences about the effect the growing populations of rabbits were having on ecosystems.

c What do you think 'enrich the country' meant to the early European settlers?

d Suggest why the rabbits were not found in plague proportions in Spain or the rest of Europe.

3 The body temperature of birds and mammals is fairly constant and changes very little even when the surrounding temperature changes greatly. Other animals have body temperatures that change with the surrounding temperature.

a Suggest why a constant body temperature might be an advantage for the survival of a particular animal.

b Which type of adaptation is a constant body temperature? Explain your answer.

c Explain the following observations.

- Snakes, frogs and insects are rarely found in places with snow and ice.
- Snakes are very slow-moving on cold mornings.
- Fish can exist in the Arctic and Antarctic regions.

4 The diagram below shows the distribution of three types of plants. Use the information in the diagram to decide, giving reasons, whether the statements are true or false.

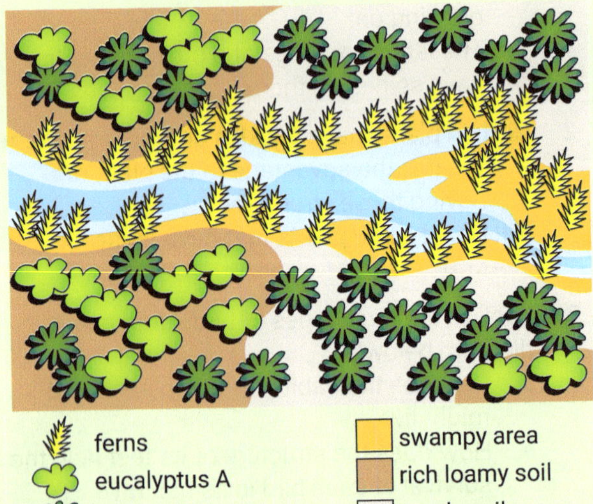

a Eucalyptus trees die in waterlogged soil.

b The distribution of ferns depends only on the type of soil.

c Eucalyptus B is adapted to different soil types.

ISBN 978 1 4202 3736 8

10.2 Matter and energy in food webs

In any ecosystem, matter in the form of solids, liquids and gases is used and recycled through food webs.

What happens to the matter as it passes through an ecosystem? How does an organism use the food it consumes? The story of Lucy and her kitten will help to explain this.

ACTIVITY

Your task

Work in a group of three or four people and read carefully the three parts of this task.

Answer the questions in each part, and be prepared to discuss them with others in the class.

Information

Lucy wanted to compare the growth of her kitten with the amount of food and water she gave it. To do this, she measured the mass of the kitten as well as the mass of the food and water it consumed each month. She also measured the mass of the wastes.

The cartoon below shows the results of Lucy's investigation.

Part A

Copy and complete the data list below using the information in the cartoon. One entry has been done for you.

Initial mass of kitten	.... g
Mass of kitten after a month	.... g
Change in the mass of kitten	.... g
Mass of food and water taken in	.... g
Mass of wastes	.... g
Mass of food and water used in body processes	.... g
Mass of new tissue of kitten	620 g
Mass of food and water used by the body but not for growth	.... g

ISBN 978 1 4202 3736 8

1 What does 'mass of food and water used by the body but not for growth' mean? What main process in the body would this refer to? What other substance is used in this process? What substances are produced?

2 Of all the food and water eaten by the kitten, what mass was used for the growth of new tissues? What mass was not used for growth?

3 What are the two main processes for which organisms use food?

Part B

Copy the matter flow diagram below into your notebook. Then use the data in the table on the previous page to fill in the missing quantities.

1 The matter flow diagram shows three possible destinations for the food eaten by an organism. What are they?

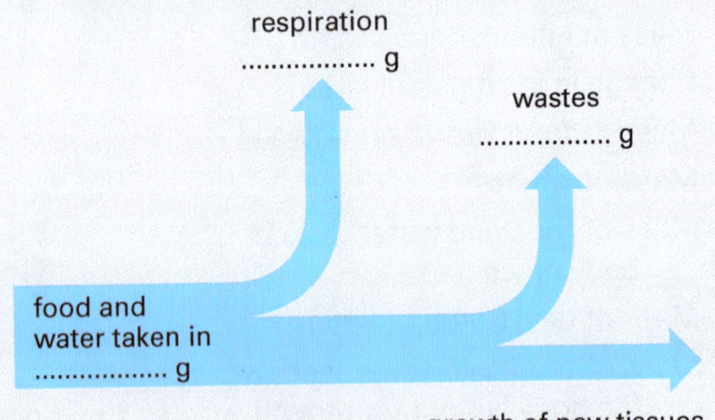

2 What percentage of the food eaten by the kitten is used for growth?

Part C

The way food is used by Lucy's kitten is an example of how food moves through a food chain. Look at the food chain below.

grass seeds ➡ mice ➡ owls

Matter flow diagrams can be combined for the three types of organisms in the food chain. The diagram at the bottom of the page is an example.

1 Matter enters the food chain through the producers. What is this matter?

2 What percentage of the matter in the raw materials is used to make new plant tissue?

3 What percentage of the plant tissue eaten by the mice is used for the growth of new tissues?

4 Explain what happens to the matter that is not used to make the tissues of the organisms.

5 Suggest what role decomposers play in a food chain like this one.

If you feel creative try this: modify the matter flow diagram below to include decomposers.

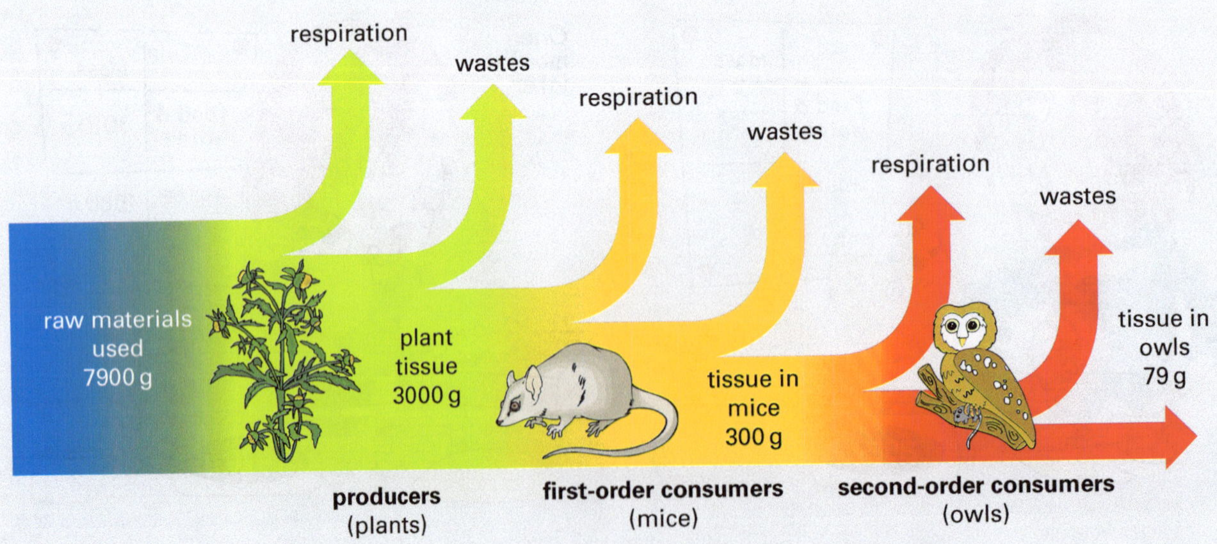

ISBN 978 1 4202 3736 8

Flow of matter in food webs

The activity on the previous pages showed that only a small percentage of the food eaten by an organism is actually used for growth. The remaining matter is given off either as products of respiration or as solid and liquid wastes (faeces and urine).

The diagram below shows the flow of matter through a food web. The size of each yellow box represents the total *mass* of organisms in each of the three feeding levels. For example, the size of the producers' box represents the mass of *all* the plants in the food web.

The diagram shows two main things about the flow of matter through a food web. First, the mass of the organisms in a feeding level decreases as you pass up through the food web from the producers to the highest order consumers. This means that the mass of food available to consumers becomes less and less at each level. Consequently, a particular mass of organisms in one level will only support a smaller mass in the next level. Second, the wastes and dead bodies are broken down by decomposers, who return the matter to the soil, air or water. This means that decomposers are vital to the functioning of the food web. The products from respiration are also recycled through the soil, air and water.

Figure 10.11 The matter cycle in a food web

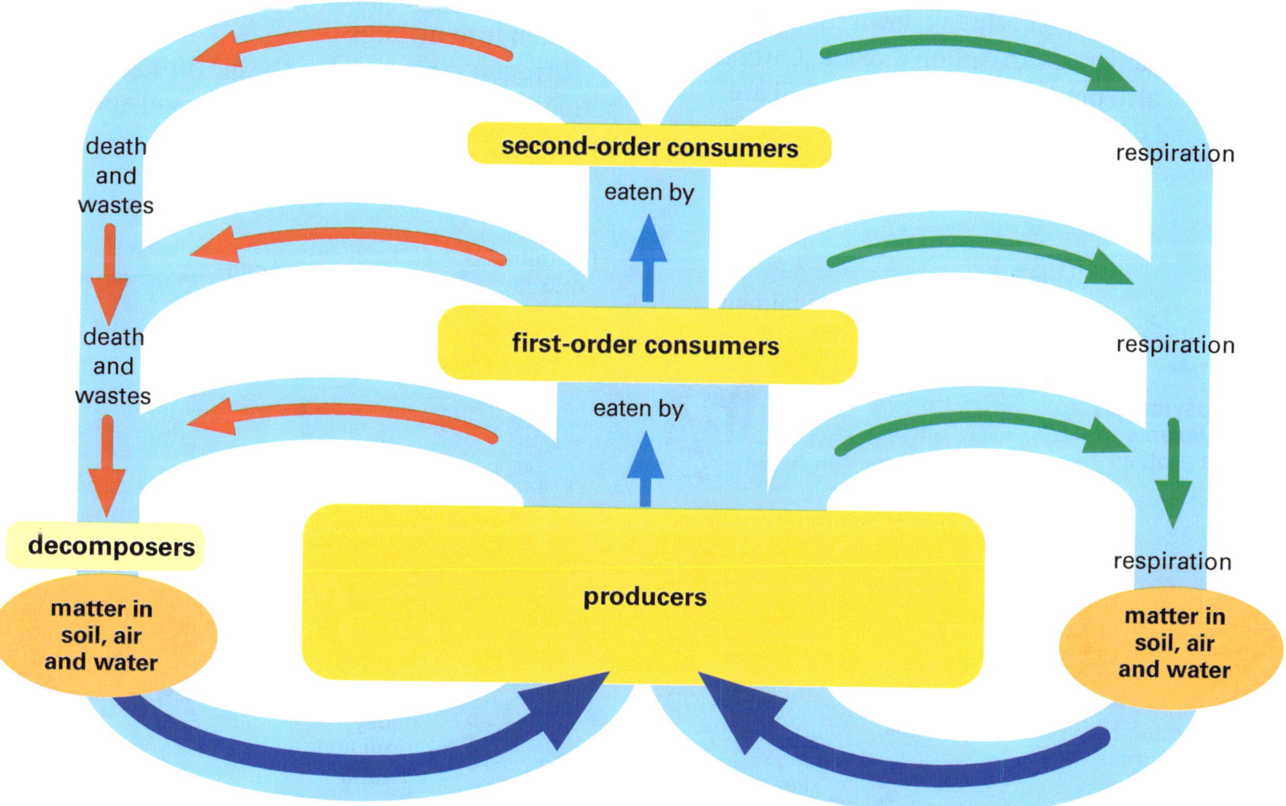

Figure 10.12 Drought reduces the mass of producers (grass), which then cannot support the mass of consumers. Animals like this cow die from starvation.

ISBN 978 1 4202 3736 8

Matter cycles

The carbon cycle

The element carbon is one of the main materials in sugars, proteins and fats in living things, and in waste products such as carbon dioxide and urea. Carbon is also found in fossil fuels—coal and oil—which are formed from decaying plants.

Sugars, proteins and fats make up the tissues of all living things. Animals obtain the raw materials to make these substances by eating other organisms. Plants and algae make these substances from carbon dioxide and water in photosynthesis.

Carbon is returned to the air or soil when decomposer organisms break down the bodies of dead organisms into carbon dioxide. Carbon dioxide is also produced when coal and oil are burnt, and during bushfires.

The nitrogen cycle

Nitrogen makes up about 78% of the Earth's atmosphere. It is also found combined with oxygen in nitrates in rocks and soil, and it is one of the elements in proteins.

Animals obtain their nitrogen from the protein in plants or other animals. The protein in their food is broken down into amino acids, which are then built into proteins in the animal's cells.

Plants and algae take in nitrogen in the form of nitrates. These compounds are soluble in water and are found dissolved in the water in aquatic habitats and in the soil of land habitats.

Nitrates are formed from the decay of dead organisms. Decomposers break down the proteins in the cells of the dead organism into amino acids and then into nitrates.

Nitrogen is returned to the air through the action of other bacteria that live in soil and in swamps.

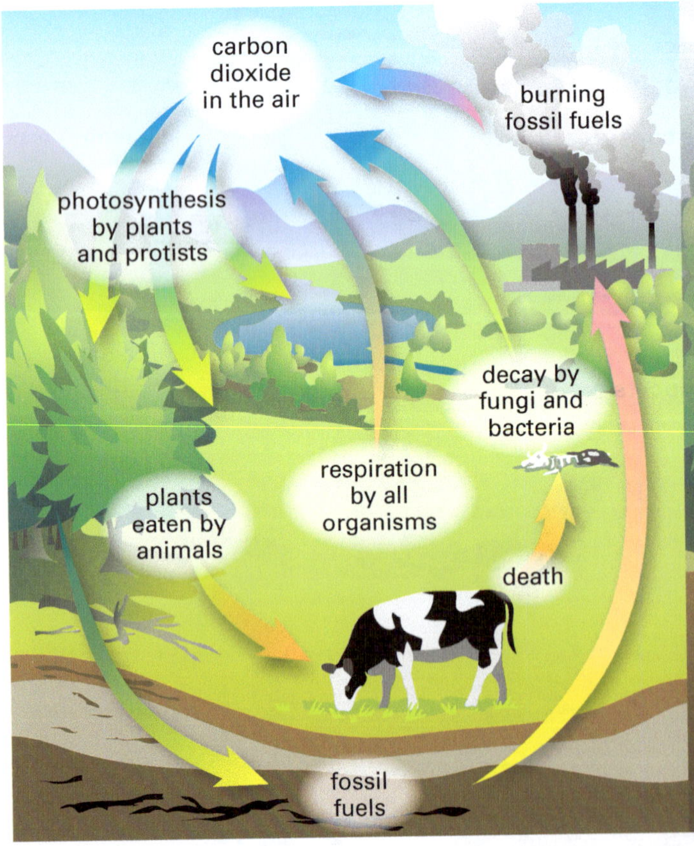

Figure 10.13 The carbon cycle

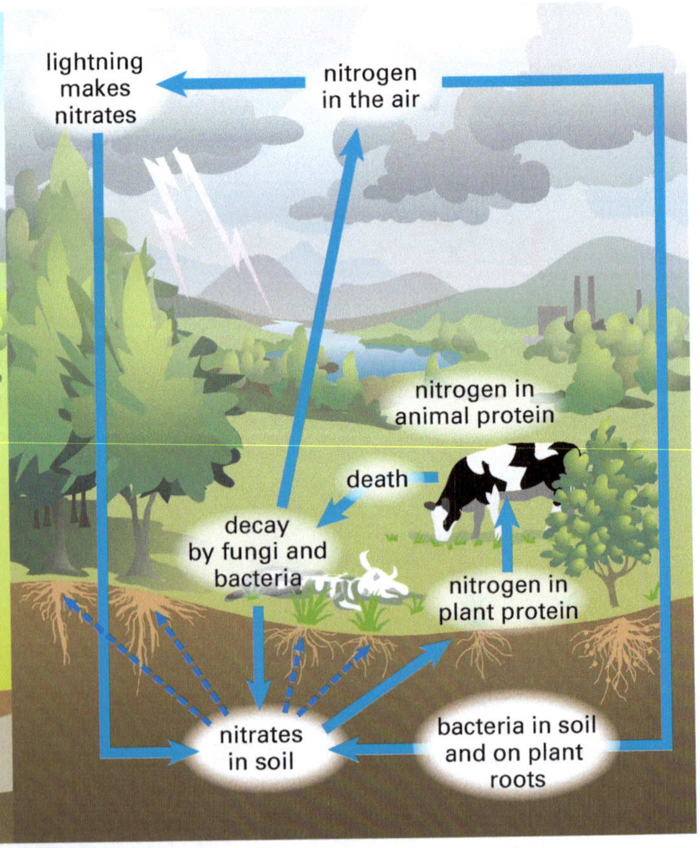

Figure 10.14 The nitrogen cycle

ISBN 978 1 4202 3736 8

Flow of energy in food webs

In a food web, producers use solar energy to manufacture large molecules of carbohydrates, proteins and fats from carbon dioxide, water and other substances from the soil.

When animals eat producers, some of the energy stored in the tissues of the plants is used for the body growth, and the remainder is released as heat energy during respiration, muscle movement and other body processes.

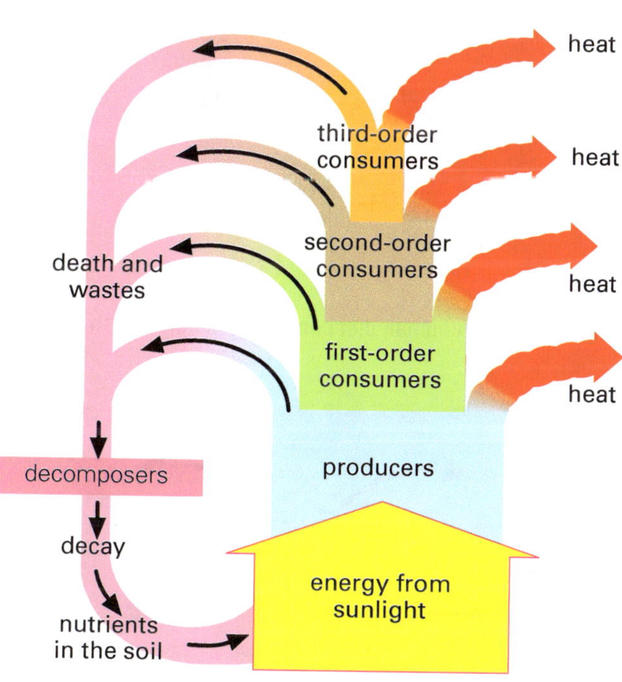

Figure 10.15 The flow of energy through a food web. Notice that only a very small amount of energy is recycled. The rest is given off as heat.

Figure 10.15 shows the flow of energy through a food web. The size of the boxes represents the amount of energy stored in the tissues of the organisms. Notice the large amount of energy that is given off as heat. This heat energy cannot be used by the organisms in the ecosystem. It is absorbed by the air and eventually radiated out into space.

Only a small amount of energy is recycled through the soil, air and water. This is in the form of energy bound up in small molecules such as carbon dioxide, nitrates and phosphates that result from the action of decomposers. When you compare the energy flow diagram with the matter flow diagram in Figure 10.11 on page 249, you can see that matter is recycled in the ecosystem but most energy is not. The energy that is lost by organisms is replaced by the energy from sunlight that is absorbed by producers.

Energy pyramids

If you stacked the energy boxes in Figure 10.15 on top of each other, an **energy pyramid** would be formed. In Figure 10.16, an energy pyramid has been constructed for the food web shown.

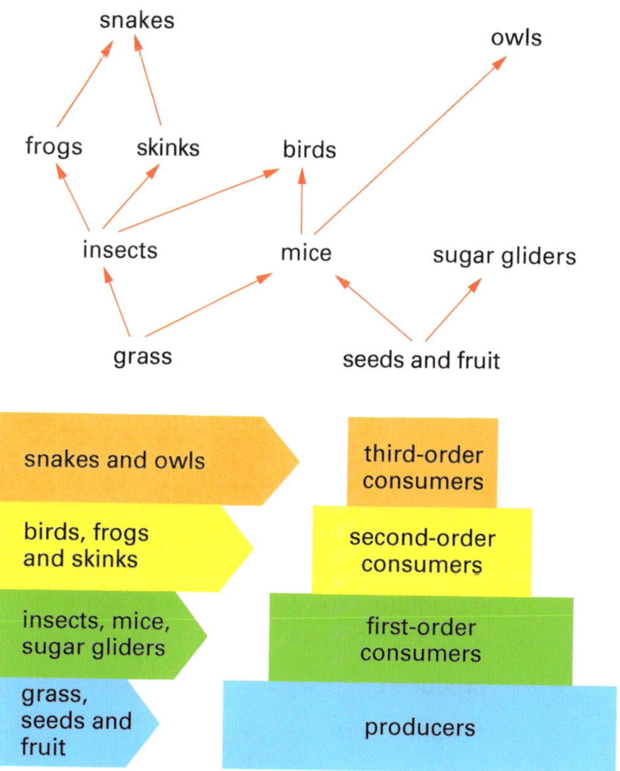

Figure 10.16 An energy pyramid of the organisms in a forest ecosystem

As there is a large loss of energy from one feeding level to the next, the food web cannot have an unlimited number of levels. For this reason, most food webs consist of producers, first-order, second-order and perhaps third-order consumers. Very rarely will a food web support fourth-order consumers or higher.

ISBN 978 1 4202 3736 8

Inferring from energy pyramids

By studying energy pyramids, a number of inferences can be made about the way humans feed themselves. Look at the two energy pyramids on the right.

Both food webs begin with producers, containing the same amount of total energy. However, in Figure 10.18, there are more feeding levels in the food web and consequently more energy lost between producers and humans. This food web supports fewer people than the vegetarian diet. This means that a mainly fish diet is less energy-efficient than a vegetarian diet.

In most cultures, the human diet is varied and includes a high proportion of foods that have come from plants. In countries where the population density is high, people have to rely on plants to supply most of their energy. In these countries, the eating of animal foods is more a luxury than an everyday occurrence.

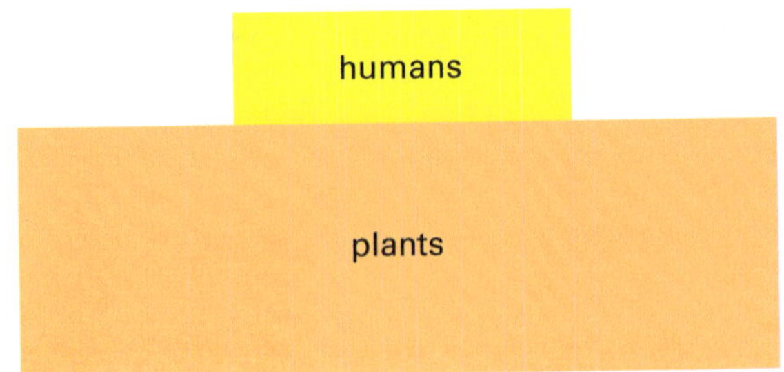

Figure 10.17 A human vegetarian diet

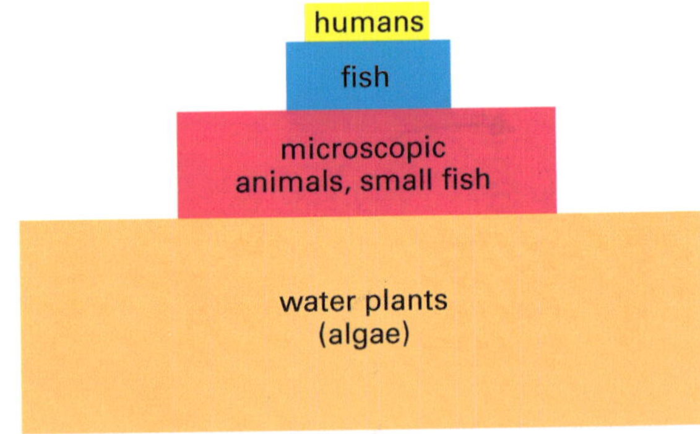

Figure 10.18 A human diet consisting mainly of fish

CHECK

1 Ecosystems are not limited by size. The Earth can be considered an ecosystem, as can a river. What do all ecosystems have in common?

2 The diagram on the right shows the flow of matter in a rocky shore ecosystem. Match the following descriptions with the letters in the diagram. (You will have to use some descriptions more than once.)

producers

decomposers

first-order consumers

food eaten

matter lost as wastes

matter lost through respiration

matter in air, water and soil

second-order consumers

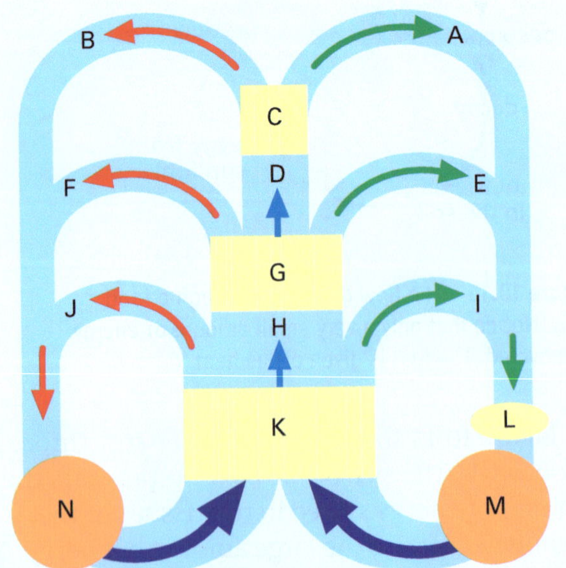

3 a In a food web, a large mass of producers supports a smaller mass of consumers. Why is this?

b Why is a vegetarian diet considered to be more energy-efficient than a diet of meat and fish?

ISBN 978 1 4202 3736 8

4 Is it possible that a carbon atom in your body could be the same as one that was part of a protein in the body of a dinosaur 180 million years ago? Give reasons for your answer.

5 In what form is energy lost from an ecosystem? Why doesn't an ecosystem run out of energy?

6 The ooze and mud at the bottom of oceans and seas contain an abundance of bacteria.
  a Why are so many bacteria found there?
  b Why are these organisms an important part of the marine food web?

7 Some parts of the labels in the cycle on the right have been left off and letters have been used in their place. For each of the letters in the cycle, choose the appropriate word.

nitrates	protein
nitrogen	decomposers
absorbed	bacteria

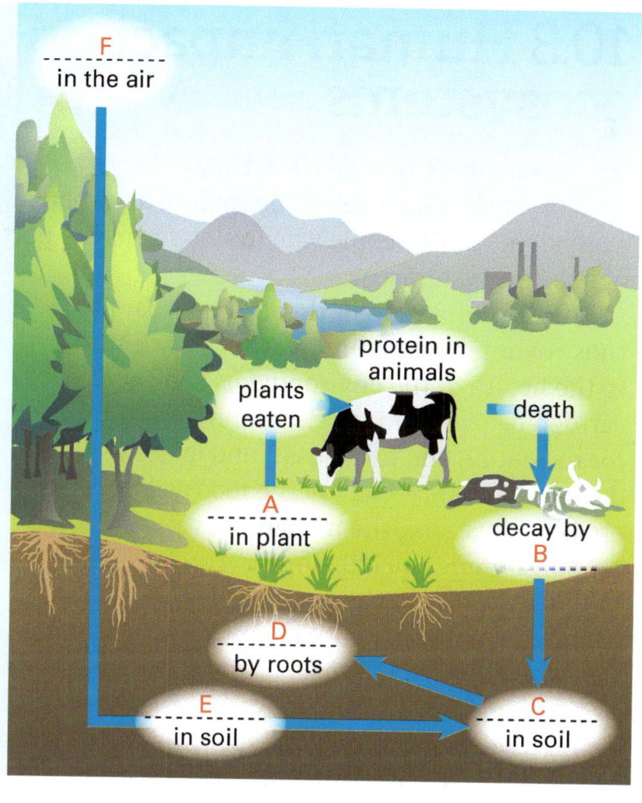

CHALLENGE

1 An ecosystem is considered productive if it contains a large mass of producers (plants and algae). Suggest why the following statements are true.
  a A rainforest is a more productive ecosystem than a desert.
  b Ecosystems in warm climates are more productive than those in colder climates.

2 a Why does a large molecule, e.g. sugar ($C_{12}H_{22}O_{11}$), contain more energy than a molecule of carbon dioxide or water?
  b You mow the lawn and dump the lawn clippings in a pile on the compost heap. A day later you notice that the pile is quite warm. Explain how this heat was generated.

3 How is it possible to increase the nitrogen content in the soil of your vegetable garden without using synthetic fertilisers?

4 Is it possible to have an energy pyramid like this one? Explain your answer.

second-order consumers
first-order consumers
producers

5 Mass pyramids are similar to energy pyramids except that the boxes in the mass pyramid represent the mass of the organisms at each feeding level. Is it possible to have a mass pyramid the shape of the pyramid in question 4? Explain your answer.

6 An area of open eucalypt forest is cleared and burnt. In its place, sugar cane is planted. Write inferences to explain the following.
  a The sugar cane is more productive over time than the native forest.
  b The sugar cane is not a self-sustaining ecosystem, whereas the eucalypt forest is.
  c There is a smaller mass of decomposers in the sugar cane ecosystem.

Test your ecological footprint.

Footprint calculator

ISBN 978 1 4202 3736 8

10.3 Human impact on ecosystems

Figure 10.19 This eco-tourism resort in Broom, WA, tries to balance human and environmental needs.

No matter where humans live, they are a part of an ecosystem. City dwellers are part of an urban ecosystem, while people who live on farms are part of rural ecosystems.

Humans often forget that their needs, like those of any other animals, are closely linked to the health of the environment. Look at the cartoon below. Humans clear land for farming, for recreation and for houses, and in doing so damage or change most of the food web in these areas. In this case, the mangroves will be cleared for houses, and the fish breeding grounds will be destroyed. Then the people who come to live where the mangroves used to be expect to catch fish when they go fishing.

Of all organisms, humans can be the most destructive. For example, the following is the story of a forest that was cleared many years ago for crops. When the forest was cleared, the natural food web in the area was destroyed. After the crops were planted, predators of the crops, mainly insects, invaded the area. However, because the original food web had been destroyed, the predators of these insects (mainly spiders, birds and frogs) had moved away from the area. The farmer, faced with losing the crops, used pesticides to kill the insect pests. These poisons affected many of the animals in the area, as well as the life in the nearby creeks.

Humans can do better to balance their activities and needs as well as conserving, protecting and maintaining the quality of the environment. The river red gum case study on the next page examines the impact of humans on this ecosystem.

Figure 10.20 Mangroves contain many complex food webs. They are the breeding grounds for fish and crabs, and their destruction affects the food webs in the ocean.

ISBN 978 1 4202 3736 8

SCIENCE AS A HUMAN ENDEAVOUR

The river red gum ecosystem

For thousands of years, large forests of river red gums have flourished along the Murray River and other large rivers that flow into it. However, over the last 200 years, huge changes have occurred to these forests.

The Murray River floodplain

The river red gums are well adapted for the floods that once occurred regularly along the Murray. Under natural conditions, the river flooded every 1.7 years for about 2–3 months, as the snow melted in the Snowy Mountains.

The floodwaters carry fertile soil, and branches and leaves from dead trees, which are caught around the roots of the trees. Over thousands of years, soil rich in nutrients has built up the floodplain.

The river red gum forest

A river red gum forest can produce 250 million seeds per hectare! Most seeds fall in spring and early summer when the floods naturally recede, and the seeds germinate in the warm moist soil during summer.

The seeds create food for ants and other insects as well as some birds. The flowers attract nectar-eating birds, insects and possums. These herbivorous animals attract echidnas, goshawks and water rats.

Human impact on the ecosystem

Farms established along the Murray River system required a dependable water supply for crops. To regulate the water flow, over 100 dams and weirs have been built along the rivers. As a result, the following changes have occurred to the natural cycle:

- Flooding now occurs only every 5–10 years.
- Flooding lasts for shorter periods instead of several months.
- The total volume of water has been reduced.

Questions

Use the information on this page and from the websites below to complete the following.

1. Describe how the river red gum is well adapted for life on the floodplain.
2. What abiotic factors have changed since agriculture was established on the Murray River?
3. Predict what effect human impact has had on the life of this ecosystem.

EXPLORE ONLINE

Follow the links to the websites to learn about river red gums.

Environment Victoria

River Red Gum Ecosystem

ISBN 978 1 4202 3736 8

Sustainable ecosystems

The river red gum forests along the Murray River flourished for thousands of years. In some years droughts may have slowed the growth of trees and the animals that lived in the forests, but over this time, the forests remained alive and productive. This is an example of a **sustainable** ecosystem.

A sustainable ecosystem is one that can exist, be productive (contain lots of producer organisms) and support a diverse group of organisms over a period of time.

Human activity, including logging, along the Murray River has affected the growth of the forests, and has reduced the numbers and types of animals that live in the forests. In general, human activity has reduced the sustainability of ecosystems.

Sustainable agricultural ecosystems

Most agricultural ecosystems are not sustainable because they do not support a diversity of plant and animal life. They generally have only one type of crop.

However, with the help of scientific research organisations such as the CSIRO, farmers have experimented with growing forests alongside their crops to increase diversity. In many cases, animals including birds, spiders and insects, have settled in the forests and have helped reduce the insect pests in the crops.

Figure 10.21 This forest has been established alongside a field used for growing crops.

Controlling populations

When crops are grown, a large population of producer organisms is established. As a result, populations of insects, birds and small mammals, whose numbers were kept in check in the natural community, grow unchecked and damage crops.

Chemical control of populations

Populations of animals and plants can be controlled by using poisons. These poisons can be obtained from natural sources, such as plants, or made synthetically in the laboratory.

Figure 10.22 Applying pesticides

There are two problems with using pesticides. First, most pesticides do not target specific organisms. This means they will kill both pests and useful organisms. For example, aphids can destroy cabbage plants, and orange and apple trees. But one ladybird can eat up to 100 aphids a day. By spraying the crops, a lot of insects, even helpful ones such as the ladybird, are killed.

The second problem is about biodegradability. A biodegradable pesticide is one that breaks down naturally to a state where it is no longer active or harmful. All pesticides are biodegradable to some degree. The natural ones break down very rapidly, but the synthetic ones can take from days to years. The now banned insecticide DDT, which was used in the 1950s to control mosquitoes, is still active in the soil in many places.

ISBN 978 1 4202 3736 8

Urban ecosystems

Urban ecosystems include cities and towns where humans live and work. But an urban ecosystem is different from other ecosystems.

First, humans are the dominant organism, not necessarily because there are more of them, but because their activities affect almost every other organism in the food webs within this ecosystem. For example, a house is sprayed with insecticide to kill cockroaches and other pests. This poison also kills spiders and predatory wasps, which control many insect populations.

Second, almost all the energy inputs in an urban ecosystem come from other ecosystems. Food, electricity and fossil fuels have to be supplied to an urban ecosystem for its survival. You notice how much your household relies on external energy supplies when there is an electrical blackout for a number of hours!

Third, an urban ecosystem produces enormous amounts of wastes that are usually not recycled. Unlike the matter cycle in Figure 10.11 on page 249, wastes are often taken outside the urban ecosystem for decomposition.

In addition, some of the wastes in an urban ecosystem are toxic. For example, oil discharged into creeks and rivers around industrial plants can affect other organisms.

ACTIVITY

Form a small group to discuss some of the following issues. Choose someone to report back to the class at the end of the discussion.

1 Most animals have predators and competitors in their habitats. Do you, as an animal, also have predators and competitors? How do you deal with these organisms? Are there any other types of organisms that affect your life? Explain your answers.

2 In a natural ecosystem, the wastes from the organisms that live there are usually broken down and recycled in that ecosystem. How do urban ecosystems deal with human wastes? What other wastes are created in urban ecosystems? What is done with these wastes?

3 Most animals in natural ecosystems move around to find food. How do humans usually find food? Does the supply of food to cities create problems for the environment? Explain your answer.

4 A natural ecosystem has to be self-sustaining to be able to survive. How could an urban ecosystem become self-sustaining? Have your group brainstorm some ideas. You might like to write your ideas on a large piece of paper to present to the class.

5 Some rural homes use a remote area power supply (RAPS) system instead of being connected to the electricity grid. Could this system be used in small urban ecosystems as well? To answer the question, use the internet to find information about the RAPS system.

ISBN 978 1 4202 3736 8

Problems in ecosystems

When Captain James Cook sailed along the east coast of Australia in 1770, he wrote in his journal that this land was a 'continent of smoke'. He was referring to the numerous bushfires he could see from his ship.

Before humans came to this land, it seems that fires, which were started by lightning strikes, occurred only very occasionally. However, Aboriginal Australians, whose ancestors arrived about 50 000 years ago, used fire for their survival and changed the natural pattern and timing of fires. This in turn changed the relationships of the organisms in certain ecosystems. However, over this long period of time the ecosystems remained sustainable.

Major changes such as bushfires, droughts and floods or large toxic chemical spills have a huge impact on the organisms in ecosystems.

Your task—impacts on ecosystems

Read the questions on this page and the next. Use the ideas in the questions to write a report about a natural or human-caused disaster and how it affects a particular ecosystem.

What to do

- Work in a small group.
- Choose a topic and decide what you are going to write about and how you are going to structure your report. You could present your report as an essay-type report, a PowerPoint presentation, a newspaper-type article, an oral presentation, a video presentation etc.
- You are not limited by the ideas listed for each topic or the order in which they appear. You can include other information as well.
- You can use information from past or recent disasters as well as models or predictions in your report.
- Document your report with the sources of your information—website addresses, book titles and authors, titles of newspaper articles and dates.

Bushfires

- How do bushfires start naturally? What weather conditions cause bushfires?
- Some bushfires are called 'low-heat fires' while others are very intense and destructive. Why does this occur?
- The Aboriginal method of burning actually protected their environment rather than destroying it. Explain what this statement means.
- What changes occur to the populations of organisms in an ecosystem as a result of a bushfire?
- What emergency services are involved in fighting bushfires?
- What methods are used to reduce the risk of bushfires and to reduce the damage caused by them?

ISBN 978 1 4202 3736 8

Droughts and floods

- How is a drought defined? Are some areas in Australia more likely to experience droughts than others? Why does this happen?
- What is the El Niño effect? Does it affect the weather in all parts of Australia? Can it be predicted?
- What changes occur to the populations of organisms in an ecosystem as a result of a drought and flood?
- Are some areas of Australia more likely to experience floods than others? Why does this happen?
- Floods cause huge losses of property, crops and livestock. However, there are benefits to the environment as a result of flooding. What are these benefits?

Oil and chemical spills

- Damage to ecosystems can occur from oil spills. This can happen when oil is being transported from the oil fields to the refineries, or oil leaks from offshore oil rigs. Find out where in Australia oil is drilled and where it is transported to.
- Give an example of a major oil spill anywhere in the world and document the damage it caused to the ecosystem. What methods are used to clean up oil spills?
- Heavy metals are very toxic to most organisms. What are heavy metals? Give examples of industrial processes that produce them. How do heavy metals get into natural ecosystems?
- The photo shows a fish kill. Why do they occur? What measures are taken to avoid fish kills in Australian waterways and seas?

ISBN 978 1 4202 3736 8

SCIENCE AS A HUMAN ENDEAVOUR

Murray River crisis

The Murray is Australia's longest river, and its basin covers 1 000 000 square kilometres of Queensland, New South Wales, Victoria and South Australia. Most of its water comes from the Great Dividing Range in the east, and it then wanders across the western plains, reaching the ocean in South Australia. An Aboriginal legend says its wandering course was formed by a giant cod thrashing as it tried to escape a hunter's spear.

Forty per cent of Australia's farms are in the Murray–Darling basin and they produce a third of Australia's agricultural production—wool, cotton, wheat, sheep, cattle, dairy products, rice, oil-seed, wine, fruit and vegetables. The water used to irrigate these farms is taken from the river. Once 25 000 gigalitres flowed down the river each year, but today the flow is less than 3000 gigalitres! Dams and reservoirs have been built to control the flow of water. This means there is not enough water in some places, and too much in others. This has affected plants such as the river red gums (page 255) and animals such as the Murray cod—Australia's largest freshwater fish. There are also blooms of blue-green algae. The algae is a sympton of low river flows and prolonged warm, dry weather.

Too much water is being taken out of the Murray River, and too much pollution is being put back in. This is a major problem for the city of Adelaide, which draws its water from the Murray. Over the last 200 years, 40% of the natural vegetation has been cleared, and shallow-rooted crops, such as wheat and pasture for livestock, have been planted. This has caused the underground water to rise, bringing salt to the surface and causing a major *salinity* problem.

Obviously the Murray River needs more water, but where will it come from, if rainfall doesn't increase? Australia suffers regular droughts and climate change could reduce the flow in the river even further. The federal and state governments are working together to try to solve the problem.

One obvious solution is to reduce the amount of water that farmers are allowed to use for irrigation. Also, large amounts of water could be saved by improved irrigation systems, as up to 85% of water is now lost due to evaporation and leaks. Perhaps farmers could switch to crops that use less water. It has even been suggested that water could be piped from high-rainfall areas into the Murray–Darling river system.

Questions

1 How would the following groups differ in their views on what should be done about the Murray?
- farmers in the Murrumbidgee Irrigation Area
- people living in Adelaide
- the federal and state governments
- the Australian Conservation Council
- people living in Melbourne

2 The Wiradjuri Aboriginal people who live along the river say, 'Look after the land and rivers, and the land and rivers will look after you.' Explain what you think this means. Try to correctly use the word *sustainable* in your answer.

ISBN 978 1 4202 3736 8

CHECK

1 a Describe in your own words what a sustainable ecosystem is.
  b What activities put pressure on these sustainable ecosystems? Give examples with your answer.

2 A lake is surrounded by a wide band of reeds. The ecosystem has been untouched by humans for over 100 years.
  a List the biotic and abiotic factors that might affect the survival of fish in the lake.
  b Suppose a farmer began to pump water from the lake for irrigation. Suggest how this might affect the lake ecosystem.
  c The farmer cleared the bush down to the reedy areas surrounding the lake. He used fertiliser on his crops and some excess fertiliser ran into the lake. Suggest how this might affect the lake ecosystem.

3 List the ways an urban ecosystem is different from a natural ecosystem, such as a eucalypt forest ecosystem.

4 On a farm that had been cleared of most of its trees many years ago, a farmer planted thousands of trees and shrubs in wide bands around her paddocks.

After some years, she noticed that the number of insect pests on the farm had dramatically decreased, without the use of pesticides. Suggest reasons for this.

5 Describe two environmental problems the use of pesticides can cause. How can these problems be solved?

CHALLENGE

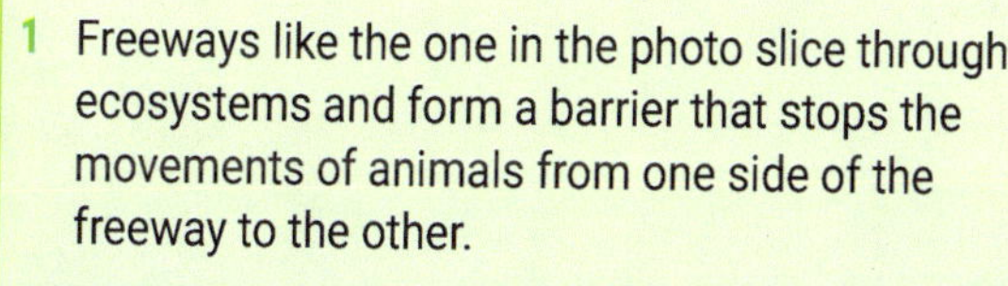

1 Freeways like the one in the photo slice through ecosystems and form a barrier that stops the movements of animals from one side of the freeway to the other.

  a What changes might occur to an ecosystem over a number of years after a freeway is built through it?
  b Suggest ways to reduce the impact of freeways forming a barrier through ecosystems.

2 Use the matter cycle in Figure 10.11 on page 249 to explain how logging can destroy a sustainable forest ecosystem.

3 In 2011, Cyclone Yasi swept in from the Coral Sea and onto the north Queensland coast. Suggest how a natural ecosystem might change as a result of extreme weather conditions such as cyclones or bushfires.

4 The photo below shows the cooling towers of a power station. The water from these cooling towers is returned to the rivers or oceans after it is used. However, the temperature of the used water can be up to 20 °C warmer than the original water. Suggest what effect this would have on the organisms in the surrounding ecosystem.

ISBN 978 1 4202 3736 8

MAIN IDEAS

Copy and complete these statements to make a summary of this chapter. The missing words are on the right.

1 An _____ is a system of relationships between organisms and the non-living parts of their environment.

2 The survival of organisms in an ecosystem depends on _____ factors (e.g. predators) and abiotic factors (e.g. availability of clean water).

3 _____ are characteristics that help an organism survive in its particular living place. They can be classified as structural, _____ or behavioural.

4 Matter is used and reused as it cycles through an ecosystem. For example, in the _____, carbon atoms are used in _____ and respiration and are recycled by _____.

5 An energy _____ shows that the energy available decreases from the producers to the highest order _____.

6 Energy from the sun is _____ and used by _____ organisms. It is then continually given off as heat by all organisms in a food web.

7 Conserving and _____ the quality of the environment should be balanced with the needs and activities of humans.

biotic
absorbed
consumers
producer
functional
photosynthesis
decomposers
pyramid
adaptations
carbon cycle
protecting
ecosystem

CH•10 REVIEW

1 In the nitrogen cycle, nitrogen in the air is:
  - **A** converted to nitrates by soil bacteria and by lightning.
  - **B** converted to amino acids by bacteria in the gut of some animals.
  - **C** absorbed directly by plants and algae.
  - **D** very reactive and forms nitrates with oxygen.

2 Which of the following would you class as a functional adaptation? (There may be more than one answer.)
  - **A** Dolphins have a layer of fat under their skin.
  - **B** Dolphins sometimes follow ships.
  - **C** Female dolphins give birth to live young and produce milk on which to feed them.
  - **D** A dolphin is able to make many sounds with its voice box.
  - **E** Dolphins have a streamlined shape.

3 In an experiment similar to the one on colour adaptation on page 241, discs (20 of each colour) were scattered over an area 3 m by 3 m. The ‘predators’ found as many they could in 10 seconds and the results are shown in the table.

Colour of disc	Number found
blue	20
green	17
yellow	4
red	8

  - **a** Draw a bar graph of the results.
  - **b** Infer the type of surroundings over which the discs were scattered.
  - **c** Explain how this experiment can be used as a model to show how important camouflage is for the survival of organisms.

ISBN 978 1 4202 3736 8

4 Which of the following would you class as a physical factor in an ecosystem? (There may be more than one answer.)
- A the number of predators in the area
- B the availability of light
- C the density of trees in the area
- D the amounts of nutrients in the soil

5 The energy pyramid below shows the organisms in a forest ecosystem. According to the pyramid, which of the following is correct?
- A Snakes, birds and possums are all second-order consumers.
- B Not all the energy in one feeding level is transferred to the one above it.
- C The energy in one feeding level is equal to the energy in the feeding level below it.
- D Insects, possums, worms and mice supply the energy for the feeding level below them.

snakes and hawks
birds, frogs and skinks
insects, worms, possums, bugs and mice
leaves, flowers, seeds and fruit

6 The photo below shows a river ecosystem. List the ways this ecosystem is different from an urban ecosystem in a large town.

7 Give reasons why a wheat field is not a sustainable ecosystem. In your answer, mention the flow of matter and energy.

8 The diagram shows some of the animals and plants that live in a pond ecosystem.

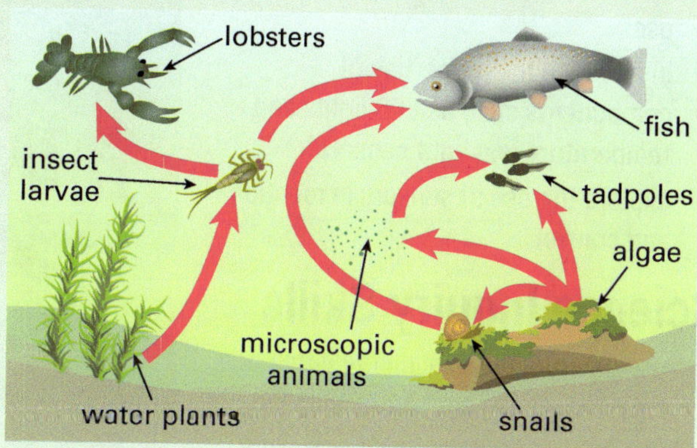

The flow of matter in the pond ecosystem can be represented in the diagram below.
- a Which organisms are producers?
- b What are the products of respiration that are reused in this ecosystem?
- c What do the letters J and K represent?
- d What do the letters L and M represent?
- e The simple diagram shows that matter is recycled in the ecosystem. In reality some matter does escape from the ecosystem. Suggest how this occurs.

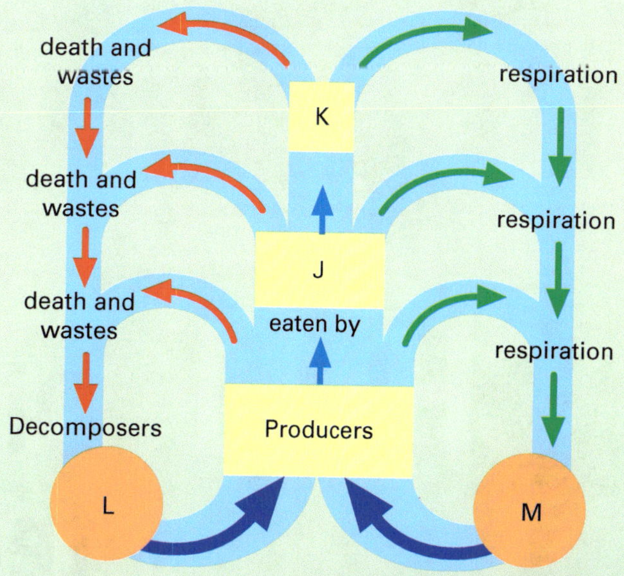

Check your answers on page 305.

ISBN 978 1 4202 3736 8

IN THIS CHAPTER

Science Understanding

- consider how communication methods are influenced by new technologies that rely on electromagnetic radiation and digital technologies
- consider how the properties of electromagnetic radiation relate to its uses
- investigate the properties of components such as LED lights and temperature and light sensors
- explore the use of sensors in robotics and control

Science Inquiry Skills

- use diagrams to explain the difference between analog and digital signals
- demonstrate how optical fibres work
- set up electric circuits using common electronic components

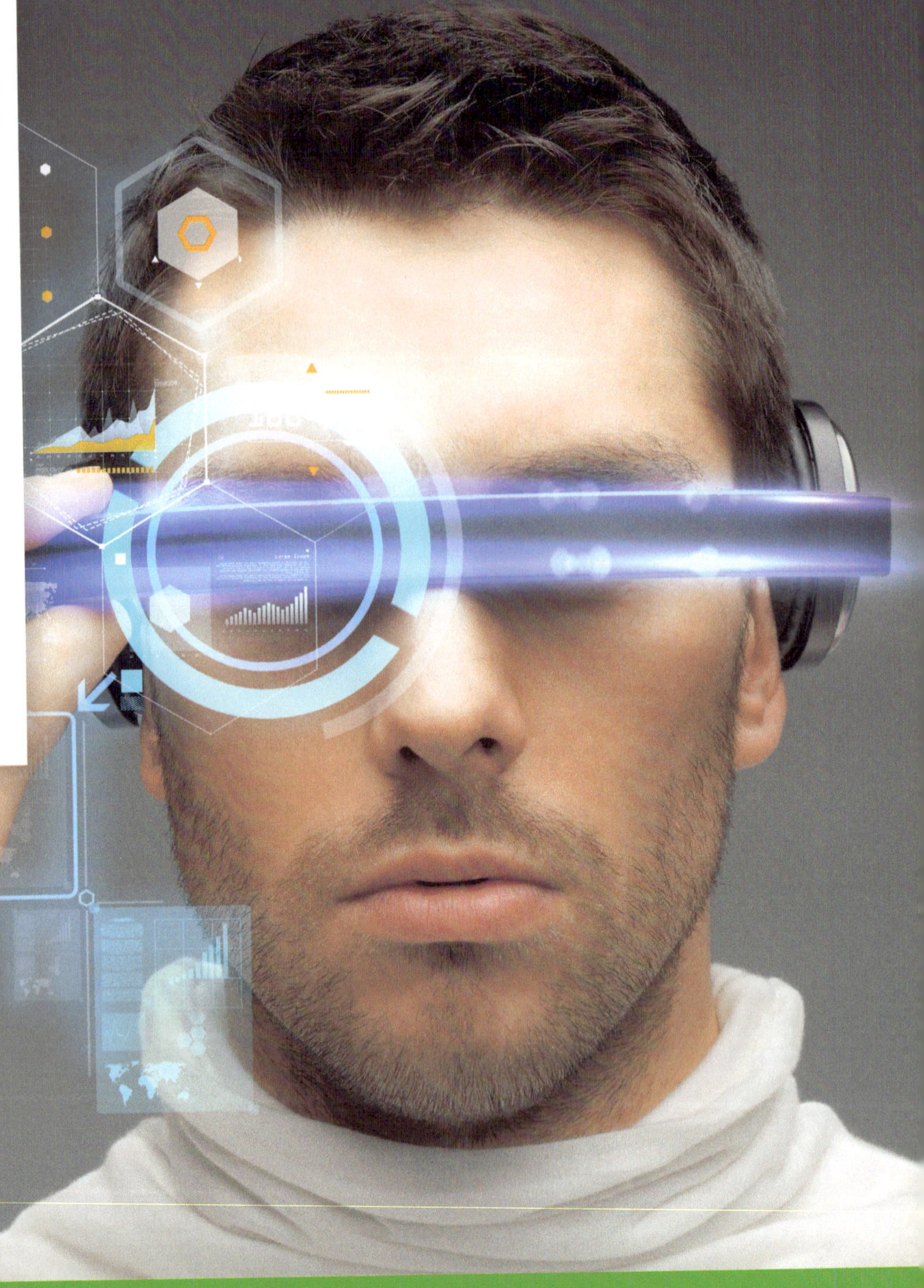

CH•11 Digital technology

ISBN 978 1 4202 3736 8

GET STARTED: *IMAGINE*

Digital technologies are expanding exponentially. New gadgets, technologies and tools are appearing all the time as computer processors get more powerful and miniaturised. Our mobile devices can do so much more than 10 years ago, and have transformed how we communicate, play and work. We are also now seeing the expansion of the 'internet of things', artificial intelligence and robotics, all of which are becoming more useful and responsive. It is said that robotics will take over many manual and repetitive tasks in coming years. Are you ready for the future? What will a school be like in 20 years' time?

1 Make a list of the many digital technologies that you currently know about.
2 Compare the technologies that your parents or guardians would have had at your age. What has changed that makes the technology so much more powerful now?
3 Highlight the technologies in your list for Part 1 that are related to communications.
4 Select one technology from your list, and explain how it influences your life.
   a How do you use it and how does it help you? Could you do without it?
   b What are the problems that this technology might also cause?
5 Imagine a digital technology you would like to have in the future.
   a Describe your technology, including the problem it solves and how it works. Use a diagram if possible.
   b Which areas of science would be required to invent your technology?
   c What do you think is currently stopping this invention from being a reality?

ISBN 978 1 4202 3736 8

11.1 Communication

Communication is the sending of a message (information) from one person to another. This message can be in written or spoken form, called *verbal communication*, or using gestures or symbols, which are forms of *non-verbal communication*. However, communicating does not only occur between humans. Communication can also involve animals.

Communication involves the transmission of information from a sender to a receiver. The sender *encodes* this information into a message suitable for transmission. That is, the message is put into a *code*. The words that you are reading now are in a code that you have learnt over a number of years. These words form sentences that have meaning. Can you make any sense of this sentence? (See Check 2 on page 274.)

To understand a message like the one above the receiver has to *decode* it—change it from this code to a code that you can understand.

The flow diagram in Figure 11.1 shows the steps in the process of communication. *Feedback* is an important part of this process because it tells the sender whether or not the message has been received and understood. *Noise* is something that might interfere with the transmission of the message; for example, someone playing loud music while you are trying to talk on the telephone. Noise can also be electronic.

The teacher in the cartoon wants the students to set up an electric circuit.

- How does she know whether her message has been understood?
- What would you need to be able to do to decode her message?
- What is the effect of noise in this situation?

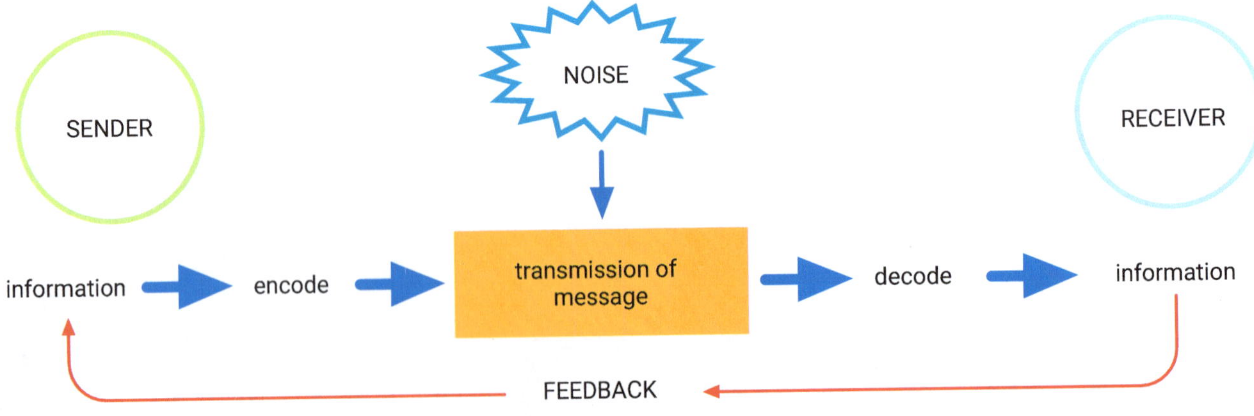

Figure 11.1 A model to show the steps in the process of communication

ISBN 978 1 4202 3736 8

Communication devices

Communication devices such as a telephone require voice to be changed into electrical signals, radio waves or light pulses. These signals are transmitted over long distances and then changed back into voice, which is heard by the receiver. Voice is changed into electrical signals by a *microphone* and the electrical signals are changed back to voice by a *loudspeaker*.

In previous studies, you learnt that a magnet induces an electric current when it moves in a coil of wire. A microphone uses this principle. Sound waves make the diaphragm (DIE-a-fram) in the microphone vibrate. The coils of wire attached to the diaphragm vibrate near a magnet. This movement then creates a current in the wires, which changes with the loudness of the voice. Soft sounds produce small currents and loud sounds produce larger currents. The pitch of the sound also affects the current.

A loudspeaker works in the opposite way to a microphone. The varying current in the wire passes through a coil near a magnet and this causes the coil to move. The coil is attached to a diaphragm that also moves. This movement causes the air next to the diaphragm to move and you hear a reproduction of the original sounds.

Digital and analog signals

In a microphone, the vibrating diaphragm produces a varying electrical signal like the one below. The size or **amplitude** of the signal determines the loudness of the sound.

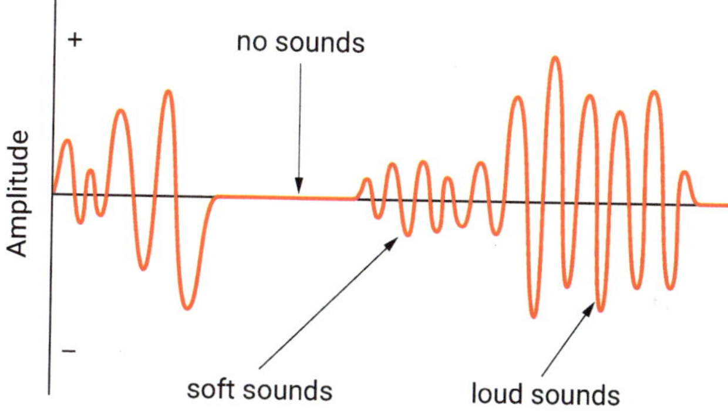

Figure 11.3 Amplitude is the loudness of sounds.

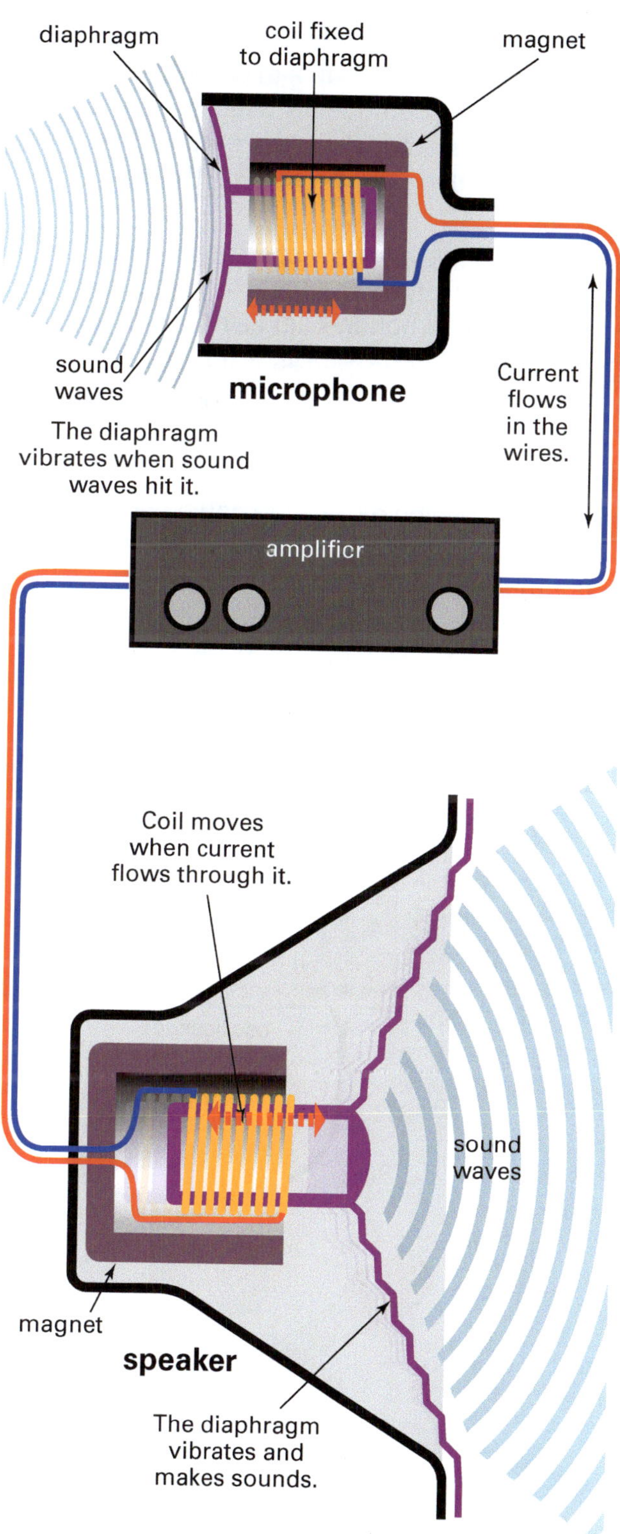

Figure 11.2 A microphone converts sound waves to electrical signals, and a loudspeaker converts electrical signals to sound.

ISBN 978 1 4202 3736 8

ACTIVITY

Teacher demonstration

Your teacher will set up a microphone attached to a cathode ray oscilloscope (CRO). The electrical signals produced by the microphone can be seen on the screen of the CRO.

Observe a variety of different sound patterns on the CRO. For example, speak into the microphone, sing a note or use a tuning fork or musical instrument.

What is the relationship between the amplitude of the waves on the CRO and the volume of the sound made?

Instead of using a CRO, you could use a sound probe connected to a datalogger and then print out the wave pattern.

You could also connect a microphone to a galvanometer and observe the movement of the pointer as you speak into the microphone.

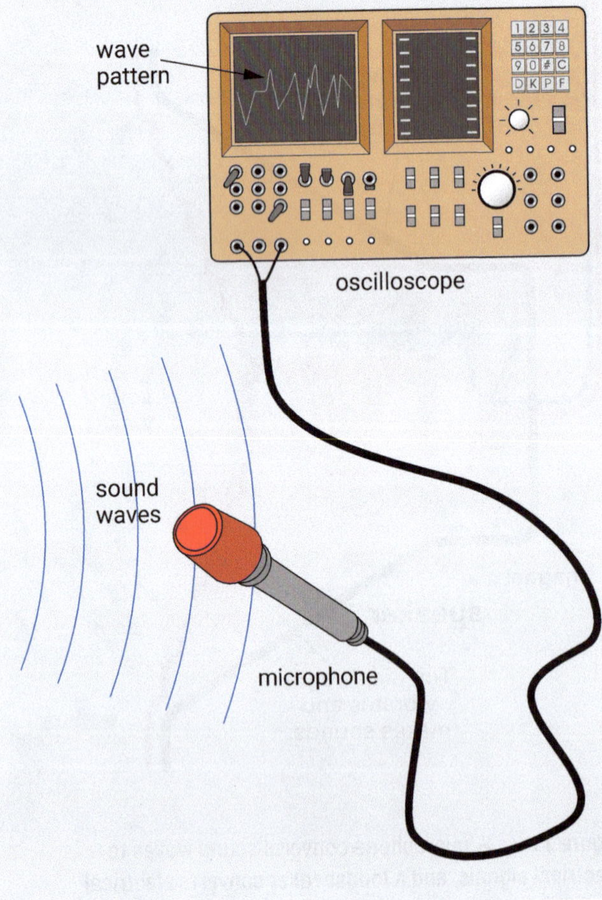

The wave pattern produced on the CRO when you speak into the microphone changes in amplitude as the volume of your voice changes. The wave pattern varies in value at different points in time. This type of signal is called an **analog** (AN-a-log) **signal**. The electrical signals that travel along the wires from the microphone in your telephone handset are similar to these waves.

Before 1980, telephone transmission in Australia was analog. Now, however, most transmissions between telephone exchanges use **digital signals**. A digital signal is made up of a sequence of *binary digits*—digits that have one of two values, zero or 1. In electronic devices, the value 1 is represented by a switch being on, and zero by the switch being off.

The two words **bi**nary dig**it** are shortened to the one word **bit**. Telephone transmissions (and computer data) are usually sent in millions of small units made up of eight bits that are called **bytes**. The digital signal in Figure 11.3 is a byte and has the value 10011010.

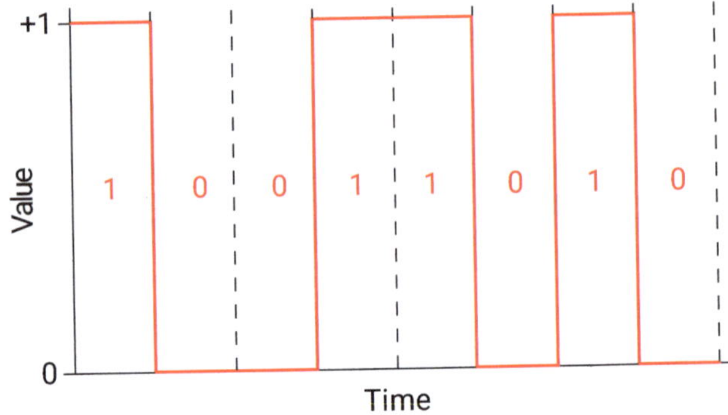

Figure 11.3 An eight-bit digital signal—each bit can have a value of 1 or 0, but nothing in between.

Digital telephone network

Figure 11.4 shows how the telephone network encodes the information in sound waves to digital electrical signals and then to optical signals. At the receiver's end, the information is decoded back again to sound waves. Instead of using cables, telephone signals can also be transmitted via microwave repeater stations or even satellites, especially in isolated areas.

ISBN 978 1 4202 3736 8

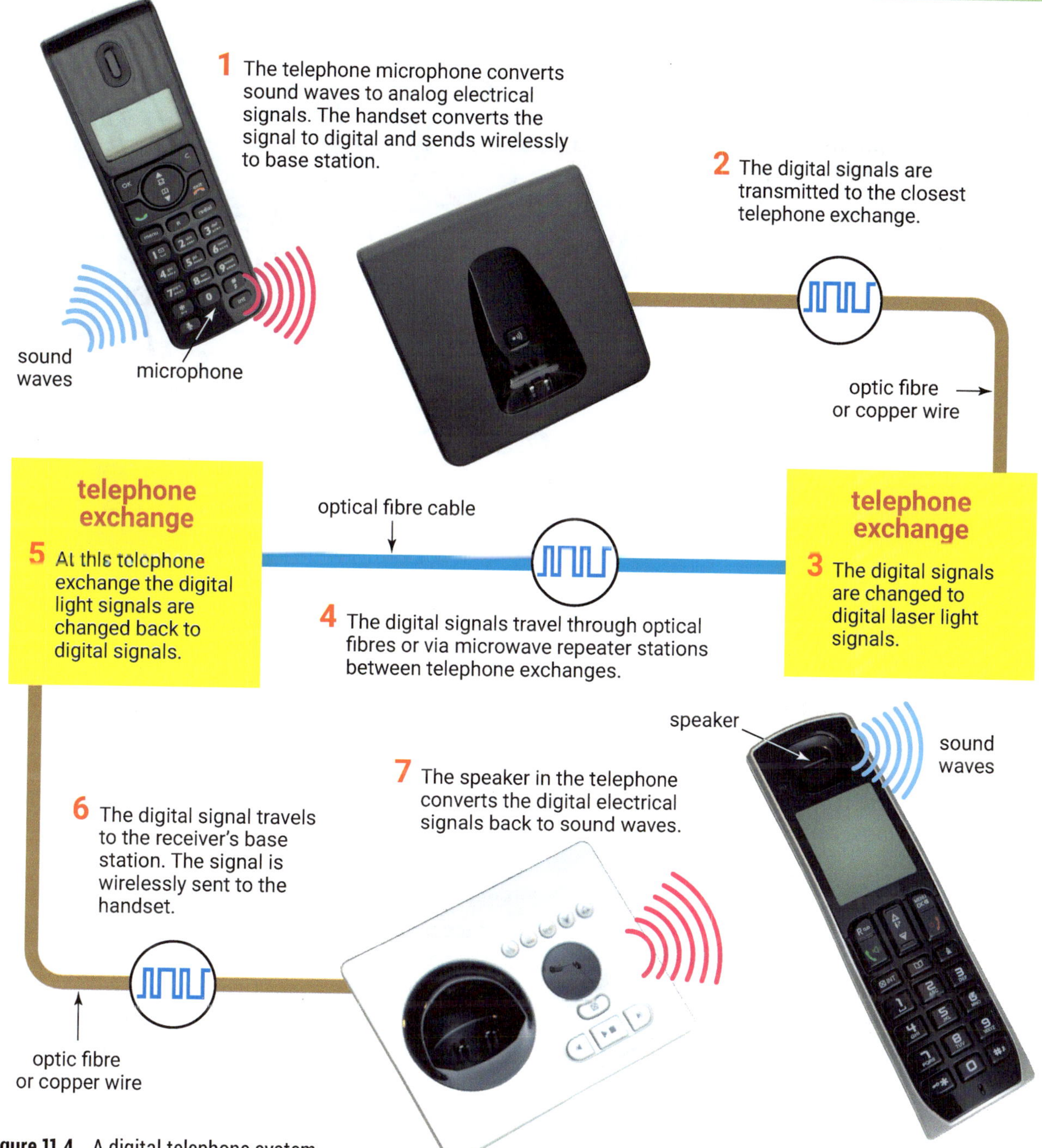

Figure 11.4 A digital telephone system

Mobile phones

Mobile phones don't need to be connected by wires. Instead they have a built-in radio transmitter and receiver. When you make a call, the mobile phone sends out a radio signal. This signal is picked up by a base station that has several antennas on top of a tower or tall building. The base station is connected to a switching centre, which switches the call to other base stations or to the fixed telephone network.

The base stations form a network of hexagonal cells, as shown in Figure 11.5. The cells range in size from 100 metres across to more than 30 km. The base stations receive mobile phone calls from the cell around them. Each call is then passed from cell to cell until it reaches its destination. The base stations also return calls to the cell around them.

ISBN 978 1 4202 3736 8

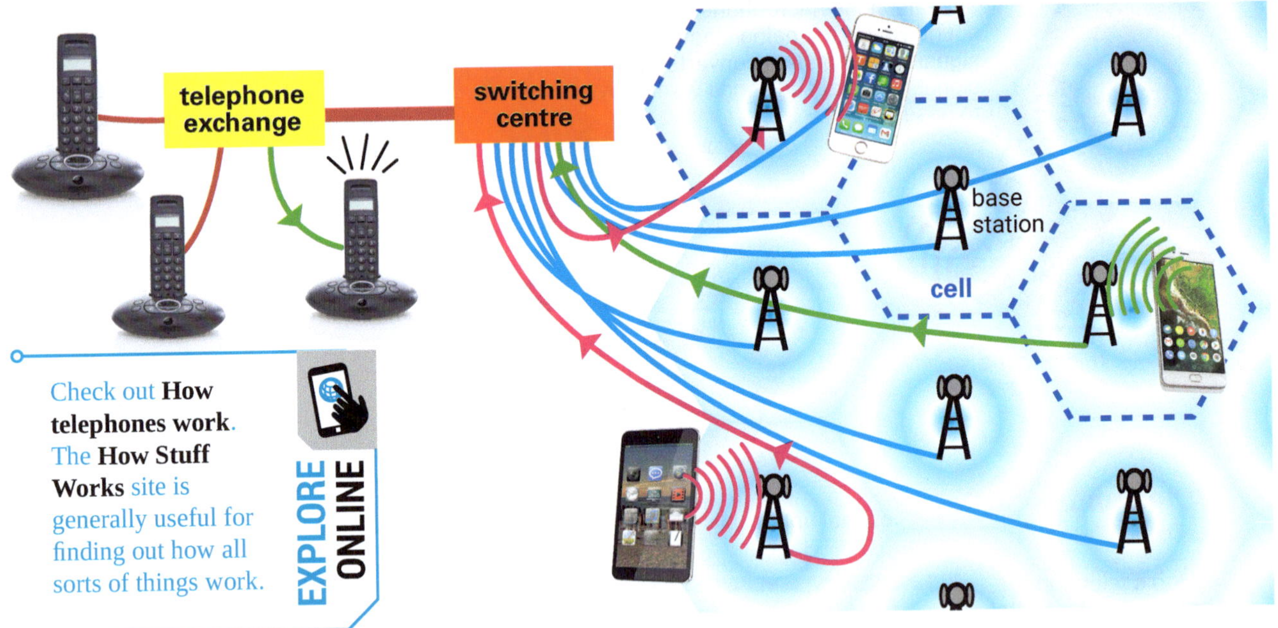

Check out **How telephones work**. The **How Stuff Works** site is generally useful for finding out how all sorts of things work.

EXPLORE ONLINE

Figure 11.5 A mobile phone network

Optical fibres

The optical fibres used in the telephone network are very thin, pure glass fibres. Each fibre consists of a glass core, a glass cladding and a protective outer jacket. The whole fibre is thinner than a human hair.

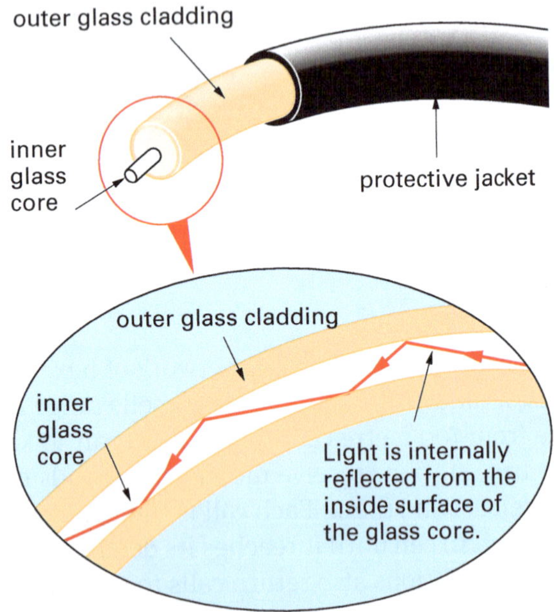

Figure 11.6 How light pulses travel through optical fibres

At telephone exchanges the electrical signals from local telephones are converted to pulses of laser light. These laser pulses are narrow high-intensity light beams of a single colour (wavelength). They are digital (on or off) and they travel through the optical fibre by *total internal reflection*. That is, the light can travel around bends and even loops by reflecting off the inside surface of the inner glass core. One optical fibre can carry up to 2 billion pulses of light per second.

The advantages of optical fibres are that they can carry much more information more quickly than copper wires do. They are lighter and cheaper to make, and they produce a better quality of communication with very low noise.

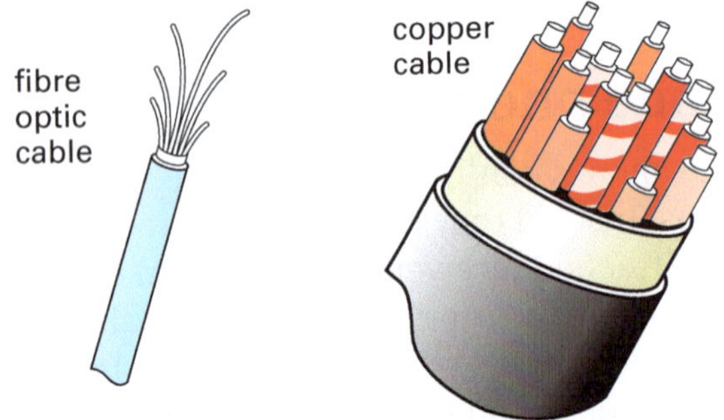

Figure 11.7 A copper cable that can carry 10 000 telephone calls is much larger and heavier than a fibre optic cable that carries the same number of calls.

ISBN 978 1 4202 3736 8

1. Set up a binocular microscope and look at the end of an optical fibre.
   Sketch what you observe.
2. The diagram below shows the equipment you need to make a model that demonstrates how optical fibres work.
3. Use the diagram as a guide to make the model. You will need to test it in a darkened room, or test it at home at night.
4. Can you improve the design of the model? Discuss your design with your partners.

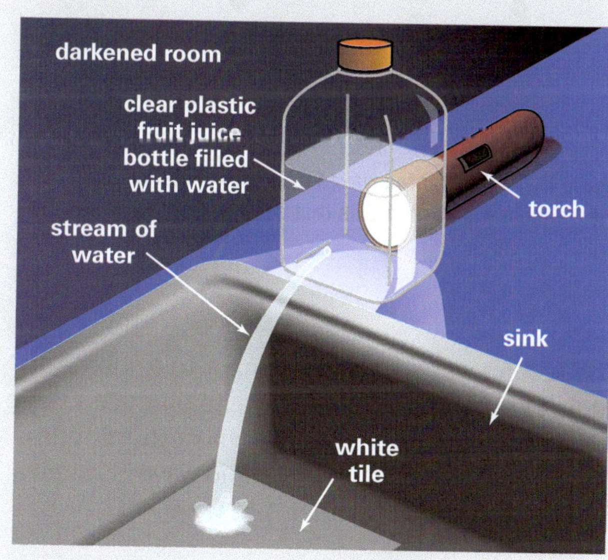

Television and radio

The information sent by television and radio is transmitted via *electromagnetic waves* of long wavelength. These waves are received by a metal aerial or antenna that converts electromagnetic waves into electric current. This current is then decoded into pictures or sound.

visible light (red light)
wavelength = 4×10^{-6} m

microwaves
wavelength = 1×10^{-6} m

radio and television waves
wavelength = 1 m to 1 000 m

Figure 11.8 Electromagnetic waves

Digital TV

Australia has now changed from analog to digital TV. Suppose the graph below shows how the electrical voltage of an analog TV signal varies during a fraction of a second. The table under the graph shows how this changing voltage can be converted to a digital signal consisting of numbers only. The voltage is sampled electronically many times per second, and these measurements are coded into binary numbers, using only zeros and 1s. The signal is then a series of pulses (pulse = 1 and no pulse = zero). The big advantage of digital TV is that you can see pictures with greater detail and clarity.

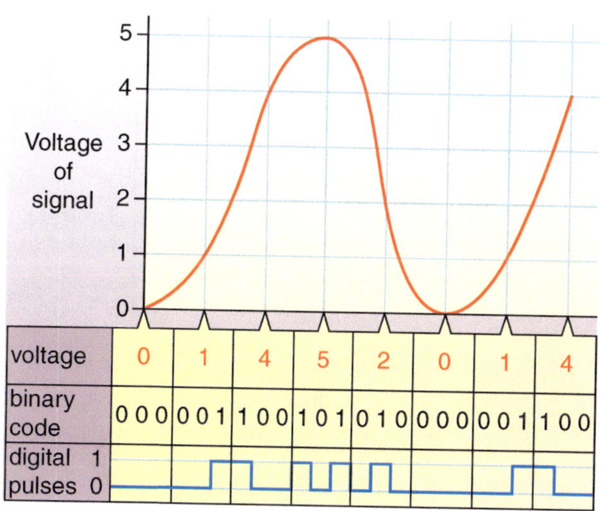

voltage	0	1	4	5	2	0	1	4
binary code	000	001	100	101	010	000	001	100

Figure 11.9 How an analog signal can be digitised

Radio and TV transmission

Your favourite radio station might have a call sign of 107.5. What does the 107.5 mean? And what is the difference between AM and FM?

To understand what frequency means, your teacher might set up a CRO as in the activity on the next page to show you different wave patterns.

A microphone converts sound waves into an electronic audio signal. These are low-frequency waves that if broadcast would travel only a few metres through the air. To overcome this problem, radio stations mix this audio signal with a much higher frequency wave with more energy, which can travel hundreds of kilometres through the air.

ISBN 978 1 4202 3736 8

This wave is called a *carrier wave*, and the combined audio signal and carrier wave is called a *modulated wave*.

Your teacher will connect an audio generator to a CRO.

1 Look at the wave pattern produced by the generator. What do you notice when the pitch of the sound is changed?

The **frequency** of a wave is the number of waves that pass a point in 1 second. Frequency is measured in hertz (Hz), where 1 hertz is one wave per second. The wave below has a frequency of 2 Hz. Two complete waves pass each point every second.

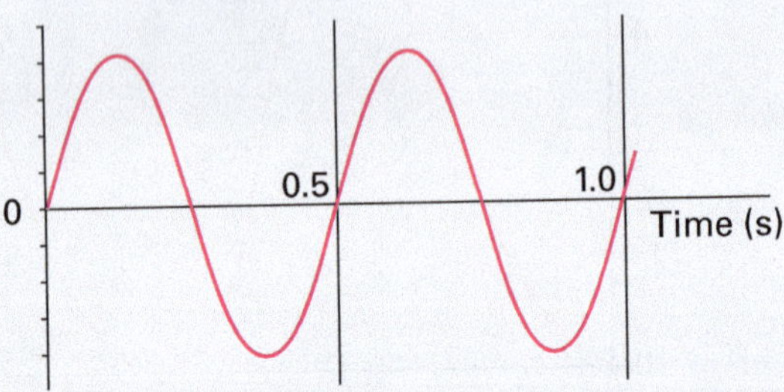

- What happens to the frequency when the pitch of a sound increases?
- What happens to the wavelength of the wave when the pitch increases?
- Make a generalisation about the pitch of sound and the frequency. Make another generalisation about the frequency of a wave and its wavelength.

2 Turn up the volume on the generator. Now turn the volume down.

- Record what happens to the shape of the wave on the screen.
- Make a generalisation about the loudness of a sound and the amplitude of the waves.
- Does changing the volume affect the frequency or wavelength of the waves?

AM

There are two ways to combine an audio signal with a carrier wave. One way results in a wave that has its **amplitude modulated** or varied. Radio stations that broadcast in this form are called **AM** stations.

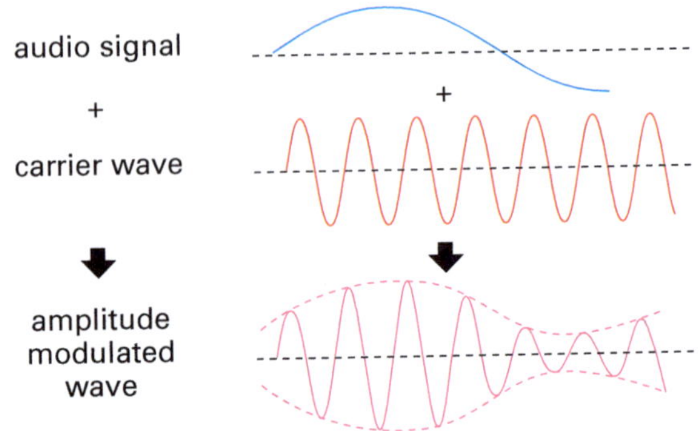

Figure 11.10 Modulated waves from an AM station. The frequency is the same as that of the carrier wave but the amplitude varies.

FM

The second way of broadcasting is to combine an audio signal with a carrier wave to produce a wave whose frequency changes but whose amplitude stays the same. This type of radio wave is called a **frequency-modulated** or **FM** wave.

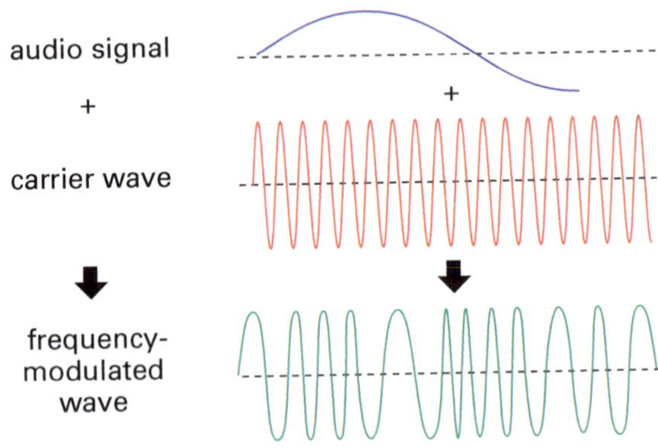

Figure 11.11 Modulated waves from an FM station. The amplitude is the same as that of the carrier wave but the frequency varies.

ISBN 978 1 4202 3736 8

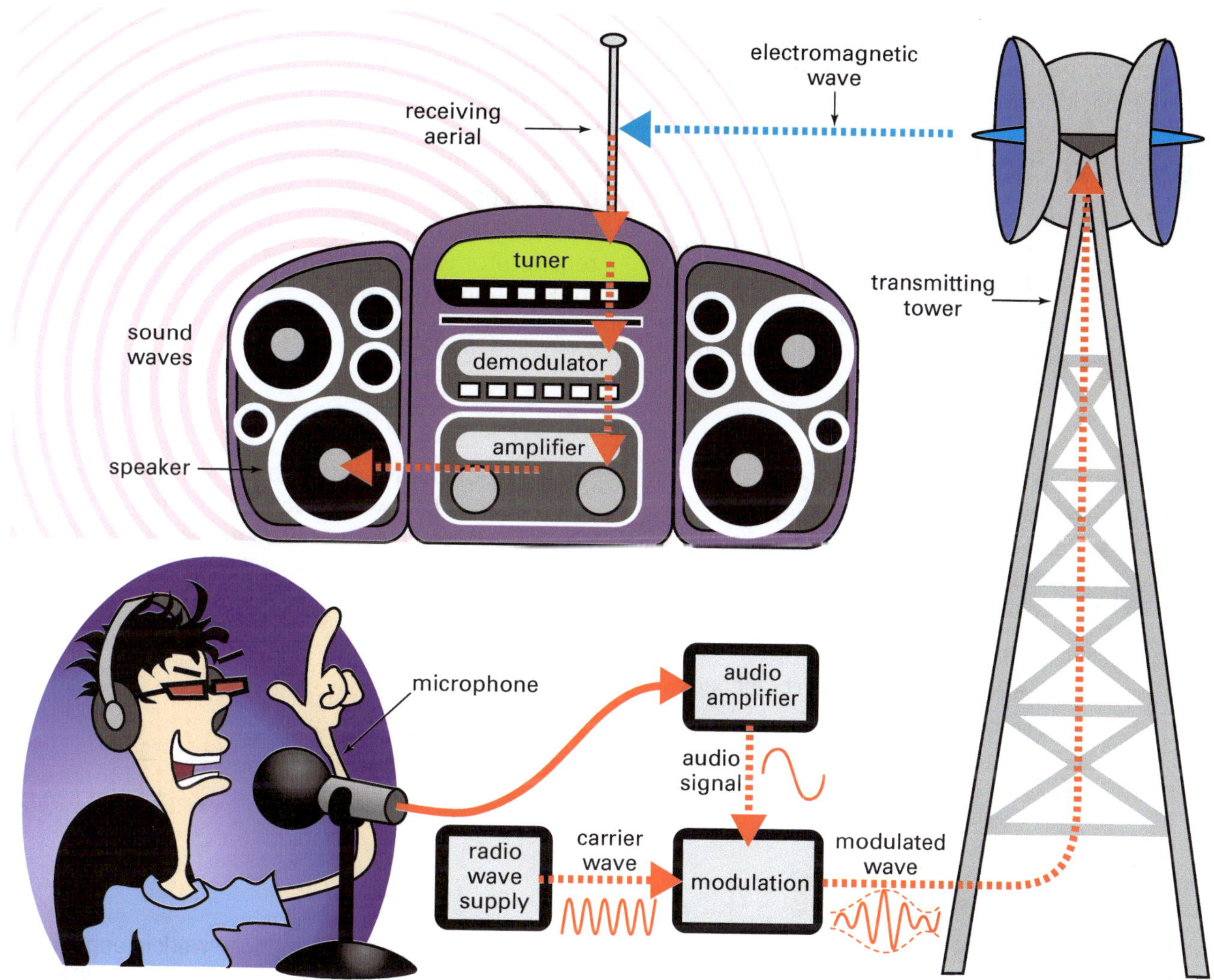

Figure 11.12 FM radio transmission

FM stations broadcast on a much higher frequency than AM stations. AM stations broadcast at frequencies between 520 kHz and 1600 kHz (1 kHz = 1 kilohertz = 1000 Hz). FM stations broadcast at much higher frequencies of between 87 MHz and 108 MHz (1 MHz = 1 megahertz = 10^6 Hz).

The steps in sending and receiving radio signals are shown in Figure 11.12.

Television transmission is much more complicated than radio because the signals have to carry both sound and pictures. The colour and the brightness of the TV picture are transmitted on a video signal that is similar to an AM signal. The sound is transmitted separately as an FM signal. A TV antenna picks up the signals and relays them to the TV set where the picture signals and sound signals are synchronised.

With cable TV, the signal travels directly to your TV without being transmitted.

Radio and TV reception

The very high-frequency waves transmitted by FM radio and TV stations suffer less interference and the sound quality is usually far superior to that of AM stations. However, these waves travel only in straight lines and are not reflected by the atmosphere. So if your antenna is behind a hill or mountain, your radio and TV reception will be poor. Another disadvantage of FM radio and TV waves is that they normally have a range of only

ISBN 978 1 4202 3736 8

about 100 km. With satellite TV, this problem is overcome by beaming the signal up to a satellite and then back down to your TV.

Digital radio

Since 2009, radio stations in Australia have been broadcasting digital radio, which is quite different from AM or FM. The audio signal is first digitised as shown in Figure 11.9 (page 271), and it is the resulting digital data that modulates the radio signal.

The big advantage of digital over AM and FM is that the quality is better. All radio signals weaken as they travel, and there is interference. However, with digital signals these effects can be corrected. Even if the incoming signals have added noise, it is still possible to tell the 1s from the zeros, as these are the only values the signal can have. A regenerator can then be used to restore the pulses to their original quality.

1 a The diagram below shows how a telephone network works. Select words from the following list to match the numbers in the diagram. You will need to use some words twice.

digital signals	sound waves
analog electrical signals	microphone
digital light signals	speaker
telephone exchange	

b Which of the numbered arrows could represent a distance of many thousands of kilometres? Explain your answer.

2 If you were given an incomplete key for the coded words on page 266, can you decipher the message?

✳ = T	❇ = I	▲ = S
❄ = E	■ = N	❏ = O

3 The CRO wave pattern below was made by sounds that were directed into a microphone.

a In which periods were there no sounds?

b Which sound was loudest?

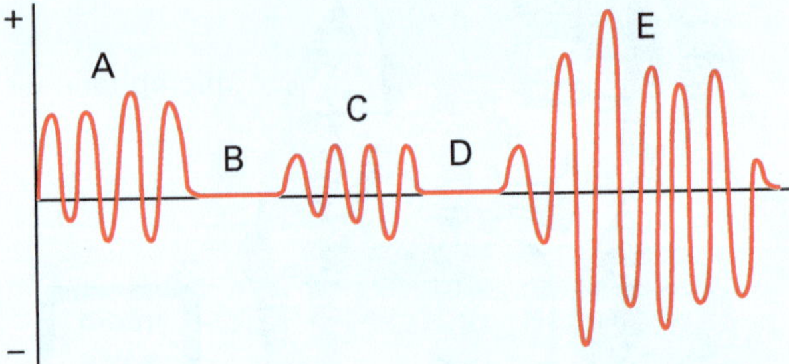

4 A speaker and a microphone work in opposite ways. Explain what this means. Use the words *sound waves, diaphragm, vibrate, coil, magnet* and *electric current* in your answer.

5 Construct an energy flow diagram that shows all the energy conversions that occur in Figure 11.2 on page 267. Use the words *sound energy, kinetic energy* and *electrical energy*. Below is the start and end of the diagram.

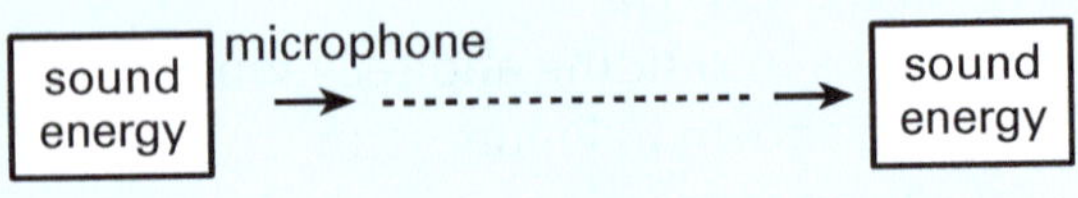

6 How are mobile phones different from fixed telephones in the way they transmit and receive voice messages?

ISBN 978 1 4202 3736 8

7 Radio station FUN broadcasts on a frequency of 690 kHz while station WIZ broadcasts on 94.5 MHz.
   a Which is the FM station?
   b What do kHz and MHz mean?
   c Which station do you predict could be heard 300 km away? Why?

8 How is it possible to receive live TV broadcasts of events taking place on the other side of the world?

9 Why do radio stations use a carrier wave to transmit their broadcasts? What is the difference between a carrier wave and a modulated wave?

CHALLENGE

1 a Why is feedback an important part of the communication process?
   b During a conversation, the person listening might say 'Yes', 'I see' or 'OK'. Explain why these responses are forms of feedback.
   c Give some examples of non-verbal feedback that might occur during a conversation.

2 Look at the telephone network diagram on page 269. Use this to construct an energy flow diagram that shows all the energy conversions that occur during a telephone conversation.

3 Use the code to decode the following message.

Code

A D G
B E H
C F I

K
J L
M

N Q T
O R U
P S V

X
W Y
Z

Message

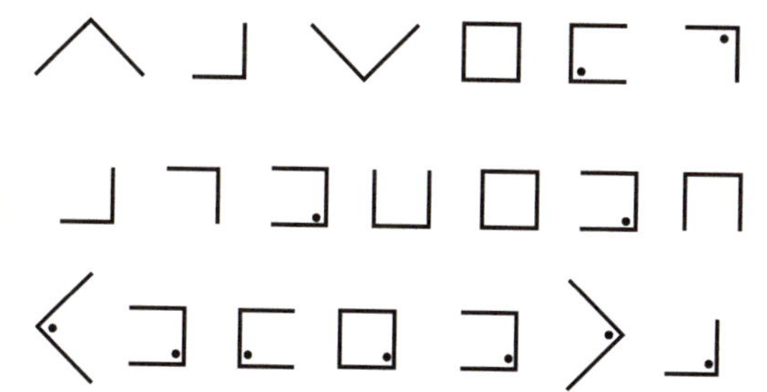

   a What would happen if the message was turned upside down?
   b Try to improve the code to overcome this problem.
   c What other problems arise using this code?

4 A digital signal is made up of eight bits having the values 11001011. Using Figure 11.9 on page 271 as a guide, draw a graph of this signal.

5 The graph below shows the current produced by a microphone when a person speaks into it. The vertical axis has both positive and negative values. The horizontal axis is measured in units of time.

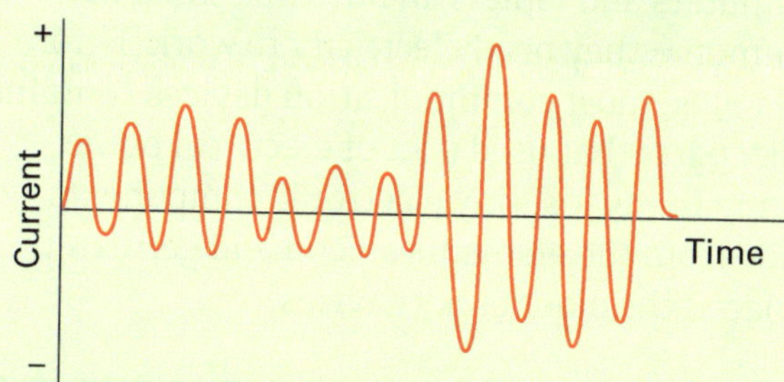

Using your knowledge of how a microphone works, explain:
   a why the current has positive and negative values
   b why the value of the current varies with time.

6 Encoding is changing one type of code to another. Decoding is the reverse process. Describe where signals are encoded during a telephone conversation and where they are decoded. In your description, include the type of device used in the processes. (Use the telephone network diagram on page 269 as a guide.)

7 Study the data in the table below.

Colour of light	Speed of light (m/s)	
	in air	in glass
red	2.988×10^8	1.983×10^8
blue	2.998×10^8	1.958×10^8

   a Describe the information in the table.
   b In optical fibres, pulses of laser light are transmitted in the glass core. If white light is used, suggest why the pulses of light become stretched out or 'smeared' after travelling through a very long optical fibre.

ISBN 978 1 4202 3736 8

c Suggest how the problem in b could be overcome.

8 The diagram on the right shows modulated radio waves from four stations.

a For each station, say whether it is AM, FM or digital. How do you know?

b Which AM station broadcasts at the highest frequency?

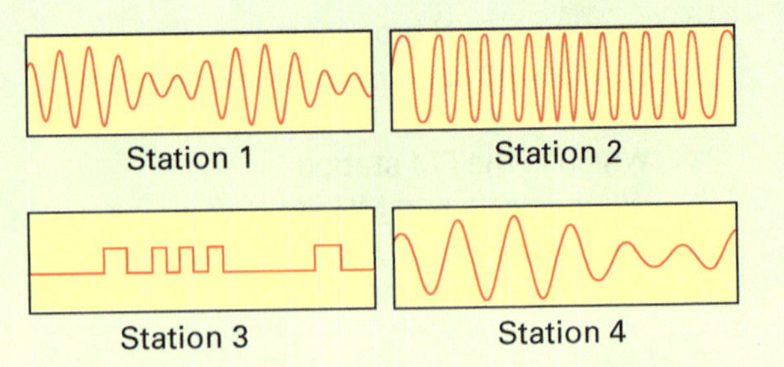

11.2 Electronics

Communication devices such as mobile phones, computers and tablets all have one thing in common—they need electricity to work. Before the 1950s, most communication devices contained bulky parts that used a lot of electrical power. During the 1950s, tiny electronic components made from the elements silicon and germanium replaced the older bulky devices.

Electronic components such as diodes, resistors and transistors are now very small and cheap to manufacture, and they also use very little electric power. This means that they can be operated for long periods using either mains power or batteries.

A great breakthrough in electronics has taken place with the development of the 'microchip' or *integrated circuit*. The microchip contains thousands of electronic components etched onto the silicon by a photographic process.

Figure 11.13 Microchips are the brains behind the many electronic devices we use every day, such as this calculator.

Resistors

Resistors control the amount of current in a circuit. The coloured stripes on the resistor indicate the size of the resistance. Resistance is measured in **ohms** (Ω).

Bands show size of resistance.

resistor symbol

Light-dependent resistors (LDR) are light- sensitive resistors. The resistance of the resistor decreases with the intensity of the light. That is, more current flows in bright light.

LDR symbol 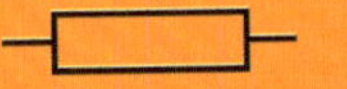

Thermistors are heat-sensitive resistors. The amount of current flowing through the resistor usually increases with the temperature.

ISBN 978 1 4202 3736 8

Diodes

Diodes allow current to flow in one direction only, making that part of the circuit a one-way street for the current. They are used in electronic circuits to stop current flowing in unwanted directions.

One end of a diode is marked with a band. This end should be connected to the negative side of the circuit.

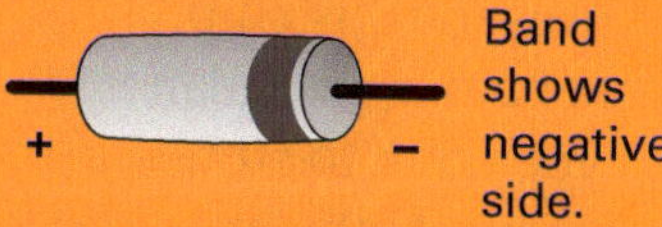

Band shows negative side.

diode symbol

Light-emitting diodes (LED) are electronic light bulbs—they glow red, green, yellow, orange or blue when electricity is passed through them. They are widely used in digital displays, traffic lights and tail-lights on cars.

LED symbol

Transistors

Transistors are devices that can act like switches, turning the current in a circuit on and off. They can also increase the size of the current. In this way, they act as amplifiers.

Transistors are made in different shapes, but each of them has three electrodes (legs), and the symbol remains the same.

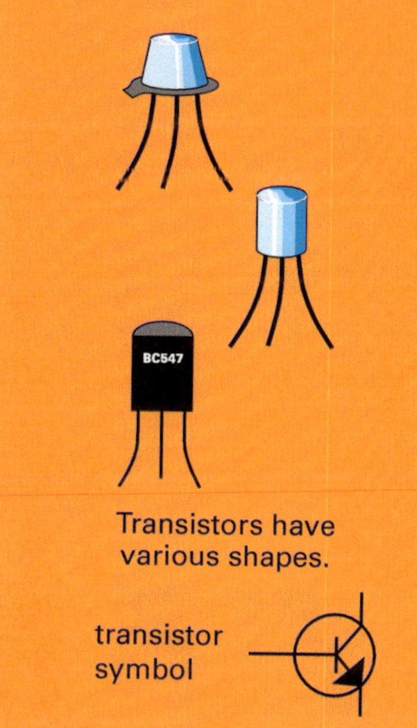

Transistors have various shapes.

transistor symbol

Capacitors

Capacitors are used in electronic circuits to store electric charge for a short time before allowing it to flow to other parts of the circuit. They are used to separate different parts of a circuit so that each can have a different current. They consist of two conducting plates separated by an insulating material called a dielectric.

The amount of charge that can be stored for each volt across a capacitor is called its *capacitance*. This is measured in farads (F), although microfarads (μF) are more commonly used in electronics.

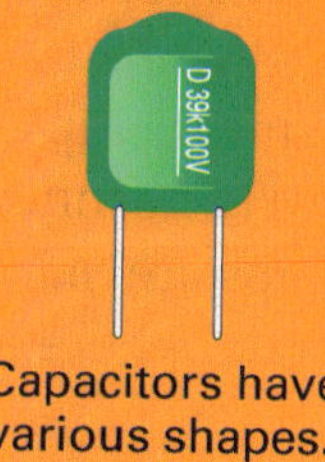

Capacitors have various shapes.

capacitor symbol

ACTIVITY

In Investigation 11.1, you will be using resistors in circuits. For this you will need to know how to tell the value of a resistor by using the coloured bands on it.

Your teacher will give you some resistors. Use the information below to work out their values in ohms.

The resistors you will use have four coloured bands on them. The code for the coloured bands is shown in the table.

Colour	Value	Colour	Value
black	0	green	5
brown	1	blue	6
red	2	purple	7
orange	3	grey	8
yellow	4	white	9

To read the code, hold the resistor with the gold or silver band on your *right*. Then start with the first colour on the *left*.

This resistor has a resistance of 560 Ω.

The colour of Band 1 gives the value of the first digit.
The colour of Band 2 gives the value of the second digit.
The colour of Band 3 tells how many zeros follow the first two digits.
The colour of Band 4 tells how precise the value of the resistor is.

ISBN 978 1 4202 3736 8

SCIENCE AS A HUMAN ENDEAVOUR

Microchips

Use the internet to research the following questions about microchips, either individually or in groups, and share your findings. Because there are so many questions, you may need to divide the task between different individuals or groups.

1 What are microchips? How are they manufactured?
2 How many transistors are there on a typical microchip?
3 Who invented the microchip? When was it invented?
4 How are microchips used?
5 Why are microchips used on credit cards?
6 Why are microchips becoming smaller?
7 What is Moore's law, and what does it have to do with microchips?
8 What are nanochips? How big are they?
9 Microchips are manufactured in special cleanrooms where the workers wear special suits, masks, caps and gloves (see Figure 11.15). Suggest a reason for this.
10 Microchips are often implanted under the skin of pets. Why?
11 In 2009 Professor Stan Skafidas from Melbourne developed a 5 mm square chip that can transmit data through a wireless connection at high speed over distances up to 10 m. Suggest uses for this chip.
12 A microchip the size of a rice grain can now be implanted under your skin to store all your medical information. Ambulance and other medical personnel can use a scanner to retrieve your medical information, enabling them to treat you more quickly and possibly save your life.
   a Discuss other ways in which human microchip implants could be used.
   b What are the pros and cons of this technology?

Figure 11.14 Fitting a microchip into a circuit board

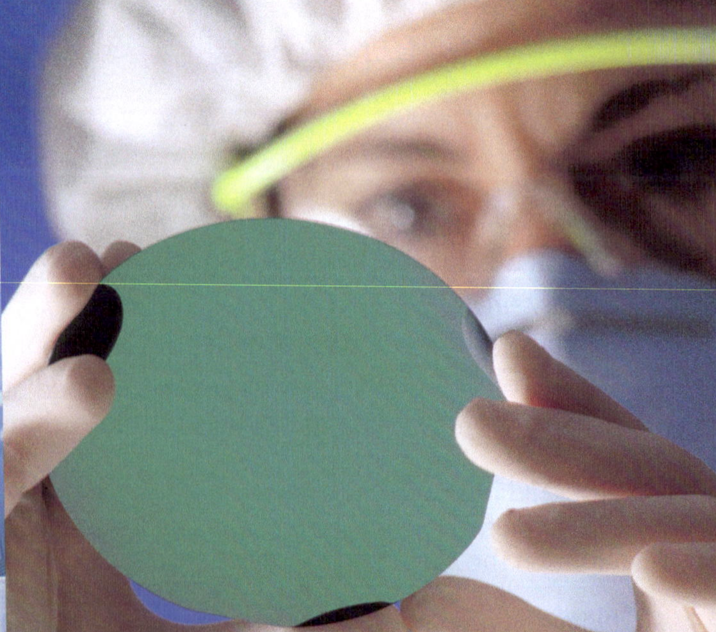

Figure 11.15 Many microchips can be made from this thin wafer of pure silicon.

ISBN 978 1 4202 3736 8

INVESTIGATION 11.1 Electronic circuits

Aim

To set up circuits using electronic components.

Materials

- resistors (1 watt) 10 Ω, 22 Ω, 56 Ω, 390 Ω, 10 000 Ω
- diode (1N4002 or similar)
- light-emitting diode
- light-dependent resistor (e.g. ORP12)
- switch
- ammeter (e.g. 1 A range) or multimeter
- power pack
- 6-volt torch bulb and socket
- 4 connecting wires with alligator clips
- two 10 cm × 10 cm pieces of cardboard, 5 drawing pins and some thin, bare wire
- clear adhesive tape

Risk assessment and planning

- Read through Part A and describe to your partner what you have to do. Swap roles and do the same for Part B (which itself has two parts).
- What precautions are necessary when using a power pack?

PART A Resistors

Method

1 Use the clear tape to fix the four lower value resistors to one of the pieces of cardboard. (Alternatively, you could use a 'breadboard'.) Write the value of each resistor next to it.

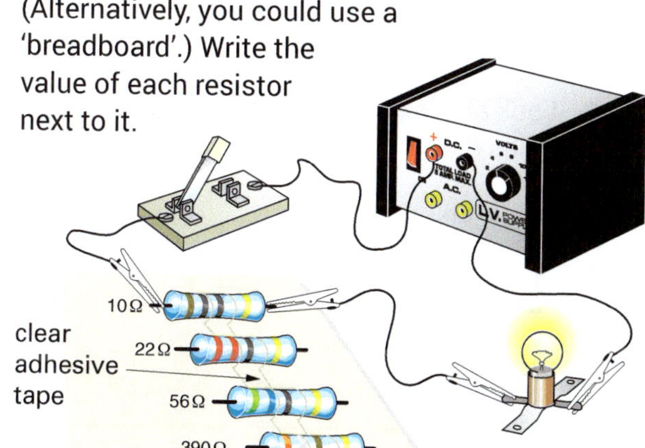

2 Connect up the circuit as shown bottom left. Set the power pack on 6 volts DC and connect each resistor in turn.

Observe the glow of the light bulb for each resistor. Record your observations.

3 Take the light bulb out of the circuit and replace it with an ammeter.

Note: Remember to connect the positive (red) terminal of the ammeter to the positive side of the power pack.

4 In turn, find the current flowing through each resistor.

Record your results in a table.

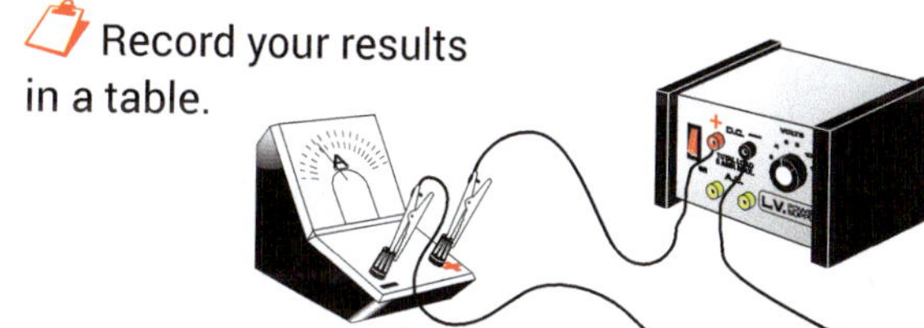

Discussion

1 Write a generalisation linking the resistance to the glow of the light bulb.

2 Write a generalisation linking the resistance to the current flowing in the circuit.

3 Predict the effect of a very large resistance (10 000 Ω) on the glow of the light bulb. Then test your prediction.

4 Why do the resistors heat up when you leave the power pack on for a while? Suggest why higher value resistors heat up more.

PART B Diodes

1 Set up the circuit as shown. Make sure you connect the banded end of the diode to the *negative* side of the power supply.

ISBN 978 1 4202 3736 8

The circuit diagram for the set-up is shown below. In this circuit, the banded end of the diode is shown by the vertical line in the symbol, and this is connected to the negative side of the power pack.

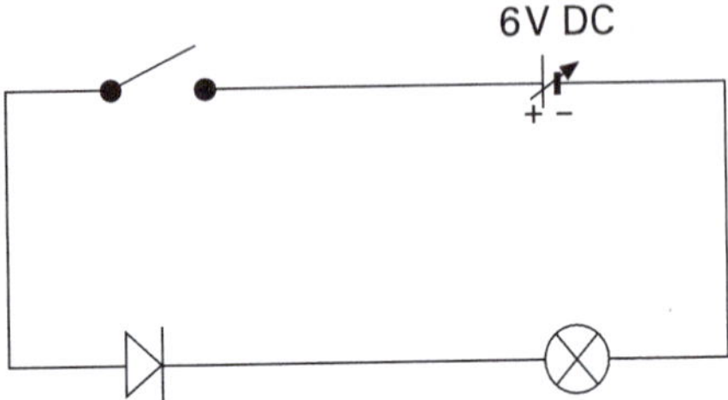

2 Set the power pack to 6 volts DC.

Record your observations when you close the switch.

3 Disconnect the diode and turn it around so that the banded side is connected to the *positive* side of the power supply.

Record what happens this time.

4 To make a puzzle for your partner, push five drawing pins into the second piece of cardboard as shown. Connect a diode between two of the pins. Then connect some bare wire between the other pins.

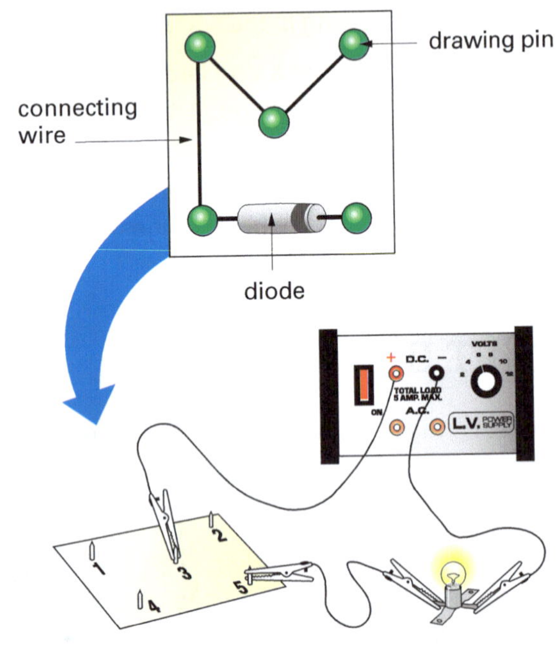

5 Turn the cardboard over and number each of the pins without your partner seeing. Now ask your partner to use the test circuit to find out where the diode is connected and which is the negative end.

Ask your partner to explain how they solved the puzzle.

6 Set up the circuit below containing a light-emitting diode (LED). The 390 Ω resistor is used to reduce the current in the circuit so that the LED does not 'burn out'.

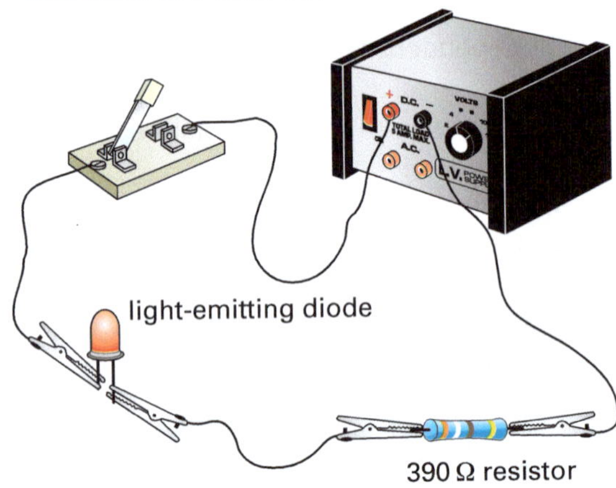

7 Experiment with the LED to find out:

a whether the LED allows current to flow in one direction only

b if the short leg of the LED is the positive or negative side

c if a current that lights an LED will light a torch bulb.

Write a report of your findings.

Discussion

1 Draw circuit diagrams using the correct symbols for the circuits in Steps 3 and 6.

2 Does an LED look brighter when viewed from the top or from the side?

3 Look at the circuits below. Will the light bulbs glow?

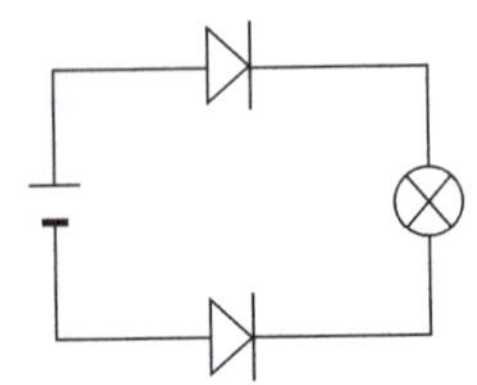

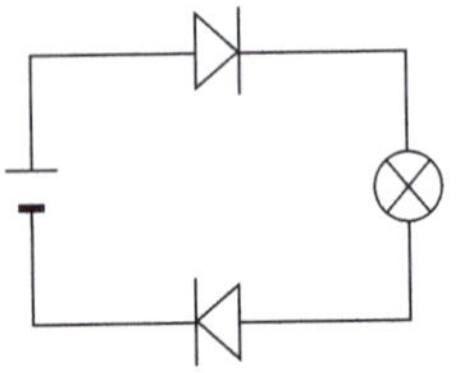

ISBN 978 1 4202 3736 8

SCIENCE AS A HUMAN ENDEAVOUR

Jim West and the electret microphone

Jim West was born in 1931 to African-American parents in Virginia, United States. Jim said that 'in those days in the South, the only professional jobs that seemed to be open to a black man were a teacher, a preacher, a doctor or a lawyer.' So his parents were disappointed when he chose to study physics instead of medicine. He went to university and began working as an intern at Bell Labs in New Jersey during his summer holidays. He joined the company full time in 1957.

A new type of microphone called a *condenser microphone* had been invented at Bell Labs in 1916. It is essentially a capacitor (see page 277) with two plates with a voltage between them. One of the plates is made of very light material and acts as the diaphragm. When sound waves hit the diaphragm, it vibrates. This changes the distance between the plates and therefore changes the capacitance, creating a small electric current. However, these microphones were not suitable for widespread use in telephones because they were expensive and required a large battery. So West and a colleague were given the task of inventing a new technology to produce a microphone that was small, high-quality and cheap to manufacture.

After several years of experimenting, West and his colleague patented an *electret microphone*. It uses a thin plastic film with a metallic coating.

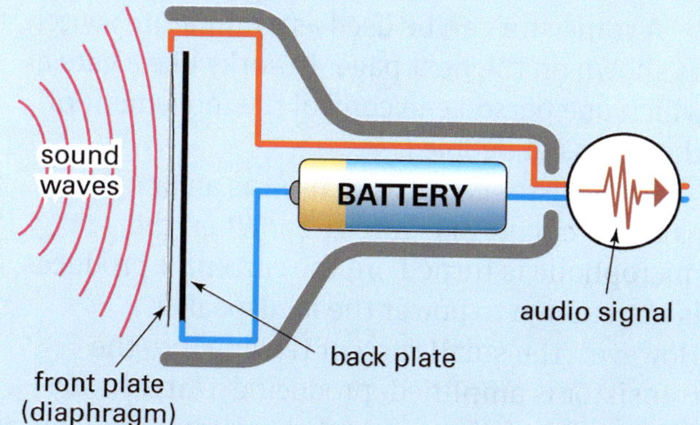

Figure 11.16 A condenser microphone

Figure 11.17 Jim West invented the electret microphone in 1962.

When exposed to a strong electric field, the film retains its electric charge, and doesn't need a battery. These electret microphones can be made very small and are now in virtually every telephone in the world.

Jim West is still working and says 'My hobby is my work. I have the best of both worlds because I love what I do'. He is active in a program aimed at encouraging more women and people from minorities to enter the fields of science, technology and engineering.

Questions

1. Why were Jim West's parents disappointed when he decided to study physics?
2. Why is the electret microphone suitable for use in mobile phones?
3. How is an electret microphone different from a condenser microphone?

ISBN 978 1 4202 3736 8

Semiconductors

Diodes and transistors are made from materials called **semiconductors**. These materials, which include the elements silicon and germanium, have properties in between conductors and insulators.

Silicon is the most important semiconductor material. It is made from sand (silicon dioxide), and it is cheap and easy to manufacture in pure form. In pure form, silicon does not conduct electricity very well. But when very small amounts of another substance, such as arsenic or boron, are added (this process is called *doping*) the silicon conducts electricity.

An atom of silicon has four outermost electrons. An atom of arsenic has five electrons, one more than silicon. When silicon is doped with arsenic and wires from a battery are placed at each end of the crystal, a current flows. The fact that the extra electrons in the arsenic atoms are relatively free to move causes the doped crystal of silicon to conduct electricity. This type of doped semiconductor is called *n-type* or negative type because of the extra electrons.

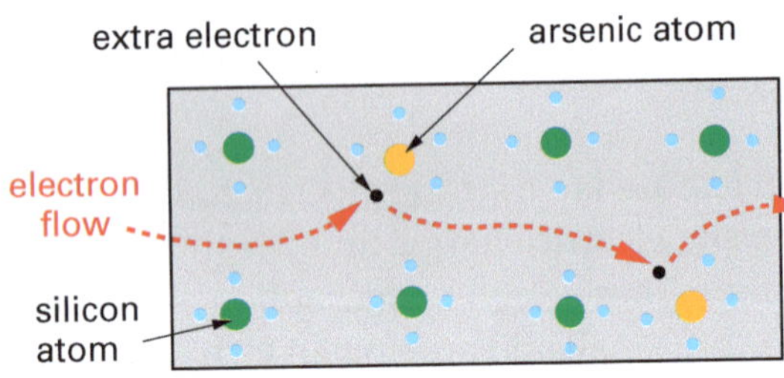

Figure 11.18 How an electric current is carried through an n-type semiconductor by mobile electrons

Boron has only three outermost electrons, one less than silicon. When silicon is doped with boron, the crystal also conducts electricity. It seems that the boron atom creates an electron space or 'hole' into which electrons from the silicon can flow, causing an electric current. This type of doped semiconductor is called *p-type* or positive type.

Diodes

A diode is made by placing an n-type crystal next to a p-type crystal. When this is connected in a circuit, the 'extra' electrons in the n-type crystal can jump across to the holes in the p-type crystal. However, if the battery terminals are reversed, the electrons cannot jump back in the other direction. This is why diodes carry current in one direction only. (By convention, the arrow in the diode symbol points in a direction *opposite* to that in which the electrons flow.)

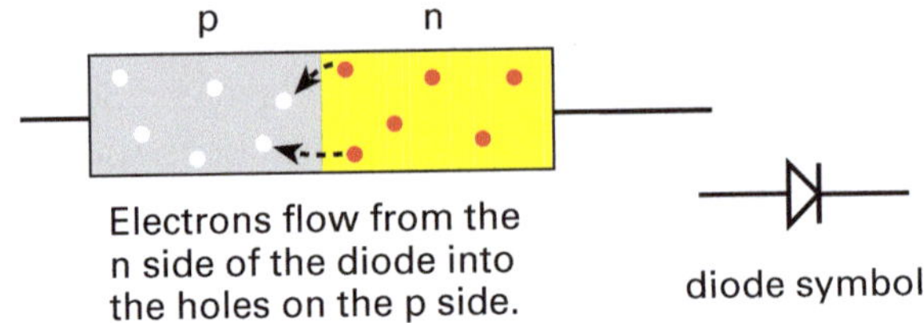

Figure 11.19 How a diode works

> **Current direction**
>
> In Figure 11.19 you will notice that the electrons flow in the opposite direction to the arrow in the symbol. When scientists first studied electricity, they thought it was a flow of positive charge—from positive to negative. It was a long time before they discovered that it was, in fact, negatively charged electrons that were moving. By this time everyone had been thinking about current flowing from positive to negative for so long that it was impossible to change. This flow from positive to negative is called 'conventional current'.

Transistors

A transistor is made of three pieces of semiconductor crystal sandwiched together. This is why transistors have three legs (electrodes).

A transistor can be used as a miniature switch, as shown on the next page. It works like a gate at which one person can control the movement of thousands of people.

A transistor can also be used as an amplifier, as in the circuit in Figure 11.22. When the microphone is turned on, the current it produces is not enough to power the loudspeaker. However, the small current flowing into the transistor is amplified, producing a larger copy of the original signal from the microphone. This amplified current is large enough to operate the loudspeaker.

ISBN 978 1 4202 3736 8

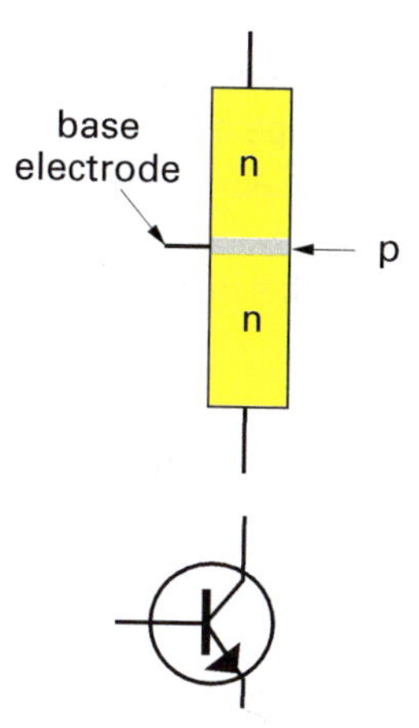

Figure 11.20 An n-p-n transistor and its symbol

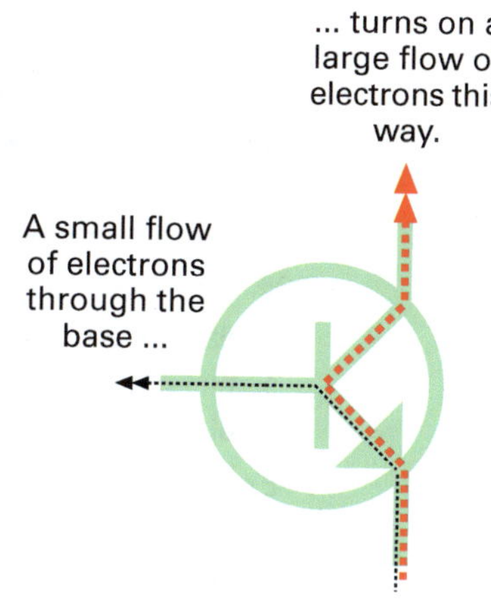

Figure 11.21 How a transistor works. When a small voltage pushes a current through the base leg, a large current can flow through the other two legs.

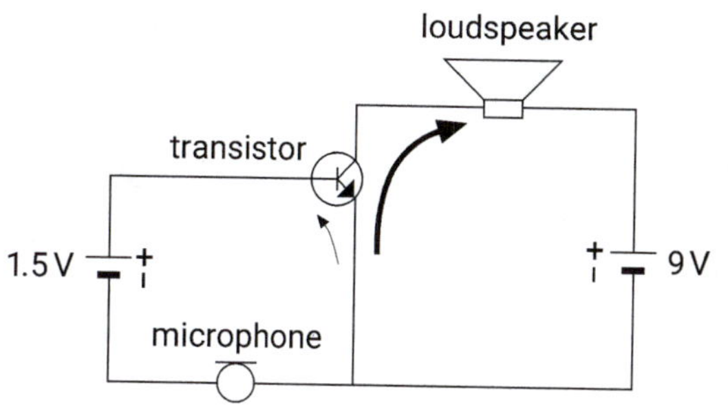

Figure 11.22 In this circuit a transistor amplifies a small microphone current to produce a large current in the loudspeaker.

1 What do the following symbols represent?

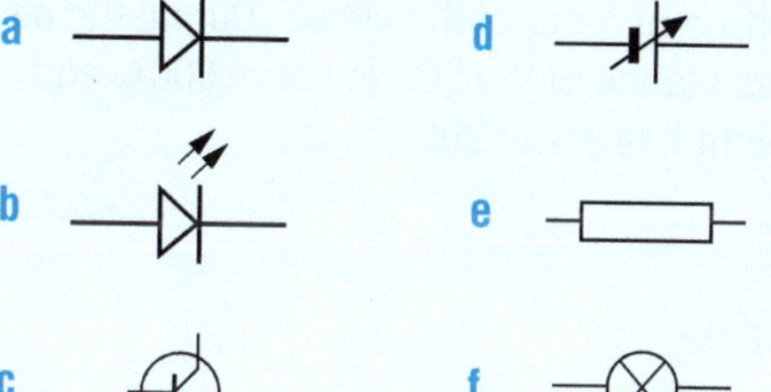

2 How is a resistor different from a diode? In what units is resistance measured?

3 Which one of the following circuits contains a battery, one resistor, one transistor and two diodes?

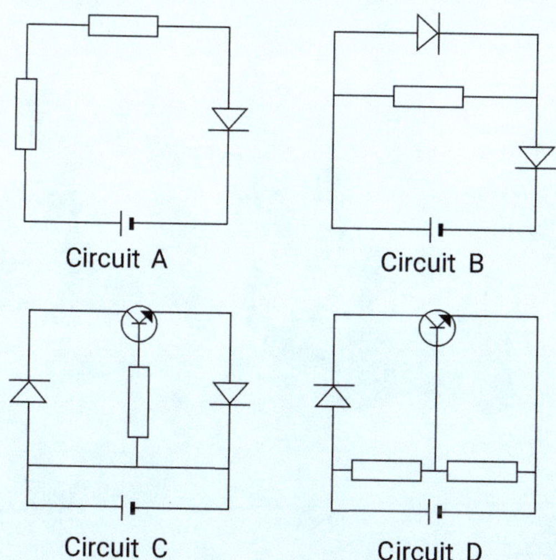

4 List the equipment you would need to build the following circuit.

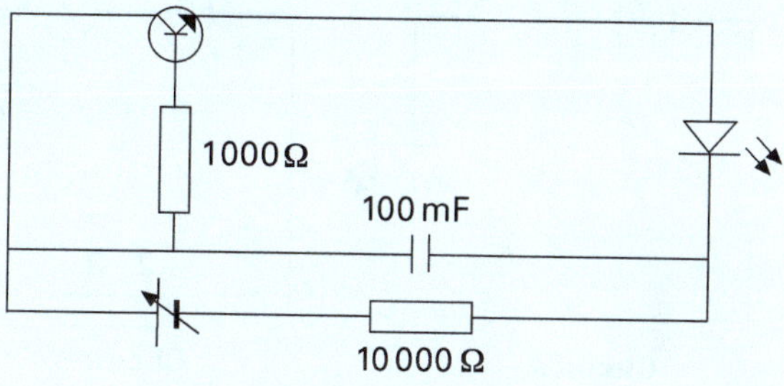

5 In which of the following circuits would you expect the light bulb to glow?

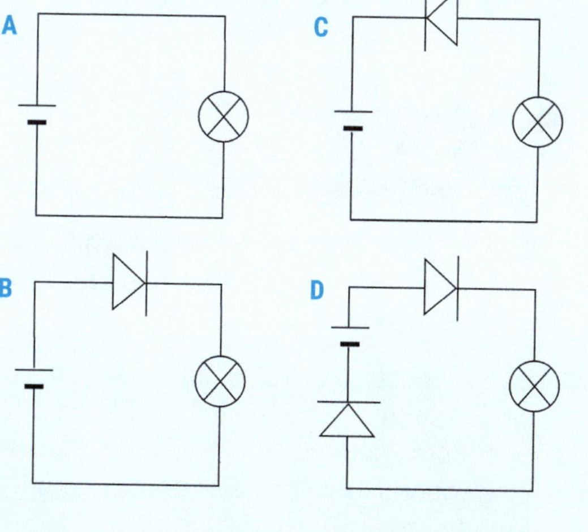

ISBN 978 1 4202 3736 8

CHALLENGE

1 Use the resistor code table on page 277 to find the values of the following resistors.

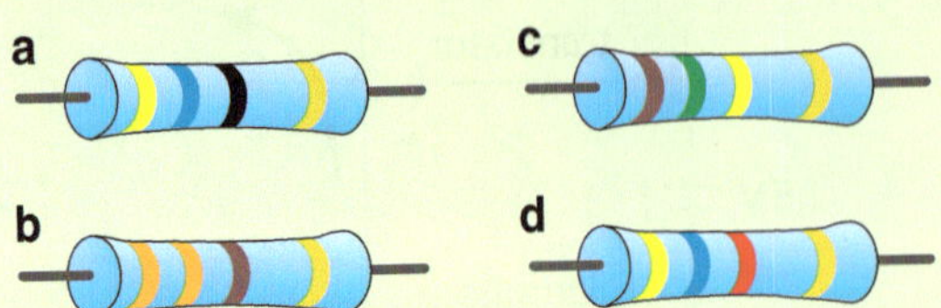

2 Why were very small portable radios not available in the 1930s?

3 Security beams in the doorways of shops sometimes use a light-dependent resistor. How do they work?

4 a In which of the two circuits below would the LED glow more brightly? Explain.

b If the LED in the circuits below has a resistance of 50 Ω, find the current flowing in each circuit.

c Suggest why LEDs, rather than light bulbs, are used in electrical appliances.

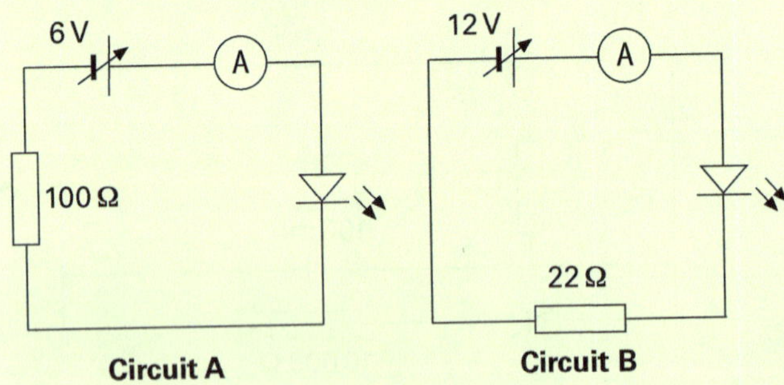

5 Suppose you are making an electronic fire alarm. Which electronic component could you use to detect the fire? Explain.

6 An undoped semiconductor such as pure silicon will not conduct electricity at low temperatures. However, as the temperature rises the ability of the silicon to carry current increases. Suggest how electronic thermometers might use this material to measure temperature.

7 a Explain in terms of electrons how semiconductors differ from conductors and insulators.

b Explain how an n-type semiconductor differs from a p-type one.

8 The circuit below can be used to switch on a light when the sun goes down.

a Explain how a light-dependent resistor (LDR) works.

b Explain what happens in the circuit during the day and when the sun goes down.

Hint: Consider the effect the two resistors and the LDR have on the electric current in the different parts of the circuit. During the day, the resistance of the LDR is about 500 Ω and at night it is about 200 000 Ω.

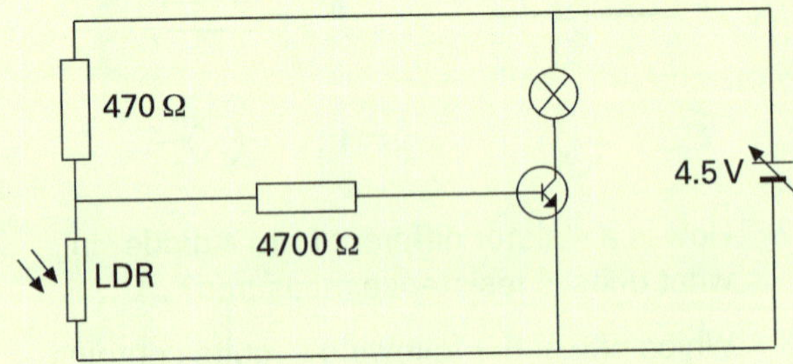

(You might like to build this circuit if your teacher can organise the components.)

EXPLORE

1 Use library resources to find out what microchips or integrated circuits are. Where are they used?

2 Use an electronics kit to build a simple everyday device such as flashing lights, a siren, a radio or a Morse code sender. You simply follow the instructions to put the electronic components together to make the device.

ISBN 978 1 4202 3736 8

11.3 Robotics and control

Home automation

Welcome home! As you step onto the verandah, the light turns on and a touch pad next to the door lights up. You place your thumb onto the pad, a green light blinks and the door swings open. You step inside and as you walk from the front door through the living area to the kitchen, the lights switch on along the way. You notice that although it was cold outside, it is warm inside. On the way home you turned the heating on using your mobile control app, and also closed the curtains. With a clap of your hands the TV turns on, and you use voice recognition to change the channel to your favourite program.

In the home, equipment and appliances can be operated and controlled remotely by using an app on a smart phone or tablet, or automatically through sensors. Usually this is done through the internet, as appliances and devices become connected to the 'internet of things'. A home automation system may control the opening of blinds, windows or vents, operate fans, coolers or heaters, or turn on ovens and entertainment equipment. A simple system may be controlled by a simple timer or there may be a programmed pattern to follow. A more elaborate system will have a computer that uses timers and programs, but will also respond to feedback from thermostats (temperature sensors), infrared motion sensors and light sensors.

A home automation system is like the nervous system that you learnt about in Chapter 7. The body has receptors that detect a change or stimulus, a central nervous system to process the information, and effectors that create a response or change. An automated house has sensors to detect change, a computer control unit or processor, and various devices that respond by changing something as instructed.

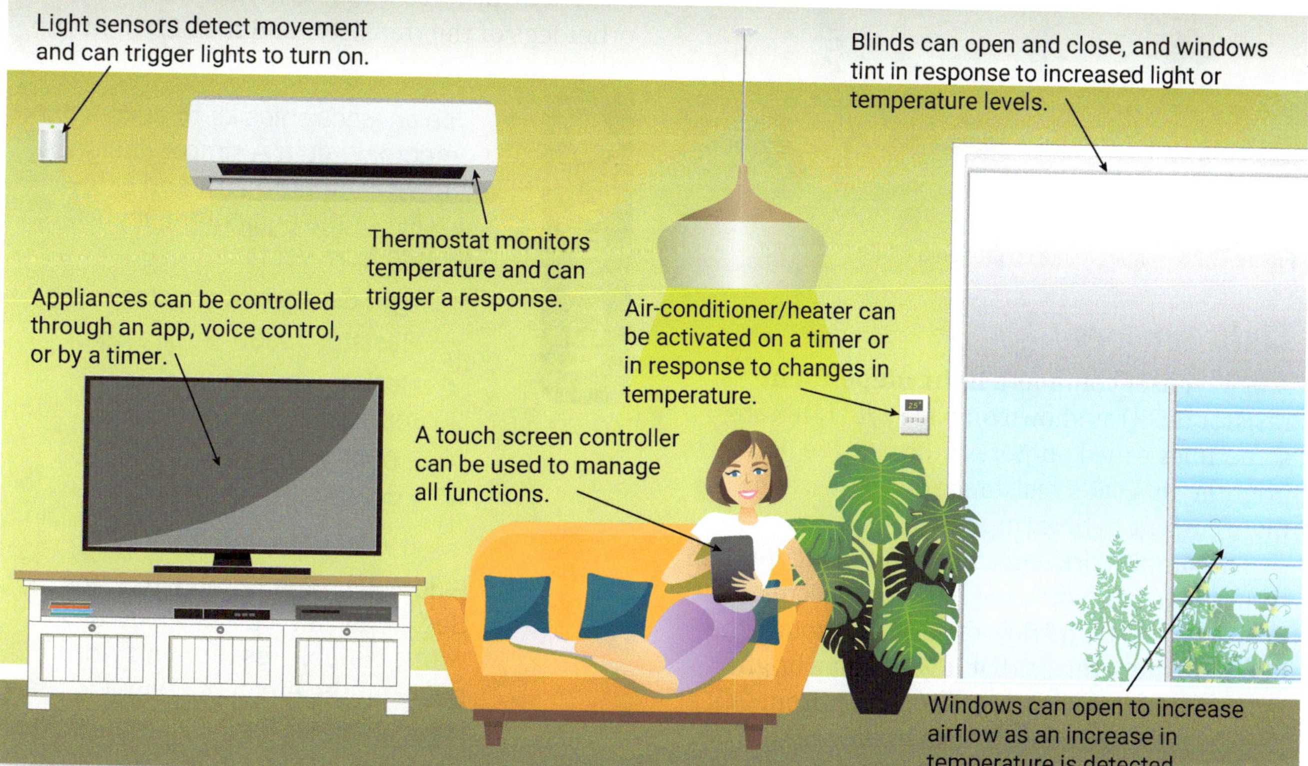

Figure 11.23 A home automation system can control many functions in the home, and in doing so can make a home much more energy efficient.

ISBN 978 1 4202 3736 8

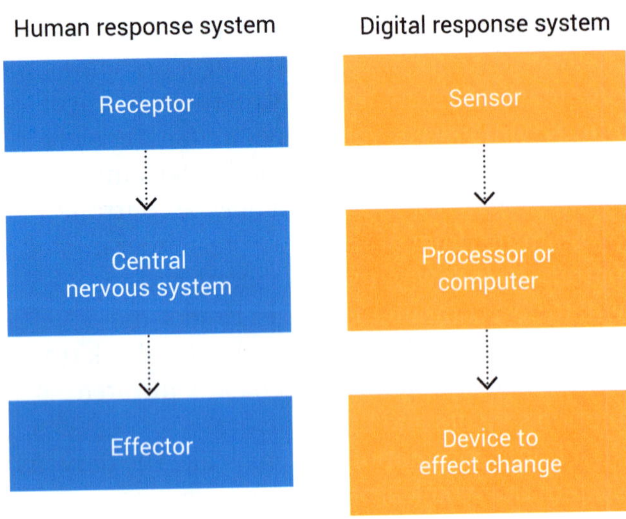

Figure 11.24 The nervous system is similar to an automated control system.

Figure 11.25 Smart home control on a tablet

Light sensors

A light sensor contains a **light-dependent** resistor (LDR) as shown on page 276. LDRs are sometimes called photoresistors. As the light gets brighter, an LDR's resistance decreases, allowing more current to flow through it.

The simple dark detector circuit in Figure 11.26 turns on an LED when the light level decreases. Resistor 1 limits the flow of current through the LED to a safe level, so that it will emit a bright light without overheating. Remember from page 277 that a transistor acts like a switch. This transistor won't switch on to let current flow through the LED until there is a small current flowing through the base leg.

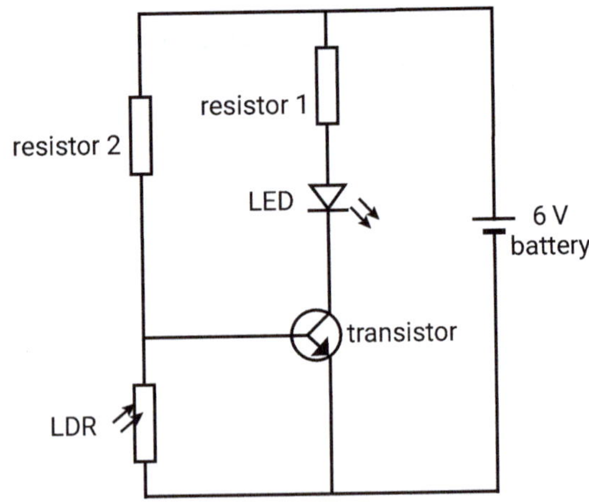

Figure 11.26 Automatic dark detector circuit

In daylight the LDR's resistance is low, so most of the 6 V provided is used up in resistor 2 and there isn't enough voltage left to switch on the transistor. As the light level decreases, the LDR's resistance increases. Now it uses up more of the voltage, while resistor 2 uses less. This increases the voltage across the transistor, creating a small current through the base leg. With the transistor switched on, current is able to flow through the other legs of the transistor and the LED lights up.

ACTIVITY

1. Use an electronics kit to build a dark detector circuit. A simple design is shown in Figure 11.26 and below is a list of the equipment you will require from your kit.
   - light dependant resistor
   - Resistor 1: 330 ohm resistor
   - Resistor 2: 100 000 ohm resistor
   - BC547 transistor
   - LED
   - 6 V DC power pack
2. Find out what a variable resistor is. See if you can replace Resistor 2 with a variable resistor and work out what the effect of changing the resistance is.
3. Research a *light* detector circuit and then try to build one.

ISBN 978 1 4202 3736 8

Motion sensors

Uses of motion detectors include:

- turning on lights when a person enters a room
- opening and closing automatic doors and gates
- turning water taps on and off and flushing toilets automatically
- activating alarm systems.

Passive infrared motion (PIR) detectors are the most widely used type in home security and lighting systems. They are passive as they only detect emitted infrared energy given off from people in the form of heat. They do not send out any signals—just wait for a change in the environment. If a change in heat is detected by a sensor it responds. PIRs work very much like photoresistors, changing their resistance as the energy detected changes, to produce an electric current.

Some motion detectors are active, sending out microwave or ultrasonic (ultra-high frequency sound) pulses, and measuring their reflection off objects. If the amount of energy reflected changes, it indicates something is moving in the area.

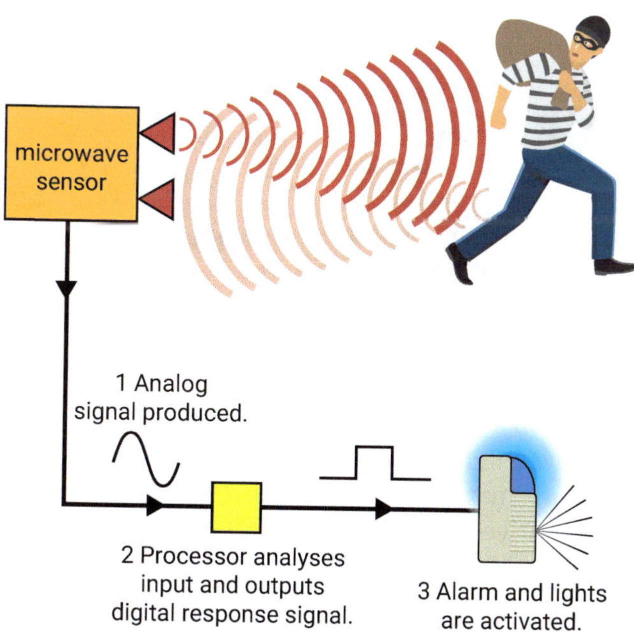

Figure 11.27 An active motion detector sends out an energy pulse, and measures any change in the energy that bounces back to the receptor.

Temperature sensors

In an automated house, temperature sensors may be used directly to turn on a cooler or fan and lower blinds if it gets too hot, or they may turn on a heater when the temperature drops. Temperature sensors are often incorporated into thermostats, which are designed to keep the temperature stable at a set temperature. They operate on a negative feedback system, which you learned about in Chapter 7 when you studied the human body.

In a thermostat, the original stimulus is the higher temperature detected by the temperature sensor. The processor responds to turn on the air conditioning to lower the room temperature. The opposite can occur, in that if a lower temperature is detected, the processor responds to turn on the heater to *increase* the room temperature.

One type of temperature sensor is the thermistor. A thermistor changes its resistance as the temperature changes. The resistance of the thermistor decreases in a predictable way as the temperature increases. The level of resistance at any point basically represents the temperature, as shown in Figure 11.28.

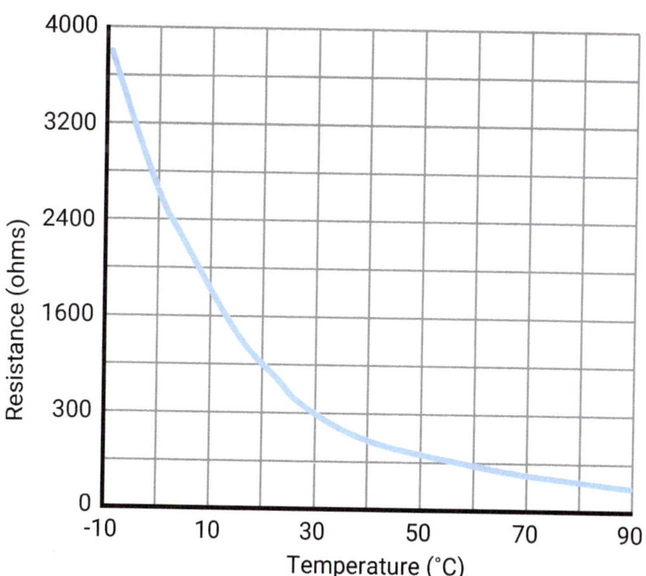

Figure 11.28 The resistance of a thermistor changes with temperature in a predictable pattern.

ISBN 978 1 4202 3736 8

SCIENCE AS A HUMAN ENDEAVOUR

Lab on a chip

Biochips are miniaturised sensors on a small piece of glass, plastic or other material, often no larger than your thumbnail. These sensors are capable of running diagnostic tests to detect various biological chemicals or other biological agents. Their uses include detecting drugs in sport and bioterrorism agents, diagnosing diseases and analysing DNA, among other things. Biochips are cheap, fast and allow for hundreds of chemical reactions to occur at the same time. Biochip sensors operate at the molecular level so require only very small samples for testing.

Biochips have led to the development of 'lab on chip'. Imagine that you go to the doctor and they take a tiny drop of blood. The blood is placed at various locations on the 'lab on a chip', and each location does a different screening test. One checks for the presence of viruses, one tests your iron levels, a third tests your white blood cell levels and another checks for cancerous cells. After a couple of minutes the results are in, and the doctor can diagnose your condition. No more waiting three days for a blood test to be sent off to the lab to get the results. One of the major applications will be diagnosing medical conditions in countries where access to medical treatment and diagnosis is currently not available for many people.

As biochip technology improves, we will start to see biochips being used much more regularly in many areas of science and medicine.

Questions

You may need to do some further research into biochips to answer the following questions.

1 Explain what a biochip is. Find out what it is made of.
2 Describe two current uses of biochips.
3 Outline three benefits of biochips in medical uses.
4 Can you think of or find out about some applications of biochips not mentioned here? Describe these applications and their benefits.

Figure 11.29 Lab on a chip

Robotics

Robots are automated systems based on digital and machine technology that can respond to their environment or undertake tasks. Robots are already all around us—drones, vacuum cleaners, self-driving cars, in factories that pack boxes or weld cars, and as space probes that you learnt about in *ScienceWorld* 7. The applications are numerous.

Robots have been used in industry for many years to complete repetitive tasks. In this situation a robot may do the same thing over and over and reduce the number of errors that humans may make, while relieving humans of a repetitive, boring or unsafe tasks. At the docks in Melbourne and Sydney where ships are unloaded, autonomous vehicles collect and move the large shipping containers around to various locations using navigation systems and various sensors, but no drivers. In future you may have a robotic house cleaner or butler that can provide you with a personalised service!

ISBN 978 1 4202 3736 8

What is a robot made of?

Robots have similar requirements to humans. The key elements of a robotic system are:

- *Body*: A robot needs a frame, or skeleton, to attach all its parts to.
- *Power source*: Energy is required to run systems and controllers, motors and sensors. Usually in a robot this energy comes from rechargeable batteries, although it could be solar power or nuclear power in a space probe or other remote robot that must be self-sufficient.
- *Sensors*: A sensory system for a robot will depend on what its task is. Sensors may include motion, light, touch (pressure), speed, tilt or gravity, distance, image (cameras) and sound (including for speech recognition). For a robot to behave like a human, it needs human-like sensors.

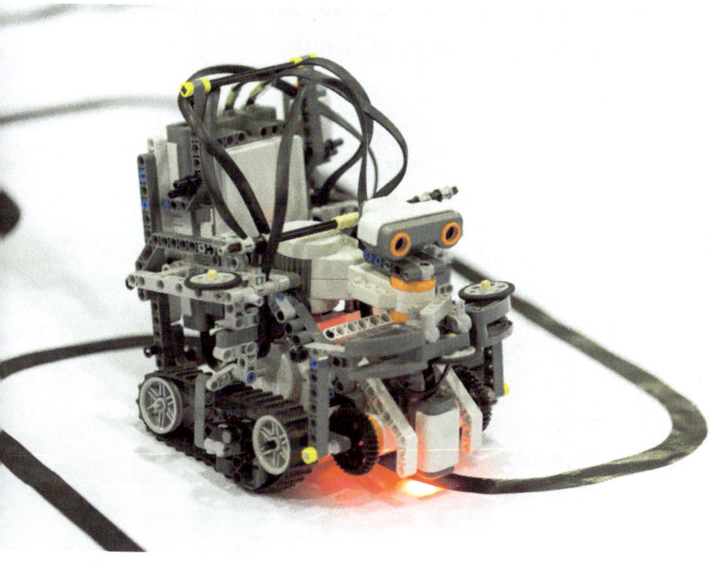

Figure 11.30 This Lego robot navigates and follows a line using a light sensor.

- *Control system* or *processor:* Like our brain, the processor in a robot is required to process incoming information from sensors, and make decisions on what to do next. Some robots, however, do not make decisions. The controller contains a pre-programmed behaviour or activity pattern that is followed. This happens with a car-welding robot that has a repetitive task. Although robots have a computer, robots are different from normal computers in that they have a body.
- *Actuators*: These are like muscles (effectors) in a human. There are many types of actuators, the most common being electric motors that spin. Linear actuators move in and out in a straight line rather than spinning. Their advantage is that they can move quickly and are very strong. Linear actuators are used widely in industrial robots, and are often powered by hydraulics or compressed air. Other actuators can be elastic or springs. Air-filled muscles have also been invented. When filled with air, the muscle shortens or contracts and when used in an arm or leg, these air muscles resemble a human muscle very closely. Actuators can also include speakers, allowing a robot to talk.

Figure 11.31 These actuators, sometimes called servos, have a motor that spins.

- *Locomotion*: 'Locomotion' means movement. There are many unusual ways in which robots have been designed to move. Many of these resemble the locomotion styles of animals. Some examples include:
  - walking on two, four or even eight legs like a spider; many legs can make it easier for a robot to navigate uneven ground
  - rolling in various arrangements, e.g. wheels
  - snaking
  - swimming (underwater robots such as deep-sea exploring craft)
  - climbing
  - flying, such as with the many drones that are currently available.

ISBN 978 1 4202 3736 8

Robots and control

Robots can be classified according to their levels of control:

- *Direct control*: If you have flown a drone, you know that you have full control. If you make a mistake and crash, it is all your fault. With direct control an operator controls the direct movements or activity of the robot. This is how a robotic operating theatre also works, with the surgeon controlling the robotic arms that undertake the operation. Many space probes are controlled this way, as operators are nervous about handing control to a robot that costs millions of dollars and is millions of kilometres away. If the robot makes a poor decision on its own, it cannot be rescued, so operators maintain control by sending programs that the robot follows to complete its mission.
- *Operator assistant or intervention*: At this level of control, the operator will assign tasks to a robot, but the robot will decide how best to achieve them. This is how some robotic landers operate on other planetary bodies. For example, the operator will send a lander instructions that it needs to move to a new location. The robotic lander will calculate the best way to get there, and use its sensors to detect and avoid obstacles along the way.
- *Fully autonomous with a fixed pattern*: This is how most industrial robots work. A robot that packs boxes is fully autonomous, but the programmed task is repetitive, allowing the robot to work for long periods of time without error or human intervention.
- *Fully autonomous*: The robot will complete all tasks on its own without human intervention. This is the most complex robot, and must be capable of problem-solving or decision-making to ensure it can successfully complete its mission or task; for example, an autonomous car is fully autonomous.

Figure 11.32 All three generations of Mars rovers have used wheels for locomotion, with legs that can move up and down to go over uneven terrain.

ISBN: 978 1 4202 3736 8

Figure 11.33 A drone is an example of direct control, while an industrial robot is fully autonomous with a fixed pattern.

Humanoid robots

There has been a lot of research into human-like, or humanoid, robots. Will we one day walk alongside humanoid robots as friends or co-workers? It is likely this will be the case as robots advance in their ability to sense, respond, communicate and move. New technologies are allowing robots to move and look more like humans than ever before.

As robots become more and more like humans they may make some people feel uncomfortable. The robots look human, but are not really human, and there appears to be something strange about them. This effect is called the 'uncanny valley', a theory that as robots become more like humans, at some point they will cause feelings of eeriness or even revulsion. Research has shown that humans feel more comfortable with a humanoid robot that is not too human looking in appearance, or is so human that you cannot tell it is a robot. It also makes people feel uncomfortable if you mix robotic and human characteristics. For example, a human with a robotic voice, or a robot with a human voice is not as easy to like as a robot with a robotic voice.

Figure 11.34 Humanoid robot

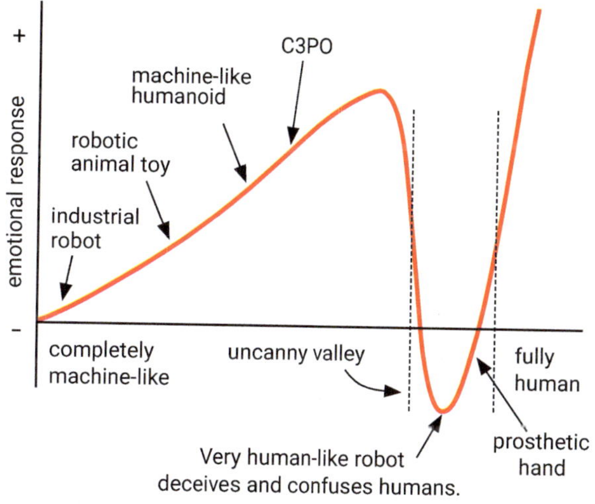

Figure 11.35 Graph of the uncanny valley

Isaac Asimov, a science fiction author proposed three rules of robotics that are widely talked about today. These rules, basically designed to protect the human race from robots, are:

1 A robot may not injure a human being or, through inaction, allow a human being to come to harm.
2 A robot must obey the orders given it by human beings except where such orders would conflict with the first law.
3 A robot must protect its own existence as long as such protection does not conflict with the first or second laws.

ISBN 978 1 4202 3736 8

SCIENCE AS A HUMAN ENDEAVOUR

Artificial intelligence (AI)

People think of **artificial intelligence** (AI) as being related to robots learning to communicate and work with humans, in human-like ways. But AI is when intelligence is shown by a machine. AI is already all around us and you will have experienced it many times in your life. Here are some of the ways you will have experienced it:

- A search engine delivers results that are tailored to your needs, because it has learnt from previous searches what you are interested in.
- Personal digital assistants like Siri (Apple) or Cortana (Microsoft) find information for you based on a verbal or typed request. These assistants collect information about you and learn about your preferences to help give you the most relevant information they can.
- Autonomous or smart cars can self-drive. Google has developed a system allows the car to learn to drive from experience, much like a person does.
- Fraud-detection software monitors your purchasing and spending patterns, and when it notes something out of the normal pattern it will send you a message to check that the purchase is valid.
- Music and movie recommendations are based on your previous behaviour and preferences.

Humanoid robots will use AI in future, and it will develop and allow such robots to learn, and respond themselves, and to improve and update their programming as they do so. The only limiting factor will be the power of computer chips themselves.

EXPLORE ONLINE

Go online and follow the links or research 'artificial intelligence' to learn more about AI. Some people think that AI will be the end of the human race; do you agree?

1. Make a list of arguments for and against the development of AI.
2. Explain your view on whether we should continue to develop AI.
3. Propose some ways that we could ensure that AI remains safe for humans.

Figure 11.36 The Google autonomous car is powered by AI software called *Chauffeur*, which is currently being refined.

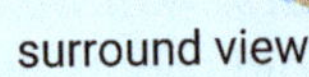

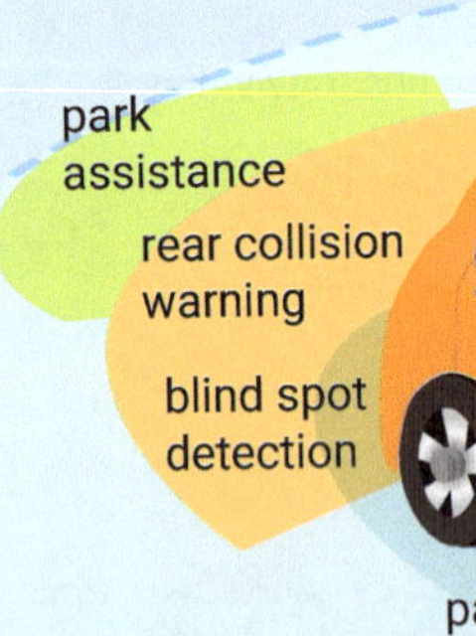

Figure 11.37 Some of the sense and response systems on a driverless car

ISBN 978 1 4202 3736 8

CHECK

1 Explain how an automated control system is similar to the human nervous system.

2 List three types of sensors that are used in automated systems.

3 Motion sensors may be active or passive. Explain the difference.

4 a Define the term *robot*.
  b List the key features that a robot requires to operate.

5 Classify each of the following robots according to its level of robotic control:
  a a lander like a Mars rover
  b a remote control toy car
  c a robot that spray-paints truck bodies while they are being produced
  d a robot vacuum cleaner
  e a landmine-clearing robot
  f a fully autonomous car
  g a robot that works in a metal smelter, pouring hot liquid metal into moulds
  h the autonomous vehicles that collect and move the large shipping containers around to various locations on the docks

6 What are the advantages of a robot that can pour hot metal into moulds?

CHALLENGE

1 a Use a diagram to explain how a thermostat keeps the temperature of a room stable.
  b Explain how this is similar to the system that maintains body temperature as explained in Chapter 7 on page 177.

2 Look at the photos of the three robots in Figure 11.38. Using the theory of the uncanny valley, describe how each robot makes you feel. Does one make you more uncomfortable than the others? Why do you think this is?

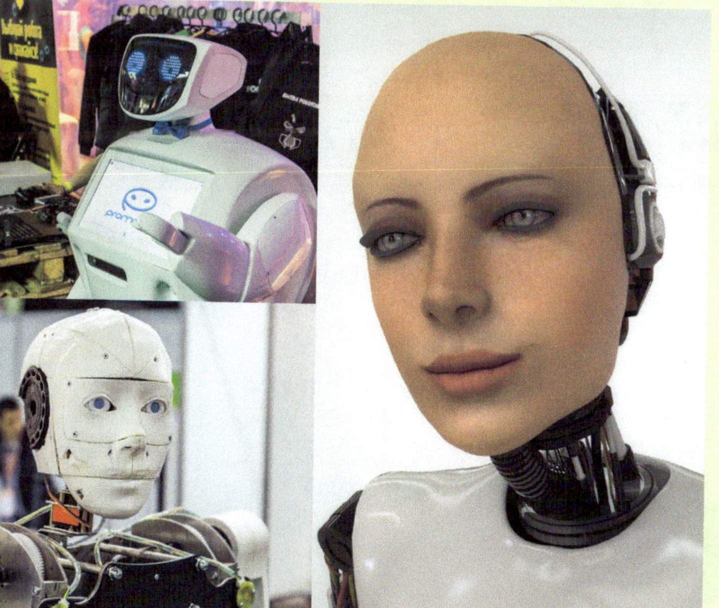

Figure 11.38 Three human-like robots

3 a Explain how an LDR works.
  b How is an LDR similar to and different from a thermistor?

4 A PIR motion sensor detects change in heat, and a thermistor responds to a change in temperature. Define the words *heat* and *temperature* and use your definitions to explain the difference between these two concepts. You may have to do some research.

5 a Think of a task you must do every day at school, home or somewhere else. Invent a robot or device that uses sensors and actuators to complete this task for you, or to help you in some way. Draw your invention and explain how it works.
  b What level of robotic control does your invention use?

EXPLORE ONLINE

If your school has robotics kits, you may be able to learn to build and code some robots to complete tasks.

There are many robotics competitions that you may be able to work towards entering. These may include robot battles, robot soccer or using your robot to solve a series of challenges.

More links are supplied online to some of the main competitions.

ISBN 978 1 4202 3736 8

MAIN IDEAS

Copy and complete these statements to make a summary of this chapter. The missing words are on the right.

1 Communication occurs when information is encoded by a sender, _____, then _____ and understood by a receiver.

2 Electronic communication devices such as telephones encode messages into _____ signals or _____ pulses which are then sent over long distances.

3 Optical fibres transmit information in the form of _____ light pulses.

4 Diodes, transistors, _____ and capacitors are electronic components used in communication devices.

5 Diodes and transistors are made from _____. These are substances that conduct electricity when doped with small amounts of another element.

6 Television and radio signals are transmitted through the air as _____ waves of long wavelength.

7 Radio signals are made up of an audio signal mixed with a carrier wave. AM radio signals have a _____ frequency than FM radio signals.

8 An _____ control system requires sensors to detect change, a computer processor, and various devices that _____.

9 Sensors are devices that detect a _____ in the environment such as light, temperature or heat.

10 A robot is a system of digital technologies that can complete a task either under _____ control or with various levels of _____.

resistors
decoded
transmitted
electrical
digital
light
electromagnetic
automated
respond
direct
lower
autonomy
change
semiconductors

CH•11 REVIEW

1 Which of the following statements is *incorrect*?
  A A microphone converts sound energy into electrical energy.
  B Noise affects the quality of the transmitted message.
  C Analog signals can only have a value of 0 or 1.
  D When the diaphragm in a microphone vibrates, an electric current is induced in the coils of wire.

2 Match the correct descriptions in list B with the electronic terms in list A.

List A	List B
resistance	1 lets electric current pass in one direction only
transistor	2 an electronic light bulb
LED	3 stores charge
diode	4 can act as a switch or as an amplifier
capacitor	5 is measured in ohms
LDR	6 is measured in amps
current	7 its resistance changes with the intensity of the light

ISBN 978 1 4202 3736 8

3 a One key feature of many robots is a locomotion system. List and describe three types of locomotion that a robot might use.
b Could a robot operate without a computer? Explain your answer.
c Define what a sensor is, and explain why they are important for use in robots and automated systems.
d Select one type of sensor and explain how it works. Use a diagram in your answer.
e Describe one application of the sensor you have selected in part d.

4 a In which of the following circuits will the light bulb glow? Explain your answer.
b If the diodes and light bulbs each have a resistance of 50 Ω, which circuit has the largest current flowing through it?

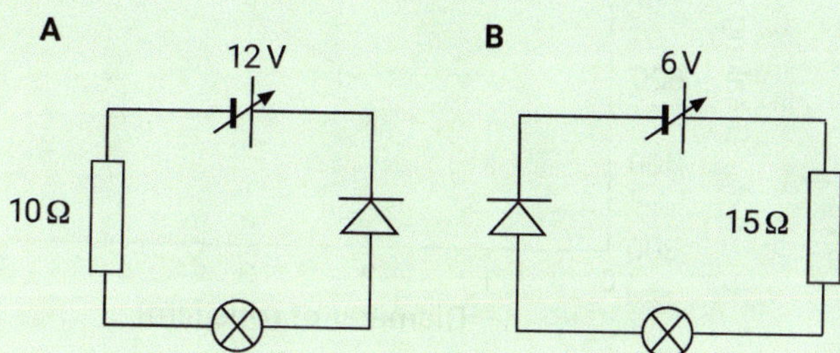

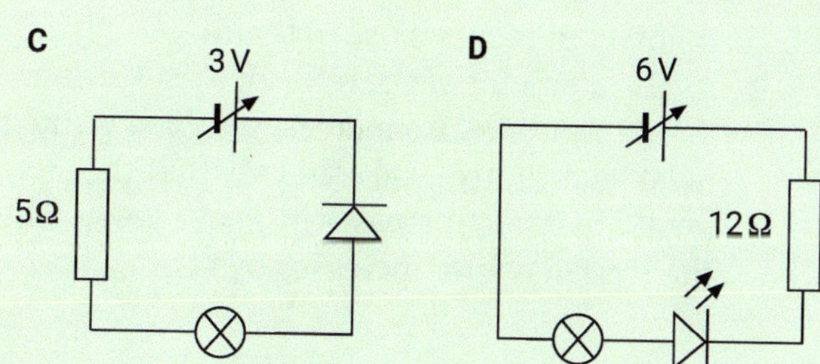

5 a How do robots help society? Outline some of the benefits and ways in which robots are useful.
b Explain what AI is.
c A robot can have different levels of control. Explain this statement and give examples.

6 Four wave patterns produced on the screen of a CRO by a sound generator are shown at top right. Which sound:
a is the loudest?
b has the lowest pitch?
c is quiet and has a high pitch?

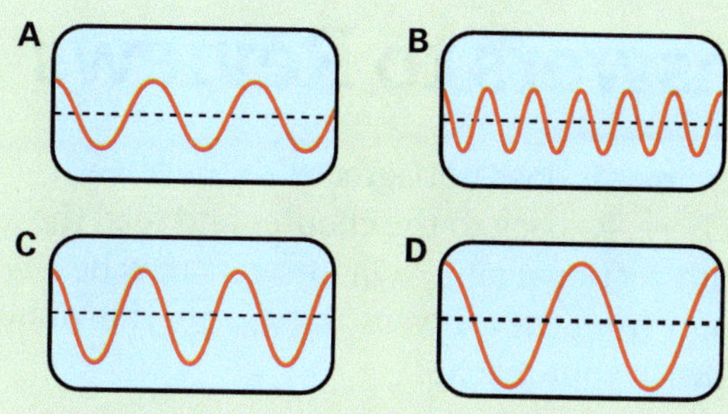

7 The wave below was produced on a CRO connected to an audio generator.

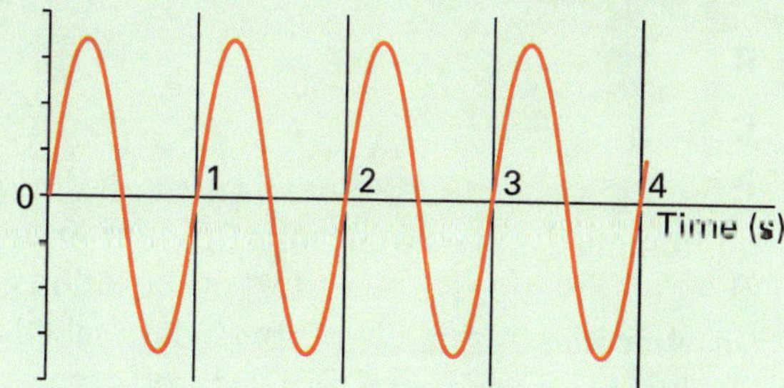

a What is the frequency of the wave?
b Sketch this wave in your notebook. On the same sketch, draw the wave produced by a sound of higher pitch but the same loudness.

8 Use the diagram below to explain how a loudspeaker works.

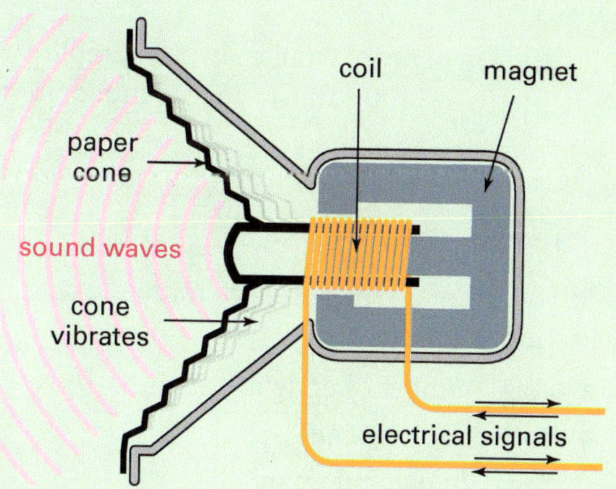

Check your answers on page 306.

ISBN 978 1 4202 3736 8

Answers to Reviews

If your answer does not agree with the answer given here, go back to the chapter and read the relevant section again. Your answers may be slightly different from the answers given here. If in doubt, check with your teacher.

CH•1 Science is investigating

1 **B**

2 **C**

3 **C**—If the researchers know who gets the lotion containing Z and who doesn't, this may influence their observations of the effect of the lotions. To overcome this problem a procedure called a double-blind experiment is used (see page 9).

4 **A**

5 You catch, count and tag a sample of mullet (e.g. 20) in that section of the river. You then release the tagged mullet back into the river. After some time you catch a second sample (e.g. 10) and count how many are tagged (e.g. 2) and untagged (e.g. 8).

Proportion of tagged mullet in 2nd sample = proportion of tagged mullet in river

So $\frac{2}{10} = \frac{20}{\text{total}}$

Therefore $\text{total} = \frac{20 \times 10}{2} = 100$

So you estimate there are 100 mullet in this section of the river.

6 **a** Total number of periwinkles = 120

b Total area = 100 m^2

Area of 1 quadrat = 1 m^2

Area of 10 quadrats = 10 m^2

$$\text{Population} = \text{no. of periwinkles in 10 quadrats} \times \frac{\text{total area}}{\text{area of 10 quadrats}}$$

$$= 120 \times \frac{100}{10}$$

= 1200 periwinkles

c The quadrat method was used because the periwinkles are not mobile—they are fixed in position on the rocky platform.

7 **a** The uncontrolled variables were the type of bicycle, the condition of its brakes, the mass of the rider, the speed of the bike and how hard the rider braked.

b It would be better for the same person to test each bike, travelling at the same speed and braking the same way on both bikes. Ideally the bikes should be the same, with different-sized wheels, but this is not possible.

8 **a**

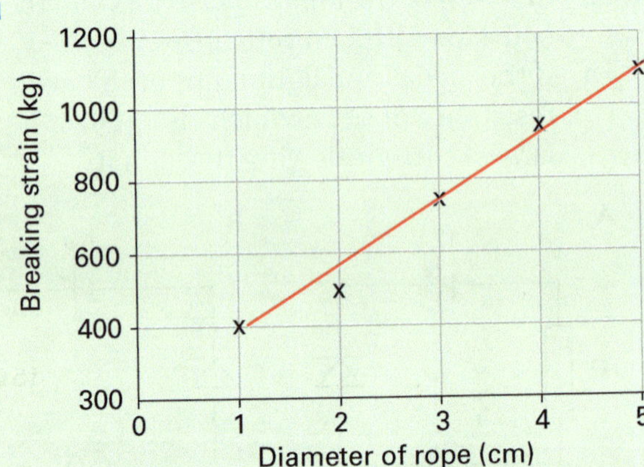

Note that the line does not go through all the points, but it goes close to them.

b As the diameter of the rope increases, the breaking strain also increases. (If you wanted to be quantitative you could say that the breaking strain increases by about 175 kg when the rope diameter increases by 1 cm.)

ISBN 978 1 4202 3736 8

CH•2 Introducing electric circuits

1 a Rods with the same charge **repel** each other.
 b Rods with opposite charges **attract** each other.

2 a If a material loses electrons it becomes **positively** charged.
 b If a material gains electrons it becomes **negatively** charged.

3

Conductors	Insulators
copper	plastic
steel	air
salt water	wood

4 a A (set-up D might or might not work, depending on whether the battery terminal touches the spring at the bottom. Because of this, torch batteries are usually put in with the + terminal nearest the bulb.)
 b The batteries are connected **in series**.

5 a The bulbs in circuit B will glow only half as brightly as the bulb in circuit A. This is because the electric current has to flow through two bulbs instead of one.
 b The bulbs in circuit C will glow as brightly as the bulb in circuit A. This is because the three bulbs are in parallel, and each bulb glows as brightly as if it were the only bulb in the circuit.
 c You would need to arrange the two bulbs in parallel. Alternatively, you could add a second battery to give the current more 'push'.

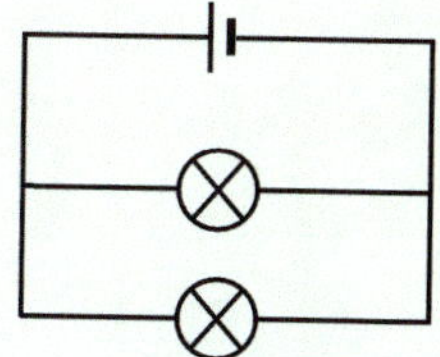
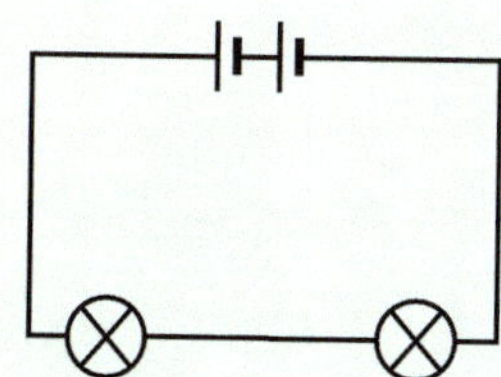

6 **C** (Current flows only in the left-hand part of the circuit—through bulbs A and B, but not through bulb C.)

7 As your shoes rub on the nylon carpet, static electricity builds up on your body. When you touch something that conducts electricity (e.g. a metal door knob), an electric current flows across your skin and you feel a slight electric shock.

8 The battery, bulb and switch need to be in parallel, as shown below. When the switch is open (off), current flows in the bottom half of the circuit, lighting the bulb. When you close the switch, virtually all the current flows through the top half of the circuit. This is because the switch has a much lower resistance than the bulb, and the current follows the path of least resistance. Hence the light goes off when the switch is closed.

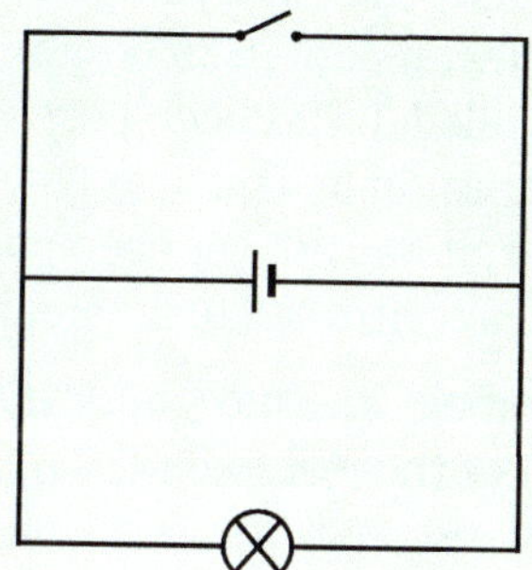

LAB REVIEW

The equipment needed is almost the same for both circuits:

- 1.5 volt battery and holder
- 3 bulbs and holders
- switch
- connecting wires (6 for the left-hand one, 7 for the right)

ISBN 978 1 4202 3736 8

CH•3 Inside the atom

1 **B**, because the number of electrons must be the same as the number of protons.

2 **B**

3 **A**—See page 70

4 **D**—See page 67

5 In the nucleus, there are protons (positive charge) and neutrons (no charge). Around the nucleus are much smaller fast-moving electrons (negative charge).

6 **a** Gamma rays, **b** beta particles, **c** alpha particles

7 **a** In nuclear fission a large nucleus splits into two smaller nuclei. In nuclear fusion two small nuclei such as hydrogen join together.

b The nuclear reactions in bombs are uncontrolled. It has been possible to control nuclear fission in a nuclear power station. However, it has not yet been possible to produce a controlled fusion reaction.

8 **a** The numbers 2, 4 and 5 indicate the mass number of the isotope (total number of protons and neutrons in the nucleus).

b All three isotopes have the same atomic number (two protons in the nucleus).

c Helium-2 has no neutrons in its nucleus, helium-4 has two neutrons, and helium-5 has three neutrons.

d You would expect helium-5 to be radioactive because there are more neutrons than protons. This makes the nucleus unstable (see page 58).

9 Alpha radiation is stopped by a few layers of paper (see diagram top right on page 69), so it probably wouldn't pass through cardboard. Therefore the alpha particles wouldn't reach the detector on the other side of the cardboard and the thickness monitor wouldn't work (see Figure 3.22 on page 74).

10 The half-life is 6 minutes, so after 6 minutes there will be 400 g left. After another 6 minutes (that is, after 12 minutes) there will be 200 g, and so on. After 30 minutes there will be only 25 g.

Now	6 min	12 min	18 min	24 min	30 min
800 g	400 g	200 g	100 g	50 g	25 g

11 If strontium-90 particles fell on the farming area, they would over time end up in the food web. If the babies' mothers then ate food from the contaminated farming area, the strontium-90 would enter their bodies and the bodies of the breast-fed babies.

12 **a** Cancer cells grow rapidly and are more sensitive to radiation than normal cells. So radiation tends to kill cancer cells rather than normal cells.

b Radiation can kill other rapidly-growing cells in the body; for example in a developing baby, in testes and ovaries, in bone marrow and in hair. This is why the reproductive organs are protected during radiation therapy and why people undergoing therapy lose their hair.

ISBN 978 1 4202 3736 8

CH•4 Magnetism and electricity

1 **D**—see page 86

2 **C**—see page 91

3 **B**

4 **a** Since X is repelled by a north pole, it is also a north pole. Y is therefore a south pole.

b

5 Use iron filings (as in Investigation 4.2 on page 89). A compass will be affected.
If you hold a piece of iron or another magnet near it you will be able to *feel* the magnetic force on it.

6 **a**

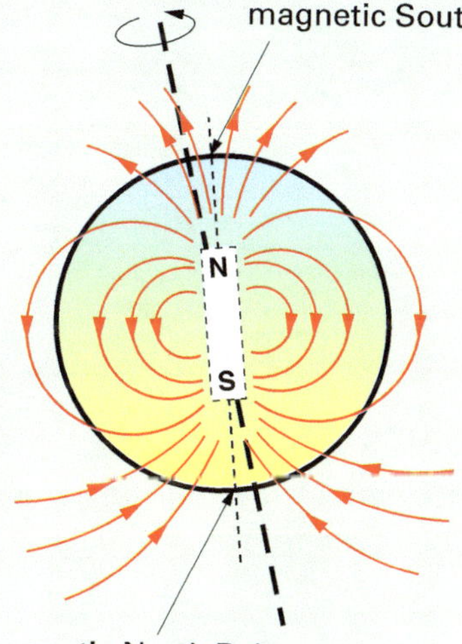

b The north pole of a compass needle would point *south* (in the direction of the magnetic field).

7 **a** Like poles repel, so the magnets will roll away from each other.

b The magnet (A) will attract the piece of iron (B). So the magnet and the piece of iron will both roll towards each other and stick together.

c Brass is not magnetic—it is not attracted to magnets—so nothing will happen.

d Nothing will happen.

8 **a** Iron filings would be attracted to the poles (top and bottom of the disc).

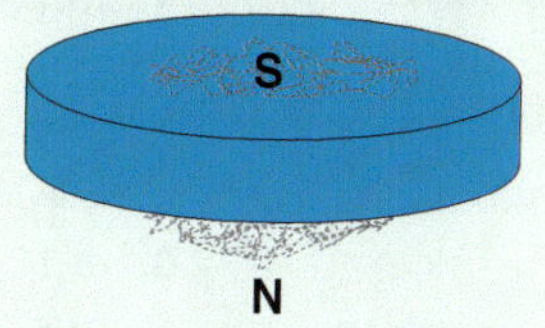

b The magnetic field goes into the south pole and out the north pole (see pages 89–90).

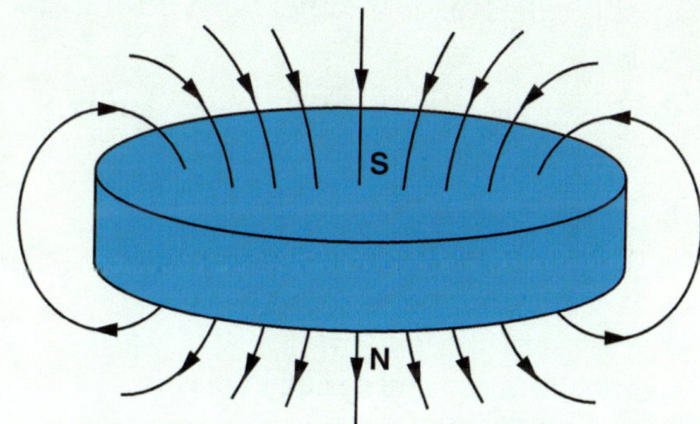

c Suspend the disc so that it is free to rotate. The north side will point north.

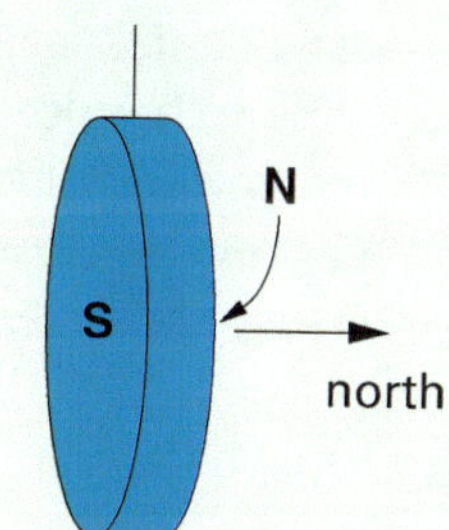

9 **C** (unlike poles attract) or **D** (B could be a piece of unmagnetised iron)

10 **a** Electromagnets 2 and 4 are the same and therefore balance each other. Electromagnet 3 is stronger than 1, so there is an unbalanced force and the ball will move towards 3.

b There is an unbalanced force towards electromagnet 3 and another unbalanced force towards 2. So, depending on the size of these unbalanced forces, the ball bearing will move towards 2 or 3 or somewhere in between.

11 **a** Electricity can be generated when a magnet is quickly moved back and forth inside a wire coil. The magnetic field causes electrons to move in the wire, creating an electric current.

b No electricity could flow as there is a break in the electric circuit.

ISBN 978 1 4202 3736 8

CH•5 Microbes

1 **A**, **B**, **D**, **E**, **F** and **H** (all except **C** and **G**)

2 **D**

3 Bacteria are usually prevented from entering the body by the skin. If they do enter the body, liquids produced in the nose, lungs and stomach trap and kill most bacteria. If bacteria get into the blood, they are destroyed by phagocytes or by antibodies made by other types of white blood cells.

4 **C**—Bacteria can reproduce independently of other organisms.

E—Antibiotics kill bacteria only.

5 **C** and **E** (Yeasts are microscopic fungi.)

6 **B**

7 sultanas—dried and sealed in plastic
mixed berries—frozen
can of chickpeas—heat sterilised
wine—chemical preservatives added
margarine—refrigerated
pasta—dried
cream cheese—refrigerated
peas—frozen
strawberry jam—heat sterilised

8 **B**

9 No carbon dioxide was produced in tube 3 because there was no sugar for the yeast to use as food.

10 Debbie wanted to find out whether sugar was needed by yeasts to produce a gas, and whether the rate of reaction depended on the temperature. Tube 3 was used as a control tube to make sure the yeast (without the sugar) did not produce any gas.

11 **a** The bacteria were added to the agar beforehand so that an even distribution of growth occurred.

b Antiseptics X and Y are effective in killing bacteria A, while only antiseptic X is effective against bacteria B.

c Josep is partly correct. The test showed that Antiseptic X is more effective than antiseptic Y in killing two types of bacteria. However, it may not be as effective in killing all types of bacteria.

12 You had chickenpox when you were younger so you have antibodies in your blood that will fight the virus. However, Belinda probably doesn't have these antibodies, and therefore has a much greater chance of catching the disease.

ISBN 978 1 4202 3736 8

CH•6 Everyday substances

1 D

2 A

3 B

4 C

5 a true
b false—Hydrocarbons contain carbon and hydrogen only.
c true
d true
e false—Petrol has a ***lower*** boiling point than diesel fuel (see Figure 6.10 on page 140).

6 a The plastic pellets are forced along by the turning screw and melted by the heaters. The molten plastic is then forced into the nozzle where it cools and takes the shape of the mould, in this case a pipe.
b B

7 a Acetone would distil first, because it has the lowest boiling point. Acetic acid would distil last.
b Propanol and water are difficult to separate. They distil together because their boiling points are very close.

8 a bridge—by painting
b bicycle chain—by coating with oil
c food cans—by coating with a thin layer of tin
d cutlery—by coating the steel with nickel or silver, or alloying it with chromium and nickel (stainless steel)

9 a A is a thermoplastic since there are no cross-links between the chains (see page 147).
b B would not melt on heating since it is a thermoset.
c B would be stronger because of all the cross-links.
d A—since only thermoplastics can be recycled.
e A is likely to be polythene (see page 146).

10 a The holes in the elbows are caused by rubbing rather than stretching. The fact that the thread is strong is not relevant in this situation.
b You would need to test the different fabrics by rubbing them, e.g. pulling strips of fabric backwards and forwards across the edge of a bench 100 times. You could then compare the wear in the different fabrics.

ISBN 978 1 4202 3736 8

CH•7 Body in balance

1 **C**

2 **B**

3 **a** Involuntary actions include heartbeat, breathing, digestion and balance (see page 163).

b The cerebellum and the brain stem coordinate involuntary actions.

4 A flash of light
B receptors in eye
C sensory neuron
D spinal cord
E motor neuron
F pupil
G pupil decreases

5 Hormones can act on the whole body, on body systems or on individual organs. Nerves act only on muscles and glands.

6 **a** Animals A, B and C are mammals or birds because they have a constant body temperature.

b 40 °C

c Animal B is probably a human because the set-point temperature is about 37 °C.

d The body temperature of animal D increases during the morning as the outside temperature increases. Its body temperature decreases when the outside temperature decreases in the afternoon.

7 **a** **Experiment 1:** The cells that produced the growth hormone were removed, so the growth of the shoot stopped.

Experiment 2: The growth hormone in the extract acted on the cells below the cut and growth continued.

Experiment 3: The black cap stopped the light getting to the cells in the tip, which in turn stopped the production of growth hormone, and growth stopped.

b There are several possible designs. Here is one way:

- Cut the tip off one plant, crush it and place the extract on the cut shoot (as in Experiment 2 on page 185).
- Cut the tip off a second plant and crush it. Then cut off another piece of the shoot to expose the cells further down. Place the extract on this shoot.
- The first plant should grow while the second one will not.

8 **a** Hormone X stimulates the kidney to reduce the amount of sodium being filtered out of the blood and therefore increases the amount of sodium in the blood.

b the receptors in the brain

c

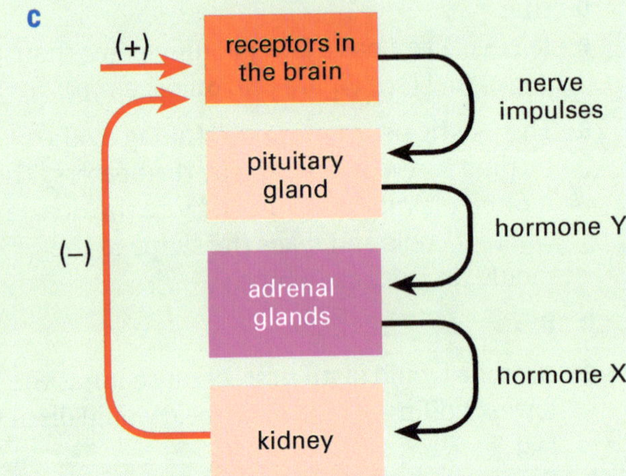

9

Sensory neuron	Motor neuron	Interneuron
• axon is attached to a receptor cell • myelin sheath present • carry information from receptors to central nervous system • usually longer than interneurons	• axon is attached to an effector such as a muscle or gland • myelin sheath present • carry information away from CNS to effectors • usually longer than interneurons	• dendrites and axons usually attached to other nerve cells • only found in the central nervous system • no myelin sheath • axon usually shorter than motor and sensory neurons • function is to process, store, retrieve and transmit information

ISBN 978 1 4202 3736 8

CH•8 Dynamic Earth

1 a B

b

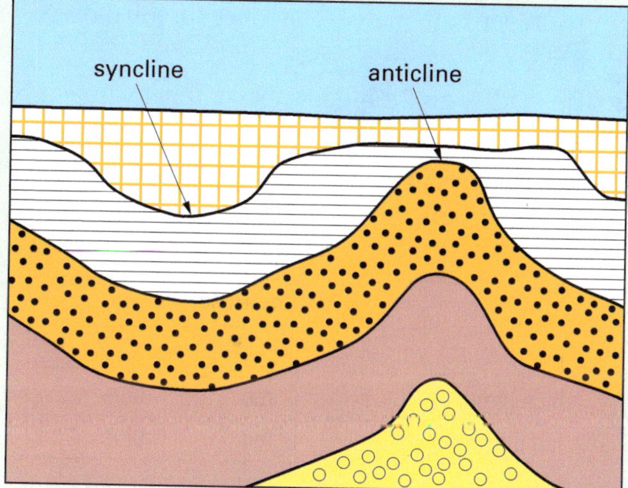

c The igneous rock probably came from lava which flowed from a nearby volcano, then later hardened to form igneous rock. Lava is liquid, so the top of the lava is flat.

d The lowest layer of rock is the oldest because it was deposited first. The other sedimentary layers were deposited on top of this one. The youngest layer is the igneous layer, which formed after the other layers had been folded and eroded.

2 Wellington is on a plate boundary. Here the relative movement of the two plates means that earthquakes are highly likely. Melbourne is well away from the edge of the plate, so the Earth there is less likely to move.

3 a The P wave is the first wave to arrive (time 3.15.00) and the L wave is the last (3.24.50). See page 195.

b P waves

c L waves have the most energy because they have the largest vibrations.

d L waves

e 4 minutes (240 seconds)

f The earthquake was about 2500 km away.

g No. You know how far away the earthquake was, but not its location. To find the epicentre you need seismograms from *three* different locations.

4 The rocks may be folded:

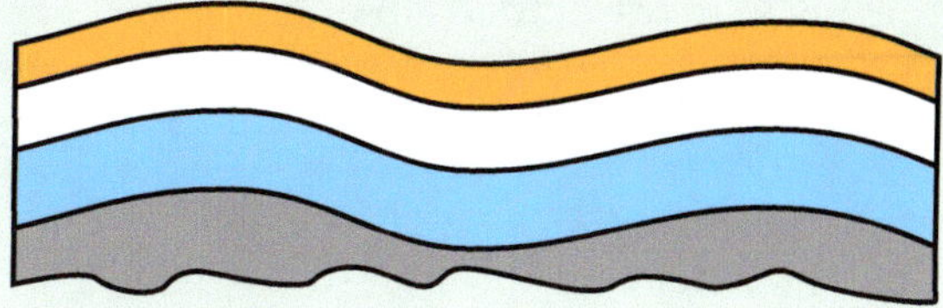

They may be faulted:

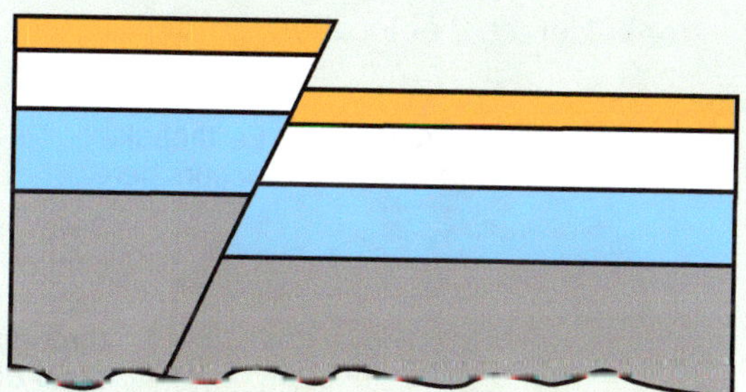

Or they may be folded *and* faulted:

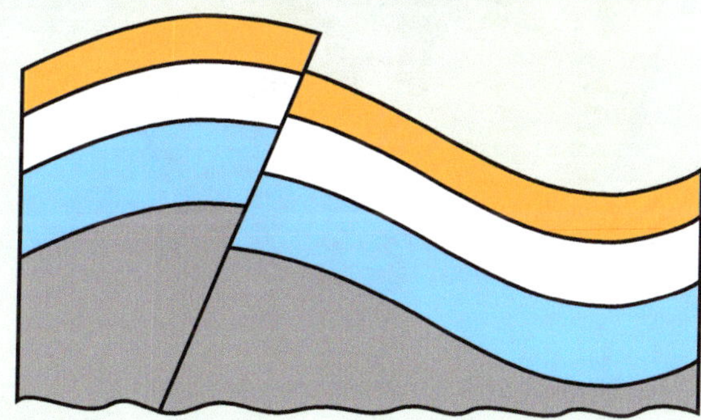

5 The African plate is moving north (upwards on the map), while the large Eurasian plate is moving south-east (downwards and to the right). The Mediterranean Sea is on the boundary of these two plates, which are moving closer together. This movement is making the Mediterranean Sea narrower.

6 a Location A earthquake, magnitude 6—most houses damaged.
Location B earthquake, magnitude 3—suspended objects swing, vibrations are like those caused by passing trucks.

b The earthquake at location A has 1000 times more energy than the one at location B (see page 196).

7 a L waves are called surface waves because they travel only on the Earth's surface

ISBN 978 1 4202 3736 8

b The fastest S waves travel at 8 km/s. The time taken to travel 1000 km is

$$\frac{1000 \text{ km}}{8 \text{ km/s}} = 125 \text{ seconds (2 min 5 s)}$$

c The outer core of the Earth is in a liquid state and S waves do not travel through it. Therefore S waves generated at location A will not be detected at location C or D.

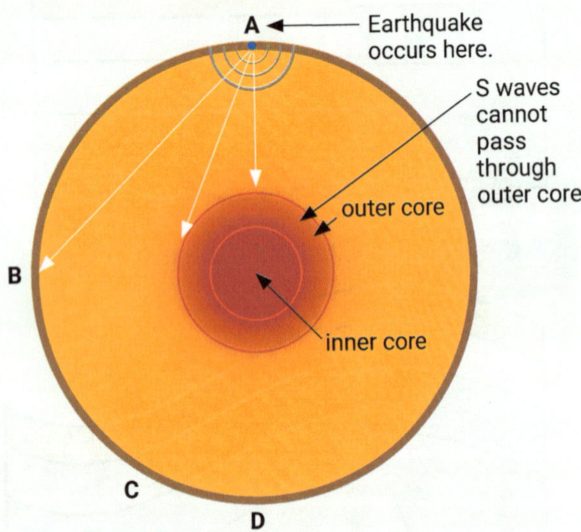

8 Indonesia and New Zealand have many earthquakes and volcanoes because they are located on the edge of tectonic plates, or plate boundaries. At these boundaries plates can move apart or towards each other, resulting in earthquakes and volcanoes. Australia is situated on the centre of a plate, away from the edges, and there are fewer volcanoes and earthquakes as you move away from the plate edges. This means Australia has had a long, stable geological history as a continent compared with countries located at plate edges.

CH•9 Everyday reactions

1 **B** and **D**

2 **A**—Vinegar reacts with baking soda (sodium hydrogen carbonate) to produce carbon dioxide gas.

3 **D**—NaCl is a salt. It does not contain hydrogen.

4 **a** exothermic
b endothermic
c exothermic
d endothermic (see page 233)

5

	Unknown solutions		
	X	Y	Z
blue litmus	blue	red	blue
red litmus	red	red	blue

6 **a** P
b R
c V
d C

7 Lemon juice is an acid and will neutralise aluminium oxide (a base) to produce a soluble salt and water. Because the coating of aluminium oxide has been removed, the saucepan is shiny. Heating the lemon juice makes the neutralisation reaction go faster.

8 Normal rainwater (pH 6) is slightly acidic because carbon dioxide in the air dissolves in raindrops to form carbonic acid (see page 228). Acid rain is even more acidic because waste gases from power stations and cars dissolve in it. Sulfur dioxide dissolves to form sulfuric acid, and nitrogen dioxide also dissolves to form acids. Distilled water (pH 7) does not contain impurities such as carbonic acid, so it is neutral.

9 Some plants grow better in acidic soils and some prefer alkaline soils. If the soil is too acidic (pH less than 7), you can add powdered limestone, and if it is too alkaline (pH more than 7), you can add compost or a soluble fertiliser. See page 217.

10 Acid A releases more H^+ ions in solution than acid C, and hence conducts electricity better—as indicated by the brightly glowing bulb. Acid B doesn't contain enough H^+ ions to make the bulb glow.

ISBN 978 1 4202 3736 8

LAB REVIEW

Use universal indicator solution or pH paper to decide which liquid is acidic (hydrochloric acid) and which is basic (caustic soda solution). The sodium chloride solution and the water are both neutral, but when you evaporate the sodium chloride solution you are left with a white solid.

CH•10 Energy in ecosystems

1 **A**

2 **C** and **D**—Functional adaptations are those that refer to the functioning or working of the organism's body.

3 **a**

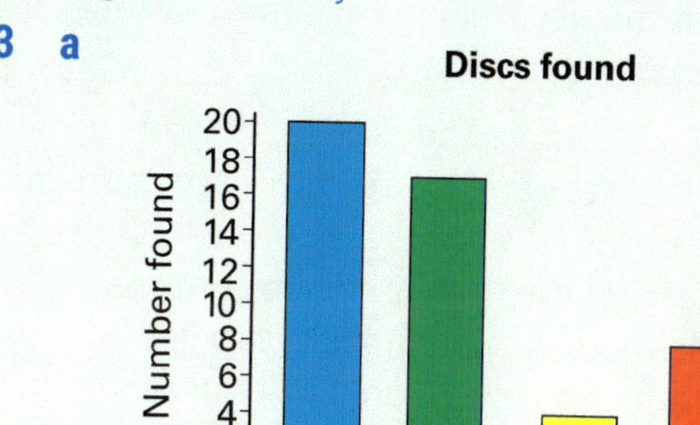

b The yellow and red discs were similar in colour to the surroundings, since fewer of them were found than blue and green. The surroundings might have been a yellow-red coloured sand or soil.

c The four different coloured discs represent the variations in a population. On this particular surface, the yellow and red discs have a better chance of survival. Over time the 'predators' will reduce the numbers of blue and green discs and the population will consist mainly of red and yellow discs.

4 **B** and **D**—The others involve living things (biotic factors).

5 **B**

6 The table shows some of the differences between the two ecosystems; you may have others. Check with your teacher if you are unsure.

River ecosystem	Urban ecosystem
• Humans are not the dominant organism. • Matter is usually recycled in the ecosystem. • Apart from solar energy, there are very few other energy inputs.	• Humans are the dominant organism. • Matter is usually taken out of the ecosystem. • There are large energy inputs from outside the ecosystem.

ISBN 978 1 4202 3736 8

7 A wheat field is not sustainable because matter is taken out of the field when the wheat is harvested and other matter in the form of fertilisers is added when the wheat is growing. So this ecosystem will always lose a large amount of matter and energy when the crop is harvested, and gain a smaller amount of matter and energy when the crop is fertilised.

8 **a** The producers are the water plants and algae.
b The products of respiration are carbon dioxide and water.
c J represents the first-order consumers—snails, microscopic animals, tadpoles (they are also second-order consumers), insect larvae. K represents fish, lobsters and tadpoles (they are also first-order consumers).
d L and M represent matter in the pond water and the soil at the bottom of the pond.
e Matter can leave (and also enter) the ecosystem when animals leave the pond. For example, the insect larvae may change into adult insects and fly away from the pond. Fish can also be caught and eaten by birds that live outside the ecosystem. Extreme weather events such as floods can wash away the organisms in the pond and the soil at the bottom of the pond.

CH•11 Digital technology

1 C

2 **1** diode
2 LED
3 capacitor
4 transistor
5 resistance
6 current
7 LDR

3 **a** Some robot locomotion systems are:
- walking: this may be on 2, 4, 6 or even 8 legs like a spider
- wheels/rolling in various arrangements, such as on the Mars rover, wheels at the ends of legs, or balls under the body of the robot. A motor to move wheels and often a balance system is required
- snaking, which involves sending snake-like movements down the body of a robot
- swimming: this may use fins, motors and propellers or other mechanisms
- climbing: requires legs or arms and some sort of grab device to hold on
- flying, with wings, motors and propellers, or gliding.

b A robot could not operate without a computer because the computer is the control unit that coordinates and controls all functions of a robot. Without this control unit a robot cannot function. The computer is like the brain in humans.
c A sensor is a device that detects a change in the environment or conditions. Sensors are important for use in robots and automated systems as they detect changes and collect data, and this allows an automated system or robot to analyse the data and calculate how to respond, and what to do next.
d Select either a light, motion or temperature sensor. See page 288–289 for details of each.
e Answers could include:
- Light sensor: to turn lights on and off in an automated home at different times of day; to turn on the headlights of a car or street lights at sunset.
- Motion sensor: burglar alarms, to turn lights on and off in an automated home

ISBN 978 1 4202 3736 8

as someone enters a room; outside home sensor lights.
- Temperature sensor: thermostat for keeping temperature in room, factory, refrigerator or other locations at a set level.

4 **a** The light bulb will glow only in circuit B where the negative end of the diode (the straight line in the symbol) is connected to the negative side of the battery (short fat stroke).

b Circuit A has the largest current. This is because it has the largest voltage and the resistance in each circuit is about the same.

5 **a** Robots help society and provide benefits such as:
- being able to explore places humans could not go, like the deep sea or space
- doing unsafe tasks such as diffusing bombs or landmines, and working in factories where there are dangerous activities such as in a metal smelter
- undertaking repetitive tasks that humans find boring, such as packing boxes or adding components to a car on a factory construction line.

b AI is artificial intelligence and is when machines show intelligence, and even learn by undertaking a task and processing how to do it better.

c A robot can have different levels of control (student examples may vary)
- direct control—An operator controls the direct movements or activity of the robot, e.g. remote control car or drone.
- operator assistant or intervention—The operator will assign tasks to a robot, but the robot will decide how best to achieve them, e.g. a shipping container robotic vehicle that moves the containers around the docks.
- fully autonomous with a fixed pattern—A robot will complete the same task over and over on its own, but without the support of the operator, e.g. factory robot packing boxes or welding, or a robot vacuum.
- fully autonomous—A robot will complete all tasks on its own without human intervention, e.g. autonomous self-driving car.

6 **a** D
b D
c B

7 **a** 1 Hz (1 wave per second)
b

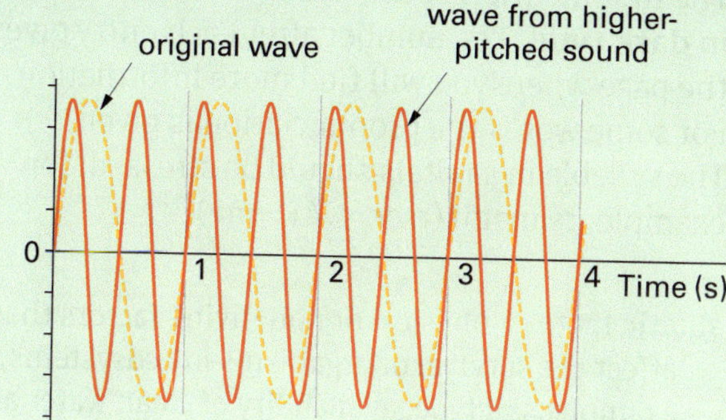

8 The sound waves cause the paper cone to vibrate. This causes the coil to move in and out of the magnet. This movement then creates an alternating current in the wires.

ISBN 978 1 4202 3736 8

Glossary

The first mention of the words in this list occurs in **dark type**. The number after each entry gives the page where you will find more information. For some words the pronunciation is given. The syllable in capitals should be stressed; for example, tsunami (tsoo-NAH-me).

abiotic factors: physical or non-living factors that affect the survival of organisms in ecosystems; for example, soil type, availability of clean water and temperature 238

acid rain: rain that is acidic due to dissolved air pollutants; it can damage plants and buildings 228

acids: substances that can form hydrogen ions (H^+) in solution; an acid can neutralise a base 210

adaptations (ADD-ap-TAY-shuns: the characteristics of an organism that enable it to survive in its habitat 240

alkalis: bases that are soluble in water 210

alloys: mixtures composed wholly or mainly of metals; they have properties different from the metals they are made from 134

alpha particle: positively charged helium nucleus (two protons and two neutrons) given off during radioactive decay 69

alternating current: electric current that changes (alternates) direction 98

ammeter: an instrument used to measure electric current, in amperes or amps (A) 38

AM (amplitude modulated): radio stations that broadcast using a type of wave whose frequency is constant but whose amplitude varies 272

amplitude: the size of a signal or the loudness of a sound, measured by the height of the wave above or below the zero point 267

analog (AN-a-log) signal: a wave signal used in communication devices that varies in value at different points in time 268

antibiotics: chemicals that stop the growth of bacteria inside the body 126

antibodies: special proteins made by the body in response to invasion by microbes and their toxins 123

anticlines: upwards-bending folds in rock layers 189

artificial intelligence (AI): when machines show intelligence and even learn by undertaking a task and processing how to do it better 292

atomic number: the number of protons in the nucleus of an atom; equal to the number of electrons 58

auxins (ORK-sins): a group of hormones responsible for the growth of cells in the stems and roots of plants 175

bacteria: unicellular organisms that have a very simple structure and no distinct nucleus 109

bases: substances that can form hydroxide ions (OH^-) in solution; a base can neutralise an acid 210

behavioural adaptation: the way an organism behaves in order to survive in its environment 241

beta particles: high-speed electrons given off during radioactive decay 69

biodegradable: able to be broken down or decayed by biological means (bacteria and fungi) 143

biotic factors: biological or living factors that affect the survival of organisms in ecosystems; for example, predators and availability of food 238

bit: a binary digit, with the value 1 (on) or 0 (off) 268

blind experiment: a controlled experiment involving people, in which the subjects do not know who is in the control group and who is in the test group 9

brain: the main organ of the nervous system, which controls all the systems in the body 162

brain stem: the base of the brain, which controls involuntary actions such as breathing and heartbeat 163

byte: a unit of information, usually eight bits, used in communications technology 268

capacitors: electronic components that store electric charge in a circuit 277

cellular respiration: the process that occurs in cells in which food is broken down in chemical reactions to release energy 232

cerebellum (ser-a-BELL-um): a small, crinkly part at the lower back of the brain that controls

ISBN 978 1 4202 3736 8

involuntary actions such as balance and coordination 163

cerebrum (ser-EE-brum): the largest part of the brain; it controls memory, speech and voluntary actions, and receives information from sense receptors 163

chain reaction: the process in which one nuclear reaction produces particles that start off a chain of similar reactions 65

circuit diagram: a standard way of drawing an electric circuit, using symbols 45

conductor: a substance that allows heat or electricity to move through it easily 38

continental drift: an early theory that suggested that the positions of the continents on the Earth's surface have changed over time 202

core: the innermost part of the Earth, made of a liquid outer core and a solid inner core 188

correlation: how closely two variables depend on each other 22

corrosion: the process in which a metal reacts with the air, water and other substances in its surroundings; the rusting of iron is one form of corrosion 136

corrosive: type of chemical that can burn or 'eat away' skin and other materials; examples include acids and alkalis 210

crust: the relatively thin, solid outer skin of the Earth 188

digital signal: a wave signal used in communication devices; it has one of two values—zero or one 268

diodes: electronic components that allow current to flow in one direction only 277

double-blind experiment: an experiment in which neither the subjects nor the experimenters know who is in the control group and who is in the test group 9

ecosystem: a system of relationships among organisms and the way they interact with the non-living things in their habitat 238

effector: a muscle, gland or organ that effects (brings out) a response to a nerve impulse 165

electrical resistance: resistance to the flow of electric current through a conductor; good conductors have low resistance 41

electric charge: results when an object gains electrons (negative charge) or loses electrons (positive charge) 30

electric circuit: a continuous path around which an electric current can flow 38

electric current: the flow of electricity around an electric circuit 98

electric generator: device that generates electricity by spinning magnets inside coils or coils inside magnets 98

electromagnet: a temporary magnet made from a coil of wire wound round a piece of iron; the electromagnet works when electricity flows through the wire 95

endocrine glands: glands found in various places in the body, which produce hormones and release them directly into the blood 169

endocrine system: system consisting of a number of endocrine glands throughout the body 162

endothermic reaction: a reaction during which energy is absorbed; energy must be supplied to keep the reaction going 231

energy pyramid: a diagram that shows the amount of potential energy in each level of a food web 251

epicentre: the point on the Earth's surface directly above where the movement of rocks occurs in an earthquake 195

exothermic reaction: reaction that releases energy 231

extrapolating (ex-STRAP-oh-late-ing): using a graph to predict a value beyond the range of a set of measurements 21

fault: a crack in the Earth's crust along which rocks move 190

fermentation (FUR-men-TAY-shun): the process in which yeasts use sugars as food, producing carbon dioxide and alcohol 114

FM (frequency modulated): radio stations that broadcast using a type of wave whose amplitude is constant but whose frequency varies 272

folds: the buckling of rocks caused by huge Earth forces 189

fractional distillation: distillation process used to separate a mixture of liquids into 'fractions', based on their different boiling points 140

frequency: the number of waves that pass a certain point in 1 second; it is measured in hertz (Hz) 272

ISBN 978 1 4202 3736 8

functional adaptation: the way an organism's body works in order to survive in its environment 241

gamma rays: very high-energy electromagnetic radiation given off during radioactive decay 69

half-life: the time required for half the nuclei in a sample of a radioisotope to break down (decay) 70

hormones: chemical messengers emitted by endocrine glands that control important processes of an organism, such as growth 162

hydrocarbons: covalent compounds containing only the elements hydrogen and carbon 140

indicator (acid–base): a substance, e.g. litmus, that turns different colours in acidic and basic solutions 211

infectious diseases: diseases caused by microbes such as viruses, bacteria and unicellular organisms 120

insulator: a substance that does not allow electricity (or heat) to move through it easily 38

interneuron: a neuron that connects a sensory neuron and a motor neuron 165

interpolating (in-TERP-oh-late-ing): using a graph to predict a value between two or more measurements 21

ion (EYE-on): an atom or a group of atoms that has a positive or negative charge, caused by the loss or gain of electrons 219

isotopes (EYE-so-topes: atoms of the same element that have the same number of protons, but a different number of neutrons 58

law of conservation of mass: the total mass of the reactants in a chemical reaction is always equal to the total mass of the products 64

light-dependent resistor (LDR): a light-sensitive resistor that decreases its resistance as the light gets brighter; sometimes called a photoresistor 286

line of best fit: a line that is closest to most of the plotted points drawn on a graph; it shows the relationship between two variables 19

magnetic field: the space in which the force of a magnet acts 88

mantle: thick layer of rock below the Earth's crust; it is partly solid and partly molten 188

mass number: the total number of protons and neutrons in the nucleus of an atom 58

microbes: microscopic organisms that include bacteria, algae, protozoans and viruses 106

motor neuron: a neuron that sends nerve impulses from the central nervous system to muscles and glands 164

mutations: changes in the DNA or genetic code of living things; they may be caused by exposure to radiation or chemicals 233

natural selection: the process in which organisms that have favourable characteristics survive in a particular habitat, and reproduce 243

negative feedback system: a system of control in the body in which the response acts as a stimulus to oppose the change caused by the original stimulus 179

nervous system: system consisting of the brain, spinal cord and nerves that run to all parts of the body 162

neuron (NEW-ron): the basic unit of the nervous system; a nerve cell 164

neutralisation: the reaction of an acid and a base to form salt and water 226

neutrons: neutral particles in the nucleus of an atom 57

nuclear fission: the splitting of the nucleus of a large atom such as uranium into smaller atoms, with the release of a large amount of energy 65

nuclear fusion: a nuclear reaction in which two small atomic nuclei (usually hydrogen) join to form a larger nucleus, with the release of a huge amount of energy; nuclear fusion occurs in the sun 66

nuclear reaction: a type of reaction that involves changes to the nucleus of the atom; these reactions release huge amounts of energy 64

nuclear reactor: a device in which nuclear reactions can be controlled 66

nucleus (atom): the positively charged core of an atom; it contains protons and neutrons 56

ohm (Ω): the unit of electrical resistance 276

ISBN 978 1 4202 3736 8

parallel connection: a method of connecting electrical components (e.g. batteries and bulbs), so that the current divides and part passes through each component 46

pH: a scale from 0 to 14 that indicates how acidic or basic something is; pH stands for 'power of hydrogen' 216

phagocytes (FAG-o-sites): large white blood cells that are able to engulf and destroy microbes and other foreign bodies 122

pituitary (pit-YOU-it-tree) gland: an endocrine gland, located on the underside of the brain; controls other endocrine glands 170

placebos (pla-SEE-bows): substances that have no chemical effect on the body, given to a subject in a blind or double-blind experiment 9

plate tectonics: a theory modified from the continental drift theory, which suggests that the Earth's crust is made up of slowly moving plates 203

poles (magnetic): the ends of a magnet where the magnetism is strongest 86

polymerisation (pol-IM-er-eyes-AY-shun): a chemical reaction in which many small molecules link up to form a giant molecule (polymer) 145

polymers: substances composed of giant molecules formed by linking many smaller molecules (monomers) together 145

protists: large group of organisms (kingdom) that have a very simple structure; the group includes algae and unicellular organisms that live in water 106

protons: positively charged particles in the nucleus of an atom 57

quadrat: a small measuring frame that can be used to sample the organisms in a particular area 13

quarks: particles even smaller that protons and electrons; scientists infer that protons, neutrons and electrons are made from quarks 90

radioisotope: an isotope that is radioactive 89

receptor: a special sense organ that can detect a stimulus 163

reflex action: an automatic response to a stimulus that does not involve the brain 166

resistors: poor conductors used to reduce the amount of current flowing in an electric circuit 276

Richter (RICK-ter) scale: a scale from 1 to 10, used to measure the magnitude or strength of an earthquake 196

robot: an automated system based on digital and machine technology that can detect changes in its environment, process data and respond in some way, or follow a predetermined set of instructions 288

salt: compound formed when an acid reacts with a base, a metal or a carbonate 225

scatter graph: a graph where you plot many points to see if there is any correlation between two variables 22

seismographs (SIZE-mo-graphs): instruments that record earthquakes and measure their magnitude 195

semiconductors: substances (e.g. silicon and germanium) that have properties between those of conductors and insulators and that are used to make diodes and transistors 282

series connection: a method of connecting electrical components (e.g. batteries and bulbs), so that the current passes through one then the other 124

sensory neurons: a neuron that sends nerve impulses from the body's receptors to the central nervous system 164

structural adaptation: a special feature of an organism's body that helps it survive in its environment 241

structural formula: a chemical formula which shows the structure of a molecule—how the atoms are linked together 138

sustainable: an ecosystem is sustainable if it supports diverse groups of organisms and can exist over a long period of time 256

synapse: a tiny gap across which a neuron can send a nerve impulse to another neuron 165

synclines: downwards-bending folds in rock layers that forms a U-shape 189

thermistor: a heat-sensitive resistor that changes its resistance as the temperature changes 287

thermoplastics: plastics that can be melted or remoulded repeatedly 147

ISBN 978 1 4202 3736 8

thermosets: plastics that cannot be remoulded after the initial heating and moulding process 147

toxin: a poison given off by a microbe 120

transistor: an electronic component that acts as a switch or an amplifier in a circuit 277

tsunami (tsoo-NAH-me): a giant wave or series of waves caused by an underwater earthquake 200

vaccination: the process in which heat-treated viruses and bacteria, or the toxins they produce, are injected into the body so that the body will make antibodies ready in case of infection by those microbes 125

viscosity: a measure of a fluids' ability to flow, sometimes referred to as the 'thickness' of a fluid 140

viruses: extremely small microbes that can only reproduce inside other living things 120

voltage: the electrical 'push' causing current to flow in an electric circuit 38

ISBN 978 1 4202 3736 8

Index

ISBN 978 1 4202 3736 8

ISBN 978 1 4202 3736 8

ISBN 978 1 4202 3736 8

ISBN 978 1 4202 3736 8

ISBN 978 1 4202 3736 8

ISBN 978 1 4202 3736 8

ISBN 978 1 4202 3736 8

Acknowledgements

The author and publisher are grateful to the following for permission to reproduce copyright material:

PHOTOGRAPHS
AAP/AP Photo/Wally Santana, **197**; (middle) ALAMY/Avalon/Photoshot, **151**, (bottom) /Kitch Bain, **269**; ANTEATER PRODUCTION/Peter Stannard, **7**, **14**, (middle left) **15**, **17**, **20**, **36**, **40**, **85**, **87**, **91**, (left) **106**, **111**, (right) **114**, **131**, (right) **151**, **163**, **171**, (top left) **173**, (bottom left) **173**, **175**, **180**, **190**, (top) **192**, **193**, **210**, **211**, **216**, **225**, (top) **237**, **238**, **242**, (left) **261**, **263**, **276**; AUSCAPE, **123**, (top) /Greg Harold, **226**, /Reg Morrison, **255**, CERN/CMS Collection/Taylor, L; McCauley, T, **61**; Coo-ee Picture Library, **152**; (bottom) CSIRO/CC BY 3.0, **256**; DREAMTIME.COM/Schall, **232**, (top) /Ali Rıza Yıldız, **149**, /Wxin, **98**; (top) FAIRFAX/Jessica Hromas, **259**; GETTY IMAGES (left) /AbleStock.com, **120**, (right) /SPL/Alex Bartel, **120**, /SPL/AGSTOCKUSA/Bill Barksdale, **221**, /Demetrio Carrasco, **117**, (bottom) /Corbis/Tim Clayton, **194**, (bottom) /Corbis/Lloyd Cluff, **192**, (top) /AFP/Fabrice Coffrini, **62**, (right) /Lonely Planet Images/Feargus Cooney, **237**, (left) /De Agostini Picture Library, **114**, /The Washington Post/Sean Gardner, **75**, (top) /SPL/Jan Hinsch, **239**, (top) /Image Source, **30**, /Corbis/Ted Horowitz, **278**, /SPL/Nancy Kedersha, **81**, /Mark Kostich, **73**, (middle) /SPL/Martin Land, **192**, (bottom) /Radius Images/Leonardo, **217**, /SPL/Dr K MacDonald, **205**, /Corbis/Perry Mastrovito, **173**, (bottom) /Matt Meadows, **109**, /AFP/Marty Melville, **201**, /SPL/Peter Menzel, **29**, (left) /Photolibrary/Stefan Mokrzecki, **240**, (bottom) /Dean Mouhtaropoulos, **62**, (bottom) /Corbis/Joe McDonald, **245**, /SPL/Claude Nuridsany & Marie Perennou, **246**, /Oxford Scientific/Richard Packwood, **228**, (top) /SPL/Alfred Pasieka, **109**, (bottom right) /UIG/Photofusion, **15**, /SPL/David Parker/IMI/Univ. of Birmingham High TC Consortium, **41**, (left) /Red Chopsticks, **151**, (top) /Corbis/VCG/Roger Ressmeyer, **187**, /SPL/S2A, **182**, /Science Photo Library, **35**, **67**, **72**, **77**, **112**, **121**, (right) **157**, **162**, **215**, (bottom) /AFP/Yomiuri Shimbun, **187**, (top) /Picture Press/Juergen & Christine Sohns, **217**, /SPL/Volker Steger, **281**; ISTOCK/BanksPhotos, **97**, (bottom) /Katarzyna Bialasiewicz, **167**, (top) /JohnCarnemolla, **245**, (top) /Jason Doiy, **292**, /dddb, **55**, (left) /dpinn, **237**, /emholk, **33**, /Furchin, **196**, /Christopher Futcher, **265**, /guvendemir, **97**, **270**, /jack0m, **139**, (bottom) /Joel Iedema, **237**, (top) /Lusoimages, **269**, **270**, /karam miri, **213**, /PamelaJoe McFarlane, **236**, (bottom) /PaulMorton, **239**, /Massimo Merlini, **118**, (bottom) /monicaodo, **292**, /marcelo pinheiro, **257**, (bottom) /Andrew Rich, **30**, (top) /Malte Roger, **105**, /ShchiGOL, **42**, /Scharvik, **126**, /titoslack, **145**, /Vershinin-M, **186**, /vjanez, **82**, /jian wang, **100**; NEWSPIX (top right) /Brianne Makin, **128**, /Michael Ross, **34**, /Adam Ward, **258**; Plantic Technologies Australia, **148**; Photodisc, **261** SCIENCE PHOTO LIBRARY (right) /Jan Hinsch, **157**, (bottom) /James King-Holmes, **149**, /Andrew Lambert Photography, **208**, /David Parker, **28**; SHUTTERSTOCK (top) /Achimdiver, **83**, (bottom) /Alila Medical Media, **105**, /alybaba, **254**, (bottom left) /asharkyu, **291**, /auleena, **135**, /belushi, **8**, (top right) /Gary Blakeley, **134**, /gualtiero boffi, **119**, (right) /bonchan, **60**, (bottom) /Nadia Borisevich, **240**, /Darren J. Bradley, **189**, (right) /Giovanni Cancemi, **291**, **293**, /ChameleonsEye, **71**, /Couperfield, **76**, /Martyn F. Chillmaid, **68**, (left) /Cylonphoto, **289**, /Andrea Danti, **160**, /Daxiao Productions, **3**, /Vereshchagin Dmitry, **168**, /DK samco, **46**, /Oleg Doroshin, **293**, /exopixel, **4**, (top) /Dirk Ercken, **167**, /Fotokostic, **256**, /goodluz, **2**, /Gino Santa Maria, **6**, (left) /Miroslav Hlavko, **244**, /imagedb.com, **90**, (middle) /Andrea Izzotti, **240**, (right) /inavanhateren, **244**, (right) /Arno Jenkins, **138**, /Kateryna Kon, **104**, (right) /Kgrif, **106**, (bottom) /ktsdesign, **83**, (bottom) /LEXXIZM, **259**, /Lucky Business, **5**, /Maksimilian, **143**, (left) /molekuul_be, **138**, /Monkey Business Images, **183**, /Constantine Pankin, **64**, (left) /PANIGALE, **278**, (top left) /Dusan Petkovic, **291**, (right) /Tanya Puntti, **240**, /Esa Riutta, **270**, /Dr Morley Read, **241**, /Fouad A. Saad, **99**, (middle) /Sam72, **114**, /science photo, **288**, /sdecoret, **270**, /seeshooteatrepeat, **249**, (right) /SerhiyHorobets, **116**, (left) /Sorbis, **60**, (right) /Peter Sobolev, **289**, /Syda Productions, **264**, /Mariusz Szczygiel, **132**, (left) /Mariusz Szczygiel, **134**, (bottom) /mTaira, **200**, (middle) /Michael J Thompson, **138**, (bottom) /Thinglass, **226**, /Vitaly Titov, **136**, /Stanisic Vladimir, **286**, /Wayhomestudio, **170**, (middle) /worradirek, **83**, /YanLev, **161**, /yevgeniy11, **74**, /zlikovec, **54**; Queensland University of Technology/QUT Marketing & Communication, **181**.

OTHER MATERIAL
The Victorian Curriculum F-10 content elements are © VCAA, reproduced by permission. Victorian Curriculum F-10 elements accurate at time of publication. The VCAA does not endorse or make any warranties regarding this resource. The Victorian Curriculum F-10 and related content can be accessed directly at the VCAA website: www.vcaa.vic.edu.au, **viii-xii.**

The author and publisher would like to acknowledge the following:
NASA, **91**, /Chris Gunn, **134**, /JPL-Caltech; **290**, Photo of cans being heat treated in factory (source unknown) **116**; (top) Courtesy of the U.S. Geological Survey, **194**.

While every care has been taken to trace and acknowledge copyright, the publisher tenders their apologies for any accidental infringement where copyright has proved untraceable. They would be pleased to come to a suitable arrangement with the rightful owner in each case.

ISBN 978 1 4202 3736 8